T0135338

Advances in Intelligent Systems and Computing

Volume 866

Series editor

Janusz Kacprzyk, Polish Academy of Sciences, Warsaw, Poland
e-mail: kacprzyk@ibspan.waw.pl

The series "Advances in Intelligent Systems and Computing" contains publications on theory, applications, and design methods of Intelligent Systems and Intelligent Computing. Virtually all disciplines such as engineering, natural sciences, computer and information science, ICT, economics, business, e-commerce, environment, healthcare, life science are covered. The list of topics spans all the areas of modern intelligent systems and computing such as: computational intelligence, soft computing including neural networks, fuzzy systems, evolutionary computing and the fusion of these paradigms, social intelligence, ambient intelligence, computational neuroscience, artificial life, virtual worlds and society, cognitive science and systems, Perception and Vision, DNA and immune based systems, self-organizing and adaptive systems, e-Learning and teaching, human-centered and human-centric computing, recommender systems, intelligent control, robotics and mechatronics including human-machine teaming, knowledge-based paradigms, learning paradigms, machine ethics, intelligent data analysis, knowledge management, intelligent agents, intelligent decision making and support, intelligent network security, trust management, interactive entertainment, Web intelligence and multimedia.

The publications within "Advances in Intelligent Systems and Computing" are primarily proceedings of important conferences, symposia and congresses. They cover significant recent developments in the field, both of a foundational and applicable character. An important characteristic feature of the series is the short publication time and world-wide distribution. This permits a rapid and broad dissemination of research results.

More information about this series at http://www.springer.com/series/11156

Pandian Vasant · Ivan Zelinka
Gerhard-Wilhelm Weber

Editors

Intelligent Computing & Optimization

 Springer

Editors
Pandian Vasant
Department of Fundamental and Applied
  Sciences
Universiti Teknologi Petronas
Tronoh, Perak, Malaysia

Gerhard-Wilhelm Weber
Institute of Applied Mathematics
METU
Ankara, Ankara, Turkey

Ivan Zelinka
Faculty of Electrical Engineering
  and Computer Science
VŠB TU Ostrava
Ostrava, Czech Republic

ISSN 2194-5357   ISSN 2194-5365   (electronic)
Advances in Intelligent Systems and Computing
ISBN 978-3-030-00978-6   ISBN 978-3-030-00979-3   (eBook)
https://doi.org/10.1007/978-3-030-00979-3

Library of Congress Control Number: 2018955576

This Springer imprint is published by the registered company Springer Nature Switzerland AG
The registered company address is: Gewerbestrasse 11, 6330 Cham, Switzerland

We also appreciate the fruitful guidance and support from Prof. Gerhard Wilhelm Weber (Poznan University of Technology, Poland; Middle East Technical University, Turkey), Prof. Rustem Popa ("Dunarea de Jos" University in Galati, Romania), Prof. Valeriy Kharchenko (Federal Scientific Agroengineering Center VIM, Russia), Dr. Wonsiri Punurai (Mahidol University), Prof. Milun Babic (University of Kragujevac, Serbia), Prof. Ivan Zelinka (VSB-TU Ostrava, Czech Republic), Dr. Jose Antonio Marmolejo (Universidad Anahuac Mexico Norte, Mexico), Prof. Gilberto Perez Lechuga (University of Autonomous of Hidalgo State, Mexico), Prof. Ugo Fiore (Federico II University, Italy), Prof. Weerakorn Ongsakul (Asian Institute of Technology, Thailand), Prof. Rui Miguel Silva (Portugal), Mr. Sattawat Yamcharoew (Sparrow Energy Corporation, Thailand), Mr. K. C. Choo (CO_2 Networks, Malaysia), and Dr. Vinh T. Le (Ton Duc Thang University, Vietnam).

Our book of proceedings provides a premium reference to graduate and post-graduate students, decision makers, and investigators in private domains, universities, traditional and emerging industries, governmental and non-governmental organizations, in the fields of various operational research, AI, geo- and earth sciences, engineering, management, business, and finance, where ever one has to represent and solve uncertainty-affected practical and real-world problems. In the forthcoming times, mathematicians, statisticians, computer scientists, game theorists and economists, physicist, chemists, representatives of civil, electrical, and electronic engineering, but also biologists, scientists on natural resources, neuroscientists, social scientists, and representatives of the humanities, are warmly welcome to enter into this discourse and join the collaboration for reaching even more advanced and sustainable solutions. It is well understood that predictability in uncertain environments is a core request and an issue in all fields of engineering, science, and management. In this regard, this proceedings book is following a quite new perspective; eventually, it has the promise to become very significant in both academia and practice and very important for mankind!

Finally, we would like to sincerely thank Dr. Thomas Ditzinger, Dr. Almas Schimmel, and Ms. Parvathi Krishnan of Springer for the wonderful help and support in publishing ICO 2018 conference proceedings in **Advances in Intelligent Systems and Computing**.

October 2018

Pandian Vasant
Gerhard-Wilhem Weber
Ivan Zelinka

Preface

The first edition of the International Conference on Intelligent Computing and Optimization (ICO 2018) will be held during October 4–5, 2018, at Hard Rock Hotel Pattaya in Pattaya, Thailand. The objective of the international conference is to bring together the global research scholars, experts, and scientists in the research areas of Intelligent Computing and Optimization from all over the world to share their knowledge and experiences on the current research achievements in these fields. This conference provides a golden opportunity for global research community to interact and share their novel research results, findings, and innovative discoveries among their colleagues and friends. The proceedings of ICO 2018 is published by Springer (**Advances in Intelligent Systems and Computing**) and indexed by DBLP, EI, Google Scholar, Scopus, and Thomson ISI.

For this edition, the conference proceedings covered the innovative, original, and creative research areas of sustainability, smart cities, meta-heuristics optimization, cyber security, block chain, big data analytics, IoTs, renewable energy, artificial intelligence, power systems, reliability, and simulation. The authors are very enthusiastic to present the final presentation at the conference venue of Hard Rock Hotel in Pattaya, Thailand. The organizing committee would like to sincerely thank all the authors and the reviewers for their wonderful contribution for this conference. The best and high-quality papers have been selected and reviewed by International Program Committee in order to publish in **Advances in Intelligent System and Computing** by Springer.

ICO 2018 will be an eye-opener for the research scholars across the planet in the research areas of innovative computing and novel optimization techniques and with the cutting-edge methodologies. This conference could not have been organized without the strong support and help from the staff members of Hard Rock Hotel Pattaya, Springer, Click Internet Traffic Sdn Bhd, and the organizing committee of ICO 2018. We would like to sincerely thank Prof. Igor Litvinchev (Nuevo Leon State University (UANL), Mexico), Prof. Nikolai Voropai (Energy Systems Institute SB RAS, Russia), and Waraporn Nimitsuphachaisin (Hard Rock Hotel Pattaya) for their great help and support in organizing the conference.

Contents

A System for Monitoring the Number and Duration of Power Outages and Power Quality in 0.38 kV Electrical Networks

Alexander Vinogradov[1], Vadim Bolshev[1(✉)], Alina Vinogradova[1],
Tatyana Kudinova[2], Maksim Borodin[2], Anastasya Selesneva[2],
and Nikolay Sorokin[2]

[1] Federal Scientific Agroengineering Centre VIM, 1-St Institutsky Proezd, 5,
109428 Moscow, Russia
{winaleksandr, alinawin}@rambler.ru,
vadimbolshev@gmail.com
[2] Orel State Agrarian University named after N.V. Parakhin, General Rodin Str.,
69, 302019 Orel, Russia
{t.kudinova77, anastasiya.selezneva.1995}@mail.ru,
maksimka-borodin@yandex.ru, sorokinnc@rambler.ru

Abstract. The proposed system for monitoring number and duration of power outages and power quality in 0.38 kV power networks makes it possible to shorten the power supply restoration time by approximately one hour by reducing the time for obtaining information about the damage and by approximately one hour by the reduction of the time for determining the location and type of damage. Besides, the effect can also be obtained by minimizing power quality inconsistency time with the standardized values. The sensors of the monitoring system are proposed to be located at customer inputs or at several network points, for example, at the beginning, in the middle or at the end of the power network as well as at the transformer substation bus bars.

Keywords: Power supply reliability · Power quality
Monitoring power supply reliability · Monitoring power quality
Power supply restoration time

1 Introduction

The power supply system efficiency can be assessed by the indices of power supply reliability and power quality. The methods and means for improving power supply reliability and power quality (PQ) [1, 2] are considered in publications of different authors. As such measures the use of the technical condition monitoring of transmission lines and network equipment operating modes are considered, which makes it possible to identify and prevent the causes of failures in the networks [3–5]. Much attention is paid to the development of technical and economic mechanisms to stimulate consumers and electric grid companies to increase power quality parameters [6]. The works of both Russian and foreign researchers are devoted to this subject [7–16].

© Springer Nature Switzerland AG 2019
P. Vasant et al. (Eds.): ICO 2018, AISC 866, pp. 1–10, 2019.
https://doi.org/10.1007/978-3-030-00979-3_1

2   Materials and Methods

Structure analysis of power supply restoration time after network failures made it possible [17, 18] to determine it by the formula:

$$t_{rest.} = t_{inf.obt.} + t_{inf.rec.} + t_{rep.} + t_{harmonize} \qquad (1)$$

where $t_{inf.obt.}$ is the time of information obtaining, h.; $t_{inf.rec.}$ is the time for information recognizing, h.; $t_{rep.}$ is the time for repairing damage, h.; $t_{harmonize}$ is the time to harmonize connection and disconnection, h.

The power supply restoration time can be reduced significantly by the implementation of a monitoring systems that controls the power outages and the voltage deviation and automatically informs the dispatcher about the outages on specific network sections.

The damage from power supply outages of consumers depends on the duration of power outage in a network supplying consumers and the type of disconnected consumers [19, 20]. The causes of outages may be wire breaks in any part of the power lines, short circuits in the line, a power failure on the 10 kV side etc. Depending on the cause, outages can occur either for all consumers connected to the network under consideration, or for a part of consumers, for example, when a wire breaks. The more sensors of outages and voltage deviations the monitoring system has, the more informative it is, the more situations in the monitored network can be recognized. The most rational option is the installation of the sensors at the input of each consumer. But this variant of sensor placement can lead to a rise in the cost of a system, therefore, in case of insufficient budget the sensors can be installed in several points of the network, for example at the beginning, in the middle and at the end of the transmission line as well as on the buses of the transformer substation. This will allow having the information about the status of the whole network and monitoring the main network parameters on its different section.

In works [6, 21] it was justified that the sensors for monitoring power quality indices were worth to install at customer inputs as well as the sensors of power supply reliability. It is proposed to control the parameters of power quality using information obtained from these sensors. The combination of monitoring of power supply outages and power quality indices along with the automated measuring and the electric power fiscal (or technical) accounting is promising. Theoretically, such an opportunity exists. At present, a rather wide range of metering devices equipped with means to monitor power outages and power quality is produced. This is a series of MAYAK meters, meters of signal frequency receivers equipped with the corresponding functions. But practically these possibilities are not used. Firstly, this is due to the impossibility to read and send the specified information remotely via AMISEPFA channels because they are used only for power consumption data transmission. The information about power outages in these meters are only stored in the meter archive. Secondly, the use of meters equipped with all the necessary capabilities is quite expensive. They are several times more expensive than meters transmitting data only about power consumption. In addition, consumer energy meters send information about power outages occurring in the internal consumer network without getting information about their reasons.

Although power outages might be caused by switching off in the external network, tripouting a switching device at the customer input or even the disconnection by the customer for servicing the wiring.

3 Results

A variant of the sensor location scheme that allows taking into account number and duration of power outages as well as the voltage deviation is shown in Fig. 1.

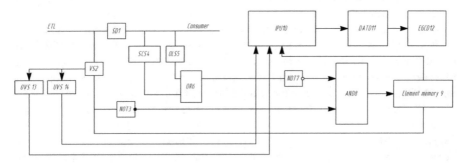

Fig. 1. Scheme of the device for monitoring number and duration of power outages and voltage deviation level at the consumer inputs.

The sensor circuit (Fig. 1) contains the switching device SD 1, the voltage sensor VS 2, the overvoltage sensor OVS 13, the undervoltage sensor UVS 14, the short-circuit sensor SCS 4, the overload sensor OLS 5, the information processing unit IPU 10, the NOT 3 element, the NOT 7 element, the OR 6 element, the AND 8 element, the element Memory 9, Data Acquisition and Transmission Device DATD 11, the data from which are transferred to the the electric grid company dispatcher EGCD 12.

In the normal operation mode there is voltage in the power transmission line supplying a consumer and there is no overload or short circuit in the internal network of the consumer. Thus the signal is present at the output of the voltage sensor VS 2 and there is no signal at the outputs of the elements SCS 4 and OLS 5. There is also no signal from the outputs of the sensors OVS 13 and UVS 14. In this case, the output of NOT7 is signaled to one of the inputs of the AND 8 element, and there is no signal at the outputs of the elements NOT 3, OR 6, AND 8, Memory 8, Memory 11. In this case, the signal from the output of the voltage sensor VS 2 is sent to reset the element Memory 9. The circuit does not start.

At the moment of failure in the transmission line, the voltage at the consumer input disappears, that is, the signal from the output of the element VS 2 disappears. Accordingly the signal appears at the output of NOT 3, which is fed to the input of the AND 8 element. If there are no signals at the sensor outputs the short-circuit current of the SCS 4 and the overload sensor OLS 5, the signal at the output of the NOT 7 element is present and fed to the second input of the AND 8 element. At both inputs AND 8 there are signals, hence a signal will appear at its output, which will be memorized by the Memory 9 element and transmitted to the information processing unit IPU 10. The IPU 10 stores the fact of power outages and its duration. The disconnection signal is transmitted via the DATD 11 to the electric grid company dispatcher (EGCD 12). When the voltage in the transmission line is restored, the VS 2 sensor will detect its presence, the signal with VS 2 will "reset" the element Memory 9, the circuit will return to its original state.

In case if a short-circuit current appears in the consumer's internal network and after its disappearing there is the voltage in the power line, the circuit will work as follows. At the moment of the short-circuit current appearance at the output of the SCS 4 element there will be a signal which will be fed to the input of the OR 6 element and from its output to the input of the NOT 7 element and also to the information processing unit IPU 10. IPU 10 unit fixes the facts of the short-circuit current and overloads in the internal network of the consumer. At the output of the element NOT 7, the signal during the period of the presence of short-circuit current will be absent. At the same time the short-circuit current in the consumer network will cause a failure in the input voltage. As a result, the signal will disappear at the output of the sensor VS 2 and at the output of the element NOT 3 the signal appears that it will feed one of the inputs of the element AND 8. But the signal at the AND 8 output will not appear because of the signal absence at its second input. After the short-circuit current has disappeared because of the switching device disconnection, the signal from the SCS 4 output will disappear as well. The voltage level at the input will return, a signal will appear at the output of the VS 2 and the signal at the output of the NOT 3 will disappear. The circuit will not start. Thus in this operation mode the block IPU 10 will detect the fact of a short circuit in the consumer network without switching off the input voltage.

In case if short circuit current occures in the internal network of the consumer and there is non-selective tripping of the switch installed in the power line, the circuit will work as follows. At the moment of the appearance of the short circuit current, at the output of the element SCS 4 there will be a signal that will be fed to the input of the OR 6 element. A signal from the OR 6 element will be sent to the input of the element NOT 7 and to the information processing unit IPU 10. unit IPU 10 fixes the facts of short circuit and overloading in the internal network of the consumer. At the output of the element NOT 7 the signal during the period of the short-circuit current will be absent. At the same time a short circuit in the consumer's network will cause a voltage drop at the input. As a result the signal will disappear at the output of the VS 2 sensor and will appear at the output of the element NOT 3 and be fed to one of the inputs of the AND 8 element. Due to the absence of a signal at its second input during the short-circuit current flow the element AND 8 will not work and the signal at its output will not appear. Also, the signal will appear at the output of the UVS 14 element and will be fed to the IPU 10. After the short-circuit current has disappeared due to the disconnection of the switching device installed in the power line, the signal from the output of the SCS 4 will disappear. But due to the nonselective disconnection of the switching device in the line, the voltage at the consumer input will disappear as well. Therefore, the signal at the output of the VS 2 will not appear and the signal at the output of the NOT 3 will not disappear. Thus, signals will be fed to both inputs of AND 8, and a signal will appear at its output. This signal, memorized by the element Memory 9, will be fed to the input of the block IPU 10. The fact and duration of the voltage outage at the consumer input will be fixed by this unit. Also, in this operation mode the block IPU 10 will detect a short circuit in the customer network. The information about these facts will be stored in the IPU and will be transmitted through the DATD and the corresponding data transmitting channel to the dispatcher.

If there is an overload in the internal network of the consumer and after its disappearance the voltage in the transmission line does not disappear, the circuit will work as follows. At the moment of the overload current appearance a signal will appear at the output of the element OLS 5 which will be fed to the input of the element OR 6. From the output of the element OR 6 the signal will come to the input of the element NOT 7 as well as to the information processing unit IPU 10, which fixes the facts of the short circuit and overloads in the consumer internal network. During the presence of the overload current the signal at the output of the element NOT 7 will be absent. At the output of the sensor VS 2 the signal will not disappear, so a signal at the output of the element NOT 3 will not appear. Because there is no signal on one of the inputs, the element AND 8 will not work and there will be no the signal at its output. After a consumer switching device get be disconnected the overload current will disappear and the signal from the output of OLS 5 disappears as well. Thus, in this operation mode the IPU 10 unit will record the fact of an overload in the consumer's network without switching off the input voltage.

The situation where there is the non-selective disconnection of a power line switching device during an overload in the consumer network is generally analogous to the situation of non-selective triggering of the switching device during a short circuit in the consumer network The difference is that the input signal of the element OR 6 will be fed from the sensor OLS 5 instead of the element SCS 4. In this case the block IPU 10 will fix a disconnection in the line as well as an overload in the consumer network.

Both the history of accounting number and duration of power outages and the facts of short circuits and overloads in the consumer network are stored in the memory of IPU 10 in the form of protocols; all these data can be transferred to the dispatching office of the electric grid company.

In case of exceeding the voltage deviation level in one or another side of the normalized value, a signal will appear at the output of the high-voltage sensors OVS 13 or the low-voltage sensors UVS 14, which will be transmitted to the IPU 10 and further to the dispatching office of the electric grid company.

Thus, the supposed device supports automatic calculation of the amount of the consumed power, accounting of the number and duration of power outages, monitoring and recording of emergency situations in the consumer network along with and voltage drops in the consumer electrical network. The information on the discrepancy of the voltage deviation is sent to the block IPU 10 and transmitted by means of the data transfer device via one of the channels (PLC, JPS, JPRS, Glonass, radio…) to the electric grid company dispatcher (EGCD 12).

Using the devices mentioned above the system for monitoring number and duration of power outages and the power quality in electrical networks of 0.38 kV can be performed as follows. The sensors for monitoring power quality indices and sensors for recording number and duration of power outages can be installed at the consumer inputs (in the simplest case only the level of voltage deviation at the customer input can be used as a monitored power quality index). Both types of sensors can be combined into one device (for example, a device for monitoring number and duration of power outages and voltage deviation - DMNDCandVD).

The information from the DMNDCandVD goes to the data processing unit and is transmitted by means of the data transfer device via one of the channels to the electric grid company dispatcher (EGCD) (Fig. 2).

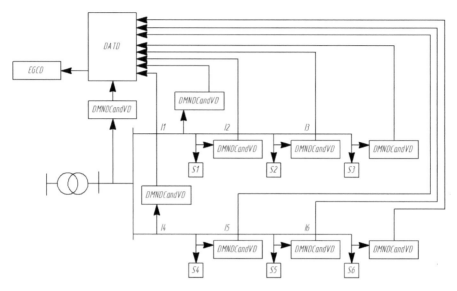

Fig. 2. System for monitoring the quantuty and duration of power outages and the power quality in 0.38 kV electrical networks

According to the Fig. 2 TS is a Transformer Substation; DMNDCandVD is the Device for Monitoring Number and Duration of Power Outages and Voltage Deviation; S1... Sn are Consumers; EGCD is the Electric Grid Company Dispatcher; DATD is the Data Acquisition and Transmission Device.

The system works as follows. The information from the DMNDCandVD of each consumer is collected in the DATD and sent to the EGCD. There on the basis of this information a company dispatcher make decisions about the need for voltage regulation or for sending a repair team. The DMNDCandVD is also installed at the TS and connected to TS buses. This device monitors voltage deviation and voltage loss on the TS buses. The information about this is transmitted to the EGCD by means of the DATD. The DMNDCandVD installed on outgoing transmission lines from the TS work similarly. They are connected after the automatic switches of outgoing transmission lines and monitor voltage disappearance, voltage deviation at the beginning of the transmission lines. If a power scheme do not has a back-up power and the monitoring task of voltage deviation is not required, the control of power outages can be made cheaper by installing voltage monitoring devices on transformer substation buses and on outgoing lines. In this case the voltage disappearance on the transmission line shows electricity supply interruption for the consumers connected to it. The voltage disappearance on the transformer substation buses shows the power supply interruption for all consumers connected to the TS. But this method is not acceptable if transmission lines are equipped with means of automatic transfer switch, partitioning means and if it is also necessary to control voltage deviation level at consumer inputs. Therefore, the system version where all consumers inputs are equipped with the DMNDCandVD are more justified since it is more functional and allows identifying some network modes other ways cannot detect. For example, if a wire (of a phase) breaks in the area between consumers S2 and S3, the signal from the DMNDCandVD installed at the consumer S3 will show the voltage failure presence and the DMNDCandVD installed at the consumer S2 will indicate that the voltage is present. Thus, the system for monitoring number and duration of power outages will find the place of the break.

4 Discussions

The economic efficiency of the system application can vary depending on the tasks assigned to it. The system application of monitoring the number and duration of power outages allows obtaining an economic effect mainly by reducing the power supply restoration time. In this case the recovery time based on the analysis of works [18, 22] can be shorten by approximately one hour by reducing the time for obtaining information about the damage and by approximately one hour by the reduction of the time for determining the location and type of damage. In total, the recovery time can be reduced by approximately 2 h. In determining the effect it is also necessary to take into account the reliability indices of the network under discussion since they can be different depending on whether the cable or overhead lines which are used in either urban or rural areas [2].

The effect of reducing the power supply restoration time can be determined as follows. First, it is required to estimate the failure probability in the considered networks during a year. In papers [23, 24] there is literature data on the failure rate for 0.38 kV networks. According to them the failure intensity for 0.38 kV power networks is 2.7 failures per 100 km during a year, for power transformers is 3.5 failures per 100 km during a year, for the 10 kV overhead lines is 35.9 failures per 100 km during a year.

For 0.38 kV overhead lines with a length L_{OL} and failure rate of 2.7 failures per 100 km during a year the probable number of failures per year is determined as follows:

$$N_{OL} = \frac{2.7 \times L_{OL}}{100} \text{ failures per year.} \tag{2}$$

For power transformers with number n_{tr} and failure rate 3,5 failures per 100 km during a year the probable number of failures per year is determined as follows:

$$N_{TR} = \frac{3.5 \times n_{TR}}{100} \text{ failures per year.} \tag{3}$$

The failures number per year for other network elements can be determined similarly.

The next step is to determine the electricity shortage per year for a given number of failures for power supply system elements.

The power undersupply is determined by the formula:

$$W_{undersupply} = N \times T_{av} \times P_{rat} \qquad (4)$$

where P_{rat} - rated power of the load connected to the faulty equipment; T_{av} - the average power supply restoration time, which according to [13] is equal to 5.86 h.

The reduction of power undersupply is defined as follows:

$$W_{red.} = W_{undersupply} - W_{undersupply.l.} \qquad (5)$$

where $W_{undersupply.l.}$ – power undersupply when implementing a monitoring system.

In the above sequence, it is possible to assess the economic damage caused by the power undersupply to consumers during power outages.

If there is voltage deviation monitoring, the effect of reducing the power supply restoration time will be added by a number of effects achieved by observing network voltage regime, which does not deviate beyond the normative values. The calculation of this effect is described in [25] in detail.

5   Conclusions

1. It is possible to increase power supply system efficiency by monitoring number and duration of power outages and power quality in 0.38 kV electrical networks. The proposed system for monitoring number and duration of power outages and power quality makes it possible to shorten the power supply restoration time and to obtain the necessary data on network operating modes and network failures.
2. The sensors for monitoring power outages and power quality are proposed to be located at the consumer inputs or at several network points, for example at the beginning, in the middle and at the end of transmission lines as well as on transformer substation buses. It makes it possible to expand the system informativeness and the possibilities of using it to diagnose failures in a controlled network.
3. Economic efficiency of the system is achieved mainly by reducing the power supply restoration time. The power supply restoration time can be reduced by approximately 1 h by reducing the time for obtaining information about the damage and approximately 1 h by reducing the time for determining the location and the type of damage. Thus, the recovery time on average can be reduced by 2 h. Besides, the effect can also be obtained by minimizing power quality inconsistency time with the standardized values.

References

1. Vinogradov, A., Vinogradova, A.: Improving the Reliability of Electricity Supply for Rural Consumers by Sectioning and Reserving 0.38 kV Transmission Lines: Monograph. Publishing House of FSBEI HE Orel State Agrarian University (2016)
2. Leschinskaya, T., Magadeyev, E.: Methodology for Choosing the Optimal Option to Improve the Reliability of Power Supply to Agricultural Consumers. Publishing House of FSBEI HPE Moscow State Agroengineering University (2008)
3. Vinogradov, A., Sinyakov, A., Semenov, A.: Computer program for choosing a monitoring system for the technical condition of overhead power lines. Sci. Pract. Mag. Agrotec. Power Supply 3(12), 52–61 (2016)
4. Kostikov, I.: CAT-1 monitoring system – increase of capacity and reliability in power lines. Power Eng. 3(38) (2011)
5. Samarin, A., Rygalin D, Shklyaev, A.: Modern technologies of monitoring of electric power networks of power lines. Natural and technical sciences, 2(58), 341–347 (2012)
6. Borodin, M., Vinogradov, A.: Increase of Efficiency of Electricity Systems Functioning by Measuring Monitoring of Electricity Quality: Monograph. Publishing House of FSBEI HE Orel State Agrarian University (2014)
7. Bolshev, V., Vinogradov, A.: Review of foreign sources devoted to increasing the efficiency of power supply systems. In: Energy Saving and Efficiency in Technical Systems: Materials of the IV International Scientific and Technical Conference of Students, Young Scientists and Specialists, pp. 372–373. Tambov State Technical University (2017)
8. Li, J., Song, X., Wang, Yu.: Service restoration for distribution network considering the uncertainty of restoration time. In: the 3rd International Conference on Systems and Informatics (ICSAI). IEEE (2016)
9. Orichab, J.: Analysis of the interrelated factors affecting efficiency and stability of power supply in developing countries. In: AFRICON 2009. IEEE (2009)
10. Gheorghe, S., Tanasa, C., Ene, S., Mihaescu, M. Power quality, energy efficiency and the performance in electricity distribution and supply companies. In: The 18th International Conference and Exhibition on Electricity Distribution, Turin. IET (2005)
11. Santarius, P., Krejci, P., Brunclik, Z., Prochazka, K., Kysnar, F.: Evaluation of power quality in regional distribution networks. In: the 23rd International Conference on Electricity Distribution. AIM, Lyon (2015)
12. Irwin, L.: Asset management benefits from a wide area power quality monitoring system. In: The 23rd International Conference on Electricity Distribution. AIM, Lyon (2010)
13. Dhapare, S., Lothe, N., Ramachandran, P.: Power quality monitoring with smart meters. In: the 23rd International Conference on Electricity Distribution. AIM, Lyon (2015)
14. Saele, H., Foosnas J., Kristoffersen, V., Nordal, T., Grande, O., Bremdal, B.: Network tariffs and energy contracts. In: The CIRED Workshop "Challenges of Implementing Active Distribution System Management". AIM, Rome (2014)
15. Mandatova, P, Massimiano, M., Verreth, D., Gonzalez C.: Network tariff structure for a smart energy system. In: The CIRED Workshop Challenges of implementing Active Distribution System Management. AIM, Rome (2014)
16. Goswami, A., Gupta, C., Singh, G.: An analytical approach for assessment of voltage sags. International Journal of Electric Power Energy Systems, 31(7–8), pp. 418-426 (2009)
17. Vinogradov, A., Vasiliev, A., Semenov, A., Sinyakov, A.: Analysis of the time of breaks in power supply for rural consumers and methods of reducing it due to monitoring the technical condition of power lines. Bulletin of All-Russian Institute of Electrification of Agriculture, 2 (27), pp. 3–11. (2017)

18. Vinogradov, A., Vasiliev, A., Bolshev, V., Semenov, A., Borodin, M.: Time factor for determination of power supply system efficiency of rural consumers. In: Kharchenko, V., Vasant, P. (eds.), Handbook of Research on Renewable Energy and Electric Resources for Sustainable Rural Development, pp. 394–420 (2018). https://doi.org/10.4018/978-1-5225-3867-7.ch017
19. Papkov, B., Osokin, V.: Probabilistic and Statistical Methods for Assessing the Reliability of Elements and Systems of Electric Power Industry: Theory, Examples, Tasks: Textbook. Stary Oskol, TNT (2017)
20. Khorolsky, V., Taranov, M., Petrov, D.: Technical and Economic Calculations of Electrical Distribution Networks. Publishing House Terra Print, Rostov-on-Don (2009)
21. Golikov, I. Vinogradov, A.: Adaptive Automatic Voltage Regulation in Rural Electric Networks 0.38 kV: Monograph. Publishing House of FSBEI HE Orel State Agrarian University (2017)
22. Vinogradov, A., Sinyakov, A., Semenov, A.: Analysis of the time of restoration of power supply to rural consumers in case of power line failures. Theor. Sci. Practical J. Innov. Agribus. Prob. Prospects **1**(13), 12–22 (2017)
23. Semenov, A., Selezneva, A., Vinogradov, A.: Comparison of reliability indicators of air and cable lines in urban and rural areas. In: The Main Directions of the Development of Technology in Agrobusiness: The Materials of the VII All-Russian Scientific and Practical Conference, Knyaginino, pp. 71–75 (2015)
24. Vinogradov, A., Perkov, R.: Analysis of the damage ability of electrical equipment in electrical networks and measures to improve the reliability of electricity supply to consumers. Bulletin of the Nizhniy Novgorod State Engineering and Economic Institute, **12** (55), pp.12–20. (2015)
25. Perova, M: Economic Problems and Prospects of Qualitative Power Supply of Agricultural Consumers in Russia, Moscow (2007)

A Novel Application of System Survival Signature in Reliability Assessment of Offshore Structures

Tobias-Emanuel Regenhardt[1], Md Samdani Azad[2(✉)], Wonsiri Punurai[2], and Michael Beer[1]

[1] Institute for Risk and Reliability, Leibniz University Hanover, Callinstrasse 34, 30167 Hanover, Germany
regenhardt@irz.uni-hannover.de

[2] Department of Civil and Environmental Engineering, Mahidol University, 25/25 Putthamonthon, Nakhon Pathom 73170, Thailand
mdsamdani.aza@student.mahidol.ac.th

Abstract. Offshore platforms are large structures consisting of a large number of components of various types. Thus a variety of methods are usually necessary to assess the structural reliability of these structures, ranging from Finite-Elements-methods to Monte-Carlo-Simulations. However, often reliability information is only available for the members and not for the overall, complex, system. The recently introduced survival signature provides a way to separate the structural analysis from the behaviour of the individual members. Thus it is then possible to use structural reliability methods to obtain information about how the failure of several constituent members of the offshore platform leads to overall system failure. This way it is possible to separate the structural from time-dependent information, allowing flexible and computationally efficient computation of reliability predictions.

Keywords: Structural reliability · Offshore platforms
Survival signature · System reliability

1 Introduction

Offshore jacket platforms are generally used for oil and gas production in shallow and intermediate water depths. Adequate performance of the platforms is ensured by designing for a service life. However, a large numbers of these steel structures are operating exceeded their design life due to high cost of replacement. Consequently, the safety of these offshore platforms creates strong reasons to develop effective methods for the reliability assessment.

For large offshore structures, reliability measures usually concern the structural reliability under the impact of external influences such as fatigue, and corrosion environment. As structural reliability concerns the behavior of an object under physical conditions, a safety assessment should prove that the risk of

© Springer Nature Switzerland AG 2019
P. Vasant et al. (Eds.): ICO 2018, AISC 866, pp. 11–20, 2019.
https://doi.org/10.1007/978-3-030-00979-3_2

structural failure is acceptable. The standard methods give some indications, such as design code, reserve strength ratio, and a probabilistic value. Design codes are claimed to be very conservative, as more knowledge of the structure is gained through some years after design thereby leading to more accurate analysis results. Methods based on reserve strength ratio can provide insight into the reserve strength of a structure [12]. The reserve strength ratio (RSR) can be obtained from the ratio of ultimate load capacity of the structure divided by the 100 year design load. But it will not cover possible failure modes that could happen to the structure as it provides information regarding the global failure phenomenon as well. Structural reliability methods typically account for the capacity versus loading, particularly deal with uncertainties of structural loads and their effects as well as resistance [14].

Reliability theories basically developed from the concepts of uncertainties (wind, wave and earthquake). An incremental loading approach till the ultimate capacity was conducted for structural reliability is delineated. However, the structural reliability methods are not sufficient measures as they are not consistent with the derivation of the reliability target levels. This is because the reliability assessments deliberates the reliability considering the intensity of environmental conditions (Loads, Corrosion) but not give enough information over time rather these provide information over fixed time. To reduce risk, a better approach is to consider all functional parts of the structure, if present (facilities on offshore platforms, the connections between platforms, pipes, dominant failure modes etc.), exploring patterns and inter-relationships within subsystems and seeing undesired events as the products of the working of the system. Some conventional tools have been used including Failure Tree Analysis (FTA) [11], Failure Mode and Effects Analysis (FMEA) [9], recently, researchers are paying more attention to the statistical techniques. For instance, grey correlation analysis [2], Bayesian Probability [19], Neural network [21], Fuzzy logic evaluation [8,20] and survival signature computing [4] have been applied to the risk assessments in engineering and related fields.

The aim of this paper is to contribute to offshore reliability assessment by using the recently developed survival signature formalism [4]. With this formalism it is possible to predict the reliability of a complex system (as, in this case, an offshore platform) from knowledge about the individual constituents (the platform members). With this, it is possible to divide the reliability assessment into two individual steps. Firstly the system structural system is analysed. This is achieved through finding the combinations of failing members that lead to total failure of the whole offshore platform. Secondly, the information about the members' reliability over time is multiplied with the corresponding entries of the survival signature to predict the overall reliability.

2 The Survival Signature

2.1 System Reliability Applying the Survival Signature

The state of a any system set together of m independent and different components can be represented by a *state vector* $\underline{x} \in \{0,1\}^m$ with $x_i = 0$ denoting a dysfunctional and $x_i = 1$ a functional component i.

The *global structure function* $\varphi : \{0,1\}^m \rightarrow \{0,1\}$ contains information whether the system is in a working state ($\varphi = 1$) or not ($\varphi = 0$) for any possible $\underline{x}$. Usually the observed systems are restricted to *coherent* systems. This refers to systems with φ not decreasing in any dimension of $\underline{x}$. This assumption is sound as most common systems are not becoming dysfunctional while gaining more functional components. Two additional assumptions are $\varphi(\underline{0}) = 0$ and $\varphi(\underline{1}) = 1$. These are intuitive, yet not necessary. However, in this paper the monotonicity of the system is assumed and thus these two conditions and the coherency of the system are assumed as well.

For more complex systems, every component belongs to one of K different types, while each set of components of type $k \in \{1, 2, ..., K\}$ consists of m_k elements and the sum of all m_k equals the number of components $\sum_{k=1}^{K} = m$. The amount of functional components of type k present in the system are denoted as l_k. This leads to $\binom{l_k}{m_k}$ possible combinations of component type k under the assumption of independent failure of all components. Then the set $S(\underline{l})$ is the collection of all state vectors that fulfil the condition that $\underline{l} = (l_1, l_2, ..., l_k)$ components are working. The system's *survival signature* Φ is now defined as the probability that the system is functional if exactly l_k components of type k are functional [23]. The survival signature is an array of K dimensions with $m_k + 1$ entries in each dimension (including the case that none of the components of that type function). For components with exchangeable random failure times the survival signature is given by

$$\Phi(l_1, l2, .., l_K) = \left[\prod_{i=1}^{K} \binom{l_k}{m_k} \right]^{-1} \times \sum_{\underline{x} \in S(\underline{l})} \varphi(\underline{x}). \tag{1}$$

The survival signature can be applied to the computation of the *survival function* of the system: $P(T_S > t)$. It provides the probability that a random failure time T_S of the system follows a specific point in time t. This provides the reliability of the system in time. Under the assumption of the failure times of the components being independently and identically distributed (*iid*), with respect to a known cumulative distribution function [15]. $F_k(t)$, the survival function of the system observed is found to be

$$P(T_S > t) = \sum_{l_1=0}^{m_0} ... \sum_{l_k=0}^{m_k} \left[\Phi(\underline{l}) \times \prod_{k=1}^{K} \binom{l_k}{m_k} F_k(t)^{m_k - l_k} [1 - F_k(t)]^{l_k} \right]. \tag{2}$$

Equations (1) and (2) indicate show that - for exact computation - many different states need to be evaluated and that the size of the survival signature itself is growing multilinearly. However, for small- and medium-sized systems the survival signature can be calculated exactly or by Monte Carlo Simulation methods. The use of signature frameworks can be useful in several ways. It seperates the information about the system in two subproblems to be solved. If the system structure is unchanged, the survival signature stays the same even if the behaviour of the components changes. Thus testing of components in simulations is computationally efficient. Additionally, additions to the framework can easily be done, for example in case of repairable systems [22] or in case of components with multiple states [5].

2.2   Obtaining the Survival Signature from Structural Simulations

The calculation of the Φ necessitate knowledge about the behaviour of the system under failure of the components. Usually due to complexity, an explicit global structure function is often not given. Instead, the states of the system can be evaluated by various means, including reliability block diagrams and cut-sets, binary decision diagrams, and failure tree analysis [17].

With reliability block diagrams, it is simple to visualize the behaviour of small systems. However, the search of cut-sets in a block diagram is NP-hard and can be very time consuming [18]. Binary decision diagrams can provide fast means to calculate the survival signature once the decision diagram data structure is available. However, the calculation of the decision diagram is also dependent on finding the cut-sets of the system and can be, inherently, slow.

For structural reliability, one is usually concerned with the behaviour of a structure under *load*. Thus the interaction of the various components is not modelled in any way described above - instead, the structure is modelled and analysed in frameworks of mechanical simulation methods (commonly, finite elements methods) concerned with the actual physical behaviour.

In this work, a bridge over this gap is presented. A large structure consists of several, possibly redundant, components. This means that the system might still be operational after the failure of some of the components. Thus, structural simulation can show various failure modes of the system under load. If a structural simulation of the structure results in a failing component, the structure is updated and the simulation started again. This is repeated until the simulation results show that the structure is failing in total. All failed components until this point are saved in a failure mode. By variation of the load parameters, all components prone to failure are identified and several failure modes are identified. These failure modes can be used as cut-sets in computation of the survival signature. Equation (1) can be evaluated using these cut-sets to compute the values of $\varphi(\underline{x})$ for all $\underline{x}$ (exact computation) or a representative sample (Monte Carlo Simulation). In this study, the Monte-Carlo approach was used as the amount of combinations to be checked is of medium size. The largest amount of combinations that is possible is for the entry placed directly in the middle of the array ($\binom{3}{2}^3 \cdot \binom{8}{4} = 1890$). Thus a sample size of 2000 samples was used.

3   Reliability of an Offshore Jacket Platform

3.1   Structural Model

The jacket platform is taken from [16]. The Jacket is designed for shallow water depth of approximately 65.31 m. It is a 4-legged jacket containing pile inside the legs. The jacket is modeled as 2×2 square grid. The overall dimensions are 8×8 m at top elevation and 21.76×21.76 m at the mud line. The total height is 81 m. Two types of bracings are used named as horizontal bracings and vertical bracings. The horizontal bracings are installed at five levels. The vertical bracings are provided as single bracings till the bottom level. At the bottom level, it was provided as K-bracings to impart more stiffness and reduce buckling. The jacket support/foundation is modelled as fixed support system. The jacket is modelled in *SAP2000* as shown in Fig. 1. Member properties of the jacket are also taken from [16]. The top mass of the oil and gas platform is simplified as a lumped mass for the easiness finite element modelling. The total weight of the topside is assumed as 1250 tons, which is equally applied over four legs where each leg, is carrying 312.5 tons at the top nodes of the jacket structures platform.

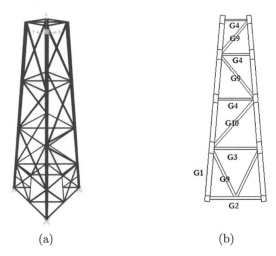

(a)   (b)

Fig. 1. (a) Three-dimensional model of the jacket (b) Grouping of the components

3.2   Generation of Failure Tress

Non-linear static (pushover) analysis is performed to understand the behaviour of structure against lateral load pattern following the procedure of FEMA356 [6,9]. The behaviour of the force displacement curve can be observed from the analysis as well. In this step, the structure is incrementally loaded over its yielding capacity and to observe the ultimate load level. The failed elements are recorded up to the ultimate load level. Here the damage level is not considered because the aim is to grasp the failure behaviour of members up to ultimate load level. Here the term failure behaviour is defined as how the member fails chronologically and which member is followed by another member. Load has been applied along three different directions comprising of 0°, 90° and 45°. The typical pushover curve is shown in Fig. 2 which is adopted from *FEMA356* [6] and the ultimate load level is the point 'C'. The failure tree can be observed in Fig. 3.

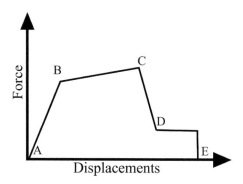

Fig. 2. Non-linear force curve

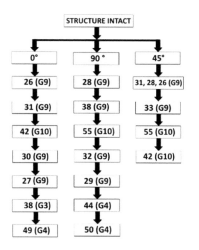

Fig. 3. Failure tree of the offshore platform under various loads

The tree is generated for three different load directions. For zero degree of direction, the first damage is observed in member 26. After that, this member is followed by members 31, 41, 30, 27, 38, and 49. When the damage initiates in member 49, the load level reached the ultimate level. For 90° direction, the first damage can be noted in member 28. The other members are 38, 55, 32, 29, 44 and 50. In case of 45 degree loading direction, the first damage is detected in 3 members in parallel labelled as 31, 28 and 26. The reason behind this is that the load is equally distributed over 2 orthogonal directions which caused three members to fail at once.

3.3   CDF of Component Types

After the successful identification of the failure-prone members of the offshore platform, it is necessary to choose proper cumulative distribution functions (CDF) for the calculation of the survival function. In this study, the components are the failure-prone members of the platform grouped in four different types according to tube total diameter (d_i) and wall thickness (τ_i). These groups can be found in Table 1.

Table 1. Overview of the failure prone component types

Group	d_i [m]	τ_i [m]	μ_i	σ_i
3	0.013	0.406	1.000	0.895
4	0.010	0.356	0.769	0.805
9	0.013	0.550	1.000	1.000
10	0.019	0.559	1.462	1.008

As many environmental factors influence each platform individually (corrosion rates, ocean movement, fatigue, and usage of protection measurements against these), in this study a general log-normal distribution is assumed. The two form factors (mean μ and standard deviation (σ) are under strong influence of the environment and the individual situation the platform is modelled or investigated in.

Thus, as a proof of concept, the mean time of failure and standard deviation is assumed to be of unity for the component group 3. To properly scale the other groups' parameters, in first order approximation, the mean follows $\mu_i \sim \tau_i$ and $\sigma_i \sim \sqrt{d_i}$, respectively. In an applicated situation, the distribution of failure times can be obtained in dependence of the individual situation. As corrosion is one of the most important influences to offshore reliability, the corrosion rate can be measured over time and compared with the thickness of the affected members to estimate the probability that a member has corroded to an unstable state at a given time. Additionally the corrosion can be modelled if precise measurements and knowledge about the situation is available (e.g. salinity, pH value and corrosion countermeasures). Also the influence of fatigue can be taken into account. In the most optimal case, a thorough study of the material behaviour can be done on similar structures already present. The repair and maintenance rates can be easily adopted to estimate values for the mean values of the failure times of a certain component type (standard deviation is then just a matter of how the data spreads).

Measurements concerning the structural properties that relate to time-dependent reliability are, as all measurements, to some degree imprecise. Additionally, many aspects of the temporal behaviour of the structural elements is only known to a certain degree beforehand. Thus imprecision has to be taken in account. To demonstrate imprecise probabilities in this case, the survival function can be computed by using an upper and lower bounds ($\overline{F}_k(t)$ and $\underline{F}_k(t)$) for the CDF together with the survival signature formalism [4]

$$\overline{P}(T_S > t) = \sum_{l_1=0}^{m_0} \cdots \sum_{l_k=0}^{m_k} \left[\Phi(\underline{l}) \times \prod_{k=1}^{K} \binom{l_k}{m_k} \overline{F}_k(t)^{m_k-l_k} \left[1 - \overline{F}_k(t)\right]^{l_k} \right], \quad (3)$$

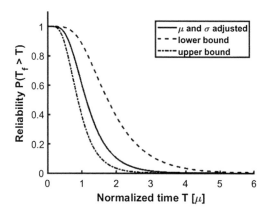

Fig. 4. Survival functions of the system

In this example, the highest and lowest value of μ_i was chosen to generate upper and lower bounds for the CDF and thus to generate a bounding p-box for the time-dependent reliability. The corresponding survival functions can be found in Fig. 4. The survival functions not only show the expected behaviour in reliability for the whole system, it becomes also clear that using only slightly different probability distributions can have grave impact on the long term reliability of the system.

4   Summary

This study shows that structural information can be obtained in order to apply the survival signature formalism - which originates from system reliability - to structural problems.

With the introduced method, it was possible to identify the members of the offshore structure whose failure leads to overall system failure. It becomes clear that under load only a selected few of the components of the platform are relevant to the failure modes of the platform. This reduces the amount of components to be considered in the system reliability measurements and makes the analysis using system reliability methods feasible. Originating from observations made during the initial state of the system a reasonable prediction of the behaviour in time can be made. For example, a maintenance cycle can be defined by setting a certain value for the reliability: as soon as the reliability is less than the defined value, maintenance has to occur.

However, individual temporal behaviour is highly dependent on the environmental situation and has to be implemented carefully. The impact of imprecise probabilities is to be taken into account properly in the future.

Acknowledgments. This project has received funding from the European Union's Horizon 2020 research and innovation under the Marie Skodowska-Curie grant agreement No. 730888. This work was also funded by the Deutsche Forschungsgemeinschaft (DFG, German Research Foundation) grants BE 2570/3-1 and BR 5446/1-1 as part of the project 'Efficient Reliability Analysis of Complex Systems'.

References

1. Bai, Y., Yan, H., Cao, Y., Kim, Y., Yang, Y., Jian, H.: Time-dependent reliability assessment of offshore jacket platforms. Ships Offshore Struct. **11**(6), 591–602 (2015)
2. Bao, X., Cao, A., Qin, F.: Safety assessment of an aging offshore jacket platform by integrating analytic hierarchy process and grey clustering method (2016). Preprints, 2016080182
3. Bilionis, D. V., Vamvatsikos, D.: Fatigue analysis of an offshore wind turbine in meditteranean sea under a probabilistic framework. In: VI International Conference on Computational Methods in Marine Engineering, MARINE (2015)
4. Coolen, P.A., Coolen-Maturi, T., Aslett, L., Gero, W.: Imprecise system reliability using the survival signature. In: Proceedings of the 1st International Conference on Applied Mathematics in Engineering and Reliability, ICAMER 2016 (2016)
5. Eryilmaz, S., Tuncel, A.: Generalizing the survival signature to unrepairable homogeneous multi state systems. Nav. Res. Logist. **63**(8), 593–599 (2016)
6. Federal Emergency Management Agency: Prestandard and Commentary for the Seismic Rehabilitation of Buildings, Chapter 3 (2000)
7. Golafshania, A.A., Tabeshpourb, M.R., Komachia, Y.: FEMA approaches in seismic assessment of jacket platforms. J. Constr. Steel Res. **65**, 1979–1986 (2009)
8. Kang, H., Han, J., Zhou, P.: Risk assessment for offshore jacket platform based on fuzzy probabilistic influence diagram. In: Proceedings of 16th International Offshore and Polar Engineering Conference, San Francisco, CA, USA (2006)
9. Kim, D., Ok, S., Song, J., Koh, H.: System reliability analysis using dominant failure modes identified by selective searching technique. Reliab. Eng. Syst. Saf. **119**, 316–331 (2013)
10. El-Din, M.N., Kim, J.: Seismic performance evaluation and retrofit of fixed jacket offshore platform structures. J. Perform. Constr. Facil. **29**(4), 040114099 (2015)

11. Kurian, V.J., Wahab, M.M.A., Kheang, T.S., Liew, M.S.: System reliability of existing jacket platform in Malaysian water. Appl. Mech. Mater. **567**(316–331), 307–312 (2013)
12. Majid, W.A., Hashim, A.R., Embong, MB.: Determination of structural reserve strength ratio (RSR) of an existing offshore structure. In: The Eighth International Offshore and Polar Engineering Conference (1998)
13. Melchers, R.E., Beck, A.T.: Structural Reliability Analysis and Prediction. Wiley, Hoboken (2017)
14. Moses, F., Stahl, B.: Reliability analysis format for offshore structures. J. Pet. Technol. **31**(03), 347 (1979)
15. Patelli, E., Feng, G., Coolen, F.P.A., Coolen-Maturi, T.: Simulation methods for system reliability using the survival signature. Reliab. Eng. Syst. Saf. **167**, 327–337 (2017)
16. Punurai, W., Azad, M.S., Pholdee, N., Sinsabvarodom, C.: Adaptive meta-heuristic to predict dent depth damage in the fixed offshore structures. In: European Safety and Reliability Conference, NTNU, Trondheim, Norway (2018)
17. Reed, S.: An efficient algorithm for exact computation of system and survival signatures using binary decision diagrams. Reliab. Eng. Syst. Saf. **165**, 257–267 (2017)
18. Supplement information to [17]. https://github.com/sean-reed/bdd-signatures
19. Ren, J., Wang, J., Jenkinson, I., Xu, D.L., Yang, J.B.: A Bayesian network approach for offshore risk analysis through linguistic variables. China Ocean. Eng. **21**(3), 371–388 (2007)
20. Ren, J., Jenkinson, I., Wang, J., Xu, D. L., Yang, J. B.: An offshore risk analysis method using fuzzy Bayesian network. J. Offshore Mech. Arct. Eng. **131**(4) (2009)
21. Uddin, Md. A., Jameel, M., Razak, H. A.: An offshore risk analysis method using fuzzy Bayesian network. Indian J. Geo-Mar. Sci. **44**(3) (2015)
22. Walls, L., Revie, M., Bedford, T.: Risk, Reliability and Safety: Innovating Theory and Practice. CRC Press, Boca Raton (2017)
23. Zamojski, W., Mazurkiewicz, J., Sugier, J., Walkowiak, T., Kacprzyk, J.: Complex Systems and Dependability. Springer, Heidelberg (2012)

Security Assurance Against Cybercrime Ransomware

Habib ur Rehman[1,3]([✉]), Eiad Yafi[2], Mohammed Nazir[1],
and Khurram Mustafa[1]

[1] Department of Computer Science, Jamia Millia Islamia, New Delhi, India
way2habibmca@gmail.com, {mnazir,kmustafa}@jmi.ac.in
[2] Malaysian Institute of Information Technology, Universiti Kuala Lumpur,
Kuala Lumpur, Malaysia
eiad@unikl.edu.my
[3] DXC Technology, Noida, India

Abstract. Cybercrime is not only a social ill but it does also pose a tremendous threat to our virtual world of personal, corporate and national data security. The recent global cyberattack of WannaCry ransomware has created an adverse effect on worldwide financials, healthcare and educational sectors, highlighting the poor state of cyber security and its failure. This growing class of cyber attackers is gradually becoming one of the fundamental security concerns that require immediate attention of security researchers. This paper explores why the volume and severity of cyberattacks are far exceeding with the capabilities of their mitigation techniques and how the preventive safety measures could reduce the losses from cybercrime for such type of attacks in future. It further expresses the need to have a better technological vision and stronger defenses, to change the picture where human cognition might be the next big weapon as a security assurance toolkit.

Keywords: Cybercrime · WannaCry ransomware · Human cognition
Cybersecurity · Cyberintrusions · Cyberattack

1 Introduction

Constant stream of news on cyber offences has contributed to a sense that cybercrime is out of control, whether it is or not, but it knocks down the false sense of growing security in the arena. Every year, millions of personal information are stolen or critical data become unavailable by encryption [30]. It may not be wrong to say that today's cyber security challenge, in a nutshell, is crafting an impression that attackers are far ahead of the defenders [25]. McAfee Net losses reports says that "Annual economy grows from $2 trillion to $3 trillion and cybercrime costs US$375 billion to $575 billion to the global economy" extort 15%–20% of the value [16]. Keeping a dollar cost on cybercrime and cyber espionage becomes the headline, but the most obvious

---

On Sabbatical leave from DXC Techology.

© Springer Nature Switzerland AG 2019
P. Vasant et al. (Eds.): ICO 2018, AISC 866, pp. 21–34, 2019.
https://doi.org/10.1007/978-3-030-00979-3_3

questions about its damage on victims are the loss of cyberspace [24]. The threat landscape is dynamic and constantly adapting new methods of exploitations [27, 39]. The security measure usually gets exposed from time to time which make traditional security measures insufficient to protect the digital assets. We studied the adapting ransomware attack incidents; they are threatening the cyber security in an ongoing fashion. Hence, we need to strengthen security with power of computation.

2   Background

Cybercrime is taking various forms to exploit financial, corporate, consumers credit card or government agencies' confidential information and usually gain attention when millions of records are stolen by the hackers. The past attack trends show the

Table 1. Major Cyberattacks of the years

Dates	Company	Description of breach
May 2017	Microsoft	A worldwide cyberattack by the WannaCry ransomware crypto worm, exploit Microsoft Windows & targeted the computers running over Windows operating system, and have encrypted the data of approx. 200,000 systems under 150 countries [6, 32], demanded ransom payments in the *Bitcoin* cryptocurrency
Dec 2016	Ukraine power grid	The spear-phishing emails hampered utility's network and malware caused denial of service attack, targeted number of critical infrastructures even blackout the power plants
Sep 2016	Yahoo	Massive data breach exposed the records of 500 million customer names, email addresses, phone numbers and even their hashed passwords
April 2016	Philippine elections voters' data	The personal information of approximately 55 million voters was made public by Lulzsec Pilipinas [7] from (COMELEC) Philippine Commission on Elections
Feb 2016	Hollywood Presbyterian Hospital	Sensational security attacks against Hospital that was compelled to pay a whopping $17,000 ransom to regain access on the files locked by ransomware
July 2015	Ashley Madison data breach	Through a bad MD5 hash implementation over 37 million customer records and their account's passwords were made vulnerable [12]
Feb 2015	Anthem data breach	Millions of records compromised in a healthcare network via attack on administrator password [27]
Dec 2014	Yahoo	The biggest data breach in history up to one billion user accounts was compromised [27]
May 2014	EBay	Massive data breach exposed records of site's 233 million customers, including names, email addresses, physical addresses, phone numbers and birthdates [14]

(continued)

Table 1. (*continued*)

Dates	Company	Description of breach
Feb 2014	Michaels stores	Under a fraudulent activity hacker attacked on the data security of Michael's stores and exposing 2.6 million records of U.S. payment cards that were used at Michaels stores [2]
Jan 2014	Neiman Marcus	Hacker exposed 1.1 million records of customer cards
Dec 2013	JPMorgan Chase	Hackers directly attacked on banking giant's network and compromised the personal information of 465 thousand card holders [24]
Nov/Dec 2013	Target	A highly specialized attack through a Malware stored on Target's checkout registers that theft of data for approx. 40 million credit & debit card accounts and personal information of up to 70 million customers [30]

exponentially rising impacts of cybercrimes in our real and digital world. Some of the prominent attack incidences are imprinted in Table 1 depicts the growing confidence and maturity of the complex cyberattacks, requires immediate attention.

There are various reasons for hackers to hack—better indigenous organizational capabilities mean a greater return from hacking [16]. Attackers see low risk from cybercrime, with ever increasing benefits as industrial, manufacturing, information communication and research capabilities improve around the world [28, 31], their returns on stealing data, IP or other critical resource will always increase [16]. Regardless of the reasons of attacks – corporate revenge extortion or simply malicious hacker's behavior – web services are at higher risk of critical data and financial losses as well as damage of their global business reputation [24]. The cybercriminals produce high returns at (relatively) low cost and low risk e.g. the ransomware attacks.

3 Study of Ransomware

The rise of ransomware cyberattack is not a new form of extortion – from its dormant introduction three decades ago. It is a type of cybercrime extortion scheme formed from **ransom** with *software*—prohibits users for accessing their critical resources until a ransom is not paid [5]. Currently, ransomware attacks hinder system operation in three ways: *locker ransomware:* by blocking accessing to the system; *crypto ransomware:* by making data unusable by means of encryption algorithms; and by combining of *locker/crypto ransomware*. The crypto ransomware can be most destructive typically using strong encryption algorithms. The cybercriminals typically trap the users to activate this malicious software through phishing email (HTML links or attachments), they encrypt the critical data with a private key owned by attacker and demands money from the users for the digital key to unlock and regain the access on their own data [6]. To escape from law enforcement, these cybercriminals use anonymous payment methods usually in cryptocurrencies e.g. *bitcoin* [15].

The rising usage of an untraceable digital *cryptocurrencies,* the swift money that can be made in ransomware scams appears for the incensed innovation in the cybercrime world. These cyberattackers have started representing *ransomware-as-a-service* model. To wider the network they are providing customizable ransomware and easy-to-follow instructions to nontechnical users. The increasing rate of return from attacks, is the attacker's incentives on the other hand defenders are lacking the incentive to do more, as they are underestimate the risks that results the growing losses from cybercrime. Attackers can buy software in the line of *ransomware-as-a-service* for collecting ransom, that generally happens in crypto-currencies. The critical resources of profiled clients are typical targets of ransomware attack. Also, they prepare the database of such type of clients. Hence paying ransom sometime may prove to be invitation to further extortion. Analyzing the data of previous attacks, we can summarize the types of ransomware.

3.1   Types of Ransomware

- BitCrypt is more refined ransomware act for the theft of cryptocurrency e.g., Bitcoin.
- Critroni or Curve-Tor-Bitcoin (CTB) locker ransomware uses the Tor network to mask its C&C communications. e.g. CTB Torrent Locker adds CAPTCHA code and redirection to a spoofed site and asks ransom in Bitcoins.
- CRIBIT malware extorts in the form of Bitcoins for unlocked files.
- Cryptorbit or CryptoDefense, malware not only encrypts database, web, Office, video, images, scripts, text, and other non-binary files but also deletes backup files to prevent restoration of encrypted files. It traps the system and demands money for decryption. It can be easily spread as compared to other malwares via removable drives hence eliminate the need of network connection or relying malware down-loader to infect the systems.
- CryptoLocker Ransomware encrypts files, rather than locking the system. It usually downloads from the spammed message attachment. Variant of CryptoLocker might abuse Windows PowerShell feature to encrypt files and remain undetected on the system and/or network.

3.2   Activities of Ransomware

A Ransomware zip's target critical file types such as .DOC, .DLL, .EXE, .XL and overwrites them keep them in password protected zip files in the user's system. It asks ransom through a note for the zip file password or the decryption key. Furthermore, a Ransomware targets Master Boot Record of a vulnerable system to keep undetected till it displays ransom notification. A Ransomware infects known critical file, such as user32.DLL and locks the screen of the infected computer thereby prevents detection by behavioral monitoring tool. e.g. the infected user32.DLL will begin a chain of routinesload ransomware and lock the screen with projecting a ransom image messages. In this paper, we carried out a detailed analysis of a severe *crypto ransomware* attack. It is known as *WannaCry* because of the name present at its application descriptor. It has caused global business loss due to the fast spread rate and encryption complexity [10]. The healthcare, educational, financial and telecommunication organizations were adversely affected [4]. We put an effort to expose the uniqueness of this attack and explore the direction of research to strengthening the security measure.

4 WannaCry Ransomware

At the peak of a harried work day on Friday May 12, 2017, and a legitimate-looking e-mail lands in the inbox: the company's security software detects a security breach, prompting to run an instant background scan for malicious code over the machines. Usually, in hurry, the common tendency among the users is just to click 'accept' without reading [9]. That swift, incensed click send over the fake e-mail drill malware itinerary into the company's network and encrypted the data and convert the machine to a zombie *botnet* ready to fires off more spam [15]. Many accessible files are encrypted over the infected machine and a ransom note appears over the screen shown in Fig. 1. The attacker seems to have a much better grasp on human psychology over the practices meant to defend [26].

Cause of attack: The new software releases, brings up fresh and important features along with unexpected vulnerabilities. The critical notifications and fixes turn up after the exploits are identified [35]. These vulnerabilities are ongoing problems for the customer facing applications and services that create storm of threats [20]. The files detailing vulnerability was leaked before it gets patched [4], which was exploited to encrypt data on affected user machines and demands ransom payment in the form of *bitcoin*. Microsoft instructed users to apply the recommended patches over services network environments as a preventive measure. The benefits of being able to swiftly update code to address newly discovered vulnerability outweigh the risks leaving millions of devices vulnerable until they are modified [35].

The impact was wide spread due to the delay in patching across the globe hence, infected over million systems within a short span. Those systems which applied the Microsoft Windows critical patch released in April 2017 were protected against the attack [32]. The *Bitcoin* activities were also increased indicating some success with the ransomware. It would encourage more variants and efforts to expand this type of attacks in near future. *Will the patches and Managed Security Services tools fully protect a customer's environment from such type of attacks in future?* The patching is just a virtual pace to move ahead with real loss – not a security approach for the organizations and is unavoidable in future at the cost of next vulnerability.

In past ransomware attacks ransom note were displayed as an image over the infected machine, but the WannaCry uses an executable file pretending to be a benign program displaying the note using *SetForegroundWindow()* adaptable in different language formats [32]. Nevertheless, the attacker's plans to execute a *ransomware* attacks to generate maximum illegal revenue, but the recent attack set comparatively less ransom starting from $300, increased to $400 after two hours and finally $600 per endpoint so on [4]. The motive appears to have widespread impact. This ransom payment application was engineered to make it easy for the victim to pay via *bitcoin* to anonymously receive payment from anyone [5]. Furthermore, Unique *bitcoin* wallet address is generated for every infected machine, however due to a race condition bug the code has not executed as intended by the attackers. Later, the *bitcoin* address is defaulted to only three hardcoded addresses, due to which it was impossible for attackers to identify which victims have paid, hence unlikely to ever get their files decryption key. On May 18, the attacker subsequently released a new version of the malware to correct the flaw. But they were comparatively unsuccessful like the original version impacted [4].

Fig. 1. The critical files gets encrypted and application wanna descriptor ransom note appears over victim system informing the instructed to pay ransom.

There were two issues associated under this exploitation, widespread patching for MS17-010 is the first challenge and secondly the malware had used a strong asymmetric encryption by employing the RSA 2048-bit cipher to encrypt the files. This was potentially the most damaging aspect of these attacks due to the strength materialize virtually impossible to break [32]. Comparatively the past ransomware codes were simplistic without modularity but the recent WannaCry architecture is modular in nature. This feature is known to be used for legitimate big software development projects, or used in various complex malware projects such as banking Trojans [32]. This indicates that WannaCry code is more likely prepared by highly organized cybercrime gangs and reminds Dridex [17] and Locky attack [22].

WannaCry spread at an alarming rate until a 'kill switch' was accidentally discovered by Marcus Hutchins, a 22-year-old self-taught cybersecurity researcher [4] who identified a hardcoded domain from a part of its code i.e.http://www.iuqerfsodp9ifjaposdfjhgosurijfaewrwergwea.com. This was unregistered and swiftly turned it into a sinkhole by the security researcher [32]. It enabled access to the sinkhole domain from the endpoints. The machines infected over the weekend carrying this domain, became reachable after weekend and malware was not activated on those endpoints [32]. However, the 'kill switch' was not helpfull for organizations whose data was already affected, leaving behind the option to rely over the backups [5]. These domain calls have only worked for the directly connected systems i.e. the attack was halted for consumer devices, which were not having proxies [32]. The workaround had failed if the endpoint proxies the traffic. It specifies attacker's intent was the corporate networks where endpoint traffic is usually through proxies. It shows how intelligently the attackers have targeted the attack vectors. The traditional security measures remain insufficient to protect the digital assets with evolving attacks.

5   From Security to Safety

Continued successes of a cyber-security attack have highlighted the weakness of security measures. When a new security measure evolves, its effectiveness and accuracy increases rapidly by the time it gets deployed, where it grabs broad exposure to the real-world scenarios and threats as depicted in Fig. 2. When the expert's suggestion and feedback of the development team with inclusion of other defense measures are incorporated, the technique further improves in accuracy. This trend continues up to a level of effectiveness by the time adversaries starts responding. After performing rigorous experiments, the attackers explore new ways to evade the defense by developing countermeasures to trim down its value.

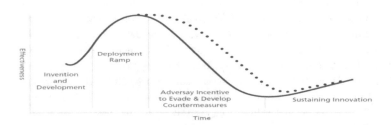

Fig. 2. Reducing effect of security measure [33]

5.1   Insufficient Security Measures

The security measure gets exposed over a period. They should regularly strengthen with power of computation. It is also desirable to learn from the past incidents and incorporate the required changes in the security measures (Table 2).

Table 2. Limitation of traditional Security measure

Attacks	Security measure*
Intrusion	As the numbers of signatures are exponentially growing with recognized threats. It is impractical that endpoint systems always have the most updated version of the latest signature, hence foolproof endpoints security is impossible in signature based detection
Zero-day	The attack surface remains open for a new exploit by the time it is identified and signature is updated in antivirus based intrusion detection systems. The cybercriminals and hackers use this window period to propagate new attacks using the botnets and various other techniques
Specific IP	The cybercriminals and hackers target to a specific system or organizations with a new method of attack or exploit where no prior information or signature exist leaving the endpoint insecure in a targeted attack [29]
The hybrid	Cybercriminals are using hybrid techniques using malware crypters, server-side polymorphism and explores other methods to disguise malwares from the antivirus packages hence fail to recognize by the signatures

The researchers have hopes from behavior analysis to overcome the limitations. Under *behavior recognition,* every program execution has certain behavioral characteristics. The behavior recognition technique allows the program to execute under a safe 'sandbox' and observe its behavior as a benign or a malware program—for example, the program execution stats accessing the email distribution lists or editing registry keys or trying to disable the anti-malware packages without user interference [9]. The analyses of these behavior patterns are complex that separates the malicious behavior from the benign behavior [34]. The threat behavior database also requires continuous update but not as instantaneously as it is required for signature base update to avoid zero-day attacks.

Journaling and Rollback: The behavior recognition might also not be enough for short-term behavior recognition [34]. The malware writers might be clever enough that program exhibit the malicious behavior with some delay. So, we need a long-term behavior analysis pooled with journaling and rollback feature [34], checkpoints based execution, file modifications, memory locations mapping, registry keys tracking i.e. journaling of all entities. This is to keep track of a before and after image of every change. If malicious behavior is identified at some later stage, its change impact can be minimized by deleting and rolled back, returning to the checkpoint at some known good state. Together, these approaches would improve performance of attack detection and mitigation techniques to a broader range of malware and will reduce the time spent in distributing signature updates and cleaning infected systems [9, 34].

5.2   Genesis

Cleaning up in aftermath of cybercrime is costlier than the cybercrime itself. The organizations that fail in protecting their networks might have rising competitive disadvantage. They might experience a reduced valuation if ever hacked. This is the time to embrace new security approach or existing measures to make sensitive enough for considering the limitations of the existing approaches. Cybersecurity is not just the problem of a techno scientist—its potential solutions need inputs from psychologists, economists and human-factors specialist [26]. The human behavior and cognitive response can play a prominent role to safeguard from unauthorized access in cyberspace [21]. The active involvement of user cognition can strengthen the security. Security and safety are open-ended subjective problems, though security becomes worse over a compromised system, with growing uncertainty about the knowledge and capabilities of attackers. The system's interaction with its environment requires subjective assessment by a designer, security administrator and user.

We cannot assure that system is 100% free from malicious programs. It is also difficult to assure that entropy of credential is free from brutal force attack from an illegitimate user. Cryptographic security is also dependent on computational limitation. Hence, security assurance against unauthorized access is difficult to evaluate under the existing access control mechanism. A critical application needs high security, privacy to assurance its digital identity. Application might have various distinguishable functionalities. Do all require same level of protection? The need of security assurance can also vary with the criticality of operations the application is performing at an instance. How do we achieve confidence about access control security as classic philosophical problem especially when it is impossible to examine every possible circumstance that assures our belief? Assurance cases are inherently defeasible, which means that there is always the possibility that something has been omitted. When we say that we have "complete" confidence in a claim, we understand that this only reflects what we know at a point in time [4].

5.3   Security Assurance

Security assurance intends to provide a degree of confidence instead of a true measure of how secure is the system [37]. An assurance case is a structured argument that system properties are as per desire i.e. safe, or reliable, or secure against attack, where as *"safety case communicates a clear, comprehensive and defensible argument about the acceptability for safe operation in a particular context."* [38]. The assurance against ransomware attack also depends on the safe design of the system, i.e. the system properties are safe (considering the threat of ransomware attack). It assures that existing vulnerability cannot be exploited by a ransomware. Furthermore, ***Implicit assurance*** *by* means of best practices for recover to acceptably safe or secure to minimize impact.

Explicit assurance: A system is acceptably safe or secure; means have an implicit assurance case. To have an explicit assurance case, it needs some changes in the system, and setup rules to assess the user, service provider for subjective assessment producing the impact on assurance case, where even a compromised system can become insensitive to exploit the full capability. In the line of explicit assurance, we propose an assurance framework, which does not strictly follow the interpretation of conventional method of accepting full range of evidences to provide either *full-access* (trust) or *no-access* (no-trust), rather it could accept partial evidence and provide different levels of assurance controls. The less critical functionality can be performed by providing the most relevant evidence. Hence if vulnerability exists in the system, the full capability of the system cannot be exploited during the attack.

5.4   Theoretical Basis for High Assurance Framework

The assurance case argument shows how a high-level claim (e.g. "the system is adequately secure") is ultimately supported by detailed evidence at specific level of confidences. An assurance case will refer to a range of evidence, and will show how various evidences combine to provide confidence in higher-level properties [38]. It captures the rationale of why the evidence we have produced because of the low-level activities provides a reason to believe the high-level claim. Evidence could be prioritized from the weakest to the strongest; the most relevant evidence only provides a gradual support for claim, hence setup connection between evidence and claims (Fig. 3).

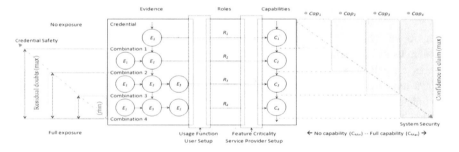

Fig. 3. Claim confidence and security assurance framework.

To explain the framework, we give an example of traditional access control methods, which provides full privilege P_{Max} to an application up to its maximum functionalities F_{Max}, depicted as follows:

$$p_i(x) := \begin{cases} 0, & x = 0, \\ F_1||F_2||F_3||F_4||F_5||F_6, & x = P_{Max} \end{cases}$$

If the application capabilities are disintegrated under different functionalities, which can be mapped with gradual variation in privilege, the traditional access control methods can be represented as $p_i(x) := \int_0^{P_{Max}} F(x)$ where $F(x) \subset (F_1 < F_2 < F_3 < \ldots < F_{Max})$. Under assurance framework there should be a threshold in privilege p as per user is need and recognition $F(x)$. It sets the benchmark for executable and non-executable functionalities i.e. below threshold functionalities are executable $p_i(x) < \int_0^p F(x)$ while above threshold remains non-executable $p_i(x) > \int_p^{P_{Max}} F(x)$, hence during an attack instant, certain functionality remains protected unlike the traditional threat over the entire system. Hence, protected up to the extent of functionality usage.

The proposed framework will utilize user cognition to use specific functionality of the application which will set the threshold of privilege; and hence meets the situational need with optimum level of privilege at each of the login instances. We suggest the functionality separation approach for the shake of application security and balance it with credential safety by means of user cognition. It is controlled by multiple credentials to provide different level of access. Hence, functionality access can include both user identity and intent. The access to the resources is stablished by setup of the multi-layer trust instead of solely user's identity (password), although user cognition should be an exemplary constituent of the overall judgment [36].

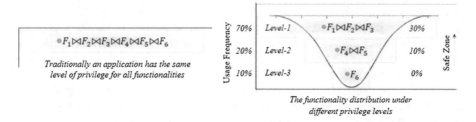

Fig. 4. Functionality vs. privilege usage threshold, and its comparison.

The Fig. 4 illustrates if there are three levels of privilege covering six different functionalities of an application. Assume the level-1 functionalities fulfil user needs in 70% occasions which would keep the 30% uncovered functionalities in safe zone (20% in level-2 and 10% in level-3). At level-2 the 10% uncovered functionalities will remain under safe zone, whereas at level-3, since none of the functionality is protected that implies entire application will behave in traditional way and no functionality will be under the safe zone. Traditionally the entire application provides full privilege for all functionalities without the concept of safe zone as illustrated in Fig. 4. Hence the entire application suffers the threat against any attack vector (Table 3).

Table 3. Three level privilege

Level	Range	User request	Usage %	Safe zone
Level-1	$0 < x < P_1$	$f_1(x) = \int_0^{P1} (F_1 < F_2 < F_3)$	70	30%
Level-2	$P_1 < x < P_2$	$f_2(x) = \int_{P1}^{P2} (F_4 < F_5)$	20	10%
Level-3	$P_2 < x < P_3$	$f_3(x) = \int_{P2}^{P3} (F_6)$	10	0

The traditional design, which does not allow privilege verses functionality subdivision, hence it cannot leverage the benefit for securing the critical functionalities. The concept of safe zone provides explicit assurance to the system. Here we need to design the application by clear subdivision among the functionalities, and the criteria of subdivision is influenced with the criticality of the functionality. It provides extra range of control to users, who can take naturalistic decision as per *situation of logics* [36] e.g. avoid performing critical functionality of official work from personal computer where no extra protection of firewall.

6 Conclusion and Future Direction

We are facing tremendous threats to our personal data and the value it represents attracts thieves. The paper covers past ransomware attacks and the detailed analysis of the recent WannaCry ransomware attack that have impacted wide range of computer systems by exploiting a security flaw of Microsoft software. In recent years *ransomware* is progressively an intensifying threat; with forthcoming new variations presenting the increasing success of *'ransomware-as-a-service'* of attackers' business model. The paper discuss how ransomware authors are constantly modifying their games, techniques to snag victims and to stay ahead of security researchers. They are trying new tricks to exploit the vulnerabilities of widespread software. The prominent causes of cyberattacks are due to the ongoing expansion of attack surface, exploitation of applications vulnerabilities, augmented attacker sophistication, the lack of integrated security technologies, the rising cost of breaches, and the shortage of skilled security talent to fight back.

The frequent occurrences of ransomware attacks, employing complex encryption algorithms, which become more challenging to protect the mission critical application. It became undoubtedly a serious challenge, as most of the time the encryption methods incarnations are unbreakable, as unable to reach the limits of modern cryptography. Enterprises and application designer should take preventative measures to design the application in such a way that can reduce the impact of the attack and can safeguard the critical functionality of an application by harden the systems among different layers. The proposed framework suggests few steps to consider while designing a mission critical application to protect against ransomware attacks. The key is to proactively deter ransomware attacks through explicit assurance. We hope this framework can suggest application designer to more proactively cope with the increasingly sophisticated threats of ransomware and try to address it at application design level.

References

1. Adham, M., Azodi, A., Desmedt, Y., Karaolis, I.: How to attack two-factor authentication internet banking. In: International Conference on Financial Cryptography and Data Security, pp. 322–328. Springer, Heidelberg (2013)
2. Arlitsch, K., Edelman, A.: Staying safe: cyber security for people and organizations. J. Lib. Admin. **54**(1), 46–56 (2014)

3. Bergman, M.K.: White paper: the deep web: surfacing hidden value. J. Electron. Publ. **7**(1) (2001)
4. Collier, R.: NHS ransomware attack spreads worldwide. CMAJ **189**, E786–E787 (2017). https://doi.org/10.1503/cmaj.1095434
5. Everett, C.: Ransomware: to pay or not to pay? Comput. Fraud Secur. **4**, 8–12 (2016)
6. Gandhi, K.A.: Survey on ransomware: a new era of cyber attack. Int. J. Comput. Appl. **168** (3), 38–41 (2017)
7. Greenleaf, G.: Philippines Appoints Privacy Commission in Time for Mass Electoral Data Hack (2016)
8. Jøsang, A., et al.: Local user-centric identity management. J. Trust. Manag. **2**(1), 1 (2015)
9. Kirlappos, I., Parkin, S., Sasse, M.A.: Learning from 'Shadow Security': why understanding noncompliant behaviors provides the basis for effective security. In: USEC Workshop on Usable Security (2014)
10. Laszka, A., Farhang, S., Grossklags, J.: On the economics of ransomware (2017). arXiv preprint arXiv:1707.06247
11. Levchenko, K., et al.: Click trajectories: end-to-end analysis of the spam value chain. In: Proceedings of IEEE Symposium on Security and Privacy, pp. 431–446 (2011)
12. Lunker, M.: Cyber laws: a global perspective. Internet Source (2005). http://unpan1.un.org/intradoc/groups/public/documents/APCITY/UNPAN005846.pdf
13. Mansfield-Devine, S.: The Ashley Madison affair. Netw. Secur. **9**, 8–16 (2015)
14. Martin, G., Kinross, J., Hankin, C.: Effective cyber security is fundamental to patient safety (2017)
15. Minkus, T., Ross, K.W: I know what you're buying: privacy breaches on ebay. In: International Symposium on Privacy Enhancing Technologies Symposium, pp. 164–183. Springer International Publishing (2014)
16. Mohurle, S., Patil, M.: A brief study of Wannacry Threat: ransomware attack. Int. J. **8**(5), 1938–1940 (2017)
17. Net Losses: Estimating the Global Cost of Cybercrime McAfee, Center for Strategic and International Studies (2014). http://go.nature.com/15nom3
18. OBrien, D.: Dridex: Tidal waves of spam pushing dangerous financial trojan. Symantec, White Paper (2016)
19. OWASP: AppSec Europe HTTP Parameter Pollution (2009). http://www.owasp.org/images/b/ba/AppsecEU09_CarettoniDiPaola_v0.8.pdf. Accessed 20 Apr 2014
20. Perlroth, N.: Hackers in China attacked The Times for last 4 months. NY Times, 30 January 2013
21. Rehman, H., Nazir, M., Mustafa, K.: Security of web application: state of the art. In: International Conference of Information, Communication and Computer Technology ICICCT 2017 likely to be appear soon in Springer CCIS series (2017)
22. Robert S., Philip S.: Client-side attacks and defense. In: Syngress (2012). ISBN: 978-1-59749-590-5
23. Rudman, L., Irwin, B.: Dridex: analysis of the traffic and automatic generation of IOCs. In: Information Security for South Africa (ISSA), IEEE 2016, pp. 77–84, August 2016
24. Scaife, N., Carter, H., Traynor, P., Butler, K.R.: Cryptolock (and drop it): stopping ransomware attacks on user data. In: 2016 IEEE 36th International Conference on Distributed Computing Systems (ICDCS), pp. 303–312. IEEE, June 2016
25. Shields, K.: Cybersecurity: recognizing the risk and protecting against attacks. NC Banking Inst. **19**, 345 (2015). http://scholarship.law.unc.edu/ncbi/vol19/iss1/18
26. Turpe, S.: Security testing: turning practice into theory. In: IEEE International Conference on Proceedings of Software Testing Verification and Validation Workshop, ICSTW 2008, pp. 294–302 (2008)

27. Waldrop, M.M.: How to hack the hackers: the human side of cybercrime. Nature **533**(7602), 164–167 (2016)
28. Walters, R.: Cyber attacks on US companies in 2014. Heritage Foundation Issue Brief, vol. 4289 (2014).
29. Web Application Attack and Audit Framework. http://w3af.sourceforge.net. Accessed 20 Apr 2014
30. Weinberger, S.: Is this the start of cyberwarfare? Nature **474**(7350), 142 (2011). Chicago
31. Wilkinson, C.: Cyber Risks: The Growing Threat (2013)
32. Zhang, H., Yao, D.D., Ramakrishnan, N.: Detection of stealthy malware activities with traffic causality and scalable triggering relation discovery. In: Proceedings of the 9th ACM Symposium on Information, Computer and Communications security, pp. 39–50, June 2014
33. https://www.mcafee.com/in/resources/reports/rp-threats-predictions-2016.pdf. Accessed 27 Jun 2018
34. https://www.webroot.com/shared/pdf/reinventing-antivirus.pdf
35. https://www.mcafee.com/in/resources/reports/rp-threats-predictions-2016.pdf. Accessed 27 Jun 2018
36. Bruza, P.D., Wang, Z., Busemeyer, J.R.: Quantum cognition: a new theoretical approach to psychology. Trends Cogn. Sci. **19**(7), 383–393 (2015)
37. Rehman, H., Khan, U., Nazir, M., Mustafa, K.: Strengthening the Bitcoin safety: a graded span based key partitioning mechanism. In: International Journal of Information Technology (selected for publication in vol. 10) (2018)
38. Alexander, R., Hawkins, R., Kelly, T.: Security assurance cases: motivation and the state of the art. High Integrity Systems Engineering, Department of Computer Science, University of York, York, UK (2011)
39. Almasri, A.H., Zuhairi, M.F., Darwish, M.A., Yafi, E.: Privacy and security of cloud computing: a comprehensive review of techniques and challenges. J. Eng. Appl. Sci. (Under Review)

SAR: A Graph-Based System with Text Stream Burst Detection and Visualization

Tham Vo Thi Hong[1,2(✉)] and Phuc Do[3]

[1] Lac Hong University, Biên Hòa, Dong Nai, Vietnam
[2] Thu Dau Mot University, Thủ Dầu Một, Binh Duong, Vietnam
thamvth@tdmu.edu.vn
[3] University of Information Technology, VNU-HCM, Linh Trung, Thu Du, Ho Chi Minh, Vietnam
phucdo@uit.edu.vn

Abstract. Smart city trend with Artificial Intelligence, Internet Of Thing and Data Science has been attracting a lot of attention. Following this trend, smart applications that help users improve their quality of life, as well as work, has been investigating by many researchers. In an era of industry 4.0, collecting and exploiting information automatically is essential so that many studies have proposed models for solving storage problems and supporting efficient data processing. In this paper, we introduce our proposed graph-based system called SAR (Smart Article Reader) that can store, analyze, exploit and visualize text streams. This system first gathers daily articles automatically from online journals. After articles are collected, keywords' frequency of existence is calculated to rank the importance of keywords, finding worthy topics and visually display the results from user requests. Especially, we present the application of Burst Detection technique for detecting periods of time in which some keywords are unusually popular. This technique is used for finding trends from online journals. In addition, we present our method for rating keywords, which share similar Bursts patterns, based on their term frequencies. We also perform system algorithm testing and evaluation to show its performance and estimate its responding time.

Keywords: Graph database · Visualization · Keyword extraction
Burst detection

1 Introduction

There are averagely more than 100 articles posted every day on an online newspaper. It takes at least 2 min to read an article and more than 3 h to read all of them. Since there are more than 20 online newspapers in Vietnam, it takes plenty of time to read and convey information in order to find newsworthy topics. Hence, we build a system that automatically collects articles from online newspapers, manages and exploits information from those articles in order to help users to find newsworthy topic quickly and visually as well as track the progression of those topics.

The purpose of our application is to answer a number of questions such as "What are keywords of an article?", "What newsworthy keywords are widely used for a

© Springer Nature Switzerland AG 2019
P. Vasant et al. (Eds.): ICO 2018, AISC 866, pp. 35–45, 2019.
https://doi.org/10.1007/978-3-030-00979-3_4

particular topic?", "What keywords have been most commonly used lately?", "What is a lifespan of a keyword?", "What keywords are used most?", "What are interesting topics during a time span?", "What periods of time a keyword occurs with unusually high rates?", etc. So, our goal is to develop a system for collecting, storing, exploiting, and visualizing text stream of daily articles from Vietnamese Online Journals. We believe that our proposed system is totally new and necessary in Vietnam and it is absolutely simple to apply to other languages in other countries. To develop this system, we research and apply some main techniques including Web Crawling, Text Stream Processing with Topic Model, Burst Detection and Text Stream Visualizing by Graphs. We use the TF_IDF technique to find popular topics of articles and also of columns (each article is divided into some categories called columns). In particular, we use the Kleinberg algorithm [1] in searching for periods of time in which keywords appear continuously. Those periods of time are known as "Burst" and this Kleinberg algorithm is called Burst Detection algorithm. In this paper, we introduce our general system design, describe how to apply Burst Detection technique to build an important system function, carry out testing, evaluation and analyze experimental results. The paper is organized as follows. The next section gives a review about document stream and Burst Detection. Section 3 presents our design, explains the implementation of Burst Detection technique and illustrates the algorithms. The system testing and evaluation are shown in Sect. 4. And finally, we draw a conclusion in Sect. 5 which discusses about the advantages and disadvantages of our proposed system and plans some future works.

2   Related Work

2.1   Document Stream

In this work, a document stream is considered as a data stream. It is defined as a sequence of articles that have a temporal order. Figure 1 shows an example of a document stream. In this figure, the documents which are articles arrive in a temporal order.

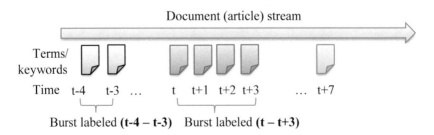

Fig. 1.   Document stream and bursts of keywords

2.2   Burst Detection

The rapid increase of a term's frequency of appearance defines a term burst in the text stream. Thus, a term is considered as bursty when its frequency is encountered at an unusually high rate. The identification of bursts is known as burst detection procedure. Burst detection has been applied by many fields such as biological and finance [2] to solve problems related to time series analysis. Among the proposed methodologies such as test-based method [3], parameter-free method [4], burst detection for events [5, 6], burstiness [4, 7–9], a widely-used one is the two-state automaton proposed by Kleinberg [1] which is used for our system implementation. Kleinberg proposed a burst model where a burst is defined as a rapid increase of a term's frequency of occurrences. If the term frequency is encountered at an unusually high rate, then the term is labeled as 'bursty'. The work is used to identify bursts in text stream and produce state labels of bursts. In this work, articles are published every day, so we identify a burst of an article by a period of time it continuously appears and we label its burst name "start time – end time" (see Fig. 1). The shorter the data arrival time interval, the higher is the degree of burst state and vice versa. The Kleinberg algorithm is applicable to various document streams [10, 11] such as e-mails [12], blogs, online publications [13], and social network [14, 15], etc.

To sum up, Burst detection is a way of identifying bursts, periods of time in which some events are unusually popular. In other words, it can be used to identify fads, or "bursts" of events over time. The idea of using burst detection in this paper is to identify trends from online journals. We carry out our experiment with Vietnamese journals which need more complex text processing tasks.

3   System Design

3.1   Context Introduction

There are averagely more than 20 online newspapers with dozens of articles posted daily in Vietnam online journals attracting a large number of readers. The most important requirement of readers is to find the significant information contained in this huge amount of data spending as less time as possible. Readers also have the need of tracking timeline and evolution of crucial topics. Thus, we consider these articles which are published continuously as a document stream. Then, our whole system goal is to collect these articles to create, store, extract and visualize important information with graphs. To extract information from collected data, we keep developing functions and Burst Detection is one of technique we apply for a function to support users find topical trends and track the evolution of the topic over time. We believe that our proposed system is currently a new system in Vietnam and Burst Detection application in extracting information from online journals is also novel and useful for further research.

4   System Architecture

Our system model is presented in Fig. 2. It can be seen that data is first collected by the Crawler and transferred to the Processor. Here, the data is organized in a Tree structure. Next, the processor implements two main groups of algorithms: the first group consists of Text Processing Algorithms, including Vietnamese Text Segmentation, Stop Word Removal, Topic Detection using the Keyword Extraction approach, which means each topic is represented by a set of keywords and Burst Detection which is our main contribution in this paper. This group of algorithms is responsible for processing, calculating and storing the results. The second group comprises algorithms that expire outdated data, eliminating meaningless data from the system, using a time sliding window, including some algorithms such as Wjoin, PWJoin, etc. In general, the main task of the Processor is to process, calculate and store the results. Finally, the Visualizer interacts with users through the visualization interface and allows users to view, organize and save the output.

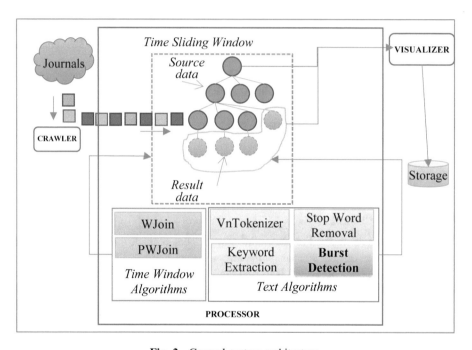

Fig. 2. General system architecture

The system collects, processes data and then stores the processed data every day. We consider each article as an object called Article (Title, Author, Description, Content). Articles come continuously over time into a large text stream. Based on the characteristics of the articles, we group them by Day, Column: Each Day has n Columns, each Column has n Articles. Collected data can be stored (also on external storage if needed) and transferred to the storage structure of a Tree. Then, the system performs a text segmentation step (Data Processor) and calculates the term importance using TF-IDF. Next, the system visually shows the results of user requests. In addition, the system also runs algorithms that eliminate expired datasets in the text stream based on a sliding window with a set time parameter.

The main goal of the system is to detect important topics as well as visualize results with graphs. Bursts detected are also results need to be visually displayed.

There are five main processing steps in the flow diagram as Fig. 3 describes. First, the crawler collects articles from online journals in forms of texts. Next, in step 2, the articles are processed by a word segmentation algorithm called Vntokenizer for creating connecting syllables of more-than-one-syllable words. Vntokenizer is claimed 96–98% accurate. Then, at step three, the articles are processed by removing stop words. These stop words are words which are less meaning getting by calculating from a large set of texts using TF-IDF algorithm. After that in step four, word term frequencies are calculated by using TF-IDF algorithm so that the keyword extraction is performed and the set of top n important keywords of each article is found out so that trustworthy topics from those keywords can be extracted (see [16] for more information about the system).

After defining topics, the set of keywords, we calculate bursts of keywords (a Burst is a period of time that a keyword occurs continuously by day) based on their timestamps. The storage structure is as follows (see Fig. 4).

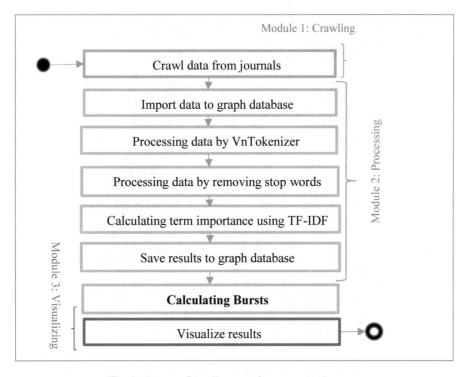

Fig. 3. Process flow diagram of our proposed system

Figure 4a shows the storage structure for Bursts calculation. Each keyword appears at different time points may belong to multiple bursts. Figure 4b shows an example of the keyword "Facebook" that appears at many time points and belongs to 4 bursts.

Figure 4c describes the Burst of the keyword "Facebook" from 2018-03-21 to 2018-03-24. The 4 numbers: 8,25,11 and 10 are the ranks (from 1 to n) of the keyword "Facebook" versus the other n-1 keywords appearing at the same time point. Then we take the sum of these rankings = 54 as the weight used for comparing the importance of this Burst to other Bursts of this keyword.

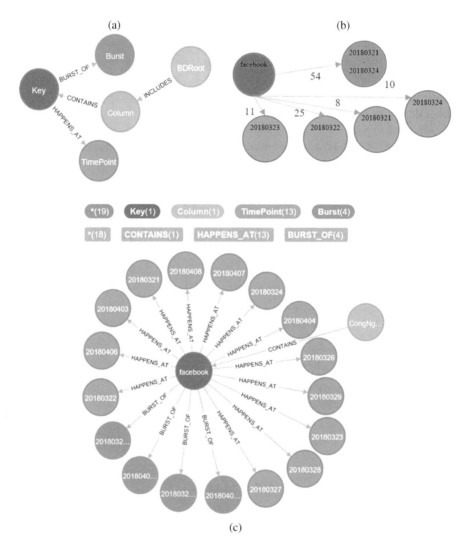

Fig. 4. a Storage structure for Burst Detection, **c** An instant of storage structure for Burst Detection.

4.1 System Algorithm

In this section, we describe the programming codes for applying Burst Detection algorithm and rating for keywords which share similar Burst patterns. Burst detection by Kleinberg's algorithm receives a set of natural numbers and return Burst patterns which are arrays of continuous numbers. We change the original algorithm to solve our problem, so the new algorithm called AdaptingBurst gets an array of date as input parameter and return Burst patterns, arrays of continuous dates (see Algorithm 1).

Algorithm 1. AdaptingBurst

```
Input: Keyword, Column
Output: List result
1: Get the list of TimePoint (TL) of the Keyword in the
Column
2: if the list is empty then exit
3:   result← kleinberg (TL, s=2, gamma=0.01)
```

Algorithm 2 is created for ranking Keywords which share similar Burst patterns by calculating the sum of Keywords frequency (weight of keywords) at every time points in its Burst pattern. The values calculated can be used for identifying the most important keywords.

Algorithm 2. KeywordImportanceIndexCal

```
Input: Keyword, Column
Output: List result
1: Get the list of Burst patterns of the Keyword in the
Column
2: if the list is empty then exit
3: else
4:     result← Sum of weights at every time points for
       each Burst pattern
```

---

5   Experimental Results and Evaluation

5.1   Programming Framework

To develop the system, we use some tools and libraries. We will mention briefly their names and versions here. They are: Scala 2.10.5, Java 8, Spark 1.6.3, Windows Utilities 2.6.x, Maven 3.3, Neo4j 3.2, Apache-tomcat-9.x, IntelliJ IDEA and Vis.js, Python 3.6 and JetBrains PyCharm 2017.3.2.

5.2   Main Functions

Table 1 shows information about the Bursts of keyword "facebook". Each Burst (BurstID, Start, End, Weight of Importance) contains an identified number that counted from 1, a starting time, an ending time, and an importance index that indicates the index of user's interest in the topic containing this keyword. This index is calculated by summing all keyword frequency values in a burst pattern. So, the higher is the weight, the higher level of interest the keyword has.

Table 1. Bursts of keyword "Facebook"

Keyword	Burst ID	Start	End	Importance index
Facebook	1	03-21-2018	03-24-2018	46
Facebook	2	03-26-2018	03-29-2018	22
Facebook	3	04-03-2018	04-04-2018	78
Facebook	4	04-06-2018	04-08-2018	69

Figure 5 illustrates how keyword "Facebook" has appeared in articles from 2018, March 21st to April 04th. It can be seen that "Facebook" become the most popular in 2 days from April 03rd to 04th. After that, people continue to mention "Facebook" from 2018, April 06th to 08th. Therefore, burst detection can help users track "hot trend" of key-words. As we all know, the reason why "Facebook" has become so highly regarded with high frequency in the above periods is due to the leak of user information related to the Trump Presidential election in the United States. Our system also supports users see the results in a visual manner (as seen in Fig. 5).

6   System Testing and Evaluation

The configuration of the computer used for testing is Intel(R) Core(TM) i5-6300HQ CPU @ 2.30 GHz, 4 GB of DDR4 memory under Windows 10 operating system.

We carry out test cases for Burst Detection function. Experimental results of Burst Detection function on articles obtained for a continuous period of 9 days are presented in Table 2. Table 2 shows the statistics on the number of articles collected, keywords analyzed, keywords occurrences, bursts found and the corresponding processing time. Through this table, the factors which affect the processing time of the system can be identified.

In general, the lowest processing time was 11770 ms in the first day when there was no Burst calculation, the processing time then changed during the following days when Bursts are detected and especially when Bursts reached the highest number of 170, the processing time almost reached the highest level at 19220 ms. Thus, it can be seen that as the number of bursts increases, the processing time increases.

Table 2. Time tesing of processing Burst Detection of articles in 19 days

Date No	Articles num	Keywords num	Occurrences num	Bursts num	Processing time (10*ms)
1	162	180	180	0	1177
2	165	179	218	39	1193
3	162	179	251	57	1408
4	159	180	288	73	1534
5	163	177	328	94	1650
6	164	180	272	48	1781
7	159	179	368	133	1620
8	159	179	351	77	1781
9	165	179	350	72	1639
10	163	179	402	111	1512
11	153	179	398	89	1169
12	155	179	455	129	1459
13	156	180	**567**	170	1922
14	159	179	442	101	1385
15	157	179	514	159	1857
16	159	179	505	115	1561
17	192	177	531	140	1740
18	155	178	454	114	1851
19	150	180	541	146	1923

Furthermore, other factors such as the number of articles, the number of keywords, the number of occurrences of keywords also has little effect on the processing time. Specifically, on day 19, although the number of Burst is not the maximum one, the processing time reaches the highest of 19230 ms when the number of keywords reaches the highest of 180.

Figure 6 is drawn from Table 2 showing the correlation between the number of articles collected per day and the Burst detection processing time. In general, processing time is directly proportional to the number of articles. Thus, when the number of keywords occurrences increases over time, the Bursts calculation takes more time. It can be seen in Table 2 that when the number of occurrences gets almost highest values (567 and 541 on 2 days 13 and 19), the processing time is also the highest one (19220 ms and 19230 ms). Therefore, solutions for storage or expiration of past data needs to be taken in the near future in order to optimize the system's processing time.

In addition, our system collects data every day and generate datasets which can be effectively reused for further relevant research. These datasets include articles, pre-processing articles, top n keywords of articles, top n keywords of columns, keywords, and bursty patterns, top n keywords in a bursty pattern, etc. The form of datasets can be freely adjusted to meet several different requirements.

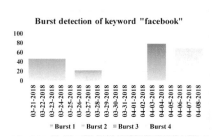

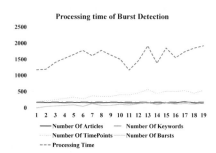

Fig. 5. Bursts detection of keyword "facebook"   **Fig. 6.** Processing time of Burst Detection

7 Conclusion and Future Work

In this paper, we have developed the function of burst detection to find important keywords in bursty patterns which are relevant to topical trends. The results of our system can answer the several questions related to a keyword such as (1) How many revisions a keyword survived, and (2) How long for a keyword appears without interruption, etc. The outcomes in forms of datasets can be effectively reused for relevant research. We keep carrying out our further studies to develop new functions for the system such as News automatic summarization, News automatic synthesis, Event detection, etc. These functions can use the outcome of existing functions as their input.

Acknowledgements. This research is funded by Vietnam National University Ho Chi Minh City (VNU-HCMC) under the grant number B2017-26-02.

References

1. Kleinberg, J., Bursty and hierarchical structure in streams. Data Mining and Knowledge Discovery, 2003. 7(4): p. 373-397
2. Kürüm, E., G.-W. Weber, and C. Iyigun, Early warning on Stock Market Bubbles via methods of optimization, clustering and inverse problems. Annals of Operations Research, 2018. 260(1-2): p. 293-320
3. Vlachos, M., et al. Identifying similarities, periodicities and bursts for online search queries. In: Proceedings of the 2004 ACM SIGMOD International Conference on Management of Data. ACM
4. Bakkum, D.J., et al., Parameters for burst detection. Frontiers in computational neuroscience, 2014. 7: p. 193
5. Weng, J., Lee, B.-S.: Event detection in twitter. In: ICWSM, vol. 11, pp. 401–408 (2011)
6. Romsaiyud, W.: Detecting emergency events and geo-location awareness from twitter streams. In: The International Conference on E-Technologies and Business on the Web (EBW2013). The Society of Digital Information and Wireless Communication (2013)

7. Fung, G.P.C., et al.: Parameter free bursty events detection in text streams. In: Proceedings of the 31st International Conference on Very Large Data Bases. VLDB Endowment (2005)
8. van Pelt, J., et al., Long-term characterization of firing dynamics of spontaneous bursts in cultured neural networks. IEEE Transactions on Biomedical Engineering, 2004. 51(11): p. 2051-2062
9. Wagenaar, D., DeMarse, T.B., Potter, S.M.: MeaBench: a toolset for multi-electrode data acquisition and on-line analysis. In: 2nd International IEEE EMBS Conference on Neural Engineering, 2005. Conference Proceedings. IEEE (2005)
10. Lee, S., Y. Park, and W.C. Yoon, Burst analysis for automatic concept map creation with a single document. Expert Systems with Applications, 2015. 42(22): p. 8817-8829
11. Lee, D., Lee, W.: Finding maximal frequent itemsets over online data streams adaptively. In: Fifth IEEE International Conference on Data Mining (ICDM'05). IEEE (2005)
12. Heydari, A., et al., Detection of review spam: A survey. Expert Systems with Applications, 2015. 42(7): p. 3634-3642
13. Zhang, Y., W. Hua, and S. Yuan, Mapping the scientific research on open data: A bibliometric review. Learned Publishing, 2018. 31(2): p. 95-106
14. Khaing, P.P., New, N.: 2017 IEEE/ACIS 16th International Conference on Adaptive methods for efficient burst and correlative burst detection. in Computer and Information Science (ICIS), IEEE (2017)
15. Yamamoto, S., et al., Twitter user tagging method based on burst time series. International Journal of Web Information Systems, 2016. 12(3): p. 292-311
16. Hong, T.V.T., Do, P.: Developing a graph-based system for storing, exploiting and visualizing text stream. In: Proceedings of the 2nd International Conference on Machine Learning and Soft Computing. ACM (2018)

Detection of Black Hole Attacks in Mobile Ad Hoc Networks via HSA-CBDS Method

Ahmed Mohammed Fahad[1(✉)], Abdulghani Ali Ahmed[1],
Abdullah H. Alghushami[2], and Sammer Alani[3]

[1] Faculty of Computer System & Software Engineering,
Universiti Malaysia Pahang, 26300 Kuantan, Malaysia
ahmedsipher2010@yahoo.com, abdulghani@ump.edu.my
[2] Information Technology Department, The Community College of Qatar,
Doha, Qatar
abdullah.alghushami@ccq.edu.qa
[3] Faculty of Electronic and Computer Engineering (FKEKK),
UTEM University, Durian Tunggal, Malacca, Malaysia
itsamhus@gmail.com

Abstract. Security is a critical problem in implementing mobile ad hoc networks (MANETs) because of their vulnerability to routing attacks. Although providing authentication to packets at each stage can reduce the risk, routing attacks may still occur due to the delay in time of reporting and analyzing the packets. Therefore, this authentication process must be further investigated to develop efficient security techniques. This paper proposes a solution for detecting black hole attacks on MANET by using harmony search algorithm (DBHSA), which uses harmony search algorithm (HSA) to mitigate the lateness problem caused by cooperative bait detection scheme (CBDS). Data are simulated and analyzed using MATLAB. The simulation results of HSA, DSR, and CBDS-DSR are provided. This study also evaluates the manner through which HSA can reduce the inherent delay of CBDS. The proposed approach detects and prevents malicious nodes, such as black hole attacks that are launched in MANETs. The results further confirm that the HSA performs better than CBDS and DSR.

Keywords: Dynamic source routing · Cooperative bait detection scheme
Harmony search algorithm · Black hole attack

1 Introduction

This Mobile ad hoc networks (MANETs) [1] have various important applications, such as in military crisis operation and emergency preparedness and response, because of the widespread availability of mobile devices [1]. Scholars have focused on investigating routing protocols and developed several of them for MANETs [2].

The security of MANETs is important because it ensures the presence of network services and the integrity and confidentiality of data [3]. Frequent security attacks on MANETs are usually caused by several factors, such as open medium features, changes in dynamic topology, and lack of a central management, clear defense mechanisms, and cooperative algorithms. MANETs consider information security as a critical element

© Springer Nature Switzerland AG 2019
P. Vasant et al. (Eds.): ICO 2018, AISC 866, pp. 46–55, 2019.
https://doi.org/10.1007/978-3-030-00979-3_5

[4]. In this event, secure communication and information transmission should be strictly observed and protected from attackers. Security information and transmission engineers must be conversant with all possible methods of attacks [5]. Examples of attacks that can be launched to MANETs include black hole attack, wormhole attack, selfish device misbehavior, Sybil attack, routing table overflow attack, flooding attack, impersonation attack, and denial of service [6].

MANETs are significantly vulnerable to attacks because of information exposure caused by the mutual trust of communication among devices [7]. The security weakness of a MANET makes it a potential destination for routing attacks [8]. For example, a black hole attack modifies the routing function of a MANET, thereby exposing it to abnormal packet transfers [9]. A black hole attack can also access and propagate through the network in an unauthorized manner [10]. If the black hole attack accesses the network, then the protocol will send a route request (RRQE) message to find the shortest path to the destination. When the black hole attack receives the message, the protocol will directly send a fake route reply (RREP) to the source node and tell the node that this path is the shortest path to the destination.

Accordingly, the source node will send the packet through this path; as such, all packets will drop. Figure 1 shows a typical scenario of a black hole attack [11].

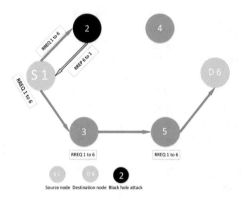

Fig. 1. Black hole attack.

Many solutions are proposed to secure MANETs against black hole attacks. Although helpful, these solutions may still be exposed to further attacks during data transmission because a malicious node may become active and may mislead the detection process. Therefore, this problem can be solved by the solution established by Chang using the CBDS scheme in 2015. This scheme uses a 90% threshold. When the packet delivery ratio (PDR) is within the threshold, the scheme will launch an alarm in the network. Thereafter, the source node, which will be placed on the blacklist, will stop sending packets and reverse tracing will begin to search for the black hole attack. The CBDS scheme has also a problem with delay [12], which must be resolved to attain an increasingly robust and efficient solution. This paper proposes a solution for black hole attacks in MANETs by using the harmony search algorithm (HSA). The solution aims to mitigate the delay limitation that exists on the method of Chang [12].

The dynamic source routing (DSR) protocol and the HSA will simultaneously begin to send the packet. When the PDR uses the threshold, the CBDS will begin to search for the black hole attack and the HSA will select other paths for sending the packet.

The rest of this paper is organized as follows. Section 2 presents the literature review. Section 3 introduces the proposed method for detecting black hole attacks. Section 4 discusses the implementation and the results of the study. Section 5 concludes the paper.

2   Related Work

Security parameters" are metrics in MANET security that are important in all attack prevention approaches. Being unaware of these parameters may render a security approach useless [13].

Reference [14] proposed a malicious node detection system (MDS) to prevent cooperative black hole attacks; the system uses detection and defense mechanisms to remove intruders by considering normal and abnormal activities. Various fake RREP parameters, such as destination sequence number, hop count, destination IP address, and timestamp, are considered in identifying attacks. MDS can improve the PDR by 76%–99%. In this method, decisions about unsafe routes are taken independently by the source node, and no additional overhead is required. Another study [15, 16] proposed a secure routing method for the DSR protocol to mitigate black hole attacks; the DSR RREP is modified to generate a plain packet when the destination replies to reverse path nodes. If the packet is sent, then the node is considered normal; otherwise, the node is considered malicious [17]. Monitoring the spatial and temporal behavior of MANETs is helpful. The performance, topography, and security of the networks should also be observed. An extensive set of experiments were used to validate the performance of the scheme proposed by Wang et al. However, this scheme may be exposed to further attacks during 27 data transmission cycles.

Chang et al. designed and tested the CBDS based on DSR [14]. The results showed that CBDS performed better than DSR, BFTR, and 2ACK protocols [18]. CBDS provided higher PDR than BFTR, 2ACK, and DSR. CBDS also presented higher throughput than that exhibited by the DSR for all simulation cases. Nonetheless, the 2ACK scheme can provide the highest routing overhead over CBDS, BFTR, and DSR, regardless of the number of malicious nodes [19]. The authors concluded that CBDS is an efficient scheme in terms of routing overhead and PDR. Figure 2 shows the random selection of a cooperative bait address (Fig. 3).

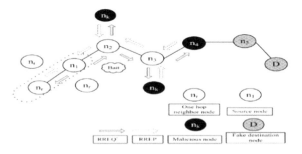

Fig. 2. Random selection of a cooperative bait address.

3   Proposed Work

HSA, which is inspired by music, is an evolutionary algorithm that mimics improvisation and is widely used by musicians. This algorithm is easy to implement and is based on a simple concept [20]. Simulation with this algorithm involves only a few parameters because its theory is based on stochastic derivative [21]. HSA was initially developed for discrete optimization but was eventually used for continuous optimization [22]. Figure 4 shows the step-by-step pseudo code of HSA.

```
HSA pseudocode

// initialize
Initiate parameters
Initialize the harmony memory
//main loop
        While (not-termination)
                for I = 1 to number of decision variables {N} do
                R1 = uniform random number between 0 and 1
                    if {R1<P∞} (memory consideration)
                X {I} wi  ll be randomly chosen from harmony
                memory
                R2 = uniform random number
                    if (R2<P) (pitch adjustment)
                X [1] = X [I] = Δ
                    end if
                else (random selection)
                X [I] = X ∈ <Φ = Value set>
                    end if
                end do
// evaluate the fitness of each vector
Fitness-X {X, fitness-X}
// update harmony memory
update-memory {X, fitness-X} if applicable
                end while
                end procedure
```

Fig. 3. HSA pseudo code.

In this minimization problem, the solution vector (2, 3, 1) for the global minimum can be easily found. However, the harmony search finds the solution vector in another way.

Figure 4a shows that harmony memory (HM) is initially structured with randomly generated values that are sorted based on the objective function. A new harmony (1, 2, 3) is then improvised after considering that HM: x1 selects (1) out of (2, 1, 5); x2 selects (2) out of (2, 3, 3); and x3 selects (3) out of (1, 4, 3). The new harmony (1, 2, 3) is then included in the HM because its function value is 9, whereas the worst harmony (5, 3, 3) is excluded from the HM Fig. 4b. Finally, harmony Search improvises the harmony (2, 3, 1), which has a function value of 3, known as the global minimum.

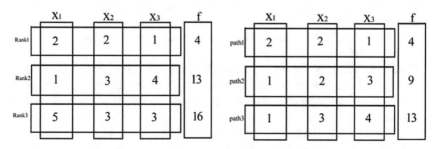

Fig. 4. a and **b.** Initially structured HM and Subsequent HM.

where n is the dimension of the solution vector; $f(\times)$ is the objective function, and HMS is the number of solution vectors hived in the HM [23].

The proposed approach can detect and prevent black hole attacks that launch malicious nodes in MANETs. This scheme can help the source node select a neighboring node that will work in tandem to identify the bait destination for a malicious node, which will then send an RREP message as a response. The scheme uses reverse tracing to identify and prevent malicious nodes, thereby helping launch an alarm in the network when the system observes a significant decrease in the PDR. In turn, the alarm launches a malicious node detection mechanism to prevent false nodes.

DSR, which is the basis for CBDS, traces all node addresses used during the routing path (i.e., from source to destination). However, DSR may not provide any information to the source node to differentiate the replies of malicious nodes (false RREP) and true nodes. This occurrence may misdirect the source node, which may send packets through the false shortest route advertised by the malicious nodes.

This scheme adds a HELLO message to the CBDS to facilitate and identify the true neighboring node and traverse the subsequent node within one hop. This paper proposes the use of HSA to reduce the delay in malicious node detection techniques, such as CBDS.

This algorithm uses the proposed HSA–CBDS scheme to send test bait in the form of address for identifying and eliminating malicious nodes. These route request (RREQ) baits are similar to true RREQ packets but can identify the address of a malicious node. The following rules should be followed:

1. The proposed method initializes the nodes and considers some of the nodes as malicious (random selection), and the remaining nodes are considered trustworthy. Fifty nodes are initially adopted, and the algorithm is run with 10% of the malicious nodes. Thus, five of the 50 nodes are malicious (fake nodes), and the remaining nodes are trustworthy.
2. The proposed algorithm subsequently computes the false and true paths by using CBDS.
3. The proposed algorithm updates the threshold by using the dynamic threshold algorithm at every run.
4. The HSA is used to optimize the end-to-end delay, thereby increasing the overhead, throughput, and PDR.
5. The delay for every route in the network (false and true paths) is computed.
6. The HSA obtains n number of HM (in this paper, we obtain n different routes). Thus, different routes, which are commonly known as HM in the HSA, are obtained as route (x1, x2, …xn).
7. The HSA considers the paths as inputs and works on them to optimize (minimize) the best solutions.
8. The fitness values of the routes (harmony) are predicted. A fitness function evaluates the fitness values for these routes. The end-to-end delay, PDR, routing overhead, and throughput is considered as input parameters to evaluate the fitness values.
9. The fitness values for the paths are computed and saved as f1, f2 …fn. Fitness values evaluate the quality of paths within the range of {0, 1}t.

10. The loop works for every harmony and computes the best harmony with every iteration by updating. A random parameter is used to add the HM and create a new solution (path).
11. After N number of iterations, the HSA finds the best solution with the minimum cost.
12. Each parameter is optimized using the HSA.

An objective function with HSA memory is required to improve the scheme of Chang et al. by improving the end-to-end traversal time.

Figure 5 shows the flowchart process of the proposed algorithm.

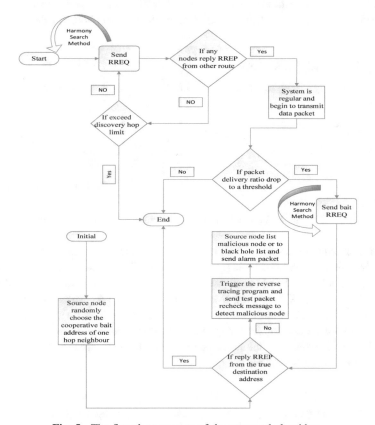

Fig. 5. The flowchart process of the proposed algorithm.

4   Results and Discussion

After simulating Scenarios 1 and 2, we evaluated the performance of the proposed algorithm in terms of average.

PDR, end-to-end delay, average throughput, and overhead. The effectiveness of the CBDS–DSR protocol using the HSA–CBDS was investigated based on the two scenarios. The percentage of malicious nodes in the first scenario varied (10%, 20%, 30%, and 40%), and the speed of nodes in the second scenario also varied (5, 10, 15, and 20 m/s).

Figure 6a and b show the variations in the PDRs of the HSA, DSR, and CBDS protocols. The HSA offered rapid convergence with the optimal fitness value and computation time. Upon introduction of the malicious nodes, the PDR, DSR, and CBDS decreased. The HSA achieved the highest PDR, indicating that fewer packets were lost during transmission because, once established, the route was maintained correctly.

Figure 7a and b show the variations in the throughputs of the three routing protocols after introducing four malicious nodes and speed. The throughputs for the HSA, CBDS, and DSR decreased. Among these protocols, the HSA exhibited the highest throughput. Based on the HSA protocol, the packets reached their destinations without any delay and loss.

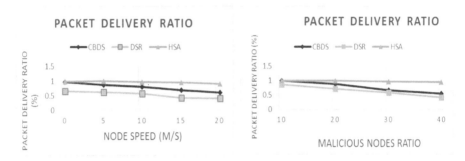

Fig. 6. a and **b.** Variations in PDR with node speed malicious nodes.

Figure 8a and b present the variations in the routing overheads of the three protocols for different malicious node ratios and speed. The routing overhead increased in the HSA and CBDS. By contrast, the routing overhead decreased in the DSR with malicious node ratio. The routing overhead of the HSA was better than those of the DSR and CBDS. The route established by the protocol was maintained efficiently, resulting in reduced end-to-end delay with high PDR and high throughput. This well-established route required minimal retransmissions for any missing segments, thereby reducing the routing overhead.

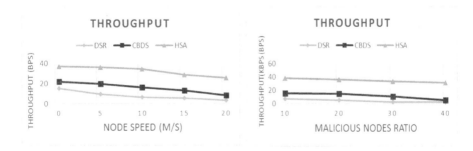

Fig. 7. a and **b.** Variations in throughput with node speed malicious nodes.

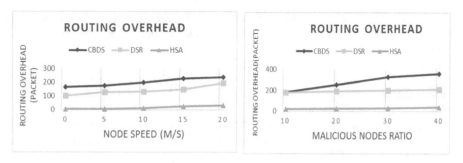

Fig. 8. a and **b.** Variations in routing overhead with node speed and malicious nodes.

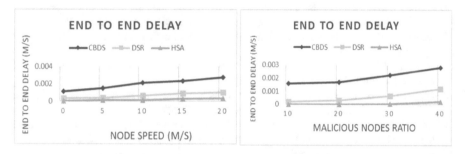

Fig. 9. a and **b.** Variations in end-to-end delay with node speed and malicious nodes.

Figure 9a and b illustrate the variations in the end-to-end delays of the three routing protocols. Upon introducing malicious and speed nodes into the network, the end-to-end delay of the HSA increased insignificantly. Moreover, the delay of the DSR increased more than that of the HSA, and that of the CBDS increased more than those of the other protocols. The HSA showed a relatively better performance than the DSR and CBDS.

5 Conclusions

This paper proposes the use of the HSA to reduce the delay in the CBDS. This approach can detect and prevent black hole attacks that launch malicious nodes in MANETs. The scheme can also help the source node select a neighboring node that will work in tandem to identify a bait destination for a malicious node, which will then send a RREP message as a response. This method adopts reverse tracing to identify and prevent malicious nodes based on the HSA. The proposed approach shows improvement in the delay, throughput, PDR, end-to-end delay, and routing overhead.

Acknowledgements. This work was supported by the Faculty of Computer System and Software Engineering, Universiti Malaysia Pahang under FRGS Grant No. RDU160106 and RDU Grant No. RDU160365.

References

1. Vimala, S., Srivatsa, S.: Security using data compression in MANETS, pp. 528–531 (2017)
2. Taneja, S., Kush, A.: A survey of routing protocols in mobile ad hoc networks. Int. J. Innov. Manag. Technol. **1**(3), 279 (2010)
3. Ahmed, A.A., Jantan, A., Wan, T.-C.: SLA-based complementary approach for network intrusion detection. Comput. Commun. **34**(14), 1738–1749 (2011)
4. Akbani, R., Korkmaz, T., Raju, G.: Mobile ad-hoc networks security. In: Recent Advances in Computer Science and Information Engineering, pp. 659–666. Springer, Berlin (2012)
5. Ahmed, A.A., Sadiq, A.S., Zolkipli, M.F.: Traceback model for identifying sources of distributed attacks in real time. Secur. Commun. Netw. **9**(13), 2173–2185 (2016)
6. Jhaveri, R.H., Patel, S.J., Jinwala, D.C.: DoS attacks in mobile ad hoc networks: a survey, pp. 535–541 (2012)
7. Ullah, I., Rehman, S.U.: Analysis of black hole attack on MANETs using different MANET routing protocols (2010)
8. Goyal, P., Parmar, V., Rishi, R.: Manet: vulnerabilities, challenges, attacks, application. IJCEM Int. J. Comput. Eng. Manag. **11**(2011), 32–37 (2011)
9. Jaiswal, R., Sharma, S.: A novel approach for detecting and eliminating cooperative black hole attack using advanced DRI table in ad hoc network, pp. 499–504 (2013)
10. Bawa, K., Rana, S.B.: Prevention of black hole attack in MANET using addition of genetic algorithm to bacterial foraging optimization. Int. J. Curr. Eng. Technol. **5**(4) (2015)
11. Tseng, F.-H., Chou, L.-D., Chao, H.-C.: A survey of black hole attacks in wireless mobile ad hoc networks. Human-Centric Comput. Inf. Sci. **1**(1), 4 (2011)
12. Chang, J.-M., Tsou, P.-C., Woungang, I., Chao, H.-C., Lai, C.-F.: Defending against collaborative attacks by malicious nodes in MANETs: a cooperative bait detection approach. IEEE Syst. J. **9**(1), 65–75 (2015)
13. Ahmed, A.A., Jantan, A., Wan, T.-C.: Real-time detection of intrusive traffic in QoS network domains. IEEE Secur. Priv. **11**(6), 45–53 (2013)
14. Chaudhary, A., Tiwari, V., Kumar, A.: A cooperative intrusion detection system for sleep deprivation attack using neuro-fuzzy classifier in mobile ad hoc networks. In: Jain, L., Behera, H., Mandal, J., Mohapatra, D. (eds.) Computational Intelligence in Data Mining, vol. 2, pp. 345–353. Springer, New Delhi (2015)
15. Bhandare, A., Patil, S.: Securing MANET against co-operative black hole attack and its performance analysis-a case study, pp. 301–305 (2015)
16. Choudhary, N., Tharani, L.: Preventing black hole attack in AODV using timer-based detection mechanism, pp. 1–4 (2015)
17. Bhardwaj, A.: Secure routing in DSR to mitigate black hole attack, pp. 985–989 (2014)
18. Wankhade, S.V.: 2ACK-scheme: routing misbehavior detection in manets using OLSR. Int. J. Adv. Res. Comput. Eng. Technol. (IJARCET) **1**(5), 1–7 (2012)
19. Arya, P., Negi, G.P., Dhiman, P.K., Kapoor, K.: CBDS (Cooperative bait detection scheme) ATTACK—a review. Int. J. Adv. Res. Comput. Eng. Technol. **4**, 3428–3434 (2015)
20. Assad, A., Deep, K.: Applications of harmony search algorithm in data mining: a survey, pp. 863–874 (2016)

21. Geem, Z.W.: Novel derivative of harmony search algorithm for discrete design variables. Appl. Math. Comput. **199**(1), 223–230 (2008)
22. Lee, K.S., Geem, Z.W.: A new meta-heuristic algorithm for continuous engineering optimization: harmony search theory and practice. Comput. Methods Appl. Mech. Eng. **194** (36), 3902–3933 (2005)
23. Geem, Z.W., Kim, J.H., Loganathan, G.V.: A new heuristic optimization algorithm: harmony search. Simulation **76**(2), 60–68 (2001)

Network Intrusion Detection Framework Based on Whale Swarm Algorithm and Artificial Neural Network in Cloud Computing

Ahmed Mohammed Fahad$^{(\boxtimes)}$, Abdulghani Ali Ahmed,
and Mohd Nizam Mohmad Kahar

Faculty of Computer Systems & Software Engineering,
Universiti Malaysia Pahang, 26300 Kuantan, Malaysia
ahmedsipher2010@yahoo.com,
{abdulghani,mnizam}@ump.edu.my

Abstract. Cloud computing is a rapidly developing Internet technology for facilitating various services to consumers. This technology suggests a considerable potential to the public or to large companies, such as Amazon, Google, Microsoft and IBM. This technology is aimed at providing a flexible IT architecture which is accessible through the Internet for lightweight portability. However, many issues must be resolved before cloud computing can be accepted as a viable option to business computing. Cloud computing undergoes several challenges in security because it is prone to numerous attacks, such as flooding attacks which are the major problems in cloud computing and one of the serious threat to cloud computing originates came from denial of service. This research is aimed at exploring the mechanisms or models that can detect attacks. Intrusion detection system is a detection model for these attacks and is divided into two-type H-IDS and N-IDS. We focus on the N-IDS in Eucalyptus cloud computing to detect DDoS attacks, such as UDP and TCP, to evaluate the output dataset in MATLAB. Therefore, all technology reviews will be solely based on network traffic data. Furthermore, the H-IDS is disregarded in this work.

Keywords: IDS · WOA · ANN · TUIDS · Cloud computing

1 Introduction

A cloud refers to a distinct IT environment that is designed to remotely provide scalable and measured IT resources [1]. This term originated as a metaphor for the Internet, which is a network of networks that provide a remote access to a set of decentralised IT resources [2]. The symbol of a cloud is commonly used to represent the Internet in various specifications and mainstream documentations of web-based architectures before cloud computing has become a formalised IT industry sector [3]. Figure 1, illustrates the importance of cloud computing in remote services and virtual desktop applications [4].

© Springer Nature Switzerland AG 2019
P. Vasant et al. (Eds.): ICO 2018, AISC 866, pp. 56–65, 2019.
https://doi.org/10.1007/978-3-030-00979-3_6

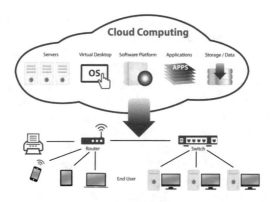

Fig. 1. Cloud computing.

A crucial aspect of cloud security is detecting DDoS attacks [5] and intrusions that disrupt the resources of end-users or organisations [6]. An N-IDS can detect these types of attacks. However, the N-IDS in cloud computing is irrelevant if the attack classifier is inaccurately written [7]. A DDoS classifier in machine learning is neural networks. Several important studies have proposed various ANN-based machine learning classifiers against DDoS in cloud computing [8]. However, considerable attention for accurate classification and reduction of false alarms remains necessary [9]. Hence, this work is aimed at proposing a new model, namely, WOA-ANN, to detect UDP/TCP flooding attacks in the N-IDS in cloud computing. To reduce the false alarm of the N-IDS system, the WOA-ANN model is used to enhance the accuracy of the N-IDS by improving the analysis of network traffic which will be generated by our cloud testbed. We compare the proposed model with existing works using MATLAB to evaluate and compare their efficiency. At the 2014 Black Hat conference, a pair of testers from Bishop Fox demonstrated the pooling of a free-tier public cloud service VM into a mini botnet that could mine bitcoin cryptocurrencies and potentially perform DDoS or password cracking [10]. Moreover, the qualities that make the public VM useful, that is, scalability, ease-of-use and stewardship by high-profile vendors, make this technology an ideal platform for staging DDoS attacks [11].

2 Proposed Framework

Whales are the largest mammals in the world. An adult whale typically measures 30 m in length [12]. Several whale species include killer, Minke, Sei, humpback, right, finback and blue. Whales, as a predator, never sleep because they breathe on the ocean surface. These animals are intelligent and show emotions. Hof and Van Der Gucht highlighted that whales have brain cells, namely, spindle cells, in which are common in humans. These cells control emotions and social behaviours in humans. The number of spindle cells in whales is twice as much as that in an adult human; therefore, whales can think, learn, judge and communicate. For example, killer whales can create their own dialect.

Most whales live in groups. A killer whale can live in a group in its lifetime [13]. Figure 2 exhibits the largest baleen whale called humpback. Its size is comparable to that of a school bus. It typical preys on krill and small fish herds [14]. Its hunting method is called bubble-net feeding in which it hunts preys that are close to the surface. This method is accomplished by creating distinctive bubbles along a circle or a '9'-shaped path. Goldbogen et al. studied this interesting behaviour using tag sensors. A total of 300 tag-derived bubble-net feeding events were captured. These authors discovered that whales use 'upward-spiral' and 'double-loop' manoeuvre patterns.

For the 'upward-spiral' pattern, humpback whales dive 12 m below the surface and create bubbles in a spiral shape around the prey. Subsequently, they swim up towards the ocean surface. The latter pattern consists of three stages: coral loop, lob tail and capture loop. In the current work, this unique spiral bubble-net feeding manoeuvre pattern was used for optimisation.

Fig. 2. WOA and bubble movement.

The mathematical models of encircling preys, spiral bubble-net feeding manoeuvre and searching for preys were outlined. The WOA algorithm was then reported. Humpback whales are aware of the location of their preys whilst hunting. The current best candidate solution is assumed as the target prey in the WOA algorithm because the location of the optimal design in the search space is unknown. The positions of other search agents are updated by defining the optimal search agent. This behaviour can be explained by the following equations:

$$\vec{X}(t+1) = \vec{X}^*(t) - \vec{B}.\vec{S} \tag{1}$$

$$\vec{S} = \left| \vec{K}.\vec{X}^*(t) - \vec{X}(t) \right| \tag{2}$$

where t is the current iteration, and are the coefficient vectors, is the position vector of the current best solution obtained and is the position vector. Here, is updated after iterations and are computed as

$$\vec{B} = 2\vec{b} \cdot \vec{r} - \vec{b} \tag{3}$$

where decreases from 2 to 0 during the iterative phase, and is a random vector between [0, 1]. The rationale behind Eq. (2). Figure 6 explain The new position (X, Y) of a search agent is updated on the basis of the current best position (X*, Y*). The locations of the optimal agent can be manipulated by adjusting the and vectors. Any position that is located within the search space is reachable by using the random vector, as displayed in Fig. 3, which simulates the encircling prey movement of a whale. The same method can be applied to high-dimensional problems.

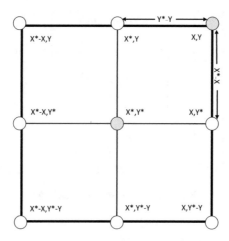

Fig. 3. position vectors and their possible next locations (X* is the best solution obtained so far).

2.1 Bubble-Net Attacking Method (Exploitation Phase)

The bubble-net strategy can be performed using the following approaches:

1. Shrinking encircling mechanism: This strategy is achieved via reducing the value of in Eq. (1) from 2 to 0 during the iterative procedure. The new position of a search agent can then be identified by setting the random values in [−1, 1].

   Figure 4, presents several possible solutions (X, Y) that can be obtained by setting $0 \leq K \leq 1$.

2. Spiral updating position: In Fig. 8, the distance between (X, Y) and (X*, Y*) is calculated first. A spiral equation is then established to represent the helix-shaped movement:

$$\vec{X}(t+1) = S.e^{ml} \cdot \cos(2\pi l) + \vec{X}^*(t). \tag{4}$$

where indicates the distance of the ith whale to the prey, m is a constant that defines the shape of the logarithmic spiral and l is a random number within the range [−1, 1]. In general, a humpback whale swims around the prey within a shrinking circle, following the spiral-shaped path. A probability of 50% is prescribed on activating either the shrinking encircling mechanism or the spiral model whilst updating the whale position to model this condition.

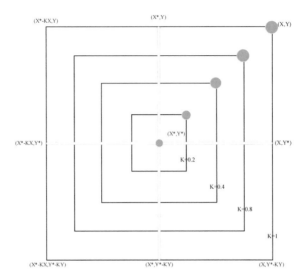

Fig. 4. Possible solutions (X, Y).

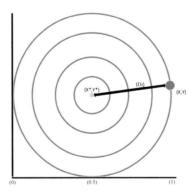

Fig. 5. Distance between (X, Y) and (X*, Y*).

The WOA algorithm is initialized using a set of random solutions. For each iteration, the positions of the search agents are updated based on a randomly selected search agent or the current best solution, depending on || (Fig. 5).

Theoretically, WOA is a global optimizer because it contains exploration and exploitation capabilities. Moreover, the current hypercube mechanism defines a search space in near the optimal solution, thus permitting other search agents to search for the current best record within the domain. The mathematical model of the adaptive version of the WOA algorithm is

$$\vec{X}(t+1) = \begin{cases} \vec{X}^*(t) - \vec{B}.\vec{S} & if \quad \ell < 0.5 \\ \vec{S'}.e^{ml}\cos(2\pi l) + \vec{X}^*(t) & if \quad \ell \geq 0.5 \end{cases} \tag{5}$$

where is a random number between [0, 1]. In certain cases, a humpback whale searches for a prey randomly.

2.2   Searching for Prey (Exploration Phase)

The method can be adopted by using a similar approach of varying the vector to search for a prey (i.e. exploration). In the random method, with random values greater than or less than 1 is used to ensure that each search agent is far from the reference. Whale. Here, the position is updated randomly in accordance with the randomly selected search agent. The global search operation in WOA can be performed by applying this mechanism and setting | | > 1. The corresponding mathematical model is:

$$\vec{S} = \left| \vec{K}.\vec{X}_{rand}(t) - \vec{X}(t) \right| \tag{6}$$

$$\vec{X}(t+1) = \vec{X}_{rand}(t) - \vec{B}.\vec{S} \tag{7}$$

3   Classifier Design

In practice, a smooth transition between exploration and exploitation is feasible. Here, several iterations are allocated exploration (| | ≥ 1), whereas the remaining iterations are dedicated for exploitation (| | < 1). In WOA, only two main adjustable internal parameters are available. The current work considers a simple version of WOA by neglecting the other evolutionary operations that mimic the real behavior of humpback whales. Therefore, hybridization with evolutionary search schemes can be further explored (Fig. 6).

The average values obtained from the fitness function are considered those of the candidate solutions. The value of this fitness function is verified during the iteration until a new best value is found. The attribute that provides a minimum error value is selected to complete the selection and evaluation processes. The flow of the proposed method is as follows:

$$E(i) = wTrain.TrainData.E + wTest.TestData.E \tag{8}$$

where wTrain and wTest are the heights to be used for training and test data. The error values obtained from the training and test data will be used to calculate the fitness function. In the proposed method, wTrain: 0.7 and wTest: 0.3 are considered. E(i) denotes the fitness values obtained at the end of three runs. Figure 8 depicts the method

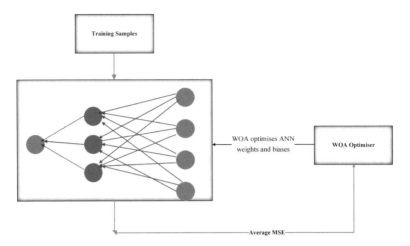

Fig. 6. Process for collecting dataset and designing a classifier.

used by WOA to feed the weights and biases and produce additional training samples efficiently.

3.1 BFGS Quasi-Newton Backpropagation

Newton's method is an alternative to conjugate gradient methods used for fast optimization. The basic step of Newton's method is:

$$X_{k+1} = X_k - A_k^{-1} g_k \tag{9}$$

where is the Hessian matrix (second derivatives) of the performance index at the current values of the weights and biases. Newton's method frequently converges faster than conjugate gradient methods. However, computing the Hessian matrix for feedforward neural networks is complex and costly. A class of algorithms is based on Newton's method and does not require calculating the second derivatives. These methods are called quasi-Newton (or secant). They update an approximate Hessian matrix at each iteration of the algorithm. The update is computed as a function of the gradient. This algorithm requires more computation in each iteration and more storage than the conjugate gradient methods, although it generally converges in few iterations. The approximate Hessian must be stored, and its dimension is n × n, where n is equal to the number of weights and biases in the network. For large networks, using the conjugate gradient algorithms is favorable.

4 Model for N-IDS In-Eucalyptus Cloud Computing

The proposed model is aimed at synthesizing the Eucalyptus cloud as the N-IDS that analyses the generated traffic and blocks the TCP/UDP flooding and Smurf DDoS attack. A TUIDS DDoS dataset is prepared using the TUIDS testbed architecture with a demilitarized zone (DMZ); hosts are divided into several VLANs, where each VLAN belongs to an L3 or L2 switch inside the network. The attackers are placed in wired and wireless networks with reflectors, but the target is placed inside the internal network. The target generates low- and high-rate DDoS traffics. We consider real-time low- and high-rate DDoS attack scenarios for both datasets during our experiments. However, a low-rate attack does not consume all computing resources on the server or bandwidth of the network that connects the server to the Internet. Hence, a real low-rate DDoS attack scenario contains attack and attack-free traffics. We mix low-rate attack and legitimate traffics during our experiment to prepare the real low-rate DDoS attack scenarios in the TUIDS DDoS dataset (Fig. 7).

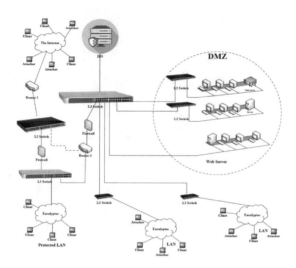

Fig. 7. New pre-processor rule structure in Eucalyptus cloud computing.

Many experiments have been conducted in this work to empirically demonstrate the impact of methodological factors, as discussed in Sect. 4.3.2. The experiments can be summarized as follows: General observations: exploring classifier performance and validation methods. Performance on original datasets: providing a benchmark that is used to compare the results obtained from subsequent experiments. Removing new attacks from the test set: this experiment is conducted to determine.

5 Classifier Process in MATLAB

We offer three specific benefits of classifier combinations. Statistical: if the amount of training data insufficiently models the hypothesis space with one classifier, then the combined knowledge of an ensemble of classifiers may reach accurate predictions. 2. Computational: algorithms can be trapped in local optima and finding the global optimum may be computationally expensive. However, executing several local search algorithms from different starting points and combining them may be favorable.

A hybrid classifier is developed in MATLAB using our dataset. First, we must perform the normalization process for the dataset. Then, the dataset will have been delivered to the training and testing sets in the following proportions: 70% training and 30% testing. The target will be TCP, UDP and Smurf attacks. The results are validated through evaluation which was derived from the machine learning metrics (Fig. 8).

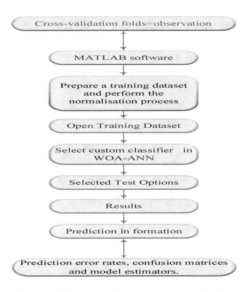

Fig. 8. WOA classifier test using MATLAB.

6   Conclusions

This work proposes a new classifier design based on a hybrid artificial neural network and whale swarm algorithm to feed the ANN weights and biases. However, the model cannot function without a derivative dataset. This dataset is derived from our testbed design based on a developed TUIDS network topology over the Eucalyptus cloud computing. In the N-IDS, sensing is used in real time for the normal and upnormal traffic DDoS attacks through Snort. The log-output from the dataset has been analyzed in MATLAB.

Acknowledgments.   This work was supported by the Faculty of Computer System and Software Engineering, Universiti Malaysia Pahang under FRGS Grant No. RDU160106 and RDU Grant No. RDU160365.

References

1. Rittinghouse, J.W., Ransome, J.F.: Cloud Computing: Implementation, Management, and Security. CRC press, Boca Raton (2016)

2. Ahmed, A.A., Jantan, A., Ali, G.A.: A potent model for unwanted traffic detection in QoS network domain. JDCTA **4**, 122–130 (2010)
3. Kaul, S., Sood, K., Jain, A.: Cloud computing and its emerging need: advantages and issues. Int. J. Adv. Res. Comput. Sci. **8**(3) (2017)
4. Sultan, N.: Cloud computing for education: a new dawn? Int. J. Inf. Manag. **30**(2), 109–116 (2010)
5. Mirkovic, J., Reiher, P.: A taxonomy of DDoS attack and DDoS defense mechanisms. ACM SIGCOMM Comput. Commun. Rev. **34**(2), 39–53 (2004)
6. Elejla, O.E., Jantan, A.B., Ahmed, A.A.: Three layers approach for network scanning detection. J. Theor. Appl. Inf. Technol. **70**(2) (2014)
7. Ahmed, A.A.: Investigation model for DDoS attack detection in real-time. Int. J. Softw. Eng. Comput. Sci. (IJSECS) **1**, 93–105 (2015)
8. Ahmed, A.A., Jantan, A., Wan, T.-C.: Filtration model for the detection of malicious traffic in large-scale networks. Comput. Commun. **82**, 59–70 (2016)
9. Rowland, C.H., Rhodes, A.L.: Method and system for reducing the false alarm rate of network intrusion detection systems. ed: Google Patents (2011)
10. Nanavati, M., Colp, P., Aiello, B., Warfield, A.: Cloud security: a gathering storm. Commun. ACM **57**(5), 70–79 (2014)
11. Ahmed, A.A., Sadiq, A.S., Zolkipli, M.F.: Traceback model for identifying sources of distributed attacks in real time. Secur. Commun. Netw. **9**(13), 2173–2185 (2016)
12. Mirjalili, S., Lewis, A.: The whale optimization algorithm. Adv. Eng. Softw. **95**(Supplement C), 51–67 (2016)
13. Mirjalili, S., Lewis, A.: The whale optimization algorithm. Adv. Eng. Softw. **95**, 51–67 (2016)
14. Aljarah, I., Faris, H., Mirjalili, S.: Optimizing connection weights in neural networks using the whale optimization algorithm. Soft Comput. **22**(1), 1–15 (2018)

The 'Smart' as a Project for the City Smart Technologies for Territorial Management Planning Strategies

Cinzia Bellone[1(✉)] and Vasiliki Geropanta[2(✉)]

[1] DIS, Università degli Studi Guglielmo Marconi, Rome, Italy
c.bellone@unimarconi.it
[2] R&D Department, Università degli Studi Guglielmo Marconi, Rome, Italy
v.geropanta@unimarconi.it

Abstract. This paper investigates the project for the city by looking at one of the latest trends in urban planning: "the smart city". It starts with a general overview of the definitions of the term, explaining how the ascendance of information technologies and the associated focus on sustainability, as a response to globalization, urbanization and climate change, can generate effective territorial management planning strategies. Building on this objective, the authors review a smart city project within a consolidated and mature urban fabric in Italy, from three perspectives: territorial planning, technology, and social practices. The study consists of a detailed analysis of the contextual and typological configuration of the case and its surrounding area, narrating on the possible transformation of the territory. The results of the study lead the authors to argue that through this instance of enhanced reality, the potentialities for urban regeneration become apparent.

Keywords: Smart city · Governance · Strategy territorial planning

1 Introduction

Last few years, new sets of relations between the state, citizens, places and the market appeared because of the shift towards an economy based on knowledge [1] and the domination of data-ism [2] as the cultural and scientific trend of contemporaneity. These relations emerged to improve the accessibility of places through a multi-scale approach, transforming the city into a porous think tank. Aiming at enhancing the efficiency and quality of living and work of individual actors by offing a sustainable future, they use new advanced technologies, and form, in this way, a new urban paradigm [3, 4].

The smart city is one of these paradigms, being first on the list of the world's venture capital distribution [5]. The digital component of the smart city collects and distributes tacit and explicit knowledge towards all channels of interest and actors revealing a new immaterial city dimension that is able to solve issues coming from the physical space. Some of these issues relate to the way we use space, some in the way the city is organized, managed and governed and some represent global challenges

© Springer Nature Switzerland AG 2019
P. Vasant et al. (Eds.): ICO 2018, AISC 866, pp. 66–75, 2019.
https://doi.org/10.1007/978-3-030-00979-3_7

(such as the lack of natural resources or the unlimited demographic expansions) that affect the city performance [3, 4, 6]. In its full completion, the smart city treats urban territory as a complex mixture of networks, places, flows, and information, in which multiple relations and activities have the possibility to generate creative synergies and respond to contemporary urban demands [7].

Among all sectors of a smart city project, mobility seems to acquire notable attention (*smart mobility*)[1] [8]. This is because smart mobility seems to offer many solutions against the dysfunction of the previous static urban infrastructure and the large energy consumption. In fact, as the necessity for improved connectivity between the different "landscapes" [9] of the city rises, so do the concerns about transportation, global greenhouse gas emissions, congestion, noise and poor air quality in cities [10]. The dense built environment, the lack of economic resources, and a degraded aged infrastructure turned its back to the traditional city planning and seeks for solutions that are sustainable, dynamic and technologically driven [3]. In fact, when supplementing urban planning and management practices with digital technologies of a human-centered approach, there is an opportunity to improve mobility services for citizens and generate wider economic and environmental value in the physical infrastructure.

In Italy for example, the results of the study "Efficient cities Siemens 2012", that analyze 54 main Italian cities, showed that urban mobility is the most financially supported area, followed afterward by the sustainability of buildings among every possible intervention via a smart city project [11]. That study indicated surprisingly the popularity and effectiveness of the project: in 5 years' time, the northern Italian cities showed remarkably improved performances in urban mobility, and the southern ones were described as "having strong potentials for improvement" [14].

Rome does not appear in the highest rankings for smart city implementation projects [12] but the city center has potentialities to be considered as a smart district [13]. A number of initiatives in the sector of tourism and the surveillance of the city have brought about the creation of dynamic platforms and applications. In fact, if we look at urban mobility, the issue of energy production, the quality of air, the strong traffic congestion are only a few of the most popular key sectors to look for instances of the application of ICT in the Italian capital [18]. The ability of ICT for delivering real-time information and re-creating any environment, enhanced, virtual or augmented in a historical, stratified and culturally rich city creates new layers of meanings and reveals new potentialities for the enhancement of the city performance.

Following this premise, this paper offers an examination of how the smart city strategy in the case of urban mobility in Rome can give new insight to the project for the city. Specifically, the paper presents a review on the smart city initiatives for ameliorating the mobility of Rome, launched by the Municipality of Rome the last 6 years, and makes a reflection about their effectiveness to bring about regeneration in the city. Since they are all on-going projects, the study uses observations coming from empirical observation and changes in social practices.

---

[1] The European Smart City Model consists of six key fields: Smart Economy, Smart Mobility, Smart Environment, Smart People, Smart Living and Smart Governance (Giffinger, Kramar, Haindlmaier & Strohmayer, 2007).

Our general research question is: How can ICT contribute to or alter the experience of the urban space in the case of smart mobility, and how can the use of ICT, therefore, generate new planning processes? A second, more specific question, regarding the case study, is: How do the projects related to the smarter mobility in the case of Rome offer possibilities for innovative urban regeneration? The main objective is the analysis of the deployment of a layer of ICT in the historical center of the city, extrapolating the possible effects on the social and urban practices. The methods applied are a detailed analysis of the projects in the center of Rome, complemented by documentary research on the urban configuration of the area and participant observation.

2   Intelligent Urban Mobility in the Smart City

Most of the objectives attached to Smart Mobility look at the implementation of ICT in the field of road transport, including infrastructure, vehicles, and users [16]. A number of sensors, installed in the city, study the urban traffic control systems, track car locations in real time and adapt intelligent transportation systems that allow administrators to manage instantly spontaneous traffic congestion issues. The information that is collected for this activity is real-time data (location, weather, and traffic raw data), generated continually by dynamic patterns of human behavior as people navigate the city. In this way, personal smartphones or city sensors become a roaming source of information that allows transport operators to make immediate interventions and create additional capacity where it is needed in physical infrastructure [15]. For the users, the advantages are various among which the changing speeds in daily life mobility and the reduction of transfer costs.

For operators of Intelligent Transportation systems, this means an immediate detection of incidents on the road, an ability to create personalized mapping services to drivers, detect blockages, surveillance, and realize remote controls, managing, therefore, and better travel conditions. For the users of public transport this means the ability to see the route and time of arrival in precision of every transport mean, the ability to plan the route and map online one's travel in the city and the ability to be updated any moment about any available means of transportation at whatever city in the world. Common examples are the basic management systems such as car navigations (using GPS), traffic signal control systems, speed cameras among others.

The overall goal of these projects is the CO_2 emissions[2] reduction. Furthermore, monitoring noise pollution offers possibilities for a better quality of life and is, therefore, an important pillar in city management. Consequently, Smart Mobility is a multifaceted topic, involving all the smart city paradigms (digital city, knowledge city, green city etc.) and generates a set of heterogeneous benefits for all the smart city stakeholders.

---

[2] Sanseverino and Orlando (2014), citing the "Expert Working group on Smart City applications and requirements, 2011", show the effects of lowering by 3 min the average time needed to find a public parking place in relation to the reduction of CO_2 emissions in Barcelona.

3   The Case Study: Smart Mobility in the City Center of Rome

The "Sustainable Mobility Plan" of Rome (*SUMP*) under the guidance and indications of EU,[3] tried to respond to the challenges for safe, sustainable and accessible transport. The objective of the plan is to "reconnect the city, reduce the use of private vehicles (cars and motorbikes, recover and redistribute public spaces and improve the environmental status" [16]. This was a project, built on the "Environmental Action Plan for achieving the objectives of the Kyoto Protocol in the city of Rome, 2001", a guideline document (supplement of the Environmental Action Plan) for sustainable development [17]. In this document, the major areas of activity are transport, housing, waste and land use, and the overall goal of the action was to reach the 6.5% reduction in CO_2 emissions by 2012. The document presented the actions taken during the first decade of the 2000s and connected the managing of motorization rates with the use of less parking spaces or garages in the city with the application of few new smarter techniques[4] [18].

SUMP was announced in 2012 and works with three objectives. The first one is the systemization of the transport system, by expanding two metro lines (B and C) and ameliorates the tramline.[5] The second regards a number of e-mobility devices, car sharing and car-pooling for the city. The last one regards the use of ICT that could have the ability, firstly control several of the traffic systems and then to allow citizens' participation [19]. The hard infrastructure of the first objective, started its realization in the first part of 2000 while in the second decade on the 2000s, the other two acquired more attention and were enriched by a focus on ICT sector.

Agenda Urbana, a national platform created by ANCI[6] and IFEL[7] presents the four key smart mobility projects of Rome that Sump brought about. The first three of them refer to a number of several sharing services for automobiles (Car2Go and Enjoy), motorcycles (eCooltra) and bicycles (O'Bike) supported by the use of GPS, GIS and a number of sensors and satellite connected devices in the city center. In total almost 2000 car sharing means are deliveries and more than 100 stalls are created for hosting

---

[3] At a European level there are highlighted the following documents: (a) 2009 Urban Mobility Action Plan - Communication from the Commission to the European Parliament, the Council, the European Economic and Social Committee and the Committee of the Regions - [COM (2009) 490 final] (b) The 2011 White Paper - Roadmap to the Single European Transport Area for a Competitive and Sustainable Policy [COM (2011) 144] (c) the 2013 Urban Mobility Package, which is the most recent and specific document on urban mobility. In the annex "The methodological reference framework for the Pums" the relevance of the latter is reiterated as a planning tool and indicates the main requirements.

[4] In Italy, the motorization rate has diminished by 5.4% in Rome between 2008 and 2012.

[5] In 2013, the tramway terminus in Venice square, new tramway line in via Botteghe Oscure and the requalification of Largo Argentina were constructed. Metro B and C are under construction, while many parts are given already to public.

[6] Anci: National Association of the Italian municipalities (Associazione Nazionale Comuni Italiani).

[7] Ifel: National Association of the Italian municipalities Foundation – Institute of Finance and Local Economy (Fondazione ANCI, Instituto per la Finanza e l'Economia Locale).

this new activity. The sharing of bicycles, as well as the promotion of this activity on behalf of the commune, led to the creation of new cycle paths of about 241 km in total [20].

These projects aimed at ameliorating the accessibility in all the areas of the city promoted sustainability by not consuming fuels and worked against urban noise. Beyond the construction of infrastructures and interchange nodes, fleets with low environmental impact as well as intelligent transport systems were financed [21]. Furthermore, new mobility manager and car-pooling create the connection among users and mobility managers [22]. The same document describes the creation of 1200 new parking lots, the replacement of traffic lights, access gates, checkpoints, and the release of 65 new ecological, electrical autobus and the installation of numerous smart technologies for monitoring the different movement activities in the city. Lastly, a number of projects supported the e-mobility diffusion and the information about the decarburization of the inner city center. Apps are now offered information on how to avoid flow traffic and are enhanced by 'AR' systems, suggested new pedestrian areas and pathways. To facilitate cycling, the municipality of Rome allow foldable bicycles onto buses and trains; and installed road signs to alert motorists to the presence of cyclists along frequently used routes. In addition, 700 small supporting colonies integrated with higher power systems (fast recharge) along the GRA (fuel distributors) and along the perimeter of the railway ring were installed that serve as reloading equipment and 15.500 free parking posts were offered to their service [20].

The fourth one refers to the project "Intervention for the monitoring of the urban penetration routes and the main urban itineraries of the intra-GRA network of the Municipality of Rome: Master Plan Phase 3 (MP3)". This project consists of the redesign, development, and implementation of a system of 10 new variable interactive and responsive panels (PMV – VMSS). It requires the re-engineering of 19 traffic light systems, 20 new measuring and monitoring stations of non-invasive nature that transmits and elaborates all data collected to Rome's mobility center. Lastly, it regards the integration of the new systems within the technological framework of the Mobility center with adjustment of the HW and SW infrastructures. It is called the Monitoring Centre for Road Safety [21].

The projects consist of a number of city interventions using ICT and a web interface. In the Municipality's document "Programma Della struttura di linea" from the Department of mobility and transport from the Municipality of Rome, there were presented all large and smaller operations to intervene to the various levels of street infrastructures. From the overall study, the following goals/objectives were highlighted: (A) the delivery of a transport network that is integrated, efficient, cost-effective and sustainable to meet the city's needs. (B) To develop and implement policies to encourage commuters to choose the most appropriate transportation mode. (C) To reduce the environmental cost and deliver therefore alternative solutions to the city's conservation of cultural heritage.

4   Experience of Using Smart Mobility in Rome

The application of these projects in the city of Rome include many different contents and variables, and it was difficult to provide a holistic and interrelated vision of these actions. Due to the complexity of urban mobility and the fact that there is no relatively large amount of feedback yet, we analyzed the contents of the smart mobility initiatives described examining the use of ICT, the actors and the goals and benefits towards the city from an empirical point of view. In relation to the experience of the city, the projects can be classified into the following types:

(a) Technologically-driven with applications that transmit real-time information. These include the initiatives carried out by companies, depend on citizen's participation, and affect citizen's behaviors (implementation of multimodality to enhance the use of public transportation instead of private one - mobile apps that facilitate the transition between public transport and car and bike sharing, online tickets, parking, etc).

(b) The ones that receive information from drivers and vehicles through technology and use this data for the better functioning of the city. This group includes the initiatives carried out by the companies or organizations that supply the local public services in the city. They are a large and heterogeneous set of applications, including the following: Info mobility, Sanctioning and fining, Monitoring Controlling Management [21].[8]

(c) The localization of services in support of the smart mobility projects including policies.

We argue here that the three categories affect the city at a spatial, social and environmental level.

In all of the projects, the presence of ICT acquires a physical dimension in the urban space. Firstly, the material/physical use of the existing space is transformed in order to host all hardware and relevant technologies. This is hidden at a first view, but all over the city, there are cameras and sensors installed covering the spectrum, of all the kilometers of the ancient city. There are also terminals to access public information, displays with traffic updates led signals and data projected on responsive panels. The presence of the new stalls and the colonies of reloading electrical vehicles have their own new spaces in the city. Lastly, all sharing vehicles and all applications are represented with specific brand symbols (mostly colors that are assigned to each vehicle – yellow for smart bikes- red for enjoying car sharing- blue for motorbike sharing etc.) that can be traced easily. This has created a new spatiality of specific aesthetics such as the yellow bicycles that are found in the edges of every monument site and are now everywhere in the city center.

---

[8] Anaphoric ally some of them are: the integrated mobility Information layer dedicated to road safety, integrated parking guidance systems; Variable Message Signs (VMS); Urban Traffic Control (UTC); Video surveillance systems for area and environment security; Integrated systems for mobility management; E-gates for Limited Traffic Zones Electronic poles of bus stop, Video surveillance cameras Traffic Measurement Stations Traffic lights, Traffic data collection systems etc.

For the first two categories, it is observed that during the application of smart ecosystems, a sort of digital surface where all data exists is created as a net that is displayed over the city scale. This surface visualizes different information about every street and human activity on virtual platforms that follow the logic of augmented reality where the mixture of digital data and physical space are recreating a new image of the city dimension. The platforms are open and accessible by everybody through smart applications, a sort of simulated version of reality where instances of our lives can be uploaded acquiring new meanings and transforming places into dynamic spaces. This allows both pedestrians and drivers/travelers to have a different experience of the city, being instantly oriented. In other words, the availability of these technologies of the municipality of Rome has given birth to a new interwoven image of the city, an augmented space that describes the city and is available to every user. This image reconstructs the facades of buildings, architectural details, and human activity. They are drawn in relevant scales, open spaces are represented as urban voids, and 2d colored lines present alternatives for moving in the city, details that previously could not see on a static map (for example the route of a bus). Time, availability of parking or stalls, costs, the presence of police and others affect in this way the urban movement in Rome and therefore change the use of urban block releasing the center from congestion.

In the second category, the interconnectivity of the multiple users stresses the role of the collective, digital space that emerges in certain activities. It is the virtual space where the users connect while acquiring the possibility of interacting as if they were in the physical space (e.g. in the applications where drivers connected remotely have the possibility to talk among themselves). This collective space opens a communication network in Rome, bringing new relations of interactions at a city level. This might solve problems of density in bus stops, or problems related to directions in driving cars. The collective space in this sense is the one where each person's car or mobile meet in a digital setting and exchange information about a specific node in Rome's infrastructure.

The interaction with these applications transforms the users of roman applications to active participants in the creation of new instances in the city fabric. For the visitors of Rome, this kind of accessibility in interactive maps and local information enhances the idea of a city that is easily accessible, protected and therefore enhances public comfort.

In addition, the applications make clear the attention in Rome's specific urban areas and highlight the easiness of moving from one place to another. This does not change the routines of users (the reason for which one might move from one place to another). In this case, the presence of ICT modifies the activities that can be done in environments that were already socially defined. In this way, people can take information on a taxi or a metro station, just as well as they can do in an office and communication is instantaneous regardless of geography. People move through the various environments with the awareness of both being a presence that sets things into motion because smart city spaces are full of sensors, and of being an agent that can command the different devices by exchanging information. Inthe long term, this could lead to an important change in human consciousness and can work better towards the society that is more inclusive.

5   Smart Mobility for Urban Regeneration

From the brief description of the experience from the smart mobility in Rome, we can draw some preliminary conclusions about how the use of ICT can open a discourse about urban regeneration. In fact, the observations are all studies that can be incorporated and managed by local governments along with the three core plans that historically particularize the urban planning: the master plan of the metropolis, the general urban plan for the residential and suburban organization of the cities along with the urban studies, implementation acts and rehabilitation studies [23].

In particular, changing human activity during rush hours at the physical space means also directing the critical mass towards new directions. This, in turn, might bring about ideas for neighborhood's intensification, the different relation between the city and the road system (including the pedestrian pathways), new ideas about relocating of activities that assist the drivers/travelers in the city territory.

The new services of the smart parking and smart ticketing, parking public spaces can be more efficiently managed by guiding drivers to the closest (proximity) parking place and can be provided on demand. On the one hand, urban planners acquire the new capacity for improving the driver's experience. On the other hand, this dynamic management of transport systems means that supply can be matched with demand in real time, creating economic incentives to travel outside of peak hours or by alternative modes/routes where possible, therefore helping to distribute the peak demand. The same happens with interactive apps (e.g., Moovit) that help citizens to exchange information and avoid congestion. The data coming from this service provider can assist urban planners to plan future infrastructure, and service provision based on demand. Sharing services, such as bicycle sharing, car sharing or taxi booking services bring to the table many different future scenarios for the city management.

The digital surface of data has no thickness but works and functions where people have accessibility to the internet. These augmented places form a sort of an 'enclave' that gathers all real-time experiences. Naturally, the enclave could be everywhere inside the city, in the periphery, or it can be inside the city but not acting as an urban condition that generates relations or mixing of functions or continues the grid of the city. The enclave does not have a specific scale; it can be any group of quarters that are monitored by ICT, a few urban blocks or even a small town. The importance is the specification of this area as a separate zone from the rest of the territory, as it has different urban functions from the ones where ICT are not installed. In other words in a territory, the enclave is a compound that has fences, boundaries and is characterized by the intensity of information that is extracted. Furthermore, the enclave acts as an instrument to transform the city into a porous and permeable structure putting in the core the human activities.

At a larger scale the ICT presence could be understood as an immaterial 'archipelago', or else a "silicon landscape" to describe the idea of the dispersed condition of digital episodes in a region [22]. The silicon landscape is a network-based dispersal digital territory of IT activity in the city grid. It has the ability to transform the 'zoning' of the enclave in a topic of interest as it presents the local cultural and infrastructural characteristics of the territory. It does not have a precise scale, but it is responsible for creating a good environment for possibilities of proliferation. Centrality, localized density, distance and proximity, and notions of relation and juxtaposition are some of the principle elements that characterize the connection of the different digital episodes. Supplemented by social media, aggregated data can also provide details of citizens' thoughts and feelings about places and experiences. With the right skills and software capabilities, this massive anonymous data bank can allow urban planners to understand the detailed use characteristics of city facilities and services, and to create places that are tailored to the people who use them. Using sensor-derived real-time data, different planning conditions can be quickly tested and simulated.

6   Conclusions

The connection of smart mobility and urban planning is a complex and still challenging field of research and practice. The set of good practices and related planning actions, presented in the previous paragraphs, require more time for feedback and a considerable increase in the level of technical competencies that are among the pillars of planning disciplines. However, following the above-mentioned empirical analysis, it emerges that the applications of smart mobility can be a useful sector for improving spatial planning and that new concepts of spatial regeneration can be predicted in the three above-mentioned categories of actions.

References

1. Komninos, N.: Intelligent Cities. Innovation, Knowledge, System and Digital Spaces. Spon Press, London, and New York (2002)
2. Harari, Y.N.: Homo Deus: A Brief History of Tomorrow. Harper, New York (2017)
3. Komninos, N.: The Age of Intelligent Cities: Smart Environments and Innovation-for-All Strategies. Routledge, London and New York (2015)
4. Morandi, C., Rolando, A., Di Vita, S.: From a smart city to smart region digital services for an internet of places. In: PoliMI Springer Briefs. Springer International Publishing (2016)
5. Larson, K.: Barcelona Smart City Expo World Congress (2017). Retrieved from https://www.youtube.com/watch?v=aXFwZZ9cXD0
6. Fusero, P.: E-City, Digital Networks, and Cities of the Future. S.A Litografia (2008)
7. Healey P.: Urban Complexity and Spatial Strategies Towards a Relational Planning for Our Times. The RTPI Library Series. Routledge Barcelona, Healey (2007)
8. Giffinger, R., Kramar, H., Haindlmaier, G., Strohmayer, F.: The Smart City Model (2015). Retrieved from http://www.smart-cities.eu/?cid=2&ver=4
9. Barth, L.: Workspace urbanism: the architecture of transformation/rethinking the civic landscape. In: INTA TALLIN Conference (2012)
10. Schneider Electric: Smart Cities Cornerstone Series, Urban Mobility in the Smart City Age (2016). Retrieved from http://smartcitiesappg.com/wpcontent/uploads/PDF/UrbanMobility.pdf
11. Riva Sanseverino, R.: Competitive urban models. In: Riva Sanseverino, E., Riva Sanseverino, R., Vaccaro, V., Zizzo, G. (eds.) Smart Rules for Smart Cities. SSISS, vol. 12, pp. 1–14. Springer, Cham (2014). https://doi.org/10.1007/978-3-319-06422-2_1

12. European Commission: Smart Mobility and Services, Expert Group Report, Studies and Reports (2017). Retrieved from http://ec.europa.eu/transparency/regexpert/index.cfm?do=groupDetail.groupDetailDoc&id=34596&no=1
13. Geropanta, V., Cornelio Mari, E.: The role of ICT in the revival of cultural heritage in Rome. In: 7th International Meeting of UNESCO, Conference Communication, City and Public Space, Lima Peru, May 2018
14. Acea: Smart City a Roma: Progetti per la Mobilità Sostenibile (2017). Retrieved from https://romamobilita.it/sites/default/files/pdf/mobilityweek2017/PPT_Acea_DEF.pdf
15. Schneider Electric: Smart Cities Cornerstone Series, Urban Mobility in the Smart City Age, page 4 (2016). Retrieved from: http://smartcitiesappg.com/wpcontent/uploads/PDF/UrbanMobility.pdf
16. Nusio, F.: Roma Capital (2016). Retrieved from https://www.polisnetwork.eu/uploads/Modules/PublicDocuments/Nussio_Mobility_plan_in_Rome.pdf
17. Tira, M., et al.: Managing mobility to save energy through parking planning. In: Papa, R. (ed.) Romano Fistola Smart Energy in the Smart City. Urban Planning for a Sustainable Future, pp. 103–115 (page 103) (2016)
18. Papa, R., Fistola, R.: Smart Energy in the Smart City: Urban Planning for a Sustainable Future, p. 261. Springer (2016)
19. Piano Urbano della mobilità sostenibile, linee guida, comune di Roma. Retrieved from https://www.pumsroma.it/download/Linee_Guida_PUMS.pdf
20. Roma servizi per la mobilità', of commune di Roma (2016). https://www.comune.roma.it/webresources/cms/documents/Mobilita_sostenibile_2015.pdf
21. Dameri, R.: Smart City Implementation Creating Economic and Public Value in Innovative Urban Systems. Springer International Publishing AG (2017)
22. Nusio, F.: Smarticipate workshop, Mobility Masterplan, ITS and Opendata in Jubilee of Mercy year. Casa della città - Rome, May (2016). Retrieved from https://www.comune.roma.it/web-resources/cms/documents/smarticipate-ws-07-nussio-slide-a.pdf
23. Anthopoulos, L.G., Vakali, A.: Urban Planning and Smart Cities: Interrelations and Reciprocities. In: Álvarez, F., et al. (eds.) FIA 2012. LNCS, vol. 7281, pp. 178–189. Springer, Heidelberg (2012). https://doi.org/10.1007/978-3-642-30241-1_16

The 'Governance' for Smart City Strategies and Territorial Planning

Cinzia Bellone[1(✉)], Pietro Ranucci[2(✉)], and Vasiliki Geropanta[3(✉)]

[1] DIS, Università degli Studi Guglielmo Marconi, Rome, Italy
c.bellone@unimarconi.it
[2] Università Roma Tre, Rome, Italy
pietro.ranucci@gmail.com
[3] R&D Department, Università degli Studi G. Marconi, Rome, Italy
v.geropanta@unimarconi.it

Abstract. This article identifies methodologies for increasing the quality of life and acquiring a more democratic and participatory (inclusive) dimension in the new configuration of cities, in the case of smart cities. The analysis presents relevant strategies and implementation cases and investigates how ICT alter the meaning/ideas of "urban planning", leading to an effective "governance", of a citizen- center approach. Additional questions examine whether increasing the technological 'networks', allowing automation and monitoring are sufficient tools for cities' regeneration or if matching technology with spatial participatory models that functionally insert the 'right' formal references in the urban planning is necessary. Public governance's success is measured based on the "listening capacity" and the facilities that are provided to citizens. As such, the paper reviews the ability in managing existing complex interrelations between facilities and urban spaces. Finally, it retraces the historical arc aiming at analyzing and providing insights into the future.

Keywords: Smart city · Territorial planning · Governance

1 "Territorial Planning" as a Premise of the Smart City

The problem of the definition of territorial transformations and of the evolutionary process. The discourse about the development processes and the state of the metropolitan territorial morphology has revealed, lately, many challenges connected with the rapid 'transformations' of the increasing large territorial ambient. Regardless of the development's stage, what seems indisputable is the continuous growth of urban agglomerates that tend to occupy and expand towards nearby neighbor territories. In this diversified landscape, phenomena of centralization, re-centralization or dispersion of economic activities (because of the introduction of ICT), tend to create "centralities" that are spread in a wider territory. In fact, since the 1970s, the territorial morphology has witnessed many transformations in demographics (population growth) and in the economy (city functional specializations), and raise "questions", related to the quality of the settlement, the insertion strategies of new economic activities, mobility, cultural

P. Vasant et al. (Eds.): ICO 2018, AISC 866, pp. 76–86, 2019.
https://doi.org/10.1007/978-3-030-00979-3_8

and leisure services. The common objective in all cases appears to be the search of quality of life and of urban spaces.

In this context, the design of new urban forms that allow the balance between space and "governance" (from a physical, economic, social, political, etc. point of view) seems to be the strategic tool. In fact, population decline and demographic stability in urban places have not seized the urbanization process, as many had predicted. On the opposite, the nature and consequences of the economic growth have determined new spatial and supply demands that satisfy both social and share capital. Specifically, the urban policies of the 1960s and 1970s, guided the emerging economic activities and integrated them in different territorial and geographical contexts (residential, economic or industrial), in new urban centers (Villes Nouvelles, and New towns) and in new neighborhoods, decentralized and autonomous (new expansions in Germany and the Netherlands). Lastly, they empowered urban poles at a regional level (Le metropoles d'equilibres).

The last decades, the same policies tend to be nuanced; perhaps, less determined and are fundamentally directed towards (a) inward interventions (reuse and recovery of urban situations); (b) interventions in interstitial areas and fringe; or (c) interventions that envisage the creation of multi-sector settlements that are specialized in research, planning, consulting, information, management and programming. Since the evolutionary process of urban areas manifests itself "also" through energy consumption increase, it is essential to adapt the perimeters of the urban territory and effectively guide and control the territorial transformations of the area. This adaptation requires a strategic management based on 'processuality', a systematic series of actions that are developed in time and have as objectives the, modified and modifiable in time and space, organization and territorial planning. Following this context, it becomes challenging the definition of boundaries that can have ample degrees of flexibility in terms of function and of construction and bring about a programmatic process of planning depending on the city objectives. They are not physical but refer to boundaries of transforming governance.

Territorial discontinuity. All urban areas "explode" even with low-density settlements, often without a predefined design, creating a discontinuity of the fabric that influences the growth of the territory. In the EU countries, the planning deregulation, the localization preferences of economic activities, the territorial specialization and the spatial segregation, which in long term, might contrast with the settlement's sustainability principles, has caused extensive sub-urbanization [1]. The territorial fragmentation does not only concern the urban structure but also the localization of urban functions, the labor market, society, and mobility. When the city is expanded, the integration of other territorial and urban systems create a new landscape, a unique dispersed but strongly interconnected reality, in which the traditional territorial organizational model of hierarchical relations among areas is now outdated.

The concept of network as a structuring element of the planning of a territory, the premise of the smart city. The network of "nodes of functional relations", which altogether constitute a system, is among the most common connotations today. Organizing various human activities at a metropolitan level has facilitated the transformation of the city from a "space of places" in "space of flows", allowing the configuration of the "networked city" [2], with services that assist mobility [3].

The network appears as an invisible but structuring branching of relations/interactions among centers. In this context, the lattice scheme of Christaller, Losch-Beckmann (1958) [4], with its strict hierarchical form that takes into account only the economic aspects (economies of scale and transport costs) becomes inadequate. Christaller's theory focuses on the 'function' and the 'market', as a network of points where the relations converge. The concepts of 'threshold' and 'scope' explain the network's contribution: the first one refers to the minimum necessary existing market activity in an area and the other to the maximum distance that a consumer is willing to travel to reach the market [5]. The hierarchical organization of the Christallerian model responds to the territorial type logic. This explains how functions increase when the importance of the center increases and how this creates competition among the urban centers of the network. The synthesis of horizontal relations between the various urban centers in the network highlights the potentialities that the planning at a network scale might bring about, particularly at an international level. They raise, therefore, the challenge to understand how the network approach can become a design tool, or if it offers possibilities only in terms of analysis and interpretation of reality.

Urban form-new theoretical and governance models for territorial complexity.
The Civitas, a conscious citizen's collectivity that, in the search for efficiency and advanced social organization, acquired a precise physical form, has a precious role in the management of the historical European city. Before the metropolis, there was the big city, which locally had its own raison d'être, economic and political; but this value has diminished, and the territory has grown immeasurably affecting deeply mobility [6]. These areas should function as "city-of-villages which, through the introduction of limits and boundaries could acquire the form of local self-governing communities [7]. In this scale, it is easier to reduce energy, resources are rebalanced, and a reciprocal relationship with their agricultural territory can be managed. At a local scale, the symbolic, aesthetic and cultural identity of the place can be enhanced. In the contemporary city, the ties among these places depend on the way they relate, which is not necessarily based on proximity. This leads to a phenomenon of dissolution of boundaries, which also affects the boundaries between the rural and urban [8]. The new settlements, cities or metropolitan areas acquire a 'nebulous' identity highlighting the necessity to search scientific, pragmatic, theoretical, political and cultural models in order to govern the complexity of the territorial reality. Since the 1990s, at a national and international level the attention of urban policies [9], is focused on the creation of a balanced and sustainable territorial development that respects the human settlements' organization and consumes less energy. This reveals the need to pursue policies that focus on the concentration, exchanges, and interactions of planning actors. Therefore, in this age of networks, new forms of governance appear which are not based anymore on planning strategies of a top-down process but respond to a context of changing networks that are in continuous evolution [10].

The concept of "city networks" is a territorial organization model that allows medium-sized cities to achieve high levels of competitiveness and create synergies, network economies, and specialization. The connection between large trans-European transport and communication networks is, therefore, the construction of a polycentric system at the continental level, a favorable condition for a balanced development, territorial equity and cohesion [11]. This new way of city "functioning", produces profound changes in the social and spatial configuration of the material and immaterial relations of the community. Consequently, the role of planning, from low economic activities that generate processes of urbanization of the rural territory to the welding of the settlements, become uncertain, while the inter-urban mobility grows in a sustained way with the increase of the complexity of the territorial relations.

As a result, in this context, a polycentric network organization of urban and territorial systems, the creation of a functional "mixité", oriented towards "city effect" and the integration of diversity and activities that favor mobility interventions for the radial connection of metropolitan polarities are useful guidelines. This new urbanity is designed through invisible networks, which transform the territory in an open system. The diverse exchanges form a dense connective tissue, with small or wide meshes depending on the different territorial sites. The networks, in this sense, carry out activities that aim at improving the physical connections of the different parts or functional diversions (visible networks), favor mobility, social and cultural interaction, the circulation of goods and information, capital and innovations. They also enhance the local resources, territory and its diversity. These interdependent relationships, once existed between the parts of a city, and then between the parts of a metropolitan area, have now expanded and this invisible connective tissue is losing its boundaries and extends on a scale no longer certain.

2 The Smart City as a New City Model

The experience of the smart City – Definitions and state of the art. Over the span of the last two decades, more than 800 places in the world have been entitled "smart" as a response of urban planning towards the production of a growing number of service and technological cities of diverse nature. In addition to that thousands of initiatives at a local, national and international levels have acquired the name "smart" in an effort to describe urban areas that use different types of sensors to collect, manage and use data and information in order to transform the urban performance and efficiency. These phenomena have their origins in the spatial clusters of the 'innovation environments' and the 'knowledge clusters', spatial models that were created because of the third industrial revolution around the 1960s and now entering in the fourth acquire new form and become more visible. Generally, the smart city is an urban intervention, constructed by top down or bottom up 'educational' processes that act as the strategic device to change modern urban production factors [12].

More specifically, the smart city promotes a concept for urban performance that does not depend only on the city's endowment of hard infrastructure, but also, on the availability and quality of knowledge communication and social infrastructure [13]. As such, its focus seems to be on the role of ICT infrastructure, on the role of city management and of the education of human capital [13]. Therefore, it examines the social, relational and environmental ambient as important drivers for urban growth. It is still metabolizing, changing over time both conceptually and physically: the sustainable city, the high-tech city, the intelligent city, etc. Difficult as it might be to theorize its present form, in the same time, is obvious that the smart city is already envisioned as the choice for a future of sustainable development and welfare that has not been yet accounted as an urban paradigm of a clear, dominating hegemonic class.

What this shift would really mean for our built world is still under observation. Is the city as we know it today deeply changing when transformed into a smart city? If so, then in which ways make it visually comprehensible? A very positive or very negative answer would sound a dangerous claim: for sure, the insertion of ICT is a process that did not exist at all some decades ago, and the same stands for spin-offs or the production of new services as a new way of thinking about work and living. Currently, many scholars, organizations, corporations or large global cities, all praise the contribution of Web O_2 and ICT to the better management and functioning of cities [14]. Financially at least, these evolutions have opened a new world of possibilities, collaborations and problem-solving strategies towards the problematic of cities. The European Union and other international institutions and think tanks believe in a wired, ICT- driven form of development-. Entire new districts are created to serve the ICT evolution, and new means of digital 'zoning' is changing the conception of the traditional urban tissue.

Many examples around the world are considered as successful smart city projects. Amsterdam Smart City Initiative, which includes more than 170 projects of bottom-up and top-down processes, wins in international rankings. Barcelona similarly, praises the application of almost 60 projects within its CITYOS strategy. Manchester, Milan and Santa Cruz used ICT for a number of services that are of a human-centered approach. New entire cities are built, such as Songdo, Masdar and use technologies for a sustainable environment. In all of these cases, there appears a framework that is divided into four dimensions: technology, human infrastructure, institutional framework and data management framework. In all cases, the key sectors of the Smart city are the mobility, economy, governance, people, living, and the environment. Actors, stakeholders and large corporations collaborate and through a four-step process establish the foundation of each smart city project. The first step is the application of infrastructure (networks, Wi-Fi availability and all technological equipment) in the city ambient. The second step is the installation of sensors and IoT to collect the big data of the city and manage the infrastructure remotely. The third one is the construction of their service delivery platforms to elaborate and evaluate the concentration of big data. The final step includes sustainable applications and value-added services to the city and to citizens. These functions describe the process of collecting data, connecting various physical devices with the diverse actors (city, citizens, services, and government) and in this way city; officials can monitor and observe how the city is evolving.

The smart city with the intelligent transportation services, the e-business services, the learning services, and the environmental services aligns to each dimension of urban planning, among which the environmental protection, the sustainable residential development, resources capitalization and support of coherent regional growth. As planning dimensions are allocated to particular frameworks (demographic distributions, land uses, transportation, green spaces, environmental protections and authorities that monitor and evaluate the planning rules), the same does the smart city project. Several elements are now available and transparent regarding urban planning in participatory smart governance. For example, sites' identification, sites' characterization through the creation of local urban land use planning databases and data that describe urban values such as connectivity, spatial patterns, and proximity. As a result, the smart government tries to optimize services in the urban space, which goes hand-in-hand with activities that improve the quality of life. The engagement of various stakeholders and the use of several pervasive means such as social media, open data, and sensors, strengthen the collaboration between citizens and urban governments. They declare, in this way, that operations and services should be citizen-centric and therefore correspond better to the planning necessities. The network of the collaboration of this framework becomes pervasive, in the sense that it finds its place in all of the activities of daily life of the smart citizens, both in the private and public spheres. Smart cities may be governed completely by the organizations that comprise the network (self-governance model), or by the local government as the centralized network broker (bureaucratic model) [15].

3   Governance (Democracy, Technology, Technocracy)

The smart city designates investments in physical (transport) and intangible (ICT) communication infrastructures, with reference to the human and share capital to achieve the quality of life and sustainability in urban development. This consideration aims at qualifying the smart city, mainly, in relation to the efficiency of the urban 'machine', and the implementation of the human capital. In reality, the goal of a new urban construction process could be also expanded to include more demanding objectives such as better relationships among decision makers and citizens with a reference to "democracy, technology, technocracy".

The keywords' debate should be directed towards a research that identifies the actual relations that exist among them because none of them remotely could be able to control a sustainable urban development and a coherent definition of the urban spaces with the desired quality of level of life. One example of this research, can be found in Italy, in the three years plan of AgID (2011–2013) [16] and the initiative of Enea (Convergence Smart City and Community) that aimed at re-organizing the management of urban and territorial processes in a digital manner, starting from a conceptual, methodological and technological convergence. Specifically, Enea emphasizes the need to share a common language when identifying strategies that ameliorate the urban system's efficiency. In this way, the main actors of urban processes collaborate and create a roadmap with the necessary tools for the creation of a 'shared' idea of a smart city [17].

Another example is one of the ForumPA's debates [18] in Italy that aims at increasing the digitalization of the public administration. In this case, the connection between the processes of governance and the data are shared among the various actors. On the "acquisition and management" of the data, the governance of "the whole process" enters into law and becomes the "frame" that is decisive for keeping urban development within "democratic" scenarios. In fact, if the smart city can bring urban welfare and improve the quality of life of citizens, then how the smart city can alter the relationships between citizens and rulers seems crucial.

In the smart city, the governors increase their guiding capacity while the citizens seem unable to carry out effective control over the public and private management of the huge amounts of data, on the basis of which, political strategies and territorial infrastructural decisions are suggested. In essence, this might bring a change in the democratic character of the traditional western 'democracy'. Parag Khanna [19] in his recent text with the title "Technocracy in America" suggests that technocracy is the key word, which instantly explicates the novelty of the topic about the governance of cities, in the USA. The author analyzes various forms of governance (representative democracy, direct or not, ideological leadership, dictatorships, technocracy, etc.) and argues that a technocratic government should be based on an experts' analysis and long-term planning, rather than on typical improvisations of populism. He emphasizes that, often, forms of government based on representation (with exhausting quarrels of an ideological mold) prevail over the ones based on the administration public affairs that could quickly meet the necessary services to citizens using certain data. In this way, he argues on the necessity to give the same weight on 'figures and democracy'.

Governments should respond to the needs of citizens effectively, with long-term scenarios, bringing together democratic inclusiveness and "technocratic" efficiency. Following this argument, the author cites Switzerland ("direct" European democracy, bottom-up) and Singapore (Asian technocracy, top-down) as two states that have the best indices in international statistics for the level of citizens' well-being. In his opinion, these two states have democracy and knowledge, where knowledge implies that those who command "know" by acquiring valid "historical" data. In the state models' description, Khanna confirms that economically and politically, the development mainly focuses on metropolitan areas and their best performances are found in what he calls "Info-State" (postmodern democracy). ICT allows the Info-State to operate better in relation to the free market allowing the two states to use a direct relationship with the citizens, in real time (referendum, surveys, inquiries, public workshops…). The datum is the abandonment of the ideological guide in favor of a strong pragmatism raising consideration whether the priority is an exasperated democracy or a form of government, which allows effective responses to their needs.

A widespread "populism" that has intercepted most of the trust already entrusted to the traditional "political class", describes Italy in this discourse. Khanna wonders if democracy guarantees a country's success or if a growing inclusion of technocratic roles in governance can assist. These steps could be launched, when extracting the best from "politics and government", "democracy and services", "process and results", selecting the sense of "procedure" from democracy and sense of "result" from the technocracy. Neither of the two prevails, but both are under penalty of mutual legitimization. The author is convinced that "on the long run, the quality of Governance is more important than the type of regime in power, claiming that: "Too much politics corrupts democracy; too much democracy hinders the policies. Politics relate to positions, while policies relate to decisions. Democracies produce compromises, technocracies bring solutions; democracy adapts itself (satisfies), technocracy seeks the best solution (optimizes)". For example, the conspicuous social and economic results of Switzerland and Singapore support this argument. Therefore, it is necessary that democracy be pervaded by a different policy, which, starting from a consistent analytical recognized and shared bases, brings suitable certified knowledge forms (quantitative and qualitative data). In this way, democracy does not succumb in front of the increasing citizen participation, which often degenerates into ideological conflicts with suspensions of procedures and variations.

In Western democracies, the phenomenon of urban governance has always been accompanied by numerous analyses and by the collection of city and territorial data, scarcely used in actual realizations. Firstly, ruling classes are less prepared and willing to change their decision-making behavior and therefore rely on traditional applications that guarantee more profitable mediation among political parties.

As a result, we suggest here, a form of governance that is characterized by a decision-making system that uses a "close relationship" between representative, political and technocratic bodies; deputies to provide qualitative-quantitative assessments of the proposed development scenarios, starting a political evolution "in the technocratic sense". Khanna argues that if governance gradually acquires contributions from the technocratic system, then the "utilitarian and meritocratic" objectives of this action would clearly emerge, demonstrating that many of the political decisions are based on ideological oppositions without any knowledge on urban phenomena. The utilitarian and meritocratic system of technocracy could, therefore, offer essentially in politics. Urban planning, marked by lengthy procedures that often impose only partial revisions, would challenge the "who and how it is possible to manage" the databases in order to use them in line with social demand. However, the risk in asking for a rigorous technological admin profile is that it cannot be guaranteed that end users will be as well capable to work with data management.

The authors argue that a part of Khanna's index on building a blend of democracy and technocracy, assisted by ICT can be useful in governance. This includes a governance which is fully democratic, leading to an equally irrevocable, incremental welfare and social debate, based on a quantitative knowledge, capable of activating decision-making processes, both of a "vertical", bottom-up type, and also "horizontal." A type of governance that is able to redefine, in a shared manner, the design of the city, through ICT. In the debate on smart and digital cities, the authors highlight the creation of a global "gigantic collective memory". The impact on tax, democracy, the political class, meritocracy etc. is incontestable. For the evolution of the phenomenon, they suggest the following hypotheses:

(1) The new "political" urban governance profile should be included in the discourse about the future design of cities;
(2) "Exploring the future" by designing the present, hypothesizing a subsequent scenario, seems an idea that can be shared [20].
(3) Geo-location tools are available (GPS, Wireless etc.) and allow real-time monitoring and visualization of urban and territorial realities;
(4) Insisting on territorial scenarios that are constructed upon improbable hypotheses and are only analogically evaluated, seems less useful. On the opposite, investigating "information flows" and "data-driven" models, using data collected on the micro and macro-themes of each urban reality, and observing "how", "by whom" and with what objectives data are used, seems more adequate.

(5) "We are" experiencing an epic moment that communication between men and communities transforms, from the first "global village" to a village where communication is more democratic in character, between thrusts that are not only "vertical" but also "horizontal" (Internet, cyberspace etc.). This space of flows is a completely new phenomenon.

(6) This space interacts with the digital, analogical and physical world, stressing their spatial effects in the city. Physical "distance" is reduced even if a "fictitious" theme may appear. To what extent the Internet makes, the localization of interconnected subjects-things irrelevant is still to be searched;

(7) In fact, cities continue to grow, a fact that cannot be generalized, because it relates above all to the c.d. Third World. However, the innovative forms of communication cannot make the phenomenon of settlement dispersion real.

(8) Man is a "collective" animal that needs to live in a real and physical community where there are contact and life in the city, without denying the opportunities of cyberspace, the "triumph of atoms and bits" [21]. Urban space is a set of "virtual and physical places" that combined give rise to changing urban "forms", on a physical and functional level. This could be the smart city in necessary symbiosis with the Digital City. In this case, men and things are "sensors", in real time, of the evolution of urban reality (ubiquitous computing) and "public and private spaces" are decisive in this. Therefore, from the analogical to the digital age, the container takes "different forms in time, even short, and flexible in functions". Moreover, the concept of "form" of the city changes and acquires a more complex meaning over time.

(9) Data-driven processes change the city because urban space can be coordinated with computer platforms (C. Ratti). In this context, is the civil- hacking ("hacking" the city, S. Sassen [22]) possible? Which Governance is desirable for a complex process where the knowledge and use of new technologies are mastering? Where can we imagine a programmable and changeable Architecture? Where does "sentient" space exist? Big data's exponential growth present new research directions for the city, its "resilience" and the political-social coherence of the governance. The current form of big data governance needs to extend based on the bottom-up and horizontal model. In particular, the "way" the public space is programmable through a governance that increasingly considers this model. We observe that the data-driven process needs to evaluate the validity of the current governance forms in order to achieve a necessary substantial democracy of the whole operation.

Without a democratic profile in the process, the smart city can be negated. The meaning of cities through increasing forms of urban efficiency and human capital is no longer sufficient to clarify the deeper meaning of democracy. "Digital transformation as an ecosystem… it cannot be done, harbinger if not governed, of… terrible discrimination…. of great violation of rights… divesting monopolies…" [23].

Without denying the validity of an "efficient" city for users only, in the distinction between these and "citizens" lies the whole meaning of an "intelligent" City.

References

1. Gibelli, M.C.: Dal modello gerarchico alla governance: nuovi approcci alla pianificazione e gestione delle are metropolitan. In: Camgni, R., Lombardo, S. (eds.) (a cura di), La città metropolitan: strategie per il governo e la pianificazione Ed. Aline (1999)
2. Beguinot, C.: L'architettura è intelligente, se è capace di (inter) connettere. In: Telèma no. 15, inverno
3. Martinotti, G.: Il vero centro si è spostato, non è più "dentro" ma in periferia. In: Telèma no. 15, inverno (1998/99)
4. Beckmann, M.J.: City hierarchies and the distribution of city size. Econ. Develop. Cult. Change **6**
5. Hannerz, U.: Esplorare la città, il Mulino, Bologna (1992)
6. Magnaghi, A.: Per una nuova carta urbanistica (1990)
7. Camagni, R.: La pianificazione sostenibile delle aree periurbane, Bologna (1999)
8. Camagni, R.: Agire metropolitano: verso forme e strumenti di governo a geometria variabile. In: Convegno DPTU-DAU pensare ed agire metropolitano: verso una nuova visione istituzionale e funzionale, Roma, 23 Aprile 1998
9. Documents: Green Paper on the Urban Environment of 1990, Europe 2000 and Europe 2000 + (1992–1994) and European Space Development Chart
10. Benveniste, G.: La pianificazione come gestione a matrice di reti decentrate. In: "Pianificazione strategica e gestione dello sviluppo urbano" a cura di Curti Fausto e Gibelli, Maria Cristina, Bologna (1999)
11. Camagni, R.: Agire metropolitano: verso forme e strumenti di governo a geometria variabile. In: Convegno DPTU-DAU pensare ed agire metropolitano: verso una nuova visione istituzionale e funzionale, Roma, 23 Aprile 1998
12. Komninos, N.: The Age of Intelligent Cities: Smart Environments and Innovation-for-All Strategies. Routledge, London and New York (2015)
13. Caragliu, A., Nijkamp, P.: The impact of regional absorptive capacity on spatial knowledge spillovers. Tinbergen Institute Discussion Papers 08-119/3, Tinbergen Institute, Amsterdam (2008)
14. Engel, J.S.: Global Clusters of Innovation: Entrepreneurial Engines of Economic Growth Around the World. Edward Elgar Publishing, Cheltenham (2014)
15. Bolívar, M.P.R.: Governance models for the delivery of public services through the web 2.0 technologies: a political view in large Spanish municipalities. Soc. Sci. Comput. Rev. **35**(2), 203–225 (2017)
16. AgID - Agency for Digital Italy. It is a public agency, established in Italy by the Monti government (2011–2013). It pursues the highest level of technological innovation in the organization and development of public administration (P.A.)
17. ENEA: The main Italian National Agency for new technologies, energy, and sustainable economic development. It promotes important research on these issues
18. Forum Pa: A company of the Digital 360 Group. The company has been working for three decades to stimulate the digital growth of the public administration, favoring the meeting between the PA, companies, researchers, and citizens
19. Parag Khanna, internationally renowned geopolitical strategist. In Italy, he published for Fazi Editore the trilogy: I tre Imperi (2009), Come si governa il mondo (2011) and Connectography (2016). Parag Khanna (2017) La rinascita delle città-stato, Fazi Editore, sett (2017)

20. Ratti, C.: An internationally renowned engineer and architect, he teaches at the Massachusetts Institute of Technology where he founded and directs the Senseable City Lab. Owner of the Carlo Ratti Associati Studio, with Italian headquarters (recourse C. Ratti, "La città di domani", with Mattew Claudel, 2017) (2017)
21. idem
22. Sassen, S.: "Hacking" the city, TED talks (2013). https://www.youtube.com/watch?v=vHuX79hgtCY
23. Mochi Sismondi, C.: FPA (2018). http://www.forumpa.it/speaker/2381-carlo-mochi-sismondi

T-MPP: A Novel Topic-Driven Meta-path-Based Approach for Co-authorship Prediction in Large-Scale Content-Based Heterogeneous Bibliographic Network in Distributed Computing Framework by Spark

Phuc Do[✉], Phu Pham, Trung Phan, and Thuc Nguyen

University of Information Technology (UIT), VNU-HCM,
Ho Chi Minh City, Vietnam
{phucdo,thucnt}@uit.edu.vn,
phamtheanhphu@gmail.com, trungphansg@gmail.com

Abstract. Recently, heterogeneous network mining has gained tremendous attention from researcher due to its wide applications. Link prediction is one of the most important task in information network mining. From the past, most of the networked data mining approaches are mainly applied for homogenous network which is considered as single-typed objects and links. Moreover, there are remained challenges related to thoroughly evaluating the content of linked objects which are considered as important in predicting the potential relationships between objects. Like a common problem of predicting co-authorship in bibliographic network such as: DBLP, DBIS, etc. There is no doubt that an author who is interesting in "data mining" field tend to cooperate with the other authors who contribute on this field only. Hence, predicting co-authorships between authors work on "data mining" with others who work on "hardware" is dull as well. Moreover, in the context of large-scaled network, traditional standalone computing mechanism also is not affordable due to low-performance in time-consuming. To overcome these challenges, n this paper, we propose an approach of topic-driven meta-path-based prediction in heterogeneous network, called T-MPP which is implemented on distributed computing environment of Spark. The T-MPP not only enables to discover potential relationships in given bibliographic network but also supports to capture the topic similarity between authors. We present experiments on a real-world DBLP network. The outputs show that our proposed T-MPP model can generate more accurate prediction results as compared to previous approaches.

Keywords: Information network · HIN · Link prediction
Meta-path-based link prediction · Topic-driven link prediction
Large-scaled HIN · Distributed graph-based computing · Spark

© Springer Nature Switzerland AG 2019
P. Vasant et al. (Eds.): ICO 2018, AISC 866, pp. 87–97, 2019.
https://doi.org/10.1007/978-3-030-00979-3_9

1 Introduction

Among information network (IN) mining tasks [1–3], link prediction is one of the most important task which support to predict missing links or new relationships in future of the given networks. Link prediction [4–7] is important for mining and analyzing the evolution of the information network. In the past, several solutions of link prediction have been proposed to support for relationship discovering over the network. However, most of the existed approaches only concentrate on evaluating the homogeneous information networks (HoIN) which mean all nodes and links in the network are treated as the same type. There is no doubt that most of the networks are heterogeneous and each object and relation type carries out the different meaning as well as, for example, such as in DBLP network, multiple object types and relationships, such as: author $\xrightarrow{\text{write}}$ paper, paper $\xrightarrow{\text{cite}}$ paper, paper $\xrightarrow{\text{submit}}$ venue/conference, etc., which normally called meta-path (Definition 2) [1, 2] have different meanings and can't consider as the same. Moreover, in complex network, the attributes of object and link are also vary, for example an "author" object has its own attribute such as "gender", "address", "affiliation", etc. which are different from "paper" object's attributes such as "topic", "keywords", etc. which are extremely difficult to capture them all, but can't skip due their importance during the network evaluation process. Therefore, recently the linking prediction task in information network faces three main challenges, which are listed as following. First of all, existing problems related to the multi-typed objects and links in HINs, most of the approaches are incapable to apply for different-typed objects and relations while extracting network topological features to feed the prediction model. Secondly, the differences in length and meaning of paths which link two objects leads to the problem in generating topological features. For example, we want to evaluate the proximity of two specific "authors" via observing all paths which connect them together, in the manner of HIN-based context, two authors might be linked via different paths' type and length, such as: author $\xrightarrow{\text{write}}$ paper $\xleftarrow{\text{write}}$ author, author $\xrightarrow{\text{work\_at}}$ affiliation $\xleftarrow{\text{work\_at}}$ author, etc. and the challenge is that how to appropriately capture these differences to form proper topological features, such as: the approach of Katz$_\beta$ metric in measuring the total number of all paths between two given nodes without considering about the path's length. Last but not least, the shortage of node's attributes evaluation also is a challenge which leads to the decrease in the accuracy of link prediction task. Like as the problem of co-authorship prediction in bibliographic network, two authors who research on the same topics absolutely tend to cooperate in the future than other authors who contribute on the other topics. The topic attribute might be identified by the content of their papers or set of keywords usage. As aforementioned challenges, we need to find the other strategy which supports to properly generate topological features for the HINs. The proposed approach must be able to distinguish the differences in the paths' meanings as well as nodes' attributes. Our contributions in this paper are three-folds, include:

- First of all, we present the approach of applying LDA topic for extracting the topic distributions over content-based object in HIN, such as "paper/article" objects in bibliographic networks. Then, these probabilistic distributions are used to evaluate the topic proximity between authors.

- After that, we present the approaches of T-MPP model which is mainly inspired from the works of Sun et al. (2011) in "PathPredict" [6] which is the meta-path-based prediction approach. We describe about improvements on topic-driven meta-path-based link prediction of proposed T-MPP model. The topological features are extended by adding the topic similarity attributes while generating the feature vectors for each pairwise node.
- Finally, we conduct experiments on the real-world DBLP and Aminer dataset in order to demonstrating the effectiveness of our proposed models which is promising to leverage the accuracy of co-author relationship prediction task in the bibliographic networks.

The rest of our paper are organized in 4 sections. In the second section, we discuss about related works and motivations of previous researches. The proposed T-MPP model and methodologies are described in the third section. In the next section, we demonstrate our empirical studies on the T-MPP model as well as discuss about experimental outputs. The last section is our conclusion and future works.

2   Related Works and Motivations

From the past, most of the link prediction approaches on information network are considered unsupervised methods [4, 8, 9] which means they mainly focused on directly analyzing the graph's structure of the network to obtain the probability of new links might be appeared, but this type of method does not gained high effectiveness as well as accuracy in the outputs in complex and dynamic networks. Subsequently, the supervised learning approaches have been proposed to leverage the accuracy of link prediction tasks. The supervised approaches [10–12] enable the system to learn from the previous network dataset to generate the predictive model, which is applied to evaluate the likelihood of two nodes to be connected in the future. However, most of the previous works are mainly concentrate only on homogeneous networks. For example, such as the closest match with our works in this paper is solving the problem of co-authorship prediction [4] via extracting several topological features from the network of a single-typed co-authorship. There is seldom appropriate solution to solve the problem of multi-typed objects and links prediction task. Recently, there are several models have been proposed to overcome challenge of HIN mining. As the challenges of multi-typed object and relationship in HIN, the definition of "*meta-path*" has been proposed as the principal concept for HIN link prediction as well as other mining tasks [6, 7, 13].

3   Methodology and System Architecture

Recently, HIN mining has attracted many researchers due to its wide applications in multiple areas. Link prediction is considered as one of the most important HIN mining tasks, which encounters two main difficulties related to complex networked data mining, include:

- The diversity of node/object and link/relation type of HIN which challenges the previous HoIN-based approaches.

- The difference and complication of semantic meanings of paths which link pairs of node leads to the failure in traditional topological feature extracting mechanisms.

3.1   Topic Similarity Evaluation Between Content-Based Nodes

Topic Similarity Evaluation Between Objects via LDA Topic Model In content-based bibliographic HIN, such as: DBLP, most of text-based object such as "papers/articles" play as core object type which appear in most of meta-paths. Therefore, the carried attributes, such as: topic/category, keywords, etc. of these objects can help to effectively enrich the co-authorship predictive model. As discussed from previous sections, in realistic, authors who work on the same fields likely tend to cooperate with each other than other authors who work on the other fields. Hence, evaluating the topic similarity between authors is very important. In order to tackle this challenge, we use the LDA [14] topic model supports to produce topic distributions over the set of paper objects in given bibliographic networks. The LDA topic model support to extract the probabilistic distribution of topics within the set of papers/documents, denoted as: $P(z_i|d_j) = \theta^{d_j}_{z_{(i,i \in |Z|)}}$, is the distribution of (i)-th topic over paper/document (d_j) with $|Z|$ number of topic. Then, now each paper object are now represented as the fixed-length feature vector, $\vec{d_j}$, with $|Z|$-dimensions, denoted as: $\vec{d_j} = \left[P(z_1|d_j), \ldots P(z_{(i,i \in |Z|)}|d_j) \right]$, or $\vec{d_j} = \left[\theta^{d_j}_{t_1} \ldots \theta^{d_j}_{z_{(i,i \in |Z|)}} \right]$. After obtaining the topic feature vectors of the set of papers, we can easily evaluate the topic similarity between these papers via off-the-shelf vector distance algorithms such as: cosine similarity, Euclid distance, etc. In this paper, we used the cosine similarity metric to compute the topic similarity between these papers. Back to the problem of co-authorship link prediction, the assumption is that: "*two authors is considered as relevant in their research topics if their papers' topics are similar*". Following that idea, we define the topic similarity score between of two authors (x) and (y), depends on the linked content-based objects, such as their published papers. For each path (p), with a specific paper of each author (x) and (y), denoted as: (x_c) and (y_c), respectively. The topic similarity score between two authors for a specific path (p), denoted as: $top\_cs\_sim_p(x, y)$, as following (shown in Eq. 1):

$$top\_cs\_sim_p(x,y) = \frac{\vec{x_c} \cdot \vec{y_c}}{\overrightarrow{x_c} \cdot \overrightarrow{y_c}} = \frac{\sum_{i=1}^{|Z|} \left(\theta^{x_c}_{z_i} \cdot \theta^{y_c}_{z_i} \right)}{\sqrt{\sum_{i=1}^{|Z|} \left(\theta^{x_c}_{z_i} \right)^2} \cdot \sqrt{\sum_{i=1}^{|Z|} \left(\theta^{y_c}_{z_i} \right)^2}} \tag{1}$$

Where,

- Z, is set of extracted latent topics which extracted via LDA topic model.
- $\theta^{x_c}_{z_i}$ and $\theta^{y_c}_{z_i}$, present for the probabilistic distributions of latent topic (i)-th over a specific paper of author (x), denoted as: (x_c) and a specific paper of author (y), denoted as: (y_c), respectively.

3.2   Topic-Driven Meta-path-Based Topological Feature Extraction

In order to apply meta-path in the process of topological feature selection, Sun et al. proposed four main measure functions, include [6]: path-count, normalized path-count, random walk and symmetric random walk. Motivating from these paradigms, we proposed an extended versions of these meta-path-based topological feature extraction mechanism, as following:

Topic-driven path-count: T-PC$_{MP}$(x ⤳ y): is the total number of path instances (p) in total paths, denoted as: (P), following the defined meta-path (MP) with link two authors (x) and (y), in a given bibliographic network. Noticing that the value of path-count between x ⤳ y and y⤳ x can be different depending on the network's structure. The TPC$_{MP}$⟨x ⤳ y⟩ metric is calculated as following equation (Eq. 2):

$$\text{T-PC}_{MP}\langle x \rightsquigarrow y\rangle = \sum_{p \in P}^{p} \left[w_p . top\_cs\_sim_p(x, y) \right] \tag{2}$$

Where,

- w_p, is the total weights of path (p), normally 1 for most of the relation types in bibliographic network which are considered as binary relations.

- $top\_cs\_sim_p(x, y)$, is the topic similarity between two author (x) and (y) following a given path (p) (as explained in Eq. 1).

Topic-driven normalized path-count: T-NPC$_{MP}$⟨x ⤳ y⟩: is defined as the normalized sum of path instances between two authors following the meta-path (P), denote as following equation (Eq. 3):

$$\text{T-NPC}_{MP}\langle x \rightsquigarrow y\rangle = \frac{\text{T-PC}_{MP}\langle x \rightsquigarrow y\rangle + \text{T-PC}_{MP}\langle y \rightsquigarrow x\rangle}{\text{T-PC}_{MP}\langle x \rightsquigarrow .\rangle + \text{T-PC}_{MP}\langle y \rightsquigarrow .\rangle} \tag{3}$$

Where,

- T-PC$_{MP}$⟨x ⤳ .⟩, presents for the topic-drive path count of node (x) to other linked nodes or considered as that node's out-degrees.

- T-PC$_{MP}$⟨y ⤳ .⟩, presents for the topic-drive path count of node (y) to other linked nodes or considered as that node's out-degrees.

Topic-driven random walk: T-RW$_{MP}$⟨x ⤳ y⟩: is the probabilistic transitional weight for the walker to travel from author (x) to author (y), denoted as following equation (as shown in Eq. 4):

$$\text{T-RW}_{MP}\langle x \leadsto y \rangle = \frac{\text{T-PC}_{\text{MP}}\langle x \leadsto y \rangle}{\text{T-PC}_{\text{MP}}\langle x \leadsto . \rangle} \tag{4}$$

Topic-driven symmetric random walk: T-SRW$_P\langle x \leadsto y \rangle$: similar to random walk metric RW$_P\langle x \leadsto y \rangle$, the symmetric random walk supports to obtain the transitional probabilistic of both two node sides, include the traversals of random walker from author (x) to author (y), $(x \leadsto y)$ and author (y) to author (x), $(y \leadsto x)$, denoted as following equation (as shown in Eq. 5):

$$\text{T-SRW}_{MP}\langle x \leadsto y \rangle = \text{T-RW}_{\text{MP}}\langle x \leadsto y \rangle + \text{T-RW}_{\text{MP}}\langle y \leadsto x \rangle \tag{5}$$

In HIN-based link prediction, with different types of meta-path, we can use any topic-driven metric to obtain a distinctive topological feature or combining all these measures to form a "hybrid" approach. In the next section, we will compare each type of proposed topic-driven approach as well as comparisons with previous models.

3.3   Distributed Computing on Meta-path Count with Spark Graph-Frames

In order to optimize the performance of proposed model on the circumstance of large-scaled information network we implement the Spark [15] graph-frames library for distributed meta-path-based traversal task. The strategy for this approach is using the motifs finding, which is considered as a powerful querying tool for sequential order pattern searching in graph-based structure. The motif finding enable us to restrict the walker to follow the predefined meta-paths strictly while calculating the number of path instances between two pairwise nodes. The process for obtaining the number of path instances between two pairwise node (x) and (y) is described in Algorithm 1.

Algorithm 1. Pseudo code for meta-path-based random walk via motifs finding in Spark with Graph-Frames library

---

Input: the heterogeneous network, denoted as: $G = (V, E)$, with pre-defined meta-path $(\mathcal{P})$, with starting node (x), ending with node (y).

Output: Number of path instance (P).

1:	**Function** Distributed_MetaPath_Walk$(x, y, G, \mathcal{P})$:
2:	**Mapping**: GraphFrame$_G \leftarrow G$
3:	**Defining**: motifs$(x \leadsto y) \leftarrow \mathcal{P}$
4:	**Do finding**: paths $=$ GraphFrame$_G$(motifs(x,y)).find()
5:	**Return** paths.distinct.count()
6:	**End function**

For example, with the meta-path A-P-V-P-A which is represented for: author $\xrightarrow{write}$ paper $\xrightarrow{submit}$ venue $\xleftarrow{submit}$ paper $\xleftarrow{write}$ author, which is used to find evaluating the similarity of authors who submit their works on the same set of venues. In order to obtaining all path instances of the A-P-V-P-A meta-path, we use a Scala program with GraphFrame as follows:

```
println("All path instances of the A-P-V-P-A meta-path")

val nodeStmt = s"MATCH (n) "+
    "RETURN ID(n) as id,n.nodetype as type,n.nodename as name"
val edgeStmt = s"MATCH (n)-[r]->(m) "+
    "RETURN id(n) as src, id(m) as dst, r.edgetype as type"

val gf = Neo4j(sc)
    .nodes(nodeStmt,Map.empty)
    .rels(edgeStmt,Map.empty)
    .loadGraphFrame

gf.find("(A1)-[w1]->(P1);(P1)-[s1]->(V); (A2)-[w2]->(P2);(P2)-[s2]->(V)")
    .filter("w1.type='WRITES' and s1.type=' PUBLISHED_AT' and "+
        "w2.type=' WRITES' and s2.type=' PUBLISHED_AT' and A1.id != A2.id")
    .show()
```

With this motifs finding, Spark graph-frames supports to find all path instances following the A-P-V-P-A meta-path based on the distributed parallel computing mechanism, in a given heterogeneous network, Table 1 shows examples of path instances for meta-path A-P-V-P-A, Table 2 shows top-k predicted co-authorship relations for author "Jiawei Han".

4 Experimental Studies and Discussions

In this section, we demonstrate the empirical studies about our proposed T-MPP which is topic-driven meta-path-based prediction model. The experimental results demonstrate that T-MPP can help to effectively improve the co-authorship prediction accuracy in content-based bibliographic network.

4.1 Experimental Dataset Usage and Setup

For testing the proposed model, we use the real-world DBLP[1] bibliographic network combined with the available AMiner[2] dataset which is the dataset of abstract content of papers in DBLP network.

---

[1] DBLP bibliographic network: http://dblp.uni-trier.de/.

[2] Aminer dataset: https://www.aminer.cn/.

Table 1. Examples of path instances for meta-path A-P-V-P-A

Author_1	b (paper)	c (venue)	d (paper)	Author_2
Jiawei Han	Discovery of multiple-level association rules from large databases	VLDB	A Framework for Clustering Evolving Data Streams	Philip S. Yu
Jiawei Han	Mining concept-drifting data streams using ensemble classifiers	SIGKDD	LCARS: a location-content-aware recommender system	Yizhou Sun
...	...	...	...	...

Table 2. Top prediction of co-authorship for author "*Jiawei Han*"

Rank	Authors
1	Christos Faloutsos
2	Bing Liu
3	Hans-Peter Kriegel
4	Churu C. Aggrawal
...	...

At this time, the DBLP network contains over 4.1 M papers, over 2 M authors and more than 6 K conferences/journals. From the DBLP network, we select 60 K top-cited (to be referred/cited by the other papers) papers and split them into two main part within two main time intervals based on the published years of these papers. The first interval is in range [1985–2010] years which are used as training set, denoted as: T_{train}. The second interval is in range [2011–2017] years, which are used as test-set, denote as: T_{test}.

We store the DBLP network in Neo4J Graph Database as a graph with 5 node types and 5 relationship types. Details of the Neo4j graph are shown in Tables 3 and 4. This graph has 11,721 vertices and 52,958 edges.

4.2 Experimental Results and Discussions

T-MPP Model Accuracy Evaluation For each topological measure which is described in Sect. 3.2, we conduct the experiment for each type of approach in comparing with the hybrid measure (combination of all topological measures) approach. These approaches are tested with different dataset's sizes (%). The experimental results are shown in Table 5 (Figs. 1 and 2).

Standalone vs. on Distributed Computing Environment We implemented T-MPP model on both standalone and Spark-cluster-based distributed computing environment with 3 nodes, and comparing the execution time-consuming. The experimental results shown that the T-MPP in distribution-based gained better performance than the standalone-based environment following the increase of dataset size. The output shown in Table 6 (Fig. 3).

Table 3. Node types of the Neo4J graph

No.	Node types	Descriptions
1	AUTHOR	Authors of papers
2	PAPER	Papers
3	TOPIC	Topics of papers
4	VENUE	Venues at which papers are published
5	WORD	Keywords of papers

Table 4. Relationship types of the Neo4J graph

No.	Relationship types	Descriptions
1	CITES	PAPER – CITES – PAPER
2	CONTAINS	TOPIC – CONTAINS – WORD
3	PUBLISHED_AT	PAPER – PUBLISHED_AT – VENUE
4	RELEVANT	PAPER – RELEVANT – TOPIC
5	WRITES	AUTHOR – WRITES – PAPER

Table 5. The experimental results for different topic-driven meta-path-based topological measures

Training & test set size (%)	Training & test set size (%)			
	T-PC	T-NPC	T-RW	T-SRW
20%	0.38271	0.36213	0.33281	0.42112
30%	0.40821	0.41987	0.38213	0.45217
50%	0.41291	0.42821	0.42231	0.47292
80%	0.44291	0.43317	0.49218	0.51211
100%	0.53281	0.53821	0.58271	0.57212

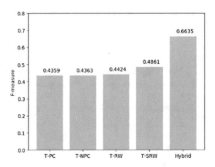

Fig. 1. Average accuracy of each T-MPP topological approach

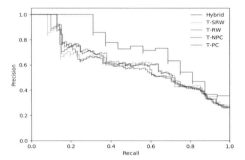

Fig. 2. Comparison of different T-MPP topological approach

Table 6. Standalone vs. on distributed computing environment

Number of node	Execution time (second)	
	Standalone	Distributed
1000	600	1.425
5000	2.850	3.213
8000	3.678	4.231
10000	5.267	4.892
50000	16.879	12.819
100000	37281	22781

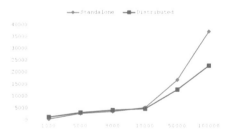

Fig. 3. Comparisons between standalone vs. distributed computing environment on time consuming

5   Conclusion and Future Work

In this paper, our works are mainly focused on solving problems related to predicting co-authorship relations between authors in heterogeneous bibliographic networks. Over experiments on the real-world DBLP bibliographic network, the T-MPP model have outperformed the traditional approaches which are applied for homogeneous networks.

The T-MPP model is designed to work on the heterogeneous networks which contain multi-typed objects and links. Inspiring from the previous proposed PathPredict model, our main contributions in this paper are the combination of thoroughly evaluating the topic similarity between authors along with the process of meta-path-based topological feature learning. This combination is promising to significantly improve the accuracy of the other relationship prediction tasks such as friendship prediction on social network via content-based objects such as: posts, comments, etc.

Acknowledgements. This research is funded by Vietnam National University Ho Chi Minh City (VNU-HCMC) under the grant number B2017-26-02.

References

1. Sun, Y., Han, J.: Mining heterogeneous information networks: principles and methodologies. Synth. Lectures Data Min. Knowl. Discov. **3**(2), 1–159 (2012)
2. Sun, Y.; Han, J.: Mining heterogeneous information networks: a structural analysis approach. ACM SIGKDD Explor. Newsl. **14**(2), 20–28 (2013)
3. Shi, C., et al.: A survey of heterogeneous information network analysis. IEEE Trans. Knowl. Data Eng. **29**(1), 17–37 (2017)
4. Liben-Nowell, D., Kleinberg, J.: The link-prediction problem for social networks. J. Assoc. Inf. Sci. Technol. **58**(7), 1019–1031 (2007)
5. Wang, P., et al.: Link prediction in social networks: the state-of-the-art. Sci. China Inf. Sci. **58**(1), 1–38 (2015)
6. Sun, Y., et al.: Co-author relationship prediction in heterogeneous bibliographic networks. In: 2011 International Conference on Advances in Social Networks Analysis and Mining (ASONAM), pp. 121–128. IEEE (2011)
7. Yu, X., et al.: Citation prediction in heterogeneous bibliographic networks. In: Proceedings of the 2012 SIAM International Conference on Data Mining. Society for Industrial and Applied Mathematics, pp. 1119–1130 (2012)
8. Adamic, L.A.. Adar, E.: Friends and neighbors on the web. Soc. Netw. **25**(3), 211–230 (2001)
9. Kuo, T.-T., et al.: Unsupervised link prediction using aggregative statistics on heterogeneous social networks. In: Proceedings of the 19th ACM SIGKDD International Conference on Knowledge Discovery and Data Mining, pp. 775–783. ACM (2013)
10. Al Hasan, M., et al.: Link prediction using supervised learning. In: SDM 2006: Workshop on Link Analysis, Counter-Terrorism and Security (2007)
11. Benchettara, N., Kanawati, R., Rouveirol, C.: Supervised machine learning applied to link prediction in bipartite social networks. In: 2010 International Conference on Advances in Social Networks Analysis and Mining (ASONAM), pp. 326–330. IEEE (2010)
12. Lu, Z., et al.: Supervised link prediction using multiple sources. In: 2010 IEEE 10th International Conference on Data Mining (ICDM), pp. 923–928. IEEE (2010)
13. Cao, B., Kong, X., Philip, S.Y.: Collective prediction of multiple types of links in heterogeneous information networks. In: 2014 IEEE International Conference on Data Mining (ICDM), pp. 50–59. IEEE (2014)
14. Blei, D.M., Ng, A.Y., Jordan, M.I.: Latent dirichlet allocation. J. Mach. Learn. Res. 993–1022 (2003)
15. Zaharia, M., et al.: Apache spark: a unified engine for big data processing. Commun. ACM **59**(11), 56–65 (2016)

Optimization of Hybrid Wind and Solar Renewable Energy System by Iteration Method

Diriba Kajela Geleta[1] and Mukhdeep Singh Manshahia[2(✉)]

[1] Department of Mathematics, Madda Walabu University, Oromia, Ethiopia
kajeladiriba@yahoo.com
[2] Department of Mathematics, Punjabi University, Punjab, Patiala, India
mukhdeep@gmail.com

Abstract. Because of depletion of fossil fuel, increasing energy demand, and increasing number of population, world has entered in to the new phase of energy extracting from alternating sources. These renewable energy sources are abundant, free from greenhouse gas and will become an alternative of fossil fuel. In this paper iteration method was involved to optimize the designed hybrid Wind and solar renewable energy system. As a result all the components are properly sized in order to meet the desired annual load with the minimum possible total annual cost.

Keywords: Hybrid renewable energy · Optimization
Iteration method

1 Introduction

Energy is one of the vital factor for the socio-economic development of societies of any country. The natural capacity of earth for supplying fossil energy will not ever lasting. Even though, these conventional energy source have been playing the leading role, 80% of the worldwide energy demand for the past many years [4]. Nowadays, the global warming, depletion of its sources and continuous increase in oil prices have got worldwide attention for the development and utilization of renewable energy sources [1]. Due to such and rapid increment of industrialization all over the world, the need for energy is exponentially increases from time to time and shortage of fossil fuels has been occurred [7].

These Conventional energy sources which include power plants using fossil fuels (natural gas, coal, etc.) have a lot of disadvantages [2]. The core disadvantages are:-

a. The issue of environmental degradation. It leads to the inevitable production of carbon dioxide (CO_2), where harmful emissions, such as carbon monoxide (CO), nitrogen oxides (NO_2), sulphur oxides (SO_2), unburned hydrocarbons (HC) and solid particles are produced.

© Springer Nature Switzerland AG 2019
P. Vasant et al. (Eds.): ICO 2018, AISC 866, pp. 98–107, 2019.
https://doi.org/10.1007/978-3-030-00979-3_10

b. It needs continuous fuel supply to operate, which contributes to the operating costs. This cost depends on various local and global parameters, such as fuel availability and type, fuel purity, world economic conditions, local prices, etc.

To overcome or at least limit some of the problems associated conventional energy sources, renewable energy resources are the solution.

Wind and Solar have abundant power which can be exploited as electric energy by the help of wind turbines and solar panels. These energy, which is renewable, environmentally clean without causing greenhouse gas and reduce the cost of electricity can be alternatives to fossil fuels [3]. The main disadvantage of these technologies is the fluctuation of their power output. To overcome these disadvantage hybrids renewable energy technology was important [1,19].

This paper is attempts to find the optimal size of hybrids of wind and solar renewable energy system base on minimization total annual cost under the power balance. The main concern is to determine the size of each components participating in the system, so that the desired load can be satisfied with minimum possible cost [4,5,7,9]. Since the problem under consideration consists of integer decision variables, numbers of wind turbines, solar panels and batteries conventional Optimization methods such as probabilistic methods, Analytical methods and Iterative method can effectively give the local extremum values [4]. But due to stochastic nature of the wind and solar system, employing nature inspired meta-heuristic Algorithms may lead to the global extremum [8,13–18]. Here the researchers apply iterative method to solve the problem and left for further research for the application and comparison of different nature inspired algorithms to solve this hybrid solar and wind renewable energy system including its cost analysis.

2   Optimization Formulation

The main Objective of the sizing Optimization problem is to minimize the total annual cost (f_{TAC}) of the system. For this problem the total annual cost is taken as the sum of initial capital cost (C_{ICC}) and annual maintenance cost (C_{Mnt}) [3,7,13]. Thus, the problem to be minimized will be taken as:

$$Minimize \quad f_{TAC} = C_{ICC} + C_{Mnt} \tag{1}$$

Maintenance cost C_{Mnt} of the system occurs during the project life time while capital cost C_{ICC} occurs at the beginning of the project. In order to compare these costs, the initial capital cost has to converted annual capital cost by the capital recovery factor (CRF) can be defined as

$$CRF = \frac{i(1+i)^n}{(1+i)^n - 1} \tag{2}$$

where i the interest rate and n denotes the life span of the system.

Now the initial capital cost of the system can be broken in the annual costs of the wind turbine, solar panel, batteries and backup generator will be given as follows:

$$C_{ICC} = \frac{i \times (1+i)^n}{(1+i)^n - 1}[N_{PV}C_{PV} + N_{WT}C_{WT}$$
$$+ (\frac{n}{LS_{Batt}})N_{Batt}C_{Batt} + C_{Backup}] \tag{3}$$

where LS_{Batt} is batteries life span, N_{PV}, N_{WT} and N_{Batt} are numbers of PV panels, wind turbine and batteries respectively, C_{PV}, C_{WT}, N_{Batt} and C_{Backup} are unit costs of PV panels, wind turbine, batteries and backup generator respectively The unit cost of solar panel C_{PV} is consists of unit cost of PV panel and its installation fee and that of wind turbine C_{WT} is also consists of unit cost of wind turbine and its installation fee as shown on Eq. 4 next.

$$C_{PV} = C_{PV,unit} + C_{inst,unit}$$
$$C_{WT} = C_{WT,unit} + C_{inst,unit} \tag{4}$$

The number of batteries N_{Batt} which, depends on the number of photovoltaic panel and number of wind turbines are decision variables and determined by the following function:

$$N_{Batt}(N_{PV}, N_{WT}) = Roundup(\frac{S_{Req}}{\eta S_{Batt}}) \tag{5}$$

where Roundup (.) is a function which returns a number rounded up to an integer number; S_{Req} is required storage capacity; η is usage % of rated capacity which guarantees batteries life span; and S_{Batt} is rated capacity of each battery.

Similar to the number of batteries, the required storage capacity S_{Req} which is defined as the number of solar panels and wind turbines in the hybrid system can be obtained by using energy curve (ΔW) defined as:

$$\Delta W = W_{Gen} - W_{Dem} = \int \Delta P dt = \int (P_{Gen} - P_{Dem})dt \tag{6}$$

Where W_{Gen} and P_{Gen} are the total energy and power generated respectively and W_{Dem} and P_{Dem} are their respective demand values.

Thus, the required storage capacity S_{Req} defined as the number of solar panels and wind turbines is given by:

$$S_{Req}(N_{PV}, N_{WT}) = \sum_{t=1}^{Maxt}(P_{PV}^t + P_{WT}^t - P_{Dem}^t)\Delta t$$
$$- \sum_{t=1}^{Mint}(P_{PV}^t + P_{WT}^t - P_{Dem}^t)\Delta t \tag{7}$$

Where Max t is the time when total energy (kwh) is highest; Min t is the time when total energy (kwh) is lowest. Δt is unit time under consideration (1 hr) here. P_{PV}^t and P_{WT}^t are the powers generated by solar panel and wind turbine at time t respectively; and P_{Dem}^t the total power demand at time t. The total power generated by the components at time t is given by:

$$P_{PV}^t = N_{PV} \times P_{PV,Eachunit}^t$$
$$P_{WT}^t = N_{WT} \times P_{WT,Eachunit}^t \tag{8}$$

where P_{PV}^t and P_{WT}^t are the total powers generated by the wind turbines and solar panels, where as $P_{PV,Eachunit}^t$ and $P_{WT,Eachunit}^t$ are the power generated from each respective components at a time t. The annual maintenance cost of the system was calculated by the following equation.

$$C_{Maint} = (C_{PV,maint} \times \sum_t^2 4P_{PV}^t \Delta t + C_{WT,maint}$$
$$\times \sum_t^2 4P_{WT}^t \Delta t) \times 365 \tag{9}$$

3   Constraints

The formulated Fitness function of the Optimization problem will be subject to the following Conditions.

1. Decision variables Constraint

$$N_{PV} \in Z, \quad N_{PV} \geq 0 \quad and \quad N_{PV} \leq N_{PV,Max}$$
$$N_{WT} \in Z, \quad N_{WT} \geq 0 \quad and \quad N_{WT} \leq N_{WT,Max} \tag{10}$$

2. Power Generated Constrain
   The total transferred power from PV and WT to the battery bank is calculated using the following Equation

$$P_{Total}^k(t) = N_{PV} P_{PV}^k(t) + N_{WG} P_{WG}^k(t)$$
$$1 \leq k \leq 365, 1 \leq t \leq 24 \tag{11}$$

The power generated from each source $P_{gen}(i)$ must be less than or equal to the maximum capacity of the source as

$$P_{gen}(i) \leq P_{gen;max}(i) \tag{12}$$

Where i is Number of sources

3. Power Balance Constraint
   The total power P_{Total} generation of the Hybrid renewable energy sources must cover the total load demand P_{demand}, the total power losses P_{Losses} and storage power $P_{Storage}$ if used.

$$P_{Total} = P_{demand} + P_{Losses} + P_{Storage}$$
$$P_{Total,supply} \geq P_{Total,demand} \tag{13}$$

4   Iterative Method

The iterative procedure selected for optimal sizing the numbers of wind turbine and PV panels needed for a stand-alone system to meet the desired specific load of a specific area. In this method all possible solutions are generated first by initializing the basic decision variables i.e. number of wind turbines and number of solar panels staring form the minimum numbers until the possible maximum numbers which optimize the system. Here under the steps used in iteration method are mentioned [5].

1. Select suitable and commercially available unit sizes for wind turbine, PV panel, and storage battery based on the data given in Table 1.
2. Since the unit cost for the wind turbine far exceeds that of a single solar panel, keep the number of wind turbines (N_{WT}) constant and increase the number of PV panels (N_{PV}) until the system is balanced, i.e. the curve of ΔP versus time for the system has an average of zero over a given period of time.
3. Repeat step 2 for different number of wind turbines, i.e. ($N_{WG} = 0, 1, 2, 3, ...$) as needed.
4. Calculate the total system annual cost for each combination of (N_{WG}) and (N_{PV}), that satisfies the requirements in step 2.
5. Choose the combination with the lowest cost under the desired conditions.

4.1   Numerical Datasets

A hybrid solar and wind renewable energy system, which is designed based on the developed Optimization formulas given (1–13) above is given here. The numerical examples used in this paper is similar to [3,7]. However, because of certain time gap, there were a change in global inflation rate and energy demand, those values under decision variables used for wind and solar system organized in Table 1 are multiplied by 3.31%, global inflation rate of 2018 and the power demand data given in Table 2 are multiplied by 1.3%, global average energy demand increment value.

Table 1. Design variables used for Solar and Wind Hybrid System

Variables	Values
Annual interest (i)	6%
Life span of the system (n)	20 years
Solar panel price (	$362/panel
Solar panel installation fee	50% of the price
Wind turbine price	$20662/Turbine
Wind turbine installation fee	25% of the price
Unit cost of the battery	$176
Cost of backup generator	$2066
Usage % of battery rated (η)	80%
Batteries rated capacity (	2.1 Kwh
Batteries life span	4 years
Unit time (Δt)	1 hr
Maintenance cost of PV array	0.5 cents/Kwh
Maintenance cost of WT	2 cents/Kwh

Table 2. Updated daily power data for wind and solar system

t	P_{Dem}	P_{WT}	$P_{PV}(W)$	$P_{PV}(kw)$	ΔP
1	1.39	0.58	0	0	−0.81
2	1.25	0.49	0	0	−0.76
3	1.19	0.48	0	0	−0.71
4	1.22	0.53	0	0	−0.69
5	1.34	0.47	0	0	−0.87
6	1.8	0.51	0	0	−1.29
7	2.66	0.46	1.6	0.002	−2.198
8	2.9	0.46	3.4	0.003	−2.437
9	2.52	0.61	10.3	0.01	−1.899
10	2.21	0.76	24.6	0.025	−1.425
11	2.05	1.1	31.7	0.032	−0.918
12	1.94	1.53	35.3	0.035	−0.375
13	1.82	1.67	36.6	0.037	−0.113
14	1.71	1.89	37.4	0.037	0.217
15	1.62	2.43	36.8	0.037	0.847
16	1.65	2.45	33.5	0.034	0.833
17	1.87	1.91	24.2	0.024	0.064
18	2.29	1.76	13.4	0.013	−0.517
19	2.58	1.57	5.6	0.006	−1.004
20	2.6	1.16	1.5	0.002	−1.438
21	2.54	0.87	0	0	−1.67
22	2.49	0.76	0	0	−1.73
23	2.28	0.74	0	0	−1.54
24	1.79	0.7	0	0	−1.09
Total	47.72	25.89		0.296	−21.534

Table 1 shows the valves of the updated decision variables for the test of the system. Here the values mentioned on [3] which related to purchase are updated. Because of life span of each battery is taken as 4 years, 5 times installations are needed during the whole systems life span. In Table 2, P_{Dem} and P_{WT} are given in kilowatts. Table 2 provides the updated valued of annual average hourly demand (P_{Dem}^{t}), generated power by each components ($P_{PV,unit}^{t}$) and ($P_{Wt,unit}^{t}$) and the difference in power in kilo watt including the total power demand versus total power generated for each unit time of the day. To illustrate this table more, the following graphs are plotted by the help of MATLAB.

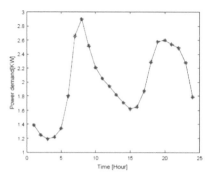

Fig. 1. Daily average of hourly power demand

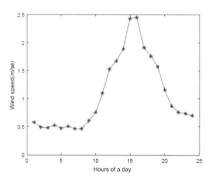

Fig. 2. Hourly average wind profile

Figure 2 above indicates the updated electrical demand and the curve shows an average hourly demand. Hourly average wind profile shown in Table 2 was illustrated in Fig. 3. The average annual power demand was plotted with Fig. 4. The power differences from each components wind turbine and solar system and also the power difference of the total generated power from the sources and the power demand used to calculate the power balance was shown by Fig. 5.

5   Results

Based on the above updated data, when iteration method was applied to Optimal sizing this problem, under the constraint the total energy was balanced, we get three different alternatives as shown in Table 3.

1. When wind Turbine alone applied ($N_{PV} = 0, N_{WT} = 2$ and $N_{Batt} = 9$)
2. When Solar panels alone applied ($N_{PV} = 162, N_{WT} = 0$ and $N_{Batt} = 17$)
3. When both Wind Turbine and Solar panels applied ($N_{PV} = 74, N_{WT} = 1$ and $N_{Batt} = 12$).

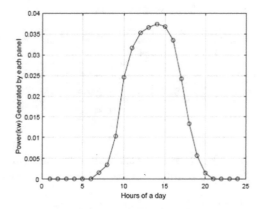

Fig. 3. Average hourly power generated by a solar pannel in a day

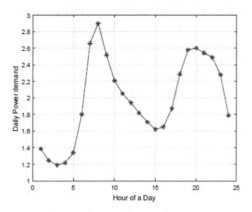

Fig. 4. Average daily annual power demand

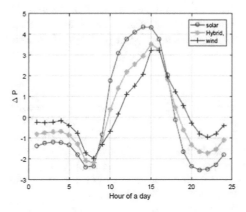

Fig. 5. Average hourly power difference in a day

Table 3. System component size

Configuration	# Tur	# Pan	# Bat	ΔP	Total annual cost
Wind alone	0	2	9	4.04 kw	$5753.09
Solar alone	162	0	17	0.216 kw	$9242.79
Hybrid	74	1	12	0.067 kw	$7085.97

In the above table, Tur, Pan and Bat stands for number of wind turbines, number of solar panels and numbers of batteries involved in the configuration respectively.

As indicated on Table 3. The optimal cost was found ($5753.09) when only Wind turbine (($N_{PV} = 0, N_{WT} = 2$ and $N_{Batt} = 9$)) was applied to generate the desired power with the possible minimum cost. But due to the stochastic nature of these renewable energy sources we recommend to select the hybrid system was selected to get continuous power with ($N_{PV} = 74, N_{WT} = 1$ and $N_{Batt} = 12$) and total annual cost $7085.97. Note that with this combination all the constraints are satisfied and the total power was also balanced with +0.067 kw to charge the battery. So, the designed backup generator will be used as contingency in case when maximum fluctuation of power occurs.

6   Conclusion

Optimal sizing of hybrid Wind and Solar renewable energy system stand alone daily average of 47.72 kw/day power generating was investigated. A simple iteration method was employed and all the constraints are taken in to account to optimal sizing of the components for the three configurations. By this optimal solution was achieved at ($N_{PV} = 74, N_{WT} = 1$ and $N_{Batt} = 12$) and total annual cost $7085.97. Hopefully, the way of organizing hybrid wind and solar renewable energy system and the updated numerical data taken to optimize the system will be come a good bench-mark for us and other researchers to apply other nature inspired algorithms and compare the their results.

References

1. Diriba, K.G., Manshahia, M.S.: Optimization of renewable energy systems: a review. Int. J. Sci. Res. Sci. Technol. **8**(3), 769–795 (2017)
2. Kosmadakis, G., Sotirios, K., Emmanuel, K.: Renewable and conventional electricity generation systems: technologies and diversity of energy systems. In: Renewable Energy Governance, pp. 9–30. Springer, London (2013)
3. Zong Woo Geem: Size optimization for a hybrid photovoltaicwind energy system. Electr. Power Energy Syst. **42**, 448451 (2012)
4. Luna Rubio, R., Trejo Perea, M., Vargas Vzquez, D., Ros-Moreno, G.J.: Optimal sizing of renewable hybrids energy systems: a review of methodologies. Solar Energy **86**(4), 1077–1088 (2012)

5. Kellog, W.D., Nehrir, M.H., Venkataramanan, G., Gerez, V.: Generation unit sizing and cost analysis for stand -alone wind, phovoltaic, and hybrid Wind/PV systems. IEEE Trans. Energy Convers. **13**(1), 70–75 (1998)
6. Askarzadeh, Alireza: Developing a discrete harmony search algorithm for size optimization of wind photovoltaic hybrid energy system. Solar Energy **98**, 190–195 (2013)
7. Kaabeche1 A., Belhamel1 M. and Ibtiouen R.: Optimal sizing method for standalone hybrid PV/wind power generation system. Revue des Energies Renouvelables SMEE, pp. 205–213 (2010)
8. Diriba K,G, Manshahia, M.S: Nature inspired computational intelligence: a survey. Int. J. Eng. Sci. Math. **6**(7), 769–795 (2017)
9. Nafeh Abd El-Shafy, A.: Optimal economical sizing of a PV-wind hybrid energy system using genetic algorithm. Int. J. Green Energy **8**(1), 25–43 (2011)
10. Koutroulis, E., Dionissia, K., Potirakis, A., Kostas, K.: Methodology for optimal sizing of stand-alone photovoltaic/wind-generator systems using genetic algorithms. Solar Energy **80**(9), 1072–1088 (2006)
11. Ashok, S.: Optimised model for community-based hybrid energy system. Renew. Energy **32**(7), 1155–1164 (2007)
12. Wang, L., Singh, C.: Multicriteria design of hybrid power generation systems based on a modified particle swarm optimization algorithm. IEEE Trans. Energy Conversion **24**(1), 163–172 (2009)
13. Yang, H., Zhou, W., Lou, C.: Optimal sizing method for stand-alone hybrid solarwind system with LPSP technology by using genetic algorithm. Solar Energy **82**(4), 354–367 (2008)
14. Yang, H., Zhou, W., Lou, C.: Optimal design and techno-economic analysis of a hybrid solarwind power generation system. Appl. Energy **86**(2), 63–169 (2009)
15. Ramoji, Satish K., Bibhuti, Bhusan R., Vijay, Kumar D.: Optimization of hybrid PV/wind energy system using genetic algorithm (GA). J. Eng. Res. Appl. **4**, 29–37 (2014)
16. Shaahid, S.M., Elhadidy, M.A.: Economic analysis of hybrid photovoltaic and dieselbattery power systems for residential loads in hot regions, a step to clean future. Renew. Sustain. Energy Rev. **12**(2), 488–503 (2008)
17. Sopian, K.A., Zaharim, Y.A., Zulkifli, M.N., Juhari, A.R., Nor, S.M.: Optimal operational strategy for hybrid renewable energy system using genetic algorithms. WSEAS Trans. Math. **7**(4), 130–140 (2008)
18. Amer, M., Namaane, A., M'sirdi, N.K.: Optimization of hybrid renewable energy systems (HRES) using PSO for cost reduction. Energy Proc. **42**, 318–327 (2013)
19. Muselli, M.N., Notton, G., Louche, A.: Design of hybrid-photovoltaic power generator, with optimization of energy management. Solar Energy **65**(3), 143–157 (1999)

Modeling of Solar Photovoltaic Thermal Modules

Vladimir Panchenko[1]($\boxtimes$), Valeriy Kharchenko[2], and Pandian Vasant[3]

[1] Russian University of Transport, Obraztsova St. 9, 127994 Moscow, Russia
pancheska@mail.ru
[2] FSBSI "Federal Scientific Agroengineering Center VIM", 1st Institutskij
Proezd 5, 109428 Moscow, Russia
kharval@mail.ru
[3] Universiti Teknologi PETRONAS, 31750 Tronoh, Ipoh, Perak, Malaysia
pvasant@gmail.com

Abstract. In the presented article the technique of creation of three-dimensional models of solar photovoltaic thermal modules in the system of computer-aided design of Compass 3D is considered. The article also considers the method of visualization of the thermal mode of operation of the water radiator of the photovoltaic thermal concentrator solar module, the model of which was created in the automated design system. A technique for manufacturing a prototype of a solar tile shell manufactured using additive technologies is proposed

Keywords: Computer-aided design system · Finite element analysis system
Additive technologies · Solar energy · Solar photovoltaic thermal module
Three-dimensional model · ANSYS · Water radiator

1 Introduction

Currently, there are a number of software complexes that, as a tool, allow you to create both three-dimensional object models, conduct various modeling of the thermal state of the modules with simultaneous visualization of the results obtained, and create prototypes of such modules using additive technologies, using relatively small resources, which is very important at the initial design stage. As a tool for creating two-dimensional and three-dimensional models of solar photovoltaic thermal modules for stationary and mobile power generation, computer-aided design systems, such as the software package of Ascon - Kompas 3D [http://kompas.ru/], can be used.

© Springer Nature Switzerland AG 2019
P. Vasant et al. (Eds.): ICO 2018, AISC 866, pp. 108–116, 2019.
https://doi.org/10.1007/978-3-030-00979-3_11

2 Method for Creating Three-Dimensional Models of Solar Photovoltaic Thermal Modules of Various Designs for Stationary and Mobile Power Generation

In the method of creating models of solar photovoltaic thermal modules, the designs of stationary and mobile power generation modules are developed, the main differences of which are the dimensions of solar cells, the number of illuminated sides of solar cells (one- and two-sided) and the dimensions of the radiator cavities due to a different solar flux to the radiation receiving surface that dimensions are optimized in the Ansys software package [http://www.ansys.com/].

The first type of solar photovoltaic thermal module for use in the concentrator system is a solar module with a two-sided beam-receiving surface. The number of components used in this type of modules is limited by the need to ensure transparency of both the outer beam-receiving sides of the module in the solar spectrum, in which the solar cell generates electricity.

The second type of solar photovoltaic thermal module for use in the planar system is a solar module with a one-sided beam-receiving side. In this type of module, the number of components used can be extended and the constructions complicated (Fig. 1).

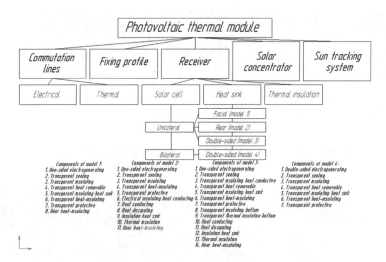

Fig. 1. The technique of creating a three-dimensional model of the receiver of a photovoltaic thermal solar module with different beam-receiving sides and types of heat removal

Both types of receivers of the photovoltaic thermal solar module are also subdivided according to the type of cooling of the radiation receiving side of the receiver (heat sink) - front, rear, two-way. Depending on the ray-receiving sides and the type of heat sink, the module is created in one of four models (Fig. 1).

The developed method for creating models of solar photovoltaic thermal modules for stationary and mobile power generation implemented in the Kompas 3D software package allows creating models of receivers of solar photovoltaic thermal modules with various planar and concentrator parameters for stationary and mobile power generation with:

– single-sided solar cells and facial heat sink (model 1, Fig. 2);

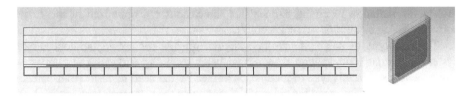

Fig. 2. Three-dimensional model of the receiver with a single-sided solar cell and facial heat sink (model 1)

– single-sided solar cells and rear heat sink (model 2, Fig. 3);

Fig. 3. Three-dimensional model of the receiver with a single-sided solar cell and rear heat sink (model 2)

– single-sided solar cells and two-sided heat sink (model 3, Fig. 4);

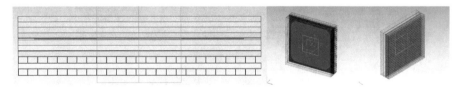

Fig. 4. Three-dimensional model of the receiver with a single-sided solar cell and a two-sided heat sink (model 3)

– bilateral solar cells and a two-sided heat sink (model 4, Fig. 5) (Panchenko et al. 2015).

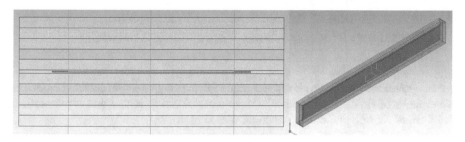

Fig. 5. Three-dimensional model of the receiver with a bilateral solar cell and a two-sided heat sink (model 4)

In the process of creating all three-dimensional components included in Model 1 with a single-sided solar cell and facial heat sink, an assembly unit is formed in the form of a solar photovoltaic thermal module (Fig. 2).

It should be noted that a two-component polysiloxane compound can be used as a sealing component in the solar photovoltaic thermal module assembly unit, which increases the term of the nominal operation of solar cells, is optically transparent, which increases the efficiency of solar cells in comparison with ethylene-vinyl acetate film and can be used in systems with concentrators, the efficiency of solar cells does not decrease either with a large positive or a large negative temperature (Panchenko et al. 2015).

In the process of creating all three-dimensional components in Models 2 with a single-sided solar cell and the rear heat sink, an assembly unit is created in the form of a solar photovoltaic thermal module (Fig. 3). Because of the rear heat sink, the quality of the thermal insulation can be improved by using a potentially larger number of components, regardless of their transparency.

In the process of creating all three-dimensional components in Model 3 with a single-sided solar cell and a two-sided heat sink, an assembly unit is created in the form of a solar photovoltaic thermal module (Fig. 4). This model combines the components used in the creation of Model 1 and Model 2, so the design becomes more complicated, but at the same time there is the possibility of more fine-tuning the cooling of the two sides of the solar cell.

For the implementation of Model 4, high-voltage solar cells with increased electrical efficiency are accepted as a bilateral solar cell in comparison with standard planar silicon solar cells used without concentrators (Panchenko et al. 2015). Along with the increase in efficiency up to 28%, the term of the rated power of solar cells also increases due to the use of a two-component polysiloxane compound. Such high efficiency can be achieved with the use of solar radiation concentrators, when working with them, high-voltage solar cells do not degrade their characteristics and the amount of solar-grade silicon used in such installations decreases.

In the process of creating all three-dimensional components in the Model 4 with bilateral solar cells and a two-sided heat sink, an assembly unit is formed in the form of a solar photovoltaic thermal module (Fig. 5). Such a solar photovoltaic thermal module is expediently used in a concentrator system with the production of warm water at the outlet.

3   Implementation of the Method for Creating Three-Dimensional Models of Solar Photovoltaic Thermal Modules

As an application of the developed technique for creating three-dimensional models of solar photovoltaic thermal modules, a three-dimensional assembly unit is presented in the form of roofing photovoltaic thermal tiles (Fig. 6).

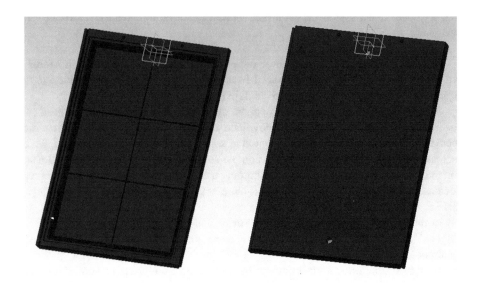

Fig. 6. Three-dimensional assembly unit of a planar photovoltaic thermal solar module in the form of a roofing panel

To create an assembly in the computer-aided design Kompas 3D, various components are created in the form of separate parts that are part of the module being developed. The assembly unit of the planar photovoltaic thermal solar module in the form of a roofing panel includes 8 components that perform various functions. The main structural element is the housing to which the other components are attached. As a cooling agent, water is adopted, which is washes an aluminum radiator of black color. Sealing of solar cells is carried out using polysiloxane two-component compound, facial thin, transparent film and black tape around the perimeter. The gas insulating region is an air layer that is bounded from the front surface of the module by an optically transparent glass.

4   Method of the Thermal Calculation of the Photovoltaic Thermal Solar Modules of Planar and Concentrator Types Using Simulation and Visualization of Processes in the Software Complex Ansys

The three-dimensional models of receivers of solar photovoltaic thermal modules with various parameters of planar and concentrator types, developed with the help of the developed method, are tested in the program of finite element analysis Ansys [http://www.ansys.com/] to optimize the design of receivers, in view of which the method of thermal calculation of photovoltaic thermal solar modules of planar and concentrator types (Fig. 7) is proposed.

Fig. 7. Method of thermal calculation of photovoltaic thermal solar modules of planar and concentrator types using process simulation in the software complex Ansys

An example of calculation in the program complex Ansys using the developed technique of the three-dimensional model of the receiver of the concentrator solar photovoltaic thermal module (Strebkov et al. 2013) is presented in Figs. 8, 9.

The definition of the boundary regions with the specification of the conditions for their interaction between of the radiator of the receiver of the concentrator solar photovoltaic thermal module is shown in Fig. 8.

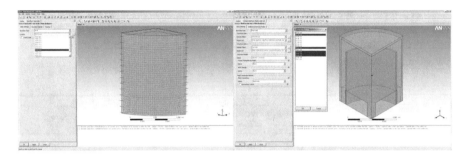

Fig. 8. Definition of boundary areas with the specification of the conditions for their interaction between of the radiator of the receiver of a concentrator photovoltaic thermal module

The derivation of thermal regimes of models and flow lines with visualization of thermal fields, coolant velocities and flow lines according to the developed method are shown in Fig. 9.

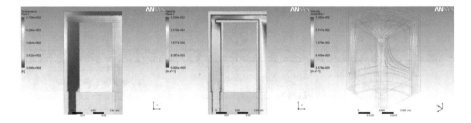

Fig. 9. The thermal fields of the model (left), the velocity of the coolant (in the middle), and the flow line (on the right) of the receiver of the concentrator solar photovoltaic thermal module

In the process of optimizing the radiator, its various three-dimensional structures were calculated, after which it is possible to judge the appropriateness of using each structure. The criterion for optimizing the radiator was the maximum temperature of the water at the outlet from the radiator and not exceeding the temperature of the lateral surface of the radiator above the maximum values at which the volt-ampere characteristic of solar cells has a rectangular shape.

With the help of the developed technique it is possible to obtain the thermal fields of the developed model, the velocity of the coolant and the current line. With the help of visualized models of the heat state of the radiator, it is possible to make decisions about the need to optimize its design to obtain the required parameters of the thermal state of the radiator itself and the coolant of the solar photovoltaic thermal module.

5   Method of Manufacturing of Three-Dimensional Solid-State Prototypes of Components of Solar Modules Using Layered Printing

As the implementation of the developed methodology method, it is proposed to use the solar roofing panel (Strebkov et al. 2014, 2015). In the developed method, much attention is paid to the development of two-dimensional and three-dimensional models of the modules being developed (Fig. 11 on the left) along with the process of layer-by-layer printing of a solid prototype that can be made from recycled polyethylene, which positively affects the environment (Fig. 10). Figure 11 on the right shows the substrate of the solar roofing panel, which is manufactured by die punching, which is too expensive due to the preparation of equipment for primary experimental modules, which may be awaited by further refinements.

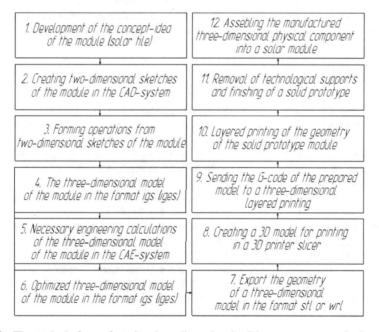

Fig. 10. The method of manufacturing three-dimensional solid-state prototypes of solar module components using layer-by-layer printing implemented in 3D printer and Kompas 3D software packages

In the case of three-dimensional prototyping using a 3D printer, any modifications of the module take place in CAD systems (three-dimensional models) and take a little time, and the sample itself is worth an order of magnitude less than production, for

Fig. 11. Three-dimensional model of a solar roofing panel prepared for 3D printing (left) and manufactured tile substrates in a standard expensive method (for experimental single samples) die punching (on the right)

example, by stamping technology. According to the manufactured solid sample, one can judge the expediency of using this design and, if necessary, make additions to the three-dimensional model with subsequent printing of the final prototype of the module.

References

Electronic resource. Access mode. http://kompas.ru/. Accessed 10 April 2018

Electronic resource. Access mode. http://www.ansys.com/. Accessed 10 April 2018

Panchenko, V.A., Strebkov, D.S., Persits, I.S.: Development of solar modules with an extended term of nominal work. In: Nanostructured Materials and Conversion Devices for Solar Energy Collection of Proceedings of the III All-Russian Scientific Conference 19–20 June 2015, Cheboksary, pp. 91–94 (2015)

Panchenko, V.A., Strebkov, D.S., Polyakov, V.I., Arbuzov, Yu.D.: High-voltage solar modules with a voltage of 1000 V. Altern. Energy Ecol. **19**(183), 76–81 (2015)

Strebkov, D.S., Kirsanov, A.I., Irodionov, A.E., Panchenko, V.A., Mayorov, V.A.: Roof solar panel. Patent of the Russian Federation for invention No. 2557272. Application: 2014123409/20, 06/09/2014. Published: 07.20.2015. Bul. 20 (2014)

Strebkov, D.S., Mayorov, V.A., Panchenko, V.A.: Solar photovoltaic thermal module with a parabolic concentrator. Altern. Energy Ecol. **1/2**, 35–39 (2013)

Strebkov, D.S., Panchenko, V.A., Irodionov, A.E., Kirsanov, A.I.: Development of a roofing solar panel. Vestnik VIESH **4**(21), 107–111 (2015)

Thermo Physical Principles of Cogeneration Technology with Concentration of Solar Radiation

Peter Nesterenkov[1]($\boxtimes$) and Valeriy Kharchenko[2]($\boxtimes$)

[1] Al-Farabi Kazakh National University, Almaty, Kazakhstan
stolkner@gmail.com
[2] Federal Scientific Agroengineering Center VIM, Moscow, Russian Federation
kharval@mail.ru

Abstract. This paper considers the thermo physical principles of cogeneration technology with the use of silicon photocells working with a low concentration of solar radiation. The efficiency of the technology is enhanced by the use of photocells at a relatively high temperature and cooling with liquid, which makes it possible to obtain high-potential heat and transmit it to the heat carrier in counter flow mode of the coolant. Transportation of heat energy to a stationary storage system is realized under the influence of the pressure head formed by the temperature gradient along the height of the circulation circuit. A mathematical model is proposed for calculating the thermal energy of linear photovoltaic modules, taking into account the experimentally determined electric efficiency of commercially available silicon photocells.

Keywords: Photocells · Optical concentrator · Heat exchanger
Energy storage system

1 Introduction

The widely available solar uncooled photovoltaic modules with silicon photocells (PV modules) which have an electrical efficiency of ≈15% emit more than 70% of the incoming solar energy into the environment in the form of heat. In coolant-cooled systems (photovoltaic-thermal modules, PVT), degree of conversion of solar energy is increased due to the utilization of thermal energy [1]. However, in such systems thermal energy is low potential, due to the negative effect of high temperature on the electrical efficiency of silicon photocells. The temperature of the coolant can be increased by using high-temperature two-junction GaAs or three-junction InGaP/InGaAs/Ge photocells [2]. At the same time, there are very few chemical elements Ga, As and In in the earth's crust, and the technology of manufacturing multi-transitional photocells is complex and expensive.

The enormous technological potential of silicon photocells is far from exhausted. In work [3] Rosell and others at a 6-fold concentration from the cooled photocells obtained a total efficiency of ≈60% and a temperature of the coolant at the outlet of about 61 °C. Engineers of the Cogenra Solar (Sun Power Corporation) at 8-fold

© Springer Nature Switzerland AG 2019
P. Vasant et al. (Eds.): ICO 2018, AISC 866, pp. 117–128, 2019.
https://doi.org/10.1007/978-3-030-00979-3_12

concentration of the sun received electricity productivity of $\approx$100 W/m^2 and heat $\approx$490 W/m^2 and for the first time proved the five-year payback of capital investments in the production of solar installations [4]. However, the high cost of installed capacity $\approx$\$1400/m^2 restrains their commercialization. The task of delivering high-potential thermal energy to the storage system without significant energy costs and heat losses remains topical. In this work, we consider the scientific and technical principles of cogeneration technology with the use of silicon photocells with a low concentration of solar radiation and the transportation of the resulting hot coolant to the storage system with minimal energy costs.

2 Methods and Calculations

The concept of the technology is based on the generation of electricity and heat by silicon photocells relatively large area with increased density of solar radiation and an elevated temperature on the surface, followed by the transport of high-grade heat released to the integrated energy storage system. The continuous generation of electric power is being realized in the technological combination of wind generation and solar generation in the space of a common carrying platform. This is facilitated by the natural phenomenon of increasing wind speed with the absense (cloudy periods, nighttime etc.) of the sun. The concept is implemented in budget solar-wind systems with concentration of solar radiation (CSWs). In all standard designs, they include a supporting structure, carrying platform, an optical concentrator, PV modules, circulation circuits with cooled photocells, a circulation circuit with a heat exchanger and a heat accumulator, wind generators, integrated energy storage system, power management and power delivery unit. The most capital-intensive are the supporting structure, carrying platform with an optical concentrator and the storage system for electrical and thermal energy.

2.1   Design of the Load-Bearing Platform

Lightweight load-bearing platform is made of a standard metal profile with the use of engineering technologies for the construction of solid truss structures. Dynamic stability of the bearing platform in conditions of turbulent wind currents is provided by using four points of support and two pass-through bearings for heat carriers. Due to the orientation of the longitudinal axis at an angle to the horizon and the counter position of the walls of the pv modules along them, a protective thermal layer is formed from the convection of the hot surfaces placed below [5]. Figure 1 presents a version of the CSWs with one section of the carrier platform, and a system for measuring technological parameters. Step-by-step tracking of the carrier platform behind the sun and seasonal adjustment for the height of the sun are carried out using trackers according to the controller program.

The total peak power of the CSWs (two sections) is $\approx$11 kW and total capacity is calculated from the condition of full provision of the average farmhouse with electricity and heat in the summer. During winter heating, solid fuel boilers can be connected to the energy storage system through integrated interface. The equipment is delivered to

Fig. 1 CSWs with one section concentrator and measuring system

the installation site in twenty-foot containers, which serve as a supporting structure and a hermetically sealed box for the energy storage system and monitoring devices.

2.2 Optical Concentrators

For mirrors, relatively thin glass with a low iron content is used, which reduces the unit cost of optical concentrators to $\approx$\$15/m^2. In further, it is planned to switch to Alanod Silver mirror films (Germany) with a reflection coefficient of $K_r \approx 0.95$ on a thin metal substrate made of aluminum or copper [6]. Design of concentrators are developed in accordance with the laws of geometric optics using computer modeling in Autodesk Inventor software. Figure 2 shows an optimal distribution of reflected solar rays in the working area of photocell surfaces.

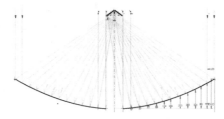

Fig. 2 Optical scheme of solar-wind systems with concentration of solar radiation

When the sun moves across the sky with angular velocity $\omega \approx 2\pi/24 \cdot 0$ rad/min during the shutdown time of the tracker engine τ, the solar beam is displaced in space from A to A$_1$. If the distance between pv modules and mirrors is denoted by R, we obtain the displacement of the reflected light spot along the wall surface $\Delta = \tau \cdot R \cdot (2\pi/24 \cdot 60)$, which is shown in Fig. 3.

At a distance R $\approx$ 1.5 m, displacement for one minute is equal to $\Delta \approx$ 6 mm. To account for the effect of displacement, the transverse size of the light spot is chosen to be larger than the size of the photocells b mm, from which the minimum transverse mirror size min $b_z \geq$ (b + Δ) is determined. The optimal reflecting surface is a

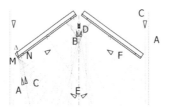

Fig. 3 Reflected solar rays on photocell surfaces

parabolic surface with a focal length f, whose analytical expression in the XY coordinate system has the form: $(X + X_1)^2 = 4 \cdot f \cdot Y$. The first mirror of the concentrators is at a distance X_1 from the coordinate axis (Table 1).

Table 1 Coordinates of mirrors

№	1	2	3	4	5	6	7	8	9	10	11	12
X, m	0.285	0.418	0.502	0.702	0.842	0.982	1.122	1.262	1.402	1.542	1.682	1.822
Y, m	0.017	0.029	0.053	0.082	0.118	0,161	0.210	0.265	0.328	0.396	0.472	0.553

Modeling allowed to determine the optimal value of the geometric coefficient - the ratio of the area of the aperture of the concentrator to the area of the mirrors $K_g \approx 0.96$. The transmission of solar radiation by the optical system depends on the total thickness of the mirrors and the protective thermal glasses. For example, with a total thickness of $\approx$10 mm, up to $\approx$8% of solar energy is absorbed [7]. Therefore, the proposed optical concentrator uses glass with a thickness of no more than 2 mm. This, in turn, limits the size of the optical elements due to the mechanical strength. As a result, the total transmission coefficient of solar radiation was increased to $K_d \approx 0.95$. Thin, relatively expensive thermal glasses have self-protection against hail and other large atmospheric precipitations. Innovative technical solutions allowed to increase optical efficiency up to $C_o = K_g \cdot K_r \cdot K_d \approx 0{,}83$. Taking into account the inevitable contamination of the optics, its real value is reduced to $\approx$0.8. Accordingly, the optical concentration for twelve mirrors is $C_{ok} \approx C_o \cdot N = 0{,}8 \cdot 12 \approx 9{,}7$.

Due to the orientation of the pv modules at an angle to the horizon and counter-angled $\varphi 0$ to each other, as shown in Fig. 3, the hot air flow is directed up along the adjacent walls and forms a thermal protective layer, which reduces the heat loss by $\approx$12% compared to conventional convection [8].

In accordance with the Stefan-Boltzmann law, the surfaces of pv modules emit heat flux into space [9]:

$$dQ = \varepsilon\sigma_0 \left[(T/100)^4 - (T_0/100)^4 \right] F \int_0^\pi d\varphi \qquad (1)$$

where - ε and T - reduced blackness and temperature of the surface of the radiating body; $\sigma_0 = 5,67 \cdot 10^{-8} W/(m^2 \cdot K^4)$ - Stefan-Boltzmann constant; T_0 - ambient temperature; F - surface area; φ - solar radiation angle.

In the case of screening of the walls of adjacent pv modules, the integration of expression (1) is carried out in the interval of angles $(\pi - \phi_0/2; \pi)$. Other things being equal, the value of the radiation loss ratio of mutually screened and freely placed surfaces is: : $\approx \varepsilon \sigma_0 (T/100)^4 (\pi /3)/ \varepsilon \sigma_0 (T/100)^4 \pi \approx 0,33$. The back-to-back inclusion of the walls with photocells working with radiation concentration leads to a reduction of thermal losses by radiation by 33%.

One of the problems of operating optical concentrators during autumn-winter period is the need for cleaning after snowfalls. In the proposed variant, unlike analogues, the working surfaces of pv modules are closed from atmospheric precipitation, and the thermal radiation from the surface of the thermal battery, pv modules and heat exchanger keeps the mirror temperature above zero. As a result, the snow does not linger and rolls down, as shown in Fig. 4.

Fig. 4 Operating during snowfall

2.3 Linear PV Modules

In the innovative method of increasing the efficiency of heat transfer, the physical effect of increasing the heat transfer coefficient with increasing medium temperature and discharging heat received from solar cells in the countercurrent mode to the second heat carrier is used.

However, it is necessary to solve the problem of ensuring reliable operation of photocells at elevated temperature and alternating thermal loads in conducting contact areas. Therefore, market photocells with a relatively low degree of degradation are selected under the specified conditions. These criteria are met by relatively cheap high-quality Maxeon solar cells from SunPower Corporation [10]. Under standard conditions, they have a peak power of ≈ 3.4 W and temperature power factors $k_P \approx 0.011 W/^0C$ and voltages $k_U \approx 0.0018 V/^0C$. At the maximum power point, voltage and current are $U_m \approx 0.57$ V and $Im \approx 5.8$ A. Under operating conditions with a low concentration (up to ten times) at a standard temperature of ≈ 25 °C, a peak power of ≈ 19 W has been obtained [11].

Silicon photocells with low light intensity within the theoretical Shockley model are described by an equivalent circuit of a diode with a photocurrent generator I_{Ph} and an internal series resistance R_S [12]. The intensive cogeneration mode changes the representation of photocells as energy converters and requires the modernization of the equivalent diode circuit, adding to it the heat generator Q and the radiation source, as shown in Fig. 5. Switches S_1 and S_2 are necessary for the transfer of photocells to short-circuit with I_{SC} current, idle mode with V_{OC} voltage, or mode with external load R_P.

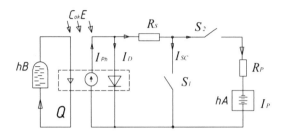

Fig. 5 Equivalent electrical circuit of CSWs

Under the influence of the radiation $C_{ok} \cdot E$, the photocell generates a photocurrent I_{Ph} directed to the chemical accumulators hA, and the heat flux transported to the thermal energy storage system hB. Balance of power released by photocells is:

$$C_{ok} \cdot E \cdot S = G \cdot Cp \cdot (t_{out} - t_{in}) + P + W \qquad (2)$$

where S - area of photocells, m^2; G - coolant flow, kg/s; C_p - specific heat of the liquid, J/(kg · K); t_{out} and t_{in} - output and inlet temperature of the coolant liquid, K; P - electric power, allocated by photocells, W; W - heat losses, W.

The operation of photocells with cogeneration is described by the electric energy η_e and the thermal efficiency η_T:

$$\eta = \eta_e + \eta_T = [I \cdot V + Q_T]/(C_{ok} \cdot S \cdot E) \qquad (3)$$

I and V - current and voltage at the maximum power point; Q_T - useful thermal power, W; C_{ok} - optical concentration; E - intensity of direct solar radiation, W/m^2; S - area of photocells, m^2.

Engineering method for calculating the efficiency of pv modules is based on experimental studies of thermal losses by cooled walls. During the transfer of solar energy to the heat carrier, the internal structure of the photocells and the shape of the heating source are not taken into account. This conclusion is the basis for the method of physical modeling of heat exchange processes. Instead of the sun and a complex tracking system, a thin electric-heated foil of nichrome is used. Between the foil and the wall is a heat-conducting adhesive, as for photocells. Thus, a convenient tool has been obtained for studying the influence of various design factors and external conditions on the thermal efficiency of pv modules. Figure 6 shows a mobile stand, which allows to standardize the testing process of the developed linear pм modules and carry out the adjustment of technical characteristics.

Fig. 6 Test bench for determining heat losses in flat channels of pv modules

The results of physical modeling studies over a wide range of changes in the external conditions of heat exchange between the channel and the coolant show that at an average coolant temperature of ≤ 45 °C, the thermal losses do not exceed $\approx 37\%$ [13]. It follows from this that the share of useful thermal power transmitted through the walls to the heat carrier is not less than $\approx 63\%$. We take it as a boundary in the mathematical model for calculating the thermal power. With the known efficiency of the photocells and the radiation value $\approx N \cdot C_{ok} \cdot E$, and $и \approx (E + E_d)$, we obtain an analytical expression for determining the useful thermal power carried away by the coolant of a two-sided pv module

$$q_1 + q_2 = 0,63 \cdot E \cdot [N \cdot C_{ok} \cdot (1 - \eta_1) + (1 + E_d/E)(1 - \eta_2)] \qquad (4)$$

The temperature of the photocells is determined from the solution of the classical thermo physical problem - temperature distribution by the wall thicknesses with boundary conditions of the second kind on the outer walls q_1 = const and q_2 = const; and boundary conditions of the third kind on the inner walls $q_1 = \alpha_1 (t_{c1} - t)$ and $q_2 = \alpha_2 (t_{c2} - t)$ [14]. The heat transfer coefficients at the inner walls α_1 and α_2 are the average along the length of the channel. Figure 7 shows the calculation scheme.

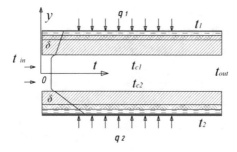

Fig. 7 The scheme of calculating photocells temperature

The thickness of the photocells and intermediate layers δ_i is not given to scale. The axis of the coordinate passes along the axis of the channel. The process of heat exchange is stationary. The specific heat of the liquid C_p is taken at the average temperature of the liquid in the channel.

The heat flow passes successively through a thermoplastic sealant of thickness δ_1, an oxidized layer δ_2, a wall δ_3 and a boundary layer along the walls δ_4. At the channels input goes coolant liquid with a mass flow rate G and a temperature t_{in}. Taking into account the accepted assumptions, we obtain the following equations, from which the local temperature of photocells on the walls is determined:

$$T_1 = t + q_1 \left(\frac{1}{\alpha_1} + \sum \frac{\delta}{\lambda} \right)$$
$$T_2 = t + q_2 \left(\frac{1}{\alpha_2} + \sum \frac{\delta}{\lambda} \right)$$

(5)

where t_1 and t_2 - temperatures of the outer walls (photocells); t - temperature of the coolant (liquid); δ/λ - thickness and thermal conductivity of the sealant, oxidized layer and metal wall; α_1 and α_2 are the heat transfer coefficients on the inner surface of the walls.

The temperature of the liquid in the channel of length L and width f varies along the flow. The amount of heat discharged by the walls into the liquid in any section of the channel f dx goes to an increase in the temperature of the liquid by an amount dt, from which we write the heat balance equation in the form:

$$q_1 + q_2 = [\alpha_1 (t_{c1} - t) + \alpha_2 (t_{c2} - t)] f dx = C_p G dt$$

(6)

Heat transfer coefficients α_1 and α_2 are averaged over the entire surface of the walls. Since the heat flux is constant, Eq. (6) with separable variables has an analytical solution:

$$t = t_{in} + \frac{q_1 + q_2}{C_p G} fX$$

(7)

From Eqs. (5) and (7) we obtain an expression for calculating the temperature of photocells at a distance X from the entrance to the channel:

$$t_1 = t_{in} + \frac{q_1 + q_2}{C_p G} fX + q_1 \left(\frac{1}{\alpha_1} + \sum \frac{\delta}{\lambda} \right)$$
$$t_2 = t_{in} + \frac{q_1 + q_2}{C_p G} fX + q_2 \left(\frac{1}{\alpha_2} + \sum \frac{\delta}{\lambda} \right)$$

(8)

The potential possibility of withdrawing the heat power emitted by photocells by the liquid is determined from the inequality:

$$0,63 \cdot E \cdot [C_{ok} \cdot (1 - \eta_1) + (1 + E_d/E)(1 - \eta_2)] \leq S \cdot \left[\frac{(t_1 - t)}{R_1} + \frac{(t_1 - t_0)}{R_2} \right] \quad (9)$$

where R_1 and R_2 - thermal resistances to the heat flux directed from photocells with temperature t_1 to the heat carrier with temperature t and into the environment with temperature t_0, respectively, equal to $R_1 = \frac{\delta_1}{\lambda_1} + \frac{\delta_2}{\lambda_2} + \frac{\delta_3}{\lambda_3} + \frac{1}{\alpha_1}$ and $R_2 = \frac{\delta_4}{\lambda_4} + \frac{\delta_c}{\lambda_c} + \frac{1}{\alpha_\beta}$; λ_1, λ_2, λ_3 and λ_c - thermal conductivity of the thermoplastic sealant, oxide film and aluminum, respectively; δ_4, δ_c and λ_4, λ_c - thickness and thermal conductivity of EVA film and heat-resistant glass; α_B - average heat transfer coefficient of the ambient air along the channel length;

Let us estimate the order of magnitude in the expression for the thermal resistance to the heat flux. When the thermal conductivity of the thermoplastic sealer is $\lambda_1 \approx 0$, 12 W/m · K and 0.1 mm thick, its thermal resistance is $\approx 0,0001/0,12 = 8 \cdot 10^{-4}$ $m^2 \cdot$ K/W. A layer of aluminum oxide with thermal conductivity $\lambda_2 \approx 2$ W/m · K and a thickness of ≈ 0.1 mm has a thermal resistance $\approx 0,5 \cdot 10^{-4}$ $m^2 \cdot$ K/W. The thermal resistance of the aluminum walls of the channel is $0,002/195 = 0,02 \cdot 10^{-4}$ $m^2 \cdot$ K/W. The flow of heat carriers in flat channels at a flow rate $G \approx 0.017$ kg/s is close to a laminar flow and the heat transfer coefficient is ≈ 370 W/m$^2 \cdot$ K. As a result, the total thermal resistance $R_1 \approx (0,8 + 0,05 + 0,002 + 2,8) \cdot 10^{-3}$ $m^2 \cdot$ K/W.

Figure 8 shows the results of the calculation of the temperature of photocells operating with low concentration of the sun. The graphs indicate the possibility of

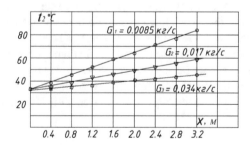

Fig. 8 Graphs of temperature variation of photocells along the channel wall as a function of liquid flow rate

flexible control of the temperature of the photocells by changing the flow rate of the liquid. In closed loop circuits, the temperature of the photocells is maintained at the required level due to the discharge of excess heat in countercurrent mode to the heat exchanger of the heat exchanger circuit.

The next step is to determine the total length of the channels of the pv modules. Denote Δt_j - difference between the temperature of the photocells on the channel walls and the standard operating temperature of 25 °C. At a temperature coefficient k_U, the voltage drop across the j_{th} photocell along the channel is determined from the expression: $\Delta U_j \approx k_U \cdot \Delta t_j$. The total value of the voltage loss of the series-connected photocells is, respectively: $U_T = \sum k_U \cdot \Delta t_j$. Taking into account the value of the voltage at the point of maximum power $U_m \approx 0.57$ and the need to maintain an optimum charge voltage of chemical accumulators $U_3 \approx 14.2$ V, the minimum number of photocells connected in series is determined: $n \approx (U_3 + U_T)/U_m$. To compensate resulting losses of voltage, additional photocells are introduced - one in the primary channels (in the direction of the coolant motion) ($n_1 = n + 1$) and three into the secondary channels ($n_2 = n + 3$). From this we get the value of the total length of the connected channels: $L_F \approx (n_1 + n_2) \cdot 0.126$.

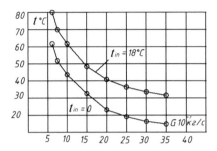

Fig. 9 Graph of the change in the outlet temperature of the liquid in the channel depending on consumption

A positive property of the serial connection of the channels along the heat carrier is the possibility of obtaining a hot liquid at the outlet. Figure 9 shows the liquid temperature at the outlet of the channel, depending on the flow rate and inlet temperature, obtained in accordance with the expression (8) at $X = L$.

In the proposed cogeneration systems, for the first time was put forward the idea of using the received heat energy (at solar energy systems) not only for its direct purpose, but also for performing mechanical work on the natural transportation of hot liquid from the heat exchanger to the energy storage system [14]. The realization of this idea is shown schematically in Fig. 10.

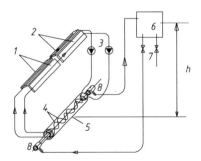

Fig. 10 Hydraulic circuit of the CSWs

The heat accumulator 6 is placed at a height h relative to the heat exchanger 5 and a temperature gradient is created between them, leading to a difference in the density of the liquid in the supply and descent pipelines of the closed circulation loop.

As a result, similar to the processes in the residential heating system, a dynamic pressure head appears that creates a natural flow of hot liquid to the heat accumulator:

$$\Delta P = gh \left(\rho_T - \rho_0 \right) \tag{10}$$

where ΔP is the dynamic head, Pa; g - free fall acceleration, 9.81 m/s^2; h - distance between the centers of the linear heat exchanger and the heat accumulator, m; ρ_0 and ρ_r - density of hot and cold coolant, kg/m^3.

With a temperature gradient of $\approx$60 °C, difference in density of the liquid is ($\rho_T - \rho_0$) $\approx$ 14 kg/m^3. At h $\approx$ 4 m dynamic pressure is $\Delta P = g \cdot h \cdot (\rho_T - \rho_0) \approx 545$ Pa which is quite enough for carrying out the transportation of thermal energy from the movable pv modules to the stationary thermal accumulator through the through-channels of the support bearings. In the steady-state regime of natural fluid flow, the dynamic head ΔP is equal to the sum of the hydraulic resistances in the channel of the heat exchanger ΔP_T and the descending and lifting pipeline ΔP_{TP}. Hot water during a sunny day accumulates in a heat-insulated thermal accumulator, the volume of which corresponds to the system capacity. The separation of the circulation circuits allowed to solve the problem of maintaining the necessary pressure in the flat channels of the pv modules and the circulation circuit with the charging of the consumable service water.

3   Conclusions

Using the example of designing low-power plants that are most in demand among farmers, the use of heat transfer laws in the processes of solar energy conversion using silicon cells of relatively large area can be traced. There is an optimal combination of the values of the concentration of solar radiation and the intensity of cooling of solar cells with liquid heat carriers, which allow the maximum amount of electrical and thermal energy to be generated per unit of aperture area of the installations, and a decrease of one value automatically leads to growth of another.

The property of the internal relationship between the electrical and thermal efficiency of photocells is particularly important in the creation of budget cogeneration plants for heating, use for energy supply of farms and desalination of water. For the first time reliable technology of transportation of high-potential heat released by photocells to the storage system under the influence of dynamic head, formed due to the temperature gradient in the heat exchanger circulation circuit, is realized. The results of physical simulation of heat exchange processes with the use of full-scale flat channels of pv modules made it possible to develop an engineering technique for calculating the productivity of linear pv modules for generation of thermal energy.

References

1. Chow, T.T.: A review on photovoltaic/thermal hybrid solar technology. Appl. Energy **2**(87), 365–379 (2010)
2. Kribus, A., Raftori, D., et al.: A miniature concentrating photovoltaic end thermal system. Energy Convers. Manag. (47), 3582–3590 (2006)
3. Al-Baali, A.A.: Improving the power of f solar panel by cooling and light concentrating. Solar Wind Technol. **4**(13), 241–245 (1986)
4. Forbes Marshall Ltd. http://www.forbesmarshall.com/solar_cogeneration.aspx. Last accessed 21 Jan 2018
5. Nesterenkov, A.G., Nesterenkov, P.A., Nesterenkova, L.A.: Innovative patent of the Republic of Kazakhstan №30003. A way of converting solar radiation into electrical and thermal energy and an installation for realizing the method (2015)
6. Alanod GmbH & Co. Homepage. http://www.alanod.com. Last accessed 21 April 2018
7. Chea, L.C., Håkansson, H., Karlsson, B.: Performance evaluation of new two axes tracking pv-thermal concentrator. J. Civil Eng. Arch. **12**(73), 1485–1493 (2013)
8. Nesterenkov, P.A., Nesterenkov, A.G., Nesterenkova, L.A.: Fundamentals of designing hybrid concentrator solar systems. In: 12th International Conference on Concentrator Photovoltaics (CPV-12), Germany (2016)
9. Blokh, A.G., Zhuravlev, Yu.A., L.N. Ryzhkov. Handbook: Heat Exchange by Radiation, 1st edn. Energoatomizdat, Moscow (1991)
10. Sunpower Corporation. Homepage. http://www.sunpower.com last accessed 21 April 2018
11. Nesterenkov, P.A., Nesterenkov, A.G.: Cogeneration Solar Systems With Concentrators of Solar Radiation. Handbook of Research on Renewable Energy and Electric Resources for Sustainable Rural Development. 1st edn. IGI-Global, Pennsylvania (2018)
12. Shockey, W., Queisser, H.: Solar concentrators. Appl. Phys. (1961)
13. Mikheev, M.A., Mikheeva, I.M.: Fundamentals of Heat Transfer. Handbook: Energy, 3rd edn. Energoatomizdat, Moscow (1977)
14. Nesterenkov, P.A., et al.: Cogeneration systems with radiation of solar concentration - a new type of equipment for solar energy. Energy of the future: innovative scenarios and methods of their implementation. In: Abykayev, N.A., Zhumagulov, B.T. (eds.) Proceedings of Word Congress of Engineers and Scientists, Astana, vol. 3, pp. 266–273 (2017)

Artificial Bee Colony Algorithm for Solving the Knight's Tour Problem

Anan Banharnsakun[(✉)]

Computational Intelligence Research Laboratory (CIRLab), Computer
Engineering Department, Faculty of Engineering at Sriracha, Kasetsart
University Sriracha Campus, Chonburi 20230, Thailand
ananb@ieee.org

Abstract. The knight's tour problem is one of the most interesting classic chessboard puzzles, in which the objective is to construct a sequence of admissible moves made by a chess knight from square to square in such a way that it lands upon every square of a chessboard exactly once. In this work, we consider the knight's tour problem as an optimization problem and propose the artificial bee colony (ABC) algorithm, one of the most popular biologically inspired methods, as an alternative approach to its solution. In other words, we aim to present an algorithm for finding the longest possible sequence of moves of a chess knight based on solutions generated by the ABC method. Experimental results obtained by our method demonstrate that the proposed approach works well for constructing a sequence of admissible moves of a chess knight and outperforms other existing algorithms.

Keywords: Artificial Bee Colony · Knight's Tour Problem · Optimization
Computational Intelligence · Combinatorial Problem

1 Introduction

Chess is a two-player game played on a square board. The standard chessboard has 64 squares arranged in eight equal rows and columns. The knight is one of the pieces used in the game of chess and is normally represented as a horse's head and neck. It is the only piece in chess that does not move in a straight line, but instead moves along an angle that is two squares away horizontally and one square vertically, or two squares vertically and one square horizontally, thus resembling the shape of the letter *L*. In addition to normal chess playing, the knight's pattern of movement on the chessboard is also very interesting to chess players. One challenge is how to produce the longest possible sequence of moves of a chess knight while visiting the squares of the board only once, which is well known as the knight's tour problem (KTP) [1].

The formal study of the knight's tour problem began with the outstanding Swiss mathematician, Leonhard Euler, in the 18th century, and one of the most ingenious methods for its solution is Warnsdorf's rule, introduced by Warnsdorf [2]. Throughout the centuries, a number of additional techniques [3–6] have been introduced to find knight's tours on a given board, such as the brute-force, the divide-and-conquer, the depth-first search with backtracking, and the neural network approaches.

© Springer Nature Switzerland AG 2019
P. Vasant et al. (Eds.): ICO 2018, AISC 866, pp. 129–138, 2019.
https://doi.org/10.1007/978-3-030-00979-3_13

Over the past two decades, recent research has shown that biologically inspired algorithms have great potential in dealing with problems in many scientific and engineering domains [7]. Specifically, there have been a number of techniques based on evolutionary computation, such as genetic algorithms (GA) [8], ant colony optimization (ACO) [9], particle swarm optimization (PSO) [10], and firefly algorithms [11], applied to solving the knight's tour problem. Gordon and Slocum [12] proposed the simple genetic algorithm (SGA) to find solutions to the problem and employed a simple repair technique that can be used to extend tours that have reached an impasse. However, their work focused on only the standard 8×8 chessboard. An ant colony optimization algorithm was presented by Hingston and Kendall [13] to enumerate knight's tours for variously sized chessboards. However, an 8×8 board is too large for a full enumeration by using their method. To initialize a wider range of potential solutions to the knight's tour problem during the search process, the variants of the angle modulated particle swarm optimization (AMPSO) were introduced by Leonard and Engelbrecht [14]. However, the specific characteristics that affect the performance of their algorithms may still be studied in more detail. Ismail et al. [15] proposed a model using the firefly algorithm to solve the knight's tour problem. Although their proposed model has the potential to be applied as a solution, the solution qualities yet need to be improved.

The artificial bee colony (ABC) method introduced by Karaboga [16] is one of the most popular biologically inspired algorithms that are used to find an optimal solution in numerical optimization problems [17]. This algorithm was inspired by the behavior of honey bees when seeking a quality food source. Previous literature [18–22] showed that the ABC can produce a more optimal solution, and thus is more effective than other methods in several optimization problems.

In this paper, an efficient and robust method for solving the knight's tour problem by using ABC strategies is proposed. The contribution of this work is to demonstrate that the ABC, a simple and efficient technique based on biologically inspired computing, can be applied and serve as a useful alternative in solving the knight's tour problem.

The remainder of the paper is organized as follows. The background and knowledge, including the knight's tour problem and the artificial bee colony algorithm, are briefly described in Sect. 2. Adopting the artificial bee colony algorithm to solve the knight's tour problem is proposed in Sect. 3. The experimental settings and results are presented and discussed in Sect. 4. Finally, Sect. 5 concludes this paper.

2 Background and Knowledge

2.1 Knight's Tour Problem

A knight's tour is a sequence of traverses of a knight on an $n \times n$ chessboard by visiting every square once and only once, where the knight moves in an L-shaped pattern as shown in Fig. 1. The knight thus has eight possible moves from the square

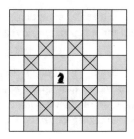

Fig. 1 The knight moves in an L-shaped pattern

where it currently resides. If the knight ends on a square that is one knight's move from the beginning square, the tour is called a closed tour; otherwise, it is considered an open tour.

Each square on the chessboard can be represented as a vertex (V) in the graph (G) and each permissible move by the knight can be represented as an edge (E) in the graph. Thus, the graph of a knight's moves for an $n \times n$ chessboard can be defined as:

$$G = (V, E) \text{ where } V = \{(i,j) | 1 \leq i, j \leq n\}, \text{ and}$$
$$E = \{((i,j), (k,l)) | \{|i - k|, |j - l|\} = \{1, 2\}\}.$$

That is, there is a vertex for every square of the board and an edge between two vertices exactly where a knight's move from one to the other occurs.

The objective of the knight's tour problem is to find the longest possible sequence of moves of a chess knight. This objective function can be expressed as:

$$f(s) = \sum_{i=1}^{(n \times n)-1} c_i \begin{cases} c_i = 1 & \text{if the move is valid for } i^{\text{th}} \text{ move} \\ c_i = 0 \text{ and } i = n - 1 & \text{if the move is not valid for } i^{\text{th}} \text{ move} \end{cases} \quad (1)$$

where s represents a candidate sequence of moves of a chess knight, and $n \times n$ is the total number of squares on the chessboard.

2.2 Artificial Bee Colony Algorithm

The artificial bee colony (ABC) is a meta-heuristic in which the artificial bees of a colony cooperate in finding good quality solutions to optimization problems. An important characteristic of the ABC is that it was inspired by nature, or more precisely by the behavior of honey bees seeking a quality food source. Essential components in the ABC that are modeled after the foraging process of bees are defined as follows:

- Food Source: This component represents a feasible solution to an optimization problem.
- Fitness Value: This value represents the quality of a food source. For simplicity, it is represented as a single quantity associated with the objective function of a feasible solution.

- Bee Agents: This component is a set of computational agents. The 'honey bees' in the ABC are categorized into three types: employed bees, onlooker bees and scout bees.

In the ABC algorithm, the colony is equally divided into employed bees and onlooker bees. Each solution in the search space consists of a set of optimization parameters that represent the location of a food source. The number of employed bees is equal to the number of food sources. In other words, there is one employed bee for each food source.

The details of the ABC algorithm are as follows. First, randomly distributed initial food source positions are generated. The process can be represented by Eq. (2) below:

$$F(x_i), x_i \in R^D, i \in \{1, 2, 3, \ldots SN\}, \tag{2}$$

where x_i is the position of a food source as a D-dimensional vector, $F(x_i)$ is the objective function that determines how good a solution is, and SN is the number of food sources. After initialization, the population is subjected to repeated cycles of three major steps: updating feasible solutions, selecting feasible solutions, and avoiding suboptimal solutions. In order to update feasible solutions, all employed bees select a new candidate food source position. The choice is based on the neighborhood of the previously selected food source. The position of the new food source is calculated by using Eq. (3) below:

$$v_{ij} = x_{ij} + \phi_{ij}(x_{ij} - x_{kj}), \tag{3}$$

where v_{ij} is a new feasible solution that is modified from its previous solution value (x_{ij}) based on a comparison with a randomly selected position from its neighboring solution (x_{kj}), ϕ_{ij} is a random number between $[-1,1]$ that is used to randomly adjust the old solution to become a new solution in the next iteration, and $k \in \{1,2,3,...,SN\} \wedge k \neq i$ and $j \in \{1,2,3,...,D\}$ are randomly chosen indexes. The difference between x_{ij} and x_{kj} is a difference of position in a particular dimension.

The old food source position in the employed bee's memory will be replaced by a new candidate food source position if the new position has a better fitness value. Employed bees will return to their hive and share the fitness value of their new food sources with the onlooker bees. In the next step, each onlooker bee selects one of the proposed food sources depending on the fitness value obtained from the employed bees. The probability that a proposed food source will be selected can be obtained from Eq. (4) below:

$$P_i = \frac{fit_i}{\sum_{i=1}^{SN} fit_i} \tag{4}$$

where fit_i is the fitness value of the food source i, which is related to the objective function value ($F(x_i)$) of the food source i.

The probability of a proposed food source being selected by the onlooker bees increases as the fitness value of the food source increases. After the food source is selected, the onlooker bees will go to the selected food source and select a new candidate food source position in the neighborhood of the selected food source. The new candidate food source can be calculated and expressed by Eq. (3) above.

In the third step, any food source position that does not have an improved fitness value will be abandoned and replaced by a new position that is randomly determined by a scout bee, which helps to avoid suboptimal solutions. The new random position chosen by the scout bee will be calculated by Eq. (5) below:

$$x_{ij} = x_j^{min} + rand[0, 1](x_j^{max} - x_j^{min}) \tag{5}$$

where x_j^{min} and x_j^{max} are the lower bound and the upper bound of the food source position in dimension j, respectively.

The maximum number of cycles (MCN) is used to control the number of iterations and is a termination criterion. The process will be repeated until the number of iterations equals the MCN.

3   Solving the Knight's Tour Problem by Using ABC

Since the ordinary ABC algorithm is designed for solving numerical optimization problems and the algorithm utilizes Eqs. (3) and (5) to find a solution in a continuous function domain, these equations cannot be used directly to solve a problem in combinatorial optimization while its solution is in a discrete domain. However, the algorithm can be expanded for this problem type with suitable modifications. In this work, we introduce a sequence of moves of a chess knight to function as the solution, which is represented as a food source in the ABC algorithm. Each dimension in a food source represents one move pattern made by the knight. The knight has eight possible move patterns from its current square position. These move patterns will be assigned an integer value of 1 to 8, as illustrated in Fig. 2.

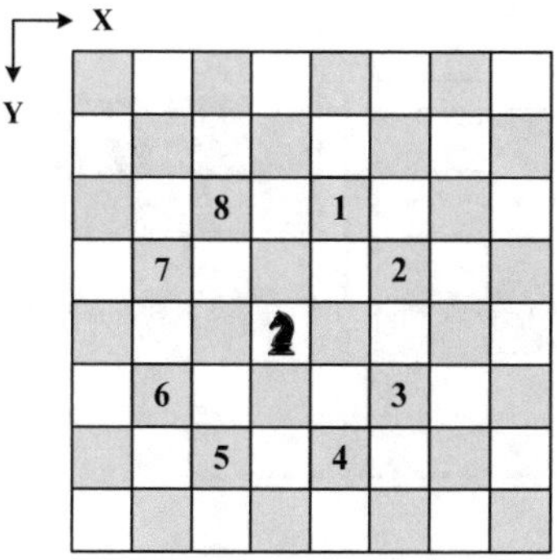

Move pattern	Change in coordinate X	Change in coordinate Y
1	+1	-2
2	+2	-1
3	+2	+1
4	+1	+2
5	-1	+2
6	-2	+1
7	-2	-1
8	-1	-2

Fig. 2 The 8 possible move patterns of the knight

Thus, for the $n \times n$ chessboard, each food source contains $(n \times n) - 1$ dimensions corresponding to $(n \times n) - 1$ move patterns. An example of a sequence of admissible move patterns for the 8×8 chessboard is shown in Fig. 3.

Fig. 3 Example of a sequence of admissible move patterns for an 8×8 chessboard

The interpretation of the example above is as follows. As we scan the sequence of admissible move patterns from left to right, the first move (S_1) of the knight will be made by using move pattern "1", the second move (S_2) of the knight will be made by using move pattern "3", the third move (S_3) of the knight will be made by using move pattern "6", the fourth move (S_4) of the knight will be made by using move pattern "5", and so on. Thus, assuming the initial position is at $(4, 5)$, this example sequence can be constructed as shown in Fig. 4.

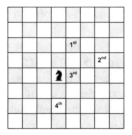

Fig. 4 Example of the knight's moves constructed from the sequence of admissible move patterns shown in Fig. 3

The ABC algorithm is then applied to optimize the sequence of admissible move patterns. In other words, we try to find the optimal sequence of admissible move

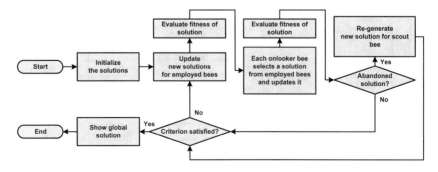

Fig. 5 ABC algorithm flowchart for finding the optimal sequence of the knight's move patterns

patterns that maximizes the objective function obtained by Eq. (1). The applied algorithm is illustrated in Fig. 5.

The solution updating process is performed by following the concept of the updating of a candidate food source based on a neighboring food source in the ABC. Three point switching (TPS) is employed to perform this task as shown in Fig. 6.

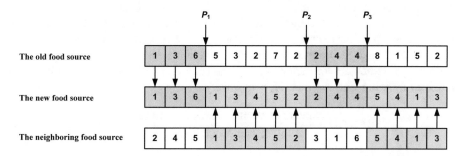

Fig. 6 Exchanging information with a neighboring food source based on TPS

In the example in Fig. 6, based on the TPS method, the points P_1, P_2, and P_3 are randomly selected points from a sequence of admissible move patterns for a 4×4 chessboard, and the dimensions in the partial sequence of the old food source and the neighboring food source are selected back and forth among these three points to produce a new food source.

The point at which the knight cannot continue to move (i.e., where the knight has jumped off the board or back onto a previously visited square) will be repaired. At that point, the move pattern will be checked to see if substituting another pattern will allow the tour to proceed.

If a substitution that extends the tour cannot be made, the scout bee will ignore this old solution and randomly search for new solutions by re-generating a new sequence of admissible move patterns. If a substitution that extends the tour can be made, this substitution can then proceed to the right solution.

4 Experimental Settings and Results

In this section, the performance of the proposed ABC method was examined and compared with other biologically inspired methods, including the GA and the PSO proposed by Gordon and Slocum [12] and Leonard and Engelbrecht [14], respectively. The aim was to evaluate the solutions obtained from the proposed approach in terms of the success rate of finding the optimal tours and the normalized objective function value. All methods used in this experiment were programmed in C++ and all experiments were run on a PC with an Intel Core i7 CPU, 2.8 GHz and 16 GB of memory. In the implementation, three chessboards of sizes 8×8, 10×10, and 12×12 were used to evaluate the performance of the proposed method. Each of the experiments was repeated 50 times with different initial positions of the knight.

For each method, the number of agents was set to 500 and the number of iterations was set to 3,000. Note that the number of function evaluations per iteration in the ABC was not equal to that in the PSO and GA algorithms. The number of function evaluations in the ABC is equal to twice per iteration (for the employed bee and onlooker bee phases), but it is only once per iteration in the ordinary PSO and GA algorithms. For an accurate comparison, the PSO and GA algorithms in this experiment were modified to perform the function evaluation twice in each iteration.

In order to evaluate the quality level of the results obtained from our proposed approach and the other aforementioned methods, the success rate of finding the optimal tours (*SRFOT*) and the normalized objective function value (*NOFV*) were used as the performance indicators for comparing the results. The *SRFOT* and the *NOFV* can be obtained by using Eqs. (6) and (7), respectively.

$$SRFOT = \frac{\text{Number of runs that found optimal tour}}{\text{Total number of runs}} \times 100\% \tag{6}$$

$$NOFV = \frac{f(s)}{n-1} \tag{7}$$

Table 1 and Table 2 show the comparative average *SRFOT* and *NOFV* results of 50 runs on the various chessboard sizes for all of the methods. Note that the range of the *NOFV* is [0,1]. The larger value of the *NOFV* indicates that the solutions are closer to the longest possible sequence of moves of a chess knight, or vice versa.

Table 1 Average *SRFOT* results of 50 runs for each chessboard size

Method	SRFOT		
	8×8	10×10	12×12
GA	90	86	82
PSO	94	90	84
ABC	**100**	**94**	**90**

Table 2 Average *NOFV* results of 50 runs for each chessboard size

Method	NOFV		
	8×8	10×10	12×12
GA	0.9946	0.9871	0.9762
PSO	0.9971	0.9891	0.9864
ABC	**1.0000**	**0.9972**	**0.9968**

As can be seen from the tables, the average *SRFOT* and *NOFV* results of the ABC method are higher than those of the GA and the PSO methods, which means that by using the proposed method, the best quantitative evaluation results have been achieved.

In addition, examples of the optimal tour results obtained from the proposed ABC algorithm are illustrated in Fig. 7.

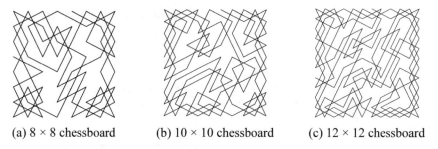

(a) 8 × 8 chessboard   (b) 10 × 10 chessboard   (c) 12 × 12 chessboard

Fig. 7 Optimal tours obtained from the proposed ABC algorithm on chessboards of various sizes

5 Conclusions

This work has proposed an approach to solve the knight's tour problem by finding the longest possible sequence of moves of a chess knight based on solutions generated by the artificial bee colony (ABC) algorithm. A detailed comparison between the proposed method and other techniques, including the GA and the PSO methods, was performed based on the criteria of the success rate of finding optimal tours and the normalized objective function value in our experiments.

The results obtained from the proposed method show that the ABC method offers good performance that outperforms other existing techniques. We can thus conclude that the ABC method is highly efficient from the perspective of both solution quality and algorithm performance. Furthermore, as a result, the proposed approach can serve as a useful alternative in solving the knight's tour problem.

References

1. Watkins, J.J.: Across the board: the mathematics of chessboard problems. Princeton University Press (2004)
2. Ball, W.W.R., Coxeter, H.S.M.: Mathematical Recreations And Essays, 13th edn. Dover, New York (1987)
3. Borrell, R.A.: Brute force approach to solving the knight's tour problem using Prolog. In: Proceedings of the 2009 International Conference on Artificial Intelligence (ICAI 2009), vol. 2, pp. 600–604 (2009)
4. Lin, S.S., Wei, C.L.: Optimal algorithms for constructing knight's tours on arbitrary n × m chessboards. Discret. Appl. Math. **146**(3), 219–232 (2005)
5. Paris, L.: Heuristic strategies for the knight tour problem. In: Proceedings of the International Conference on Artificial Intelligence (IC-AI 2004), pp. 1121–1125 (2004)
6. Takefuji, Y., Lee, K.C.: Neural network computing for knight's tour problems. Neurocomputing **4**(5), 249–254 (1992)
7. Yang, X.S., Cui, Z., Xiao, R., Gandomi, A.H., Karamanoglu, M.: Swarm Intelligence And Bio-Inspired Computation: Theory And Applications. Newnes, Massachusetts (2013)
8. Goldberg, D.E.: Genetic Algorithms. Pearson Education, India (2006)

9. Dorigo, M., Birattari, M., Stutzle, T.: Ant colony optimization. IEEE Comput. Intell. Mag. **1**(4), 28–39 (2006)
10. Poli, R., Kennedy, J., Blackwell, T.: Part. Swarm Optim. Swarm Intell. **1**(1), 33–57 (2007)
11. Yang, X.S.: Firefly algorithm, stochastic test functions and design optimisation. Int. J. Bio-Inspired Comput. **2**(2), 78–84 (2010)
12. Gordon, V.S., Slocum, T.J.: The knight's tour - evolutionary vs. depth-first search. In: Proceedings of the 2004 IEEE Congress on Evolutionary Computation (CEC 2004), vol. 2, pp. 1435–1440 (2004)
13. Hingston, P., Kendall, G.: Enumerating knight's tours using an ant colony algorithm. In: Proceedings of the 2005 IEEE Congress on Evolutionary Computation (CEC 2005), vol. 2, pp. 1003–1010 (2005)
14. Leonard, B.J., Engelbrecht, A.P.: Angle modulated particle swarm variants. In: Proceedings of the International Conference on Swarm Intelligence (ANTS 2014), pp. 38–49 (2014)
15. Ismail, M.M, et al.: Solving knight's tour problem using firefly algorithm. In: Proceedings of the 3rd International Conference on Engineering and ICT (ICEI 2012), pp. 5–8 (2012)
16. Karaboga, D., Gorkemli, B., Ozturk, C., Karaboga, N.: A comprehensive survey: artificial bee colony (ABC) algorithm and applications. Artif. Intell. Rev. **42**(1), 21–57 (2014)
17. Karaboga, D., Basturk, B.: A powerful and efficient algorithm for numerical function optimization: artificial bee colony (ABC) algorithm. J. Glob. Optim. **39**(3), 459–471 (2007)
18. Banharnsakun, A., Tanathong, S.: Object detection based on template matching through use of best-so-far ABC. Computational Intelligence and Neuroscience 2014, article no. **7** (2014)
19. Banharnsakun, A.: A MapReduce-based artificial bee colony for large-scale data clustering. Pattern Recognit. Lett. **93**, 78–84 (2017)
20. Banharnsakun, A.: Hybrid ABC-ANN for pavement surface distress detection and classification. Int. J. Mach. Learn. Cybern. **8**(2), 699–710 (2017)
21. Banharnsakun, A.: Feature point matching based on ABC-NCC algorithm. Evol. Syst. **9**(1), 71–80 (2018)
22. Banharnsakun, A.: Artificial bee colony approach for enhancing LSB based image steganography. Multimed. Tools Appl. **77**(20), 27491–27504 (2018)

The Model of Optimization of Grain Drying with Use of Eletroactivated Air

Dmitry Budnikov$^{(\boxtimes)}$ and Alexey N. Vasilev

Federal State Budgetary Scientific Institution "Federal Scientific
Agroengineering Center VIM" (FSAC VIM), 1-st Institutskij 5,
Moscow 109428, Russia
{dimml3, vasilev-viesh}@inbox.ru

Abstract. The development of equipment for energy saving grain drying is not losing its relevance. For its effective implementation it is reasonable to apply the most promising solutions in electrotechnologies, such as the use of air ions. The article describes the receiving the model describing application of electro-activated air. To obtain the sought for model, the foundations of similarity theory, as well as the dimensional method were used. The electroactivation coefficient introduced to consider air ions influence in drying process, was described. In the result, the mathematical model has been obtained, that can be used both in the description of drying process and optimization of drying equipment parameters to intensify drying process. The results of experiments of grain drying with use of the electro-activated air are given in the article and the possibility of optimization of drying process is described.

Keywords: Electroactivated air · Aeroions · Similarity theory
Drying of grain · Drying agent

1 Introduction

The necessity for grain drying of agricultural materials is determined by the necessity to ensure their quality and the period of safe storage. In order to achieve satisfactory level for the manufacturers' economic performance and maintain reasonable consumer prices, it is necessary to use energy saving equipment. The development of such technological equipment currently involves the use of computer simulation tools, building scalable models and prototyping of the required equipment.

Forced ventilation is a "soft" method of grain drying when atmosphere air is blown through grain layer. It may be heated up but only to reduce relative humidity down to equilibrium moisture content. As air heating by 1 °C reduces its relative humidity by 5%, atmosphere air supplied to grain layer is heated by 7...8 °C at most. Therefore, drying rate in forced ventilation units is not high, which reduces efficiency of postharvest treatment lines.

The research [1–5] on the use of electroactivated air for intensifying grain drying through forced ventilation has demonstrated this method efficiency. However, to calculate the process and simulate the operation modes of forced ventilation units using

© Springer Nature Switzerland AG 2019
P. Vasant et al. (Eds.): ICO 2018, AISC 866, pp. 139–145, 2019.
https://doi.org/10.1007/978-3-030-00979-3_14

electroactivated air, it is necessary to have analytical description of drying process. To obtain it, let us turn to known methods.

2 Main Part

V. I. Aniskin conducted thorough research of grain drying process in dryeration bins [6]. To describe grain drying process in thick layer with the use of forced ventilation, he developed and experimentally tested criteria equations for grain drying through forced ventilation in the conditions of both vertical and radial feed of drying agent.

$$Ho = A \cdot Ko^{\beta 1} \cdot Gu^{\beta 2} \cdot Re^{\beta 3} \left(\frac{d}{L}\right)^{\beta 4}. \tag{1}$$

Each criterion describes the mechanism of internal or external heat and moisture exchange.

The homochronicity criterion, Ho, describes duration of grain layer drying at a constant rate of drying agent velocity. The homochronicity criterion changes over time at a constant velocity of drying agent and fixed thickness of grain layer.

The Kossovich criterion, Ko, reflects the relationship between heat spent to evaporate moisture, and heat spent to warm up grain. Considering the use of electroactivated air, the Kossovich criterion would be reduced, as specific heat of evaporation decreases due to lower viscosity of moisture contained inside caryopses.

The Guchman criterion, Gu, characterizes air potentials as a drying agent. In the Guchman criterion drying potential is contained in changing difference between drying agent temperature and wet thermometer temperature. Wet thermometer temperature is in direct relationship to air humidity.

The Reynolds criterion, Re, reflects a hydrodynamic mode of drying agent movement.

In the result of experimental data processing V. I. Aniskin [6] obtained two criterion equations – for vertical and radial feed of drying agent, that reflect the process of grain drying through forced ventilation in thick layer:

$$Ho = 40,5 \cdot 10^{-5} Ko^{0,95} Gu^{-1,9} Re^{0,31} \left(\frac{d}{L}\right)^{-0,07}, \tag{2}$$

$$Ho = 50 \cdot 10^{-5} Ko^{0,95} Gu^{-1,9} Re^{0,31} \left(\frac{d}{R - r_0}\right). \tag{3}$$

2.1 Research Method

As has been noted before, one of the fundamental problems of the use of electrotechnologies in grain drying is the absence of mathematical description of processes occurring in grain layer. The thermodynamic criteria used by V. I. Aniskin for this process description, reflect them to the full extent, with the exeption of processes with the use of electroactivated air. To a certain degree the Kossovich criterion can reflect processes occurring in caryopsis when affected by electroactivated air. However, that can prove to be insufficient. Therefore, it is reasonable to apply an additional criterion that could reflect changing of processes in caryopsis. It is the Lykov criterion that can be accepted as such thermodynamic criterion.

A. I. Lykov criterion provides connection between the intensity of the development of moisture and temperature fields inside material in the process of moisture transfer [7]:

$$Lu = \frac{a_m}{a}, \qquad (4)$$

where a_m – heat exchange rate, m²/s; a – moisture diffusion coefficient, m²/s.

At Fig. 1 nomogram chart of dependence of temperature conductivity on moisture content is presented. Temperature conductivity coefficient tends to increase at up to 20% moisture content or up to 16.67% grain layer humidity, and decreases as moisture content of grain layer humidity is increased.

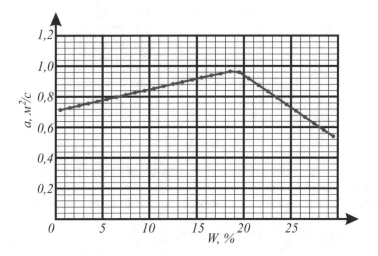

Fig. 1. Dependence of temperature conductivity on moisture content

The determination of moisture diffusion coefficient, a_m, in grain is quite a problem [8]. In various scientific sources data on this coefficient value sometimes varies considerably. It has been established [8] that the value of moisture diffusion in grain at is about 10^{-11}, m²/c. If grain moisture exceeds 10%, the value of moisture diffusion coefficient are reduced practically in accordance with linear dependence, but the order of value remains constant.

2.2   Calculation of Criteria

To consider the effect of electroactivated air on grain drying process let us add the so-called electroactivation criterion, Q_e, reflecting the process of drying agent saturation with air ions and the conditions of its interaction with grain.

At cyclic saturation of drying agent with air ions (Fig. 2) electroactivation criterion should reflect its mode peculiarities expressed in the ration of drying time under the effect of air ions and without them. In this case electroactivation criterion will be as follows:

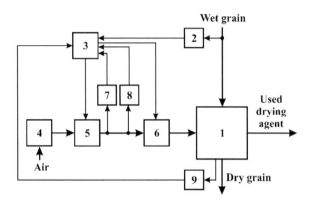

Fig. 2. Schematic view of the laboratory setup

$$Q_e = Q_1 \cdot Q_2, \tag{4}$$

where Q_1 – air ions concentration when they enter grain layer, $1/m^3$; Q_2 – air volume per one processing cycle, m^3.

The full physical equation describing the process being researched [8]:

$$Q_1 = f(E, T_a, V_a, L_T, \alpha, V_0). \tag{5}$$

where E – intensity of electric field generated by the electric activator, V/m; T – drying agent temperature, °C; V_a – drying agent velocity, m/s; L_T – distance from the electric activator to the entry to grain layer, m; α – volume recombination coefficient, m^3/s; V_0 – the velocity of air ions going to the air duct walls, s^{-1}.

It was suggested to carry out processing by air ions cyclically in accordance with the binary signal with a period of 10 min and a filling factor of 1/2. Considering the obtained expression for air ions concentration at the entry to grain layer, the expression (4) will be as follows:

$$Q_e = C \cdot \frac{\alpha \cdot V_a}{L_T^7 \cdot V_0^2} \cdot (v \cdot S \cdot T_{ae}). \tag{6}$$

where C – coefficient of reduction, determined experimentally; v – drying agent velocity, m/s; S – area of the cross section of the processing chamber for grain drying, m^2; T_{ae} – semi-oscillation of air ions concentration in drying agent, s.

Shared writing with boundary limitations would gave us the mathematical model of the process of grain drying with the use of electroactivated air:

$$A' \cdot Lu^{\beta 0} \cdot Ko^{\beta 1} \cdot Gu^{\beta 2} \cdot Re^{\beta 3} \cdot Q_{\ni}^{\beta 4} \to min;$$

$$\begin{cases} 2\pi h(R_1 - R) \leq D, \\ S_1 = \pi R^2, \\ \frac{v \cdot S_1 \cdot P}{1000 \cdot \eta} \leq N, \\ R, R_1, S_1, v > 0. \end{cases} \tag{9}$$

This model allows to optimize the parameters of the dryerationbins and ventilator performance, as well as the parameters of the criterion of electro-activation to minimize the period of grain drying through the use of electroactivated air.

2.3 Laboratory Setup

The objective is solved by that enrichment of the drying agent by aero ions is carried out periodically during which the enrichment periods and also concentration of aero ions depend on humidity of processed material, moisture content and temperature of the drying agent and can be determined by two criteria: either minimum energy consumption, or maximum speed of drying.

It is need a temperature control of the drying agent, initial humidity control of the processing material and a change of humidity of the grain during drying process to realize the method.

Figure 2 shows the structural diagram of a laboratory installation. The method is carried out as follows: the operator selects one of the realized drying variants, namely the minimum energy consumption or the maximum speed of drying, there could be specified the most admissible drying time. The operator can change the realized variant during the drying process by means of the operating controller (personal computer) in case of need. The operator also sets the required final humidity on reaching which the drying process will be stopped, and the processed grain will get further to the technological line. Crude grain comes into the technological line at the drying zone 1, at the same time the sensor 2 from which information arrives on the operating controller (personal computer) 3 controls its initial humidity. At the same time the fan 4 feeds with air the heat generator 5 and further via the ionizer 6 the zone of drying 1. At the exit of the heat generator 5 the sensors 7 and 8 control the temperature and moisture content of the drying agent respectively. Information from sensors 7 and 8 arrives on the operating controller which controls the operation of the heat generator 5 and ionizer 6. During drying process the sensor 9 controls the humidity of the processed material in a drying zone 1. Information from the sensor 9 arrives on the operating controller. Because the drying is a long process the minimum period of poll of sensors is not less than once in 15 min, however this period much lesser and doesn't exceed once in a minute.

During the process the operations of the operating controller are based on the loaded into it statistical data of experiments which example is shown on Fig. 3. There is a heating of the agent of drying (ensuring minimum admissible sorption properties in the mode of economic drying or heating up to the temperature at which is carried out the maximum moisture, according to the loaded statistical data) and an operating mode of the ionizer (switched off, switched on constantly or switching on-off cyclically) are selected using this data and according to the realizing mode of drying.

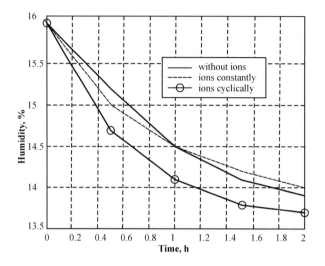

Fig. 3. Drying curves

Drying continues until the conditional humidity or set humidity is reached, after which the grain goes further to the processing line.

Realized drying variants are determined by the temperature of the drying agent and concentration of ions, but can't break the technological requirements of the drying agent (heating above the allowed temperature) and depend on the type of drying material (culture), its initial characteristics, and characteristics of the ambient air, in addition it depends on the type of installation in which drying is carried out. To realize the sensor readings and the optimal control algorithm, was used the SCADA system implemented on MasterSCADA.

3   Conclusions

Based on the foregoing, we can conclude the following:

1. The use of the method of dimensional analysis of similarity theory made it possible to obtain dimensionless criterion that can be used for the description of the process of changing concentration of air ions in their transportation from a radiating unit to grain layer.

2. The application of the method of putting the equations of physical process to dimensionless form made it possible to develop the criterion equation for the description of processes of heat and moisture exchange in grain layer during its drying through the use of electroactivated air, that can be used as the basis in building simulation models of drying of grain thick layer by electroactivated air.
3. The obtained mathematical model allows to optimize the parameters of dryeration bins and ventilator performance, as well as the parameters of the criterion of electroactivation to minimize the period of grain drying through the use of electroactivated air.

References

1. Vasil'ev, A.N., Gracheva, N.N.: Obosnovanie vozmozhnosti ispol'zovanijaj elektroaktivirovannogo vozduha dlja intensifikacii sushki zerna [Justification possibility of using electroactivated air to intensify grain drying]. Metody i sredstva povyshenijaj effektivnosti ispol'zovanijaj elektrooborudovanija v promyshlennosti i sel'skom hozjajstve: sbornik nauchnyh trudov [Methods and means of improving the efficiency of electrical equipment in industry and agriculture: collection of scientific papers]. Stavropol 2010, pp. 291–293. (In Russian)
2. Ranjbaran, M., Zare, D.: Simulation of energetic- and exergetic performance of micro-wave-assisted fluidized bed drying of soybeans. Energy (2013). https://doi.org/10.1016/j.energy.2013.06.057
3. Vasilyev, A., Budnikov, D., Gracheva, N.: The mathematical model of grain drying with the use of electroactivated air. Res. Agric. Electr. Eng. **1**(5), 32–37 (2014)
4. Vasiliev, A.N., Budnikov, D.A., Gracheva, N.N., Smirnov, A.A.: Increasing efficiency of grain drying with the use of electroactivated air and heater control. In: Kharchenko, V., Vasant, P. (eds.) Handbook of Research on Renewable Ener-gy and Electric Resourcesfor Sustainable Rural Development, USA, PA, Hershey: IGI Global, pp. 255–282, (2018). ISBN 9781522538677. https://doi.org/10.4018/978-1-5225-3867-7.ch011, https://www.igi-global.com/chapter/increasing-efficiency-of-grain-drying-with-the-use-of-electroactivated-air-and-heater-control/201341
5. Stuart, N.: Dielectric Properties of Agricultural Materials and Their Applications, p. 229. Academic Press (2015)
6. Aniskin, V. I., Rybaruk, V.A.: Teorija i tehnologija sushki i vremennoj konservacii zerna aktivnym ventilirovaniem [Theory and technology of drying and temporary preservation of grain active ventilation]. Moscow, Kolos Publishers, p. 190 (1972)
7. Lykov, A.V.: Teorija sushki [Theory of drying]. Moscow, Energiia Publication, p. 472 (1968)
8. Venikov, V.A.: Teorija podobijai modelirovanija (primenitel'no k zadacham jelektrojenergetiki): uchebnik dlja vuzov po spec. « Kibernetika j elektr. sistem » -e izd., pererab. I dop. Moscow, Visshaja shkola Publishers, p. 439 (1984)

Model of Improved a Kernel Fast Learning Network Based on Intrusion Detection System

Mohammed Hasan Ali[✉] and Mohamed Fadli Zolkipli

Faculty of Computer Systems and Software Engineering,
University Malaysia Pahang, Gambang 26300, Malaysia
Mhl80250@gmail.com

Abstract. The detection of network attacks on computer systems remains an attractive but challenging research scope. As network attackers keep changing their methods of attack execution to evade the deployed intrusion-detection systems (IDS), machine learning (ML) algorithms have been introduced to boost the performance of the IDS. The incorporation of a single parallel hidden layer feed-forward neural network to the Fast Learning Network (FLN) architecture gave rise to the improved Extreme Learning Machine (ELM). The input weights and hidden layer biases are randomly generated. In this paper, the particle swan optimization algorithm (PSO) was used to obtain an optimal set of initial parameters for Reduce Kernel FLN (RK-FLN), thus, creating an optimal RKFLN classifier named PSO-RKELM. The derived model was rigorously compared to four models, including basic ELM, basic FLN, Reduce Kernel ELM (RK-ELM), and RK-FLN. The approach was tested on the KDD Cup99 intrusion detection dataset and the results proved the proposed PSO-RKFLN as an accurate, reliable, and effective classification algorithm.

Keywords: Fast learning network · Kernel extreme learning machine
KDD Cup99 · Particle swarm optimization algorithm
Intrusion detection system

1 Introduction

Both network security and computer security systems collectively make up cybersecurity systems. Each of the security systems basically has an antivirus software, a firewall, and an IDS. The IDSs is involved in the discovery, determination, and identification of unauthorized access, usage, alteration, destruction, or duplication of an information system [1]. The security of these systems can be violated through external (from an outsider) and internal attacks (from an insider). Until now, much efforts are devoted to studies on the improvement of network and information security systems, and several studies exist on IDS and its taxonomy [1–4]. Machine learning has recently gained much interest from different fields such as control, communication, robotics, and several engineering fields. In this study, a machine learning approach was deployed to address the issues of intrusion detection in computer systems. It is a challenging task to automate ID processes, as has earlier been ascertained by Sommer and Paxson who applied ML techniques to ID systems and outlined the challenges of automating

© Springer Nature Switzerland AG 2019
P. Vasant et al. (Eds.): ICO 2018, AISC 866, pp. 146–157, 2019.
https://doi.org/10.1007/978-3-030-00979-3_15

network attack detection processes [5]. The specific approaches of using ML techniques for network intrusion detection and their and challenges have been previously outlined [6]. Some of the major problems of the current network ID systems such as high rates of false-positive alarms, false negative or missed detections, as well as data overload (a situation where the network operator is overloaded with information, making it difficult to monitor data) have been discussed [6].

Several ML algorithms have been used to detect anomalies in the behavior of ID systems. This is achieved by training the ML algorithms with the normal network traffic patterns, making them capable of determining traffic patterns that differ from the normal pattern [5]. Although some ML techniques can effectively detect certain forms of attack, no single method has been developed that can be universally applied to detect multiple types of attack. Intrusion detection systems can be generally divided into two system (anomaly and misuse) based on their mode of detection [6]. The anomaly-based detection system flags any abnormal network behavior as an intrusion, but the misuse-based detection system relies on the signature of established previous attacks to detect new intrusions. Several anomaly-based detection systems have been developed based on different ML techniques [6, 9, 11]. For instance, several studies have used single learning techniques like neural networks, support vector machines, and genetic algorithms to design ID systems [5]. Other systems such as the hybrid or ensemble systems are designed by combining different ML techniques [10, 11]. These techniques are particularly developed as classifiers for the classification or recognition of the status of an incoming Internet access (normal access or an intrusion). One of the significant algorithms of machine learning is the ELM first proposed by Huang. The ELM has been widely investigated and applied severally [12]. Several ID systems have been proposed based on the use of ELM as the core algorithm [6, 13, 14]. Furthermore, there is a heavy influx of network traffic data through the ID system which needs to be processed [7]. This study, therefore, focuses on the development of a scalable method that can improve the effectiveness of network ID systems in the detection of different classes of network attack.

2 Overview of Fast Learning Network

In the past few decades, the demand for even the high performing single hidden layer Feedforward Neural Network (FNN) has waned due to some application challenges [4]. To solve these issues, Guang Bin Huang proposed the Extreme Learning Machine (ELM) [3] whose major function is the transformation of a single hidden layer FNN into a linear least square solving a problem; it then, calculates the output weights through the Moore–Penrose (MP) generalized inverse. There are several advantages of ELM, first, it avoids repeated calculation of iteration, has a fast learning speed; cannot be trapped at the local minimum, ensure output weights uniqueness, has a simplified network framework, presents a better generalization ability and regression accuracy. Several scholars have successfully implemented the ELM learning algorithm and theory [5, 6] in pattern classification, function approximation [7–9], system identification and so on [10, 12]. The other issue is the handling of information incorporation in the ELM when multiple varying data sources are available [15]. Therefore, the kernel-based ELM (KELM) has been proposed by comparing the modeling process between SVM and ELM [16]. The results show that KELM performs better and is more robust compared to the basic ELM [15] in solving non-separable linearly samples. The KELM also performed better than ELM, KELM in solving regression prediction tasks. It achieved a comparative or better performance with a faster learning speed and easier implementation in several applications, including 2-D profiles reconstruction, hyperspectral remote sensing image classification [17, 18], activity recognition, and diagnosis of diseases [19, 20]. KELM has also been used for online prediction of hydraulic pumps features, location of damage spot in aerospace, and behavior identification [21, 22]. However, [15] the training of KELM is an unstable process; the learning parameters must be manually adjusted; and it utilizes randomly generated hidden node parameters. The adjustment of the learning parameters requires human input, and could influence the classification performance. Its kernel function parameters also need a careful selection process to achieve an optimal solution. There are many works that provide optimization methods based on KELM parameters. The meta-heuristics have been suggested for tackling the problems of parameter setting in KELM. Some of the suggested metaheuristics are genetic algorithm (GA) [18], and AdaBoost framework [23]. [24] used adaptive artificial bee colony (AABC) algorithm for parameter optimization and selection of KELM features. The features were evaluated based on Parkinson's disease dataset. In [25], the authors proposed the chaotic moth-flame optimization (CMFO) strategy to optimize KELM parameters. Also, an active operators particle swam optimization algorithm (APSO) was proposed in [15] for obtaining an optimal initial set of KELM parameters. The evaluated model (APSO-KELM) based on standard genetic datasets show the APSO-KELM to have a higher classification performance compared to the current ELM and KELM. However, the results of this work show KELM to have a better accuracy compared to ELM, showing the need to introduce the kernel function. In other words, the optimize kernel parameter results showed no fluctuation and an increasing coverage with iteration. Meanwhile, there are some issues with ELM such as the need for additional hidden neurons in some regression applications compared to the conventional neural network (NN) learning algorithms. This may cause the trained ELM to require more reaction time when presented with new test samples. Furthermore, any increase in the number of hidden layer neurons also results to an exponential increase in the number of thresholds and weights of random initialization. These values may not be the optimized parameters [26, 27]. In 2013, Li et al. suggested a novel ELM-based artificial neural network for fast learning network (FLN) [13]. FLN is a double parallel FNN made up of a single layer FNN and a single hidden layer FNN. The received information at the input layer is transmitted to the hidden and output layers (first, the message gets to the neurons of the hidden layer before being transmitted to the output layer). Therefore, the FLN can perform nonlinear approximation like other general NN. Contrarily, the information is directly transferred from the input layer to the output layer, giving the FLN the ability to establish the linear relationship between the input and the output. Hence, the FLN can handle linear problems with a high accuracy, and can also infinitely approximate nonlinear systems. FLN can also solve the issue associated with the conventional NN which does not demand iterative calculation. This work start with Sect. 2 provides the introduction. Section 3 explain the data set KDD details. Section 4 provides overview of the methodology. Section 5 provides the experiments and analysis of the results.

3   Overview of the Methodology

3.1   Fast Learning Network

The FLN was presented by [37] as a novel variant of the ELM [38]. It is structured as a combination of two NNs, the first one as an SLFNN and the second an MLFNN. The FLN depends on three layers, namely, input, hidden, and output layers. The FLN structure is shown in Fig. 1,

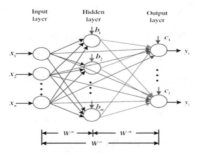

Fig. 1. Structure of FLN

The equations of deriving the output of the FLN based on the provided matrices and vectors and they are presented in the following equations.

$$y_j = f\left(w^{oi}x_j + c + \sum_{k=1}^{m} w_k^{oh} g\left(w_k^{in}x_j + b_k\right)\right) \tag{1}$$

Where

- $w^{oi} = \left[w_1^{oi}, w_2^{oi}, \ldots, w_i^{oi}\right]$ is the weight vector connecting the output nodes and input nodes.
- $w_k^{in} = \left[w_{k1}^{in}, w_{k2}^{in}, \ldots, w_{km}^{in}\right]$ is weight vector connecting the input nodes and hidden node
- $w_k^{oh} = \left[w_{1k}^{oh}, w_{2k}^{oh}, \ldots, w_{ik}^{oh}\right]$ is weight vector connecting the output nodes and hidden node, a more compact representation is given as follows

The matrix $W = \begin{bmatrix} W^{oi} & W^{oh} & c \end{bmatrix}$ can be called output weights, and G is the hidden layer output matrix of FLN, the ith row of G is the ith hidden neuron's output vector with respect to inputs $x_1.x_2.\ldots.x_N$. To solve the model, the minimum norm least-squares solution of the linear system can be written as follows: A more compact representation is given as follows:

$$Y = f(w^{oi}x + w^{oh}G + c) = f\left(\left[w^{oi} w^{oh} c \right] \begin{bmatrix} X \\ G \\ I \end{bmatrix} \right) \tag{2}$$

$$= f\left(W \begin{bmatrix} X \\ G \\ I \end{bmatrix} \right) \tag{3}$$

Where

$$H = \left[X\,G\,I \right]^{T} \tag{4}$$

$$G(W_1^{in}, \cdots, W_m^{in}, b_1, \cdots, b_m, \cdots, X_N) \tag{5}$$

$$= \begin{bmatrix} g\left(W_1^{in} X_1 + b_1 \right) & \cdots & g\left(W_1^{in} X_N + b_1 \right) \\ \vdots & \ddots & \vdots \\ g\left(W_m^{in} X_1 + b_m \right) & \cdots & g\left(W_m^{in} X_N + b_m \right) \end{bmatrix}_{m \times N}$$

$$W = \left[W^{oi} W^{oh} C \right]_{1 \times (n+m=1)} \tag{6}$$

$$I = \left[11 \cdots 1 \right]_{1 \times N} \tag{7}$$

In order to resolve the model, a Moore Penrose based equation is given as follows.

$$\hat{w} = \mathbf{f}^{-1}(\mathbf{Y}) \left(\left[X^T G^T I^T \right] \begin{bmatrix} X \\ G \\ I \end{bmatrix} \right)^{-1} \mathbf{H}^{T} \tag{8}$$

$$\hat{w} = \mathbf{f}^{-1}(\mathbf{Y}) \left(\mathbf{X^T X} + \mathbf{G^T G} + \mathbf{I^T I} \right)^{-1} \mathbf{H^T} \tag{9}$$

An algorithm that explains the learning of the FLN is presented in the flowchart depicted in Fig. 2. The algorithm starts by random initialization of the weights between the input and hidden layers and the biases of the hidden layer. Then, the G matrix is determined depending on the input-hidden matrix. This matrix represents the output matrix of the hidden layer. Next, the input-output matrix w^{oi} and w^{oh} are determined based on the Moore–Penrose equations. As a result, a complete FLN model is formulated

3.2   Kernel Fast Learning Network

A kernel function in machine learning is a measurement of the closeness between input sample data defined over a feature denoted as $K(x,x')$ [39]. In a recent study, [16] suggested that the hidden layer of an SLFN does not need to be formulated as a nodes

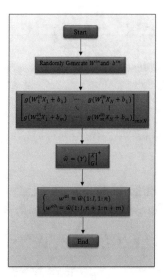

Fig. 2. Flowchart of the FLN learning model

of single layer; instead, a mechanisms of feature mapping can be a wide range used to replace the hidden layer.

This approach has been exemplified in this work as previously reported by [16] to convert an FLN model to a kernel-based model. By recalling the output of Eq. (3), replacing the output weight W with the output weight based on the Moore–Penrose generalized inverse ŵ as Eq. (8), and replacing H with Eq. (4), we can obtain Eq. (13) as follows:

$$W^{\wedge} = Y^{\wedge} H^T (H H^T)^{-L}$$

$$Y = f\left(Y^{\wedge} H^T (H H^T)^{-1} H^T\right)$$

$$= f\left(Y^{\wedge} (H H^T)^{-1} H^T H\right)$$

$$Y = f\left(f(Y)^{-1}\left([X^T\ G^T\ I^T]\begin{bmatrix}X\\G\\I\end{bmatrix}\right)^{-1}[X^T\ G^T\ I^T]\begin{bmatrix}X\\G\\I\end{bmatrix}\right)$$

$$Y = f\left(f(Y)^{-1}(X^TX+G^TG+I^TI)^{-1}(X^TX+G^TG+I^TI)\right)$$

$$Y = f\left(f(Y)^{-1}\left([X^T\ G^T\ I^T]\begin{bmatrix}X\\G\\I\end{bmatrix}\right)^{-1}[X^T\ G^T\ I^T]\begin{bmatrix}X\\G\\I\end{bmatrix}\right) \tag{10}$$

$$Y = f\left(f(Y)^{-1}\left(X^{T}X + G^{T}G + I^{T}I\right)^{-1}\left(X^{T}X + G^{T}G + I^{T}I\right)\right) \tag{11}$$

Moreover, through the addition of a small positive quantity (i.e., the stability factor) $1/\lambda$ to the diagonal of $\mathbf{H^{T}H}$, thus yielding a more "stable" solution, we can represent Y as [16] follows:

$$Y = f\left(f(Y)^{-1}(k_{1}(x.x') + k_{2}(x.x') + k_{3}(x.x') + 1/\lambda)^{-1}(k_{1}(x.x') + k_{2}(x.x') + k_{3}(x.x'))\right) \tag{12}$$

Generally, the selection of the output neurons' active function $f(\cdot)$ is often linear, such that $f(x) = x$. Then, Eq. (12) can be written as follows:

$$Y = Y \ (k_{1}(x.x') + k_{2}(x.x') + k_{3}(x.x') + 1/\lambda)^{-1}(k_{1}(x.x') + k_{2}(x.x') + k_{3}(x.x')) \tag{13}$$

Substituting in the location of k_1, k_2, k_3 three kernels, and in the location of λ a regularization factor, we obtain a MKFLN model. The power of this model is with using 3 kernels for performing the separation which is expected to outperform the classical one kernel ELM variant. However, there are two problems to be addressed. The first one is the computational concern when using kernels calculation, especially if the size of the dataset is huge. The second one is the criticality of selection suitable kernels for classification due to the sensitivity of the performance to the kernel type. Therefore, a framework for to make the developed MKFLN feasible for practical applications is designed.

However, the kernel-based learning methods usually uilize a large memory system for learning dilemmas with large data sets and are therefore modified to reduced kernel extreme learning machine (RKELM) (Deng et al. 2013). A modification to RKFLN called reduced kernel FLN has also been proposed. The kernel-based and basic ELMs have shown superior generalization and better scalability for multiclass problems with superior generalization, and considerable scalability for multiclass classification problems with much lower training times than those of SVMs [16]. These issues create the ELM an appealing learning paradigm for applied in large-scale problems, such as the IDSs.

Nevertheless, kernels are utilized in learning approaches, particularly those with potentially large amounts of memory for ML problems, with huge datasets such as IDS, which requires the collection of wide data in traffic network. To treat this issue, RKELM takes Huang's kernel-based ELM and, instead of computing $k(x,x)$ over the entire input data, computes $k(\tilde{x}, x)$, where $\tilde{x}$ is a randomly chosen subset of the input data. [40] adapted the method of reduced kernel by selecting a small random subset $\tilde{X} = \{x_i\}_{i=1}^{\tilde{n}}$ from the original data points $X = \{x_i\}_{i=1}^{n}$ with $\tilde{n} \ll n$ and using $k(x,\tilde{x})$ in place of $k(,X,X)$ to cut the problem size and computing time. As mentioned previously, FLN is better than ELM; to address this issue, RKELM is swapped out for RKFLN. Hence, RKFLN is expected to be better than RKELM. An assumption supposes that multiplying each kernel in RKFLN by a weight provides better results.

3.3   Particle Swarm Optimization

The PSO was first introduced by Li et al. [23] as a parallel evolutionary computation technique which was insired by the social behavior of swarm. The performance of PSO can be significantly affected by the selected tuning parameters (commonly called exploration– exploitation tradeoff). Exploration is the ability of an algorithm to explore all the segments of the search space in an effort to establish a good optimum, better known as the global best. Exploitation on the other hand is the ability of an algorithm to concentrate on its immediate environment and within the surrounding of a better performing solution to effectively establish the optimum. Irrespective of the research efforts in recent times, the selection of algorithmic parameters is still a great problem [41]. In the PSO algorithm, the objective function is used for the evaluation of its solutions, and to operate on the corresponding fitness values. The position of each particle is kept (including its solution), and its velocity and fitness are also evaluated [42]. The PSO algorithm has many practical applications [43–46]. The position and velocity of each particle is modified to establish an best solution for each iteration using the following relationship:

$$v_i(k+1) = wv_ik + c1r1(xbest, local - x_i) + c2r2(xbest, global) - x_i) \tag{14}$$

$$x_i(k+1) = xk + v(k+1) \tag{15}$$

Each particle's velocity and position are denoted as the vectors $v_k = (v_{k1}, \ldots, v_{kd})$ and $x_i = (x_{i1}, \ldots, x_{id})$, respectively. In (14), x vectors is the best local and best global positions; $c1$ and $c2$ represents the acceleration factors referred to as cognitive and social parameters; $r1$ and $r2$ represents randomly selected number in the range of 0 and 1; k stands for the iteration index; and w is the inertia weight parameter [47]. x_i of a particle is updated using (15).

4   Experiments and Analysis

4.1   Implementing of Multi Kernel Based on Fast Learning Network

As it has been mentioned in the previous section, RKFLN has two problems to be solved before we can apply it, the first one is the computational complexity of applying three kernels at the same time, and the second one is the selection of the kernels and its parameters. The initial one is solved through using reduced kernel approach, and the other one is solved through preparing a set of kernels and selecting out of them. In order to make the model more optimized, three weighting factors of using the kernels are imported in the model $\alpha_1, \alpha_2, \alpha_3$ and λ. The parameters α_1, α_2, and α_3 are weighting factors for the kernels k_1, k_2, and k_3, the parameter λ is the regularization parameter. The new model after optimizing is written as

$$Y = f(Y^\wedge \left(\alpha_1^* K_1 + \alpha_2^* K_2 + \alpha_3^* K_3 + \frac{1}{\lambda^*} \right)^{-1} (\alpha_1^* K_1 + \alpha_2^* K_2 + \alpha_3^* K_3)$$

where the vector $(\alpha_1^*, \alpha_2^*, \alpha_3^*, \lambda^*)$ denotes the optimal parameters of the model, they have the constraint

$$\alpha_1^* + \alpha_2^* + \alpha_3^* = 1, \alpha_1, \alpha_2, \alpha_3 \in .[1\,0]\lambda^* \in [0\,1]$$

In order to determine convenient values for these weights that multiplying each kernel in RKFLN. PSO will be used. Herein, k_1, k_2, and k_3, are multiplied by the weights α_1, α_2, and α_3 respectively, the optimization goal is to find the best weights values that give the highest testing accuracy over validation samples. Beside the searching for the weights values, the optimization process will also look for the best value for the regularization coefficient.

4.2   Experiments of Optimization Multi Kernel Fast Learning Network

This ssection describe our experimental also provide classification performanc results. In order to evaluate the PSO-RKFLN developed model, this work cover teasting results with the KDD Cup99 dataset. Optimize the RKFLN parameters to enhance the accuracy of IDS, we proposed several models such as RKFLN, RKELM, and PSO-RKELM as bunchmarks. Figure 3 shown classification evaluation measures, and Table 1 shown the comparestion results between the models.

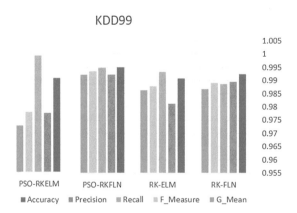

Fig. 3. Evaluation measures of classification

For PSO parameters formulae such as $c1 = c2 = 1.42$, $w = 0.75$ and number of particles $= 5$. The KDD 99 training set with duplicates removed for the first set of sets with a dataset consisting of 145,585 split into a training set of 72,792 training examples and 72,792 testing samples. For this study we used all 41 attributes of the data.

Table 1. Results of evaluatiom measures of different models

Models	G_Mean	F_Measure	Recall	Precision	Accuracy
RK-ELM	0.9863	0.98773	0.99344	0.981103	0.99073
RK-FLN	0.9868	0.98905	0.98856	0.98955	0.99243
PSO-RKELM	0.98841	0.98973	0.99513	0.97971	0.99105
PSO-RKFLN	0.99223	0.99358	0.99492	0.99224	0.99509

5 Summary

In this work, using Machine learning based on the intrusion-detection system, it's attractive for many researchers. This work provides model based on optimize of a multi kernel of fast learning network. The derived model was rigorously compared to four models, including basic ELM, basic FLN, Reduce Kernel ELM (RK-ELM), and RK-FLN. The approach was tested on the KDD Cup99 intrusion detection dataset. The accuracy of our model (PSO-RKFLN) is slightly higher than other models. For a future work, we recommend checking this model with a different number of neurons to measures and evaluate the complexity of the model.

References

1. Buczak, A., Guven, E.: A survey of data mining and machine learning methods for cyber security intrusion detection. IEEE Commun. Surv. Tutorials, vol. PP, no. 99, p. 1, 2015
2. Patel, A., Taghavi, M., Bakhtiyari, K., Celestino Jr J.: An intrusion detection and prevention system in cloud computing: a systematic review. J. Netw. Comput. Appl. 36(1), 25–41 (2013)
3. Liao, H.-J., Lin, C.-H.R., Lin, Y.-C.: Intrusion detection system: a comprehensive review. J. Netw. Comput. Appl. 36(1), 16–24 (2012)
4. Liao, H.J., Richard Lin, C.H., Lin, Y.C., Tung, K.Y.: Intrusion detection system: a comprehensive review. J. Netw. Comput. Appl. 36(1), 16–24 (2013)
5. Tsai, C., Hsu, Y., Lin, C., Lin, W.: Expert systems with applications intrusion detection by machine learning: a review. Expert Syst. Appl. 36(10), 11994–12000 (2009)
6. Fossaceca, J.M., Mazzuchi, T.A., Sarkani, S.: MARK-ELM: Application of a novel multiple kernel learning framework for improving the robustness of network intrusion detection. Expert Syst. Appl. 42(8), 4062–4080 (2015)
7. Mishra, P., Pilli, E.S., Varadharajan, V., Tupakula, U.: Intrusion detection techniques in cloud environment: a survey. J. Netw. Comput. Appl. 77, pp. 18–47, October 2016
8. Jaiganesh, V., Sumathi, P.: Kernelized extreme learning machine with levenberg-marquardt learning approach towards intrusion detection. Int. J. Comput. Appl. 54(14), 38–44 (2012)
9. Udaya Sampath, X.W., Perera Miriya Thanthrige, K., Samarabandu, J.: Machine learning techniques for intrusion detection. IEEE Can. Conf. Electr. Comput. Eng. 1–10 (2016)
10. Aslahi-Shahri, B.M., et al.: A hybrid method consisting of GA and SVM for intrusion detection system. Neural Comput. Appl. 27(6), 1669–1676 (2016)
11. Atefi, K., Yahya, S., Dak, A.Y., Atefi, A.: A Hybrid Intrusion detection system based on differen machine learning algorithms. In: Proceedings of the 4th International Conference Computing Informatics, no. 22, pp. 312–320 (2013)

12. Ding, S., Xu, X., Nie, R.: Extreme learning machine and its applications. Neural Comput. Appl. **25**(3–4), 549–556 (2014)
13. Singh, R., Kumar, H., Singla, R.K.: An intrusion detection system using network traffic profiling and online sequential extreme learning machine. Expert Syst. Appl. **42**(22), 8609–8624 (2015)
14. Ali, M.H., Zolkipli, M.F., Mohammed, M.A., Jaber, M.M.: Enhance of extreme learning machine-genetic algorithm hybrid based on intrusion detection system. J. Eng. Appl. Sci. **12** (16), 4180–4185 (2017)
15. Lu, H., Du, B., Liu, J., Xia, H., Yeap, W.K.: A kernel extreme learning machine algorithm based on improved particle swam optimization. Memetic Comput. **9**(2), 121–128 (2017)
16. Huang, G.-B., Zhou, H., Ding, X., Zhang, R.: Extreme learning machine for regression and multiclass classification. IEEE Trans. Syst. man, Cybern. Part B, Cybern. **42**, (2), 513–529 (2012)
17. Pal, M., Maxwell, A.E., Warner, T.A.: Kernel-based extreme learning machine for remote-sensing image classification. Remote Sens. Lett. **4**(9), 853–862 (2013)
18. Liu, B., Tang, L., Wang, J., Li, A., Hao, Y.: 2-D defect profile reconstruction from ultrasonic guided wave signals based on QGA-kernelized ELM. Neurocomputing **128**, 217–223 (2014)
19. Deng, W.Y., Zheng, Q.H., Wang, Z.M.: Cross-person activity recognition using reduced kernel extreme learning machine. Neural Netw. **53**, 1–7 (2014)
20. Chen, H.L., Wang, G., Ma, C., Cai, Z.N., Bin Liu, W., Wang, S. J.: An efficient hybrid kernel extreme learning machine approach for early diagnosis of Parkinson's disease. Neurocomputing **184**, 131–144 (2016)
21. Chen, C., Li, W., Su, H., Liu, K.: Spectral-spatial classification of hyperspectral image based on kernel extreme learning machine. Remote Sens. **6**(6), 5795–5814 (2014)
22. Fu, H., Vong, C.-M., Wong, P.-K., Yang, Z.: Fast detection of impact location using kernel extreme learning machine. Neural Comput. Appl. 1–10 (2014)
23. Li, L., Wang, C., Li, W., Chen, J.: Hyperspectral image classification by AdaBoost weighted composite kernel extreme learning machines. Neurocomputing **275**, 1725–1733 (2018)
24. Wang, Y., Wang, A.N., Ai, Q., Sun, H.J.: An adaptive kernel-based weighted extreme learning machine approach for effective detection of Parkinson's disease. Biomed. Signal Process. Control **38**, 400–410 (2017)
25. Wang, M., et al.: Toward an optimal kernel extreme learning machine using a chaotic moth-flame optimization strategy with applications in medical diagnoses. Neurocomputing **267**, 69–84 (2017)
26. Li, X., Niu, P., Li, G.: An adaptive extreme learning machine for modeling NOx emission of a 300 MW circulating fluidized bed boiler (2017)
27. Li, G., Niu, P.: Combustion optimization of a coal-fired boiler with double linear fast learning network (2014)
28. Abadeh, M.S., Mohamadi, H., Habibi, J.: Design and analysis of genetic fuzzy systems for intrusion detection in computer networks. Expert Syst. Appl. **38**(6), 7067–7075 (2011)
29. Chen, T., Zhang, X., Jin, S., Kim, O.: Efficient classification using parallel and scalable compressed model and its application on intrusion detection. Expert Syst. Appl. **41**(13), 5972–5983 (2014)
30. Bolón-Canedo, V., Sánchez-Maroño, N., Alonso-Betanzos, A.: Feature selection and classification in multiple class datasets: an application to KDD Cup 99 dataset. Expert Syst. Appl. **38**(5), 5947–5957 (2011)
31. Mitchell, R., Chen, I.-R.: A survey of intrusion detection techniques. Comput. Secur. **12**(4), 405–418 (2014)

32. Tavallaee, M., Bagheri, E., Lu, W., Ghorbani, A.A.: A detailed analysis of the KDD CUP 99 data set. In: IEEE Symposium on Computational Intelligence for Security and Defense Applications CISDA 2009 (June 2009)
33. Engen, V., Vincent, J., Phalp, K.: Exploring discrepancies in findings obtained with the KDD Cup'99 data set. Intell. Data Anal. **15**(2), 251–276 (2011)
34. Hu, W., Gao, J., Wang, Y., Wu, O., Maybank, S.: Online adaboost-based parameterized methods for dynamic distributed network intrusion detection. IEEE Trans. Cybern. **44**(1), 66–82 (2014)
35. Weller-Fahy, D.J.: Network intrusion dataset assessment, p. 114 (2013)
36. Chou, T.-S., Fan, J., Fan, S., Makki, K.: Ensemble of machine learning algorithms for intrusion detection. In: 2009 IEEE International Conference System Man and Cybernetics, pp. 3976–3980 (2009)
37. Li, G., Niu, P., Duan, X., Zhang, X.: Fast learning network: a novel artificial neural network with a fast learning speed. Neural Comput. Appl. **24**(7–8), 1683–1695 (2014)
38. Guang-Bin, H., Qin-Yu, Z., Chee-Kheong, S.: Extreme learning machine: a new learning scheme of feedforward neural networks. In: Neural Networks, 2004. Proceedings. 2004 IEEE International Joint Conference, vol. 2, pp. 985–990. August 2004
39. Smola, A.J., Schölkopf, B.: Learning with Kernels. February 2002
40. Flynn, H., Cameron, S.: Proceedings of the 8th International Conference on Computer Recognition Systems CORES 2013, vol. 226 (2013)
41. Trelea, I.C.: The particle swarm optimization algorithm: Convergence analysis and parameter selection. Inf. Process. Lett. **85**(6), 317–325 (2003)
42. J. Blondin, "Particle swarm optimization: A tutorial," ... *Site Http//Cs. Armstrong. Edu/Saad/Csci8100/Pso Tutor. ...*, pp. 1–5, 2009
43. Sengupta, A., Bhadauria, S., Mohanty, S.P.: TL-HLS: Methodology for Low Cost Hardware Trojan Security Aware Scheduling with Optimal Loop Unrolling Factor during High Level Synthesis. IEEE Trans. Comput. Des. Integr. Circuits Syst. **36**(4), 660–673 (2017)
44. Mishra, V.K., Sengupta, A.: Swarm-inspired exploration of architecture and unrolling factors for nested-loop-based application in architectural synthesis. Electron. Lett. **51**(2), 157–159 (2015)
45. Sengupta, A., Bhadauria, S.: User power-delay budget driven PSO based design space exploration of optimal k-cycle transient fault secured datapath during high level synthesis. In: Proceedings of the International Symposium Quality Electronic Design ISQED, vol. 2015, no. 6, pp. 289–292 (2015)
46. Mishra, V.K., Sengupta, A.: MO-PSE: Adaptive multi-objective particle swarm optimization based design space exploration in architectural synthesis for application specific processor design. Adv. Eng. Softw. **67**, 111–124 (2014)
47. Shi, Y., Eberhart, R.: A modified particle swarm optimizer. In: Proceedings of the IEEE International Conference on Evolutionary Computation, IEEE World Congress on Computational Intelligence (Cat. No.98TH8360), pp. 69–73 (1998)

Vehicular Ad Hoc Network: An Intensive Review

Ayoob A. Ayoob[1], Gang Su[1(✉)], and Muamer N. Mohammed[2]

[1] School of Electronic Information and Communications, Huazhong University of Science and Technology, Wuhan 430074, People's Republic of China
{I201522060,gsu}@hust.edu.cn
[2] Information Technology Department, The Community College of Qatar, Doha, Qatar
muamer.scis@gmail.com

Abstract. Vehicular Ad-hoc Network (VANET) is a new emerging wireless technology concept that supports communication amongst various nearby vehicles themselves and enables vehicles to have access to the Internet. This networking technology provides vehicles with endless possibilities of applications, including safety, convenience, and entertainment. Examples of these applications are safety messaging exchange, real-time traffic information sharing, route condition updates, besides a general purpose Internet access. The goal of vehicular networks is to provide an efficient, safe, and convenient environment for vehicles on the road. In this paper some wireless access standards for Vehicular Ad hoc Network (VANET) and describe was present, this paper starts with the basic architecture of networks, then discusses some of the recent VANET trials, also briefly present some of the simulators currently available to VANET researchers. Finally, discusses the popular research issues and general research methods, and ends up with the analysis of challenges and future trends of VANETs.

Keywords: Vehicles to roadside (VRC) · Vehicle to infrastructure (V2I)
Vehicular networks · Vehicular Ad-hoc network (VANET)

1 Introduction

In Vehicle Ad hoc networks (VANET), data can be communicated among vehicles or between vehicles and roadside units (RSUs), [1, 2]. VANET utilises different Ad-hoc networking technology, such as Wi-Fi, IEEE802.11 b/g, WiMAX IEEE 802.16, Bluetooth, IRA, and ZigBee for an easy, accurate, effective, and simple communication between vehicle under dynamic mobility [2].

These VANET have several properties like dynamic topology, limited bandwidth, limited energy and many more. Vehicular ad hoc network (VANET) is a subclass of MANET with some unique properties. VANETs have emerged out these days due to the need for supporting the increased number of wireless equipment that can be used in vehicles [3, 4]. Some of these products are global positioning system, mobile phones and laptops. As mobile wireless equipment and networks become increasingly

© Springer Nature Switzerland AG 2019
P. Vasant et al. (Eds.): ICO 2018, AISC 866, pp. 158–167, 2019.
https://doi.org/10.1007/978-3-030-00979-3_16

important, the demand for Vehicle-to-Vehicle (V2 V) and Vehicles-to-Roadside (VRC) or Vehicle-to-Infrastructure (V2I) Communication will continue to grow [5, 6].

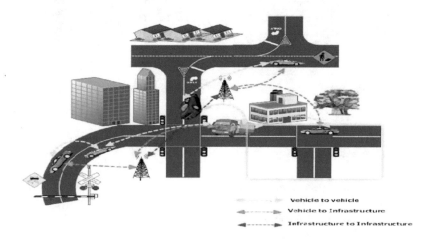

Fig. 1. A vehicular Ad-hoc network

VANETs have some dissimilar properties then MANETs like road pattern restrictions, no restriction on network size, dynamic topology, mobility models, and infinite energy supply, localization functionality and so on [7]. All these characteristics made VANET environment challenging for developing efficient routing protocols. The major factor in it is the fast moving mobile nodes [8].

1.1 VANET Architecture

There are two methodologies can be used to establish the feasible connection for VANET. as shown in Fig. 1a, the first methodology called the layered approach the provides the Required capacities and conventional layers with considerably charac-terized interfaces between them [3]. In these approach framework functionalities are adjusted to satisfy the necessities of VANET correspondence framework such as those in conventional layers for single-jump and multi-bounce Correspondence for instance, each layer is executed as an independent module with interfaces, such as SAPs (service access points) [9], above and beneath the layers. Thus conventions cannot smoothly reach the state or metadata on an alternate layer and thus generates complex infor-mation some of the particular capacities of VANETs are unsuitable for a customary layered OSI model, such as models for system strength and control [10] such capacities cannot also be particularly distributed to a specific layer. Each layer should also be converted to outer data independently without a regular interface. The un-layered approach is the result of fitting a radically new framework to the requirement of the fundamental center of VANET, that is, well-being application.In this manner, all application and then combined with outer sensors as shown in Fig. 2b [11].

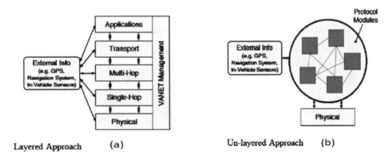

Fig. 2. Layered and un-Layered approaches for VANET

1.2   VANET Application

(A) *Efficiency and Safety Application*

The efficiency and safety Application are the two important requirements that can be used to classify VANET application by their primary purpose [12]. However, efficiency and safety are not completely separated from each other. Conversely, those and other aspects should be considered together in the design of VANET application [13]. For instance, an engine failure or an accident involving two or more vehicle can lead to a traffic jam. A message reporting this event conveys a safety warning for nearby drivers who use it to increase their awareness as shown in Fig. 3.

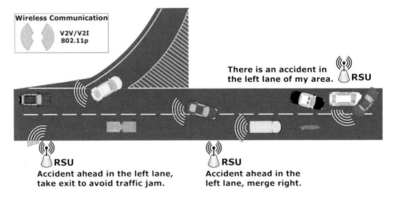

Fig. 3. VANET usage application for sent alarm car accident

While, all correspondence frameworks comprise a complex convention that relies on the assignment of correspondence frameworks, such as a web surfing mechanical control circle and cellular phone framework. The MAC layer is a sub layered of the joint information layer of the OSI reference convention on the layer is available in most correspondence system such as wireline and remote system [14]. A wide range of MAC convention has been developed, and such convention is described as dispute-based or strife-free convention. In contention-free convention,time division multiple access (TDMA) and frequency division multiple access (FDMA) are involved [15]. Despite their unique configuration, these conventions are limited by the requirement of focal instruments such as a base station or an enhance point that can share assists among client to designate time space or recurrence groups to the client.

(B) *Current Application of VANET*

Vehicle system has numerous applications, such as collision avoidance and agreeable driving. About 21,000 of the 23,000 vehicle that U.S roadside base station to a vehicle may warn the convergences [16]. The information transmitted from a roadside base station to a vehicle may warn the driver about the dangers of ending a crossing point. The correspondence between vehicle and between vehicle and roadside can save many lives and help drivers anticipate accident [17]. The worst automobile collisions involve numerous vehicles colliding with one another as a result of a solitary accident at the front of the line. In this case, by gradually decreasing its speed, the vehicle at the front can communicate effectively with its neighbouring vehicle and prevent future mishap [18, 19]. The initial recipients of this message will then handoff the message further by warning the vehicle behind them about the mishap ahead. Nevertheless, in Fig. 4 shown the VANET standard application classification.

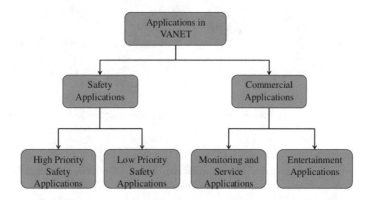

Fig. 4. Standard VANET applications

1.3 VANET Infrastructure

In the V2I protocol the infrastructure plays a coordination role by gathering global or local information on traffic and road conditions and then suggesting or imposing certain behaviors on a group of vehicles [20].

When two electronic systems communicate autonomously, That is to say without human intervention, the process is described as Machine-to-Machine (M2M) communications [20]. The main goal of M2M communication is to enable the sharing of information between electronic systems autonomously. Given the rapid reception of remote advancements, the omnipresence of electronic control frameworks, and the multifaceted nature of programming frameworks, remote M2M as attracted much education from the industry and the academic community [21, 22]. M2M correspondence incorporate wired interchanges, but recent studies have only focused on remote M2M interchanges [11]. The quantity of remote gadgets(excluding cellular phone) that work without human communication (e.g., climate station and power meters) has reached 1.5 billion in 2014 [12]. Recent studies on M2M correspondence have explored the regions that incorporate system Vitality effectiveness and green system administration, the tradeoff between gadget power utilisation and gadget insight or handling power, the institutionalisation of interchange, information collection and data transfer capacity, protection and security, and system versatility [23] (Fig. 5).

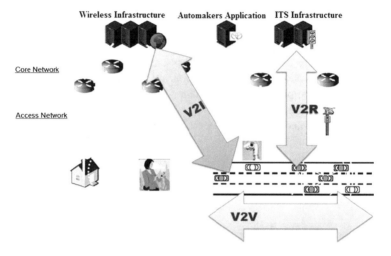

Fig. 5. VANET Infrastructure usage

Although M2M engineering alludes to the number of imparting machine (European Telecommunication Standards Institute, ETSI). M2M standards specifically apply to those systems that use countless application, even to 1.5 billion remote gadgets, without bounds. The M2M application is often discussed at a national or worldwide scale, and a large number of sensors are halfway planned [24]. Therefore, the vehicle system is assessed as M2M designs with numerous associated gadgets, some of which may not be identified with vehicle systems. The occurrence of traffic delays continues to increase infrequently, and individuals waste approximately 40 h every week stuck in traffic. Vehicle system can help reduce the frequency of these delays. The vehicle can act as information authorities that transmit activity condition data to the vehicle system. Transportation offices can use these data to case the activity clog. Specifically, the vehicle can recognize if the quantity of neighboring vehicle is excessively numerous and if their speed is too moderate. They can also transfer these data to that vehicle that is approaching the area. These data can also transfer to those vehicles that are heading in other directions to accelerate the spread of information near the area of the accident [25, 26]. Therefore, those vehicles that approach the clog area will have enough time to change course. This vehicle can also gather information about climate, street surface, development zones, roadway rail convergence, and Crisis vehicle signal appropriation, and then transfer these data to different vehicles (Fig. 6).

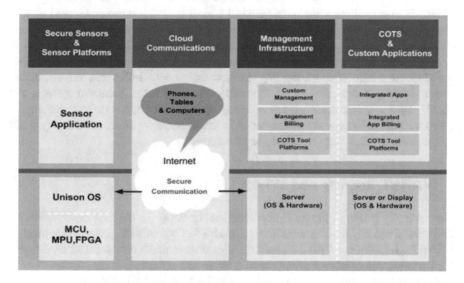

Fig. 6. A generic M2M architecture and M2M application area

It has noticed on this paper Vehicular systems can influence the M2M worldview to bolster vehicle correspondences as shown in Fig. 7. We displayed a brief review of

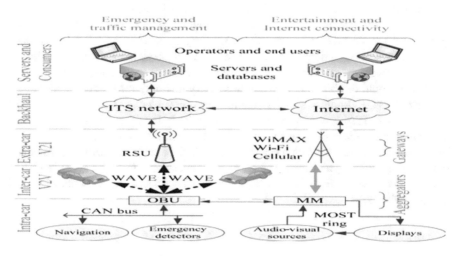

Fig. 7. M2M communication layers in vehicular network

probably the latest application ranges of M2M, to be specific brilliant framework innovation, home systems administration, medicinal services, and vehicular systems administration [27].

At the point when two electronic frameworks are imparted autonomously, i.e., without human mediation, the procedure is depicted as M2M correspondences [26, 27]. The primary objective of M2M interchanges is to empower the sharing of data between electronic framework automatically because of this rise and quick reception of remote advancements, the omnipresence of electronic control framework, and the expanding multifaceted nature of programming framework, remote M2M has been drawing attention from the industry and the educated community users. Vehicular systems and M2M interchanges are correlative developing fields of exploration that have developed. Late mechanical patterns show that vehicular correspondences could essentially profit by the advancement made in M2M interchanges.

2 VANET Security and Privacy

Since security is one of the most important requirements for VANET communication [28] several security requirements have to be satisfied before widespread deployment of VANETs. Life-critical VANET characteristics always make new security challenges. Furthermore, safety-related driving applications and services require stricter security requirements than other services and applications in VANETs, VANETs require a strong authentication method that preserves the anonymity of the drivers while the authorities should be able to identify the real identity of a message sender whenever necessary. This type of privacy methods is called conditional privacy. Moreover, this ability should not be given to a single entity since the authority could abuse its rights to violate others' privacy. Hence, this conditional tracing ability should be distributed over several authorities, and they should collectively be able to reveal the required identification information of the source vehicle when some disputes happen, security requirement for VANET is non-repudiation, in which none of the vehicles can deny their communication actions. Hence, some kinds of mechanisms should be implemented to prevent vehicles from denying their messages. Furthermore, the recipient of a message should be able to convince a third party that only the specific sender could send the message. Another security requirement for vehicular communication is the ability to detect false communication messages. One of the main challenges in secure vehicular communication is properly balancing authentication and privacy [28]. Figure 8 Shown the classification categorize an attack based the security requirement it tries to compromise. The major categories are threat and attacks on Confidentiality, Integrity and Availability. Other categories include attacks on authentication and accountability [28].

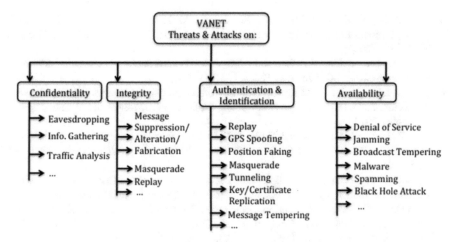

Fig. 8. VANET attacks classification and example

3   Conclusion

In this paper, we presented ideas that are key to the development of M2M interchange frameworks. We provided a brief diagram of four notable M2M applications, specifically in human services, home organizing, effective framework, and vehicular systems administration. We then focused on vehicular systems administration by presenting and M2M perspective on inter-vehicular correspondence and highlighted the particular application, requirement, and convention related to the M2M-based vehicle system. To set up where M2M interchange could increase the value of vehicle systems, we reviewed the literature exhaustively to the extent that both advances is met. Five range where distinguished namely, multiplicity, autonomy, connectivity, visualisation, and security. Lastly, we investigated the issues encountered by vehicle M2M framework. The best test is the institutionalization of correspondence interface in a system with high versatility and variability of parts. Ensuring security and protection in such a dynamic system requires further consideration.

References

1. Mohammed, M.N., Hammood, O.A.: Hybrid LTE-VANETs based optimal radio access selection. In: Recent Trends in Information and Communication Technology: Proceedings of the 2nd International Conference of Reliable Information and Communication Technology (IRICT 2017), vol. 5, p. 189. Springer (2017)
2. Singh, S., Agrawal, S.: VANET routing protocols: issues and challenges. In: 2014 Recent Advances in Engineering and Computer Sciences RAECS 2014, pp. 6–8 (2014)
3. Kim, T.M.C.Y., Kang, D.M., Lee, J.H.: A performance evaluation of cellular network suitability for VANET. World Acad. Sci. Eng. Technol. Int. Sci. Index **64**(6), 1023–1026 (2014)

4. Nemoto, Y., Taleb, T., Sakhaee, E., Jamalipour, A., Hashimoto, K., Kato, N.: A stable routing protocol to support its services in vanet networks. In: IEEE Trans. Veh. Technol. 3337–3347 (2007)
5. Chen, S., Hu, J., Shi, Y., Zhao, L.: LTE-V: a TD-LTE-based V2X solution for future vehicular network. IEEE Internet Things J. **3**(6) (2016)
6. He, Y., Li, C., Lin, H., Zhu, L.: Accident driver model for vehicular ad-hoc network simulation. In: 2013 IEEE Vehicle Power and Propulsion Conference, vol. 21, no. 8, pp. 1–5, October 2013
7. Mir, Z.H., Filali, F.: LTE and IEEE 802.11p for vehicular networking: a performance evaluation. EURASIP J. Wirel. Commun. Netw. **89** (2014). https://doi.org/10.1186/1687-1499-2014-89
8. Sepulcre, M., Gozalvez, J., Coll-Perales, B., Lucas-Estañ, M.C., Gisbert, J.R.: Empirical performance models for V2 V communications. In: Proceedings-15th IEEE International Conference on Computer and Information Technology, CIT 2015, 14th IEEE International Conference on Ubiquitous Computing and Communications, IUCC 2015, 13th IEEE International Conference on Dependable, Autonomic and Secure, October 2015, pp. 737–742 (2015)
9. Iera, A., Molinaro, A., Araniti, G., Campolo, C., Condoluci, M.: LTE for vehicular networking: a survey. IEEE Commun. Mag. **51**(2), 148–157
10. Uhlemann, E.: Introducing connected vehicles [connected vehicles]. IEEE Veh. Technol. Mag. **10**, 23–31 (2015)
11. Füßler, H., Torrent-moreno, M., Transier, M., Festag, A., Hartenstein, H.: Thoughts on a protocol architecture for vehicular ad-hoc networks. In: 2nd International Workshop on Intelligent Transportation, WIT, TR-02–003, pp. 1–5 (2005)
12. Hemakumar, V., Nazini, H.: Optimized traffic signal control system at traffic intersections using vanet. IOSR J. Comput. Eng. **15**(3), 36–43 (2013)
13. Mohammad, M.N., Sulaiman, N.: A new broadcast algorithm to optimize routing protocol in mobile ad hoc networks. J. Appl. Sci. **13**, 588–594 (2013)
14. Hoydis, M.D.J., Kobayashi, M.: Green small-cell networks. IEEE Veh. Technol. Mag. **6**(10), 37–43 (2011)
15. Mehmood, A., Khanan, A., Mohamed, A.H.H.M., Song, H.: ANTSC: an intelligent naïve Bayesian probabilistic estimation practice for traffic flow to form stable clustering in VANET. IEEE Access, **3536**, 1–1 (2017)
16. Mohammed, M.N., Kadhim, N.S., Ahmed, W. Kh.: An energy efficient multipath routing protocol based on signal strength for mobile ad-hoc network. ARPN J. Eng. Appl. Sci. **11**, 11 (2016)
17. Kadhim, N.S., Mohammed, M.N., Majid, M.A., Mohamd, S.Q., Tao, H.: An efficient route selection based on AODV algorithm for VANET. Indian J. Sci. Technol. **38**(9), 1–6 (2016)
18. Kohls, S., Scholz-Böttcher, B.M., Teske, J., Rullkötter, J.: Isolation and quantification of six cardiac glycosides from the seeds of Thevetia peruviana provide a basis for toxological survey. Indian J. Chem. Sect. B Org Med. Chem. **54B**(12), 1502–1510 (2015)
19. Pieroni, A., Scarpato, N., Brilli, M.: Performance study in autonomous and connected vehicles a industry 4.0 issue. J. Theor. Appl. Inf. Technol. **96**(4) (2018)
20. Pieroni, A., Scarpato, N., Brilli, M.: Industry 4.0 Revolution in autonomous and connected vehicle a non-conventional approach to manage big data. J. Theor. Appl. Inf. Technol. **96**(1) (2018)
21. Elmangoush, A., Coskun, H., Wahle, S., Magedanz, T.: Design aspects for a reference M2M communication platform for Smart Cities. In: 2013 9th International Conference on Innovations in Information Technology (IIT), pp. 204–209. IEEE (2013)
22. He SAFESPOT project. www.safespot-eu.org

23. Chen, Y, Wang, W.: Machine-to-machine communication in LTE-A. In: Vehicular Technology Conference Fall (VTC 2010-Fall), 2010 IEEE 72nd, pp. 1–4. IEEE (2010)
24. Swetina, Jorg, Guang, Lu, Jacobs, Philip, Ennesser, Francois, Song, JaeSeung: Toward a standardized common M2M service layer platform: introduction to one M2M. IEEE Wirel. Commun. **21**(3), 20–26 (2014)
25. Institute of Electrical Electronics Engineers 1609 Working Group Public Site. http://vii.path. berkeley.edu/1609/wave/
26. Nasir, M.K., Hossain, A.S.M.D., Hossain, S., Hasan, M., Ali, B.: Security challenges and implementation mechanism for vehicular ad hoc network, vol. 2, no. 4 (2013)
27. Isaac, J.S.C.J.T., Zeadally, S.: Security attacks and solution for vehicular ad hoc networks. IET Commun. **4**(7), 894–903 (2015)
28. Terroso-sáenz, F., Valdés-vela, M., Sotomayor-martínez, C., Toledo-moreo, R., Gómez-skarmeta, A.F.: Detection with complex event processing and VANET. IEEE Trans. Intell. Transp. Syst. **13**(2), 914–929 (2012)

Optimization of the Parameters of the Elastic Damping Mechanism in Class 1,4 Tractor Transmission for Work in the Main Agricultural Operations

Sergey Senkevich[1(✉)], Vladimir Kravchenko[2], Veronika Duriagina[3], Anna Senkevich[2], and Evgeniy Vasilev[1]

[1] Federal Scientific Agroengineering Center VIM, Moscow, Russia
sergej_senkevich@mail.ru, evgvasilev2008@yandex.ru
[2] Azov-Black Sea Engineering Institute of Federal State Budgetary Establishment of Higher Education, Don State Agrarian University, Zernograd, Rostov Region, Russia
a3v2017@yandex.ru, anna-senkev@mail.ru
[3] Southern Federal University, Taganrog, Russia
vepanuka@mail.ru

Abstract. The article deals with finding the optimum parameters of the elastic damping mechanism (EDM) in transmission of class 1,4 tractor. The tractor was used in structure of three various machine-tractor units and carried out the main agricultural operations: plowing, cultivation and cropping. EDM is intended for smooth start-off of the unit, decrease in dynamic loadings in transmission, protection of the engine against fluctuations of external loading. The indicator – "degree of the transmission transparency" is used for estimating the protective quality of the mechanism. The research was conducted by methods of experiment planning. The central composite plan of the second order for five factors has been chosen. The regression model expressing influence of key parameters of EDM on function of a response ("degree of transmission transparency") is received. The dependence of a response function from each factor, their mutual influence on the studied process is treated apart in details. The system of the differential equations in private derivatives has been received for finding the optimum values of factors and response function. The conclusion that the optimum value of the EDM parameters allow to increase quality of functioning of MTU on the main agricultural operations was made by results of the analysis of a computing experiment.

Keywords: Elastic damping mechanism · Transmission
Volume of hydropneumatic accumulator · The optimal parameters

1 Introduction

When the tractor performs agricultural operations: plowing, cultivation and sowing, the most significant factor affecting the operation of machine-tractor units (MTU) is the traction load. Vibrations of the traction loads lead to transient processes in the motor.

© Springer Nature Switzerland AG 2019
P. Vasant et al. (Eds.): ICO 2018, AISC 866, pp. 168–177, 2019.
https://doi.org/10.1007/978-3-030-00979-3_17

Scientific studies [1, 2, 4–6] prove the effectiveness of the inclusion of elastic coupling in the individual mechanisms of the tractor. The use of elastic damping mechanism (EDM) in the transmission of the tractor leads to absorption and scattering of the oscillations energy of the traction load. This reduces the dynamic loading of the drive, towing movers and so on.

One of such mechanisms is EDM [1, 3]. This mechanism is intended for smooth moving of MTU, reducing of dynamic loads in the transmission, protecting of the engine against fluctuations of external load [3]. The mechanism is developed at the "Tractors and cars" department of Azov-blacksea engineering institute. Currently, the team of scientists conducts research to improve the functioning of this device [3].

The indicator P – "degree of transmission transparency" is proposed to assess the protective qualities of the mechanism, in the form of the ratio of the current amplitude of the vibration of the motor shaft speed to its maximum value [4]. When $N = 1$ the gear unit (gearbox) is absolutely "transparent", the engine remains unprotected against fluctuations of the traction load (this happens in serial transmissions). When $N = 0$, the gear unit is absolutely "opaque" and will completely extinguish the vibrations transmitted to the engine. Professor A. B. Lurie, and professor N. M. Bespamyatnova [6, 7], considered frequency and traction characteristics as the additional influencing factor in the researches of the units working at the main agricultural operations.

2 Main Part

Purpose of Research. The aim of this work is to obtain the optimal parameters of the elastic damping mechanism in the MTU transmission of a class 1,4 tractor on plowing, cultivation and sowing. The following tasks were solved to achieve this goal:

1. identification of significant factors affecting the work of EDM;
2. the choice of the experiment plan and obtaining experimental results;
3. obtaining a regression model expressing the influence of the main parameters of EDM on the "degree of transmission transparency";
4. finding the optimal values of significant factors.

Materials and Methods. The main factors for the research were selected based on the works [1–4] and the conducted by the authors of this article.
The central composite plan of the second order meeting the requirement of rotatability was chosen for research. In contrast to the non-composite plan the selected plan for the five factors can reduce the number of conducted experiments [8].

The description and operation of the device installed in the transmission of the tested tractor is described in the source [3].

The research was carried out using the system of automatic accumulation and processing of metrological information of mobile execution (SAAP), developed at the Federal scientific center Donskoy (Zernograd). The system consists of a set of hardware and software and includes: an on-board computer, an analog-to-digital conversion board "code-figure" (ADC), an interface card, an instrumentation amplifier. All the equipment was installed in the cabin of the mobile strain-gauge laboratory TL-2 on the basis of the all-wheel drive truck car. The set of sensors (primary converters) installed on the unit under test allows measuring instantaneous values of the following energy parameters: traction resistance; torque on the axis of the driving wheel; frequency of the shaft rotation of the tractor generator; impulses of turns of a tractor driving wheel; impulses of revolutions of a tram tire wheel; oil pressure in the hydrolysis line (up to the throttle); rotation frequency of a gear wheel of the oil pump drive; fuel consumption.

The results (electronic oscillograms) of the experiment were processed on a personal computer using a software package for conversion.

The Result and Discussion. The levels and intervals of variation given in Table 1 were selected based on the results of the search experiments.

Table 1. Levels and intervals of factors variation for the plowing unit

N	Factor name	Factor identification	Code mark	Variation interval	Natural values corresponding to coded levels of the factors		
					Upper (+1)	Base (0)	Lower (−1)
1	The throttle cross-sectional area	S_{th}, m^2	X_1	$1,1 \cdot 10^{-4}$	$3,524 \cdot 10^{-4}$	$2,424 \cdot 10^{-4}$	$1,324 \cdot 10^{-4}$
2	Volume of hydropneumatic accumulator (HPA)	V_{hpa}, m^3	X_2	$1,23 \cdot 10^{-3}$	$5,079 \cdot 10^{-3}$	$3,848 \cdot 10^{-3}$	$2,617 \cdot 10^{-3}$
3	The air pressure in HPA	P_a, Pa	X_3	$1 \cdot 10^5$	$5 \cdot 10^5$	$4 \cdot 10^5$	$3 \cdot 10^5$
4	Inertia moment of additional load	J_{th}, $kg \cdot m^2$	X_4	$2,32 \cdot 10^{-3}$	$6,96 \cdot 10^{-3}$	$4,64 \cdot 10^{-3}$	$2,32 \cdot 10^{-3}$
5	The oscillation frequency of the traction load	f, Hz	X_5	$0,3$	$1,2$	$0,9$	$0,6$

Modeling was carried out under steady-state load and accelerating modes to study the influence of elastic damping mechanism on the performance of MTU during plowing, cultivation and sowing. The field plot was horizontal with a slope of not more than 2 degree. The movement of the unit was strictly rectilinear. A tractor with a tool was moving in sixth gear of the main range of speeds gearbox. The indicator P (response function Y) was calculated on the basis of the experimental data.

Expressions from (1) to (16) are standard statistical tools for research in the field of performance parameters optimization. Statistical methods for forecasting is a comprehensive, readable treatment of statistical models used to produce forecasts. The detailed technique is described in the sources [8–10]. The optimum region with sufficient accuracy is most often possible to describe by a polynomial of the second degree. The scheme of experiment planning was chosen to determine the optimization parameter. The object of study was preliminary studied on the basis of a priori information. The priori information was obtained by not only studying the literary data, but also analyzing the results of previous work.

A polynomial of the second degree adequately describes the optimal region, was used for the approximation of the response function:

$$y = b_0 + \sum_{1 \leq i \leq k} b_i x_i + \sum_{1 \leq i < l \leq k} b_{il} x_i x_l + \sum_{1 \leq i \leq k} b_{ii} x_i^2. \tag{1}$$

Regression analysis and mathematical statistics methods were used to check the adequacy of the model used and to find the optimal value of the response function.

The coefficients of the regression Eq. (1) were calculated by the formula:

$$B = \left(X^T \cdot X \right)^{-1} \cdot X^T \cdot Y, . \tag{2}$$

X — matrix based on the used experimental plan [8], Y — response matrix.

The found coefficients of the regression Eq. (2) were checked for statistical significance. For this purpose the Student's t-test was used with the construction of confidence intervals Δb_i based on the found values of the variance of the experiment reproducibility Dy and the dispersion-covariance matrix D_k

$$Dy = \frac{\sum_{i=1}^{n_0} (Y_i - \bar{Y})^2}{n_0}. \tag{3}$$

$\bar{Y}$ — average response value for points in the center of the plan, $n_0 = 6$ [5].

$$Dk = Dy \cdot \left(X^T \cdot X \right)^{-1}, . \tag{4}$$

$$\Delta b_i = \sqrt{Dk_{i,i}} \cdot t, . \tag{5}$$

t — tabular value of the Student's coefficient at the significance level $\alpha = 0{,}05$ and the number of degrees of freedom $n_0 - 1$ [8].

The absolute values of the coefficients b_1, b_2, b_4, $b_7 - b_{10}$, b_{13}, b_{16}, b_{17}, b_{19} and b_{20} were lower than the corresponding confidence intervals, so they can be considered statistically insignificant and excluded from Eq. (1).

The mathematical model has next form taking into account the obtained data:

$$y(x_1, x_2, x_3, x_4, x_5) = 0.06538x_3 + 0.06613x_5 + 0.04968x_3^2 - 0.06994x_1x_2$$
$$+ 0.04868x_2x_4 - 0.03494x_2x_5 - 0.04581x_3x_5 + 0.6212, . \tag{6}$$

The value and sign of the coefficients show the contribution of the relevant factors in the overall result – an indicator of the degree of transparency – in the transition to another level.

The resulting mathematical model was tested for adequacy using the Fisher criterion [5]

$$Fp = \frac{Da}{Dy}, \tag{7}$$

D_a – adequacy variance

$$Da = \frac{\sum_{i=1}^{N}(MD)^2 - \sum_{i=1}^{n_0}(Y_i - \bar{Y})^2}{N - \acute{k} - (n_0 - 1)}, \tag{8}$$

$$MD = Y - Yr, \tag{9}$$

$$Yr = X \cdot B, \tag{10}$$

Yr – calculated values of the response matrix taking into account the significant coefficients.

Calculations established that the Fisher coefficient for Eq. (6) is equal to $F_p = 4,217$. This value is less than table value $F_t = 4,44$ [8]. Therefore, the model is adequate.

Analyzing the Eq. (6), we conclude that the x_3 factor corresponding to the air pressure in GPA exerts the greatest influence on the degree of transmission transparency.

Figure 1 presents graphs of the dependence of the degree of transmission transparency from the traction load vibrations and the air pressure in the GPA. The response surface is shown in the left graph and the level lines in the right graph. Each level line is an equal response line that corresponds to the projection of the response surface section with the plane y = const.

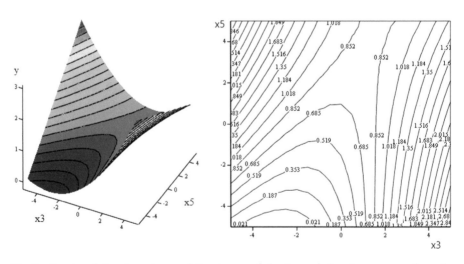

Fig. 1. Graphs of the dependence of the degree of the transmission transparency from the traction load vibrations and the air pressure in the GPA for plowing unit (surface response on the left, level lines on the right)

The system of partial differential equations was obtained to find the optimal values of the factors:

$$\frac{\partial}{\partial x_1} y(x_1, x_2, x_3, x_4, x_5) = -0.06993 x_2$$

$$\frac{\partial}{\partial x_2} y(x_1, x_2, x_3, x_4, x_5) = 0.04868 x_4 - 0.03494 x_5 - 0.06994$$

$$\frac{\partial}{\partial x_3} y(x_1, x_2, x_3, x_4, x_5) = 0.09937 x_3 - 0.04581 x_5 - 0.06537 x_1, \qquad (11)$$

$$\frac{\partial}{\partial x_4} y(x_1, x_2, x_3, x_4, x_5) = 0.04868 x_2$$

$$\frac{\partial}{\partial x_5} y(x_1, x_2, x_3, x_4, x_5) = -0.03494 x_2 - 0.04581 x_3 + 0.06613$$

We determined the optimal values of each factor solving the system (11). Converting coded values $x_1 - x_5$ to the natural values of the factors was produced by the formulas:

$$x_1 = \frac{S_{th} - 2.424 \cdot 10^{-4}}{1.1 \cdot 10^{-4}}; x_2 = \frac{V_{hpa} - 3.848 \cdot 10^{-3}}{1.231 \cdot 10^{-3}}; x_3 = \frac{P_a - 4 \cdot 10^5}{1 \cdot 10^5};$$
$$x_4 = \frac{J_{ew} - 4.64 \cdot 10^{-3}}{2.32 \cdot 10^{-3}}; x_5 = \frac{f - 1.3}{0.5}, \qquad (12)$$

Similar studies have been conducted to find the optimal parameters of the cultivator and seeder.

The regression equation for the cultivator unit has the form:

$$y(x_1, x_2, x_3, x_4, x_5) = 0.578 + 2.29 \cdot 10^{-3} x_1 - 3.72 \cdot 10^{-3} x_2 +$$
$$+ 0.053 x_3 + 4.51 \cdot 10^{-3} x_4 + + 0.021 \cdot x_5 - 1.34 \cdot 10^{-3} x_1 x_2 -$$
$$- 3.14 \cdot 10^{-3} x_1 x_3 - 2.41 \cdot 10^{-3} x_1 x_4 + 3.32 \cdot 10^{-4} x_1 x_5 -$$
$$- 1.25 \cdot 10^{-3} x_2 x_3 + 1.37 \cdot 10^{-3} x_2 x_4 - 0.044 \cdot 10^{-3} x_2 x_5 -$$
$$- 6.04 \cdot 10^{-5} x_3 x_4 - 0.035 x_3 x_5 + 8.09 \cdot 10^{-3} x_4 x_5 +$$
$$+ 0.03 x_1^2 + 0.031 x_3^2 + 0.03 x_4^2 + 0.095 x_5^2. \qquad (13)$$

$$x_1 = \frac{S_{th} - 2.424 \cdot 10^{-4}}{1.1 \cdot 10^{-4}}; x_2 = \frac{V_{hpa} - 3.848 \cdot 10^{-3}}{1.231 \cdot 10^{-3}}; x_3 = \frac{P_a - 4 \cdot 10^5}{1 \cdot 10^5};$$
$$x_4 = \frac{J_{ew} - 4.64 \cdot 10^{-3}}{2.32 \cdot 10^{-3}}; x_5 = \frac{f - 0.9}{0.3}, \qquad (14)$$

The system of differential equations was obtained taking partial derivatives of the Eq. (13). The optimal values of the factors of the elastic damping mechanism on cultivation were found after solving the system and moving from the coded values

$x_1 - x_5$ to the natural values of the factors according to the formulas (14). We obtained the response surface shown in Fig. 2 substituting (14) in (13) and analyzing the resulting equation.

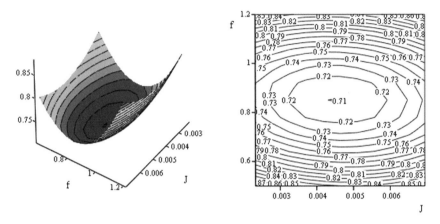

Fig. 2. Graphs of the dependence of the degree of transmission transparency from the traction load vibrations and the moment of inertia of the EDM drive for cultivation (surface response on the left, level lines on the right).

It is necessary to change the indicator x_5 to find the optimal parameters of the elastic damping mechanism for the sowing unit in Table 1. Indicators $x_1 - x_4$ remained unchanged. The values are given in Table 2 [6, 7].

The regression equation for the sowing unit has the form

$$
\begin{aligned}
y(x_1, x_2, x_3, x_4, x_5) = {}& 0.722 - 6.325 \cdot 10^{-3} x_1 - 6.657 \cdot 10^{-3} x_2 \\
& + 0.033 x_3 - 5.526 \cdot 10^{-3} x_4 - 3.072 \cdot 10^{-3} x_5 + 6.981 \cdot 10^{-3} x_1 x_2 \\
& + 6.851 \cdot 10^{-3} x_1 x_3 + 5.709 \cdot 10^{-3} x_1 x_4 \\
& - 7.045 10^{-3} x_1 x_5 + 8.462 \cdot 10^{-3} x_2 x_3 + 7.245 \cdot 10^{-3} x_2 x_4 \\
& - 5.554 \cdot 10^{-3} x_2 x_5 + 7.115 \cdot 10^{-3} x_3 x_4 + 2.725 \cdot 10^{-3} x_3 x_5 \\
& - 6.826 \cdot 10^{-3} x_4 x_5 + 2.905 \cdot 10^{-3} x_1^2 + 2.466 \cdot 10^{-3} x_2^2 \\
& - 7.031 \cdot 10^{-3} x_3^2 3.445 \cdot 10^{-3} x_4^2 + 0.038 x_5^2
\end{aligned}
$$

$$(15)$$

The transition from coded values $x_1 - x_5$ to natural values of factors is carried out by formulas according to the known method

Table 2. Levels and intervals of variation of factors of the seed unit

N	Factor name	Factor identification	Code mark	Variation interval	Natural values corresponding to coded levels of the factors		
					Upper (+1)	Base (0)	Lower (−1)
1	The throttle cross-sectional area	S_{th}, m^2	X_1	$1,1 \cdot 10^{-4}$	$3,524 \cdot 10^{-4}$	$2,424 \cdot 10^{-4}$	$1,324 \cdot 10^{-4}$
2	Volume of hydropneumatic accumulator (HPA)	V_{hpa}, m^3	X_2	$1,23 \cdot 10^{-3}$	$5,079 \cdot 10^{-3}$	$3,848 \cdot 10^{-3}$	$2,617 \cdot 10^{-3}$
3	The air pressure in HPA	P_a, Pa	X_3	$1 \cdot 10^5$	$5 \cdot 10^5$	$4 \cdot 10^5$	$3 \cdot 10^5$
4	Inertia moment of additional load	J_{th}, $kg \cdot m^2$	X_4	$2,32 \cdot 10^{-3}$	$6,96 \cdot 10^{-3}$	$4,64 \cdot 10^{-3}$	$2,32 \cdot 10^{-3}$
5	The oscillation frequency of the traction load	f, Hz	X_5	$0,5$	$1,8$	$1,3$	$0,8$

$$x_1 = \frac{S_{th} - 2.424 \cdot 10^{-4}}{1.1 \cdot 10^{-4}}; x_2 = \frac{V_{hpa} - 3.848 \cdot 10^{-3}}{1.231 \cdot 10^{-3}}; x_3 = \frac{P_a - 4 \cdot 10^5}{1 \cdot 10^5};$$
$$x_4 = \frac{J_{ew} - 4.64 \cdot 10^{-3}}{2.32 \cdot 10^{-3}}; x_5 = \frac{f - 1.3}{0.5}, \tag{16}$$

We obtain the response surface substituting (16) in (15) and analyzing the resulting equation. The data is presented in Fig. 3.

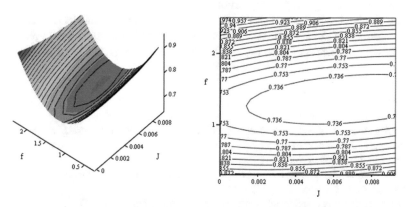

Fig. 3. The degree of transmission transparency depending on the traction load fluctuations and moment of inertia of the EDM drive for sowing (surface response on the left, line level on the right).

We used Mathcad program (section - vector and matrix operations) to solve the described equations. The use of this simple and convenient program helped in solving systems of equations and improved the clearness of the presented results.

3 Conclusions

The made computing experiment has shown the feasibility of using the EDM in the drivetrain of the class 1,4 tractor at the main agricultural operations. We conclude that the optimal values of EDM parameters can improve the quality of the MTU operation on the main agricultural operations according to the results of the analysis of the computational experiment.

The optimal value of the parameter P was obtained by varying the factors values $x_1 - x_5$.

Optimal factors values for plowing: the throttle cross-sectional area $S_{TH} = 1.49$ 10^{-4}, the volume of the hydropneumatic accumulator $V_{HPA} = 3.85 \cdot 10^{-3}$, the air pressure in the hydropneumatic accumulator $P_A = 6.77 \cdot 10^5$, the moment of inertia of the extra weight of the drive mechanism $J_{EW} = 4.64 \cdot 10^{-3}$, the most probable frequency of the traction load fluctuations $f = 1.41$, this corresponds to the optimal value of the degree of the transmission transparency equal to P = 0.63.

Optimal values of factors for cultivation $S_{TH} = 2.35 \cdot 10^{-4}$, $V_{HPA} = 4.23 \cdot 10^{-3}$, $P_A = 2.93 \cdot 10^5$, $J_{EW} = 4.56 \cdot 10^{-3}$, $f = 0.79$. This corresponds to the optimal value of the degree of the transmission transparency equal to P = 0,73.

Optimal parameters at the most probable frequency of the traction load fluctuations during sowing $f = 1.25$ Hz have the following values: $S_{TH} = 2.16 \cdot 10^{-4}$ m^2; $V_{HPA} = 4.04 \cdot 10^{-3}$ m^3; P = $5.745 \cdot 10^5$ Pa; $J_{EW} = 2.093 \cdot 10^{-3}$ kg·m^2; the value of the response function in this case is P = 0.754.

It should be noted that the found optimal values of factors for the studied EDM and response functions are applicable only to similar operating conditions and are not applicable to other agricultural operations.

The obtained data shows that the use of an elastic damping mechanism can reduce the fluctuations of the external traction load transmitted to the engine, with plowing up to 37%, with cultivation on average up to 27%, with sowing, on average up to 25%.

References

1. Shekhovtsov, V.V., Sokolov-Dobrev, N.S., Potapov, P.V.: Decreasing of the dynamic loading of tractor transmission by means of change of the reactive element torsional stiffness. Procedia Eng. **150**, 1239–1244
2. Janulevičius, A., Giedra, K.: Analysis of main dynamic parameters of split power transmission. Transport 23(2), 112–118 (2008)
3. Kravchenko, V.A., Senkevich, S.E., Senkevich, A.A., Goncharov, D.A., Duriagina, V.V.: Patent 2398147 Russian Federation, C1 F 16 H 47/04. Device to reduce the rigidity of the transmission of the machine-tractor unit. Applicant and patent holder: FGOU VPO AChGAA. – № 2008153010/11; appl. 31.12.2008; publ. 27.08.2010, Bul. № 24. – 7 p.: fig

4. Kravchenko, V.A.: The results of studies of arable unit on the basis of the class 1,4 tractor with the UDM in the transmission. Vestnik VIESH **2**(27): 87–91 (2017, in Russian)
5. Traction and Tractor Performance, Zoz, F.M., Grisso, R.D.: Agricultural Equipment Technology Conference: conference materials. Louisville, Kentucky, USA, pp. 1–47 (2003)
6. Bespamiatnova, N.M.: Nauchno-metodicheskie osnovy adaptatcii pochvoobrabaty-vaiushchikh i posevnykh mashin. Rostov n/D: OOO «Terra» , IPK «Gefest», 176 p (2002)
7. Lure A.B.: Statisticheskaia dinamika selskokhoziaistvennykh agregatov. 2-e izd., pererab. i dop. Kolos, 382 p (1981)
8. Spiridonov, A.A.: Planirovanie eksperimenta pri issledovanii tekhnologicheskikh protcessov, 184 p. Mashinostroenie, Moskva (1981). (Planning an Experiment in the Study of Technological Processes, 184 p. Mashinostroenie, Moscow (1981))
9. Box, G.E.P., Draper, N.R.: Empirical Model-Building and Response Surfaces. Wiley, New York (1987)
10. Box, G.E.P., Hunter, W.G., Hunter, S.J.: Statistics for Experimenters: An Introduction to Design, Data Analysis, and Model Building. Wiley, New York (1978)

Energy-Efficient Pasteurizer of Liquid Products Using IR and UV Radiation

Dmitry Tikhomirov[1](✉) ⓘ, Alexey Kuzmichev[1],
Sergey Rastimeshin[2], Stanislav Trunov[1], and Stepan Dudin[1]

[1] Federal State Budgetary Scientific Institution "Federal Scientific
Agroengineering Center VIM" (FSAC VIM), 1-st Institutskij 5, Moscow,
Russia109456
{tihda, alkumkuzm, s-razin-dudin}@mail.ru,
alla-rika@yandex.ru
[2] Moscow Timiryazev Agricultural Academy, Russian State Agrarian
University, 127550 Timiryazevskaya street, 49, Moscow, Russia
resurs00@mail.ru

Abstract. The article deals with the problems of development of energy-efficient technical means for heat supply of agricultural objects. Shown the energy-efficient method of heat treatment of liquid products using infrared (IR) and ultraviolet (UV) spectrum of electromagnetic waves. Much attention is drawn to method of electrical and structural calculation of pasteurization chamber for infrared radiators of coaxial type is developed. Data are given about the parameters and modes of operation of the combined installation at the influence of UV and IR radiation on the properties of milk, to reduce bacterial contamination.

Keywords: Energy saving · Electrical irradiator · UV radiation
Pasteurization chamber · Pasteurizer · Heat treatment · Disinfection

1   Introduction

Heat treatment of liquid foodstuffs (milk, juice, etc.) refers to the number of common and energy-intensive processes of primary processing with a view to their conservation. At the same time, the task is to preserve the maximum nutritional and taste qualities of the product. The effect of infrared radiation on a thin layer of a liquid product is one of the energy-efficient ways of heat treatment. The energy transfer takes place directly from the radiation source to the processed liquid product in the absence of contact between them. IR radiation penetrates to a certain depth, which eliminates local overheating and unwanted structural changes in the surface layer [1].

Short - term exposure to IR radiation with a high density simultaneously across the thickness and surface of the thin layer of liquid creates the necessary conditions for the elimination of toxic and ballast microflora, allows you to keep useful biological and physic-chemical components (proteins, vitamins, enzymes) that determine the nutritional and organoleptic properties of the product [2].

P. Vasant et al. (Eds.): ICO 2018, AISC 866, pp. 178–186, 2019.
https://doi.org/10.1007/978-3-030-00979-3_18

In thin-layer devices, all the liquid is processed at once, so the product does not overheat above the set temperature. IR pasteurizers have lower metal consumption, low heat losses, increased productivity and provide a reduction in energy consumption up to 20% compared to traditional plate pasteurization-cooling units and long-term pasteurization baths.

Figure 1 shows a thin-layer electric pasteurizer of liquid products using electromagnetic waves of infrared and ultraviolet range [3].

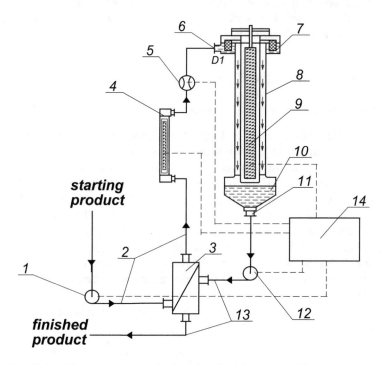

Fig. 1. Installation Diagram for pasteurization of milk with the use of IR and UV radiation: 1 – milk supply pump, 2 – pipeline hearth, 3 – heat exchanger, 4 – UV-radiator, 5 – flowmeter, 6 – nozzle input IR pasteurizer, 7 – chamber forming a thin layer, 8 – chamber pasteurizer, 9 – IR-radiator, 10 – receiving chamber, 11 – outlet pipe, 12 – pump issue, 13 – pipeline issue, 14 – controller wits GSM module.

Devices with a thin-layer flow of liquid can be made according to different schemes [4]. There are vertical, inclined and horizontal flow of liquid depending on the location of the energy source and its shape. Energy and structural calculation of the radiator and pasteurization chamber causes certain difficulties in justifying the parameters of the IR pasteurizer, associated with the dynamics of heating the liquid during the heat exchange by radiation.

The power of the radiator is usually calculated according to the well-known energy balance equation without taking into account the conditions of the heat exchange process between the heater and the liquid product [3], which in certain cases can lead to the malfunction of the designed apparatus [5, 6]. The method proposed for the calculations takes into account the process of heat exchange between the IR heater and the formed thin layer of the heated liquid product. This is its peculiarity. Most of the considered units have a similar geometric design of the pasteurization chamber. The camera contains two cylinders arranged coaxially: first – IR radiator, second – camera body. The heated liquid product flows along the inner surface of the chamber housing with a thin layer (2…3 mm).

The development of a unified methodological approach to the problem of determining the thermal and geometric parameters of the radiator and pasteurization chamber as a whole is an urgent task.

2 Methods and Results of Research

2.1 Method of Calculation of Pasteurizer

The purpose of the calculation is to determine the power of the IR radiator, geometric and structural parameters of the pasteurization chamber.

The initial data are the performance of the installation G, the surface temperature of the radiator T_1, the initial temperature of the liquid product when it is fed to the pasteurization chamber T_L, some physical properties of the processed liquid.

Thin-film flow along the vertical wall is possible only if the liquid moistens the surface. Otherwise, the film begins to disintegrate under the action of surface tension on the individual drops [7]. If the liquid moistens the wall of the working surface, on which it flows, then its surface layer, directly in contact with the wall, must have a zero-velocity value [1]. Than further away from the surface is the liquid particle, then with more speed it moves down, all other things being equal.

The diameter of the cylinder d_2 on which a thin layer of liquid flows is determined by the formula:

$$d_2 = \frac{G}{\pi \cdot \rho \cdot \upsilon \cdot \mathrm{Re}} \tag{1}$$

G – pasteurizer performance kg/s; $\mathrm{Re} = 8500$ – Reynolds number, characterizing the movement of the liquid; $\rho = 976$ kg/m^3 – milk density; $v = 0{,}43 \cdot 10^{-6}$ – coefficient of kinematic viscosity of milk at 70 °C m^2/s.

The thickness of the layer of flowing fluid, m:

$$\delta = \sqrt[3]{\frac{3Gv}{\pi D \rho g}} \tag{2}$$

Let us consider the dynamics of the process of heat transfer by radiation between the radiator and the liquid flow (liquid product), flowing a thin layer on the inner cylindrical surface of the pasteurization chamber [8]. The outer wall of the chamber is heat-insulated, so we assume that there is no heat loss to the outer space, and the heat loss from the liquid to the outer cylindrical wall is negligible. The temperature gradient at heat transfer by thermal conductivity in the radial and axial direction of the fluid flow is not taken into account. We also take the one-dimensional distribution of heat in the direction of fluid flow, i.e. the x axis (Fig. 2). Consider the element of the heat exchange surface length dx.

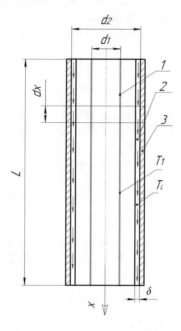

Fig. 2. Diagram of the pasteurization chamber: 1 – IR radiator, 2 – a thin layer of liquid (δ), 3 – the case of the pasteurization chamber.

The heat balance equation describing the change in the temperature of the flowing liquid for the dx element during dt has the following form:

$$cF_L\rho_L dxdT_L = qdxd\tau \tag{3}$$

q – the resulting radiant flux per unit length of the radiator, W/m; ρ_L – liquid density, kg/m^3; c – specific heat of the liquid, J/kg °C; F_L – the cross-sectional area of the liquid layer, m^2; T_L – liquid temperature, °C, τ – time, s.

Taking into account that the velocity of the fluid in the chamber $\omega_L = G/F_L$ and $\omega_L = dx/dt$, and that:

$$q = Q/L = \frac{\varepsilon_{\Pi p}C_0 F_1 \cdot 10^{-8}(T_1^4 - T_L^4)}{L}, \tag{4}$$

receive

$$cGdT_L = \frac{\varepsilon_r C_0 F_1 \cdot 10^{-8}(T_1^4 - T_L^4)}{L} dx, \quad 0 \le x \le L, \tag{5}$$

ε_r – emissivity factor of the system; T_1 – temperature of the radiator, K; C_0 – blackbody radiation, $W/m^2 \, K^4$, L – length of the radiator, m, G – flow rate, m^3/s, Q – heat flow from the radiator to the liquid product, W.

Emissivity factor for a closed system of two coaxial cylinders:

$$\varepsilon_r = \frac{1}{\frac{1}{\varepsilon_1} + \frac{d_1}{d_2}\left(\frac{1}{\varepsilon_2} - 1\right)}, \tag{6}$$

ε_1, ε_2 – emissivity factor of the radiator and the heated liquid, d_1 – the outer diameter of the radiator, d_2 – the inner diameter of the cylindrical surface through which the liquid flows, m.

Since $F_1 = \pi d_1 L$ the Eq. (5) can be reduced to:

$$\frac{dT_L}{T_1^4 - T_L^4} = \frac{\varepsilon_r C_0 d_1 \pi \cdot 10^{-8}}{cG} dx, \tag{7}$$

Integrating Eq. (7), we obtain the change in the liquid temperature along the length of the pasteurization chamber $T_L = f(x)$ at $0 \le x \le L$. the General solution is:

$$\frac{1}{2}\frac{\text{arc}\, tg(\frac{T_L}{T_1})}{T_1^3} - \frac{1}{4T_1^3}\left[\ln(T_1 - T_L) - \ln(T_1 + T_L)\right] + C = \frac{\varepsilon_r C_0 d_1 \pi \cdot 10^{-8}}{cG} x, \tag{8}$$
$$0 \le x \le L,$$

C – arbitrary constant.

At the time of the liquid supply to the pasteurization chamber at $x = 0$ (initial conditions), its temperature $T_L(0)$ can be taken equal to the temperature of the liquid leaving the heat exchanger-recuperator. The temperature of the radiator T1 $\approx$ 1173 K should be considered a constant over the entire length of the radiator. The diameter of the radiator and the inner diameter of the outer cylinder d_2 at the initial stage of calculation is set based on the adopted structural and technological scheme of the installation ($d_1 = 0{,}03\ldots0{,}1$ m). Moreover, the diameter d_2 is selected based on the thickness of thin layer is δ, the unit capacity G (kg/h) product density ρ_L and speed of movement of the liquid ω_L [9]:

$$d_2 = \frac{G}{3600\pi\rho_L\delta\omega_L} + \delta. \tag{9}$$

Under the established initial conditions from Eq. (8) we determine the unknown constant C. Solving the obtained partial equation, we find the temperature of the liquid product T_L at the output of the pasteurization chamber and specify its length.

The power of the radiator P_r is calculated from a well-known expression:

$$P_r = cG(t_{Lf} - t_{Li})/\eta_c \tag{10}$$

t_{Lf}, t_{Li} – final and initial liquid temperature, °C, η_c – the efficiency of the pasteurization chamber.

Equation (8) allows to analyze the influence of variables d_1, d_2, G on the heating level $\Delta t = (t_{Lf} - t_{Li})$ of the liquid product along the length L of the pasteurization chamber.

As a material for the working part of the radiators are often used alloys based on chromium and nickel (nichrome) with high electrical resistance in the form of wire spirals, zigzags, etc. The choice of design parameters of the emitting body is to determine the diameter, length of the heating wire, the method of its laying. The values of electrical power P_r, voltage U and temperature T_1 of the radiator are usually set [9].

2.2   Experimental Data

Table 1 shows the results of calculation of the pasteurization chamber of the electric IR pasteurizer (Fig. 1) for heat treatment of milk capacity up to 1000 l/h.

Table 1. The results of the calculation of the IR radiator and the pasteurization chamber.

Parameter	Value
Plant capacity, l/h	500…1000
The initial temperature of the milk after the heat exchanger, t_{Li}, °C	65
Temperature of pasteurization, °C	78
Supply voltage, U, V	380/220
The power of the radiator, P_r, W	12800
Electric current consumption per phase, I, A	19,4
Diameter of the radiator, d_1, m	0,07
Pasteurization chamber diameter, d_2, m	0,12
Number of partitions of the radiator, n	12
Electric current in the heater section, I_c, A	4,8
The estimated diameter of the heating wire, d_w, mm	0,74
Length of nichrome wire in section, L_w, m	14,0
Diameter of the nichrome wire, d, m	0,007
Number of turns in a section, m	1060
Length of the radiator section, L_s, m	0,85
Length of the pasteurization chamber, L, m	0,85

The design of the radiator contains a number of vertical quartz tubes of length L, inside each of which spirals of nichrome wire in the form of separate sections are laid. Quartz tubes with spirals are arranged in a circle, forming a cylinder with a diameter $d_1 = 0,07$ m.

The Eq. (8) is solved in the MathCAD system and is presented in a graphical form in Fig. 3, and the mathematical expression for the considered boundary conditions is:

$$T_{\mathrm{L}} = 15x + 338. \tag{11}$$

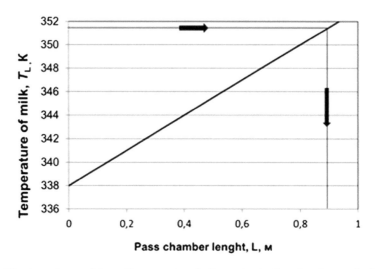

Fig. 3. The dependence of the milk temperature in the pasteurization chamber on the length of the beam.

The length of the pasteurization chamber L should be equal to 0.85 m to ensure a predetermined pasteurization temperature of milk 78 °C (351 K)

When using the proposed method, the calculated parameters of a new modular installation for IR pasteurization of liquid products were obtained [9]. The calculated thermal power and design indicators of the IR radiator and pasteurization chamber were tested by laboratory studies of the experimental sample of the installation, which confirmed a sufficiently high convergence of theoretical and practical results.

Cold pasteurization method attracts attention with low energy intensity of the process [10]. However, the effect of UV radiation on milk has a number of limitations [11]. For UV treatment of milk, it is possible to use bactericidal lamps of low pressure without ozone formation with the power of 100–300 W. In the experimental equipment (installation) (Fig. 1) [12, 13] was used bactericidal amalgam lamp type DB-145 power 145 W. The bactericidal flow of the lamp is 45 W. The area of irradiated surface $S = 600$ cm^2. The effective effect of UV radiation for the destruction of bacteria by this lamp is in the spectral region with $\lambda_{\mathrm{max}} = 253{,}7$ nm.

Theoretical and experimental studies justified the effective modes and parameters of milk disinfection by UV irradiation device: the normalized dose of radiation – 16 mJ/cm^2, milk absorption coefficient – 0,38…0,47 cm^{-1}, irradiation – 8,1 W/m^2, the layer of processed milk – up to 1 mm, processing time - about 2 s [12, 14]. The greatest spectrum of absorption of ultraviolet radiation by milk is in the wavelength range 180–220 nm, and the smallest in the range of 250–370 nm.

3   Conclusion

The results showed that the combined UV and IR effects on the milk, the pasteurization temperature can be reduced to 70 … 72 °C, which allows to reduce the cost of electricity up to 25% compared to infrared installations [15]. The values of microbiological contamination of microorganisms CFU (colony forming units)/cm^3 are $0{,}3{\cdot}10^{-5}$ (Fig. 4).

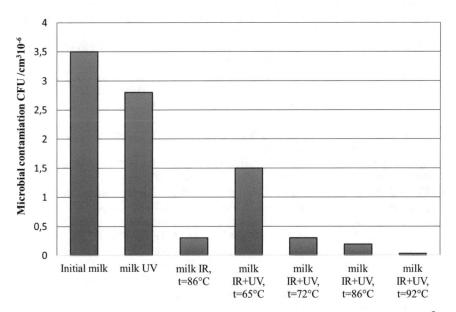

Fig. 4. Average values of microbiological contamination by microorganisms CFU/cm^3 at different methods of milk processing

UV treatment does not lead to significant deviations from the requirements for pasteurized milk. The main parameters of milk meet the hygienic requirements of safety and nutritional value of products [14]. Processing of milk with a bactericidal stream of 20–40 W did not affect the content of fat, protein, lactose [14]. When processing milk in the mode of 50 W there was an extraneous smell peculiar to the saponification of fats, there were changes in the ratio of fatty acids compared to their content in raw milk.

UV treatment of milk with bactericidal flow 49 W and with a capacity of 100 l/h increased the content of vitamin D3 in milk. For milk disinfection, it is necessary to take a normalized dose of 16 mJ/cm^2 [15, 16] and ensure the minimum possible layer of processed milk of 0,2 cm. Water for the technological washing of the pasteurizer is subjected to ultraviolet treatment as well.

A sample of pasteurizer successfully passed laboratory and economic production tests on a dairy farm of 200 cows.

References

1. Eliseev, N., Karumidze, G.: Methodical recommendations on the calculation settings for electro-thermal processing of liquid media. VIESH, 25 p (1979)
2. Fokin, M.: Nanotechnology in milk pasteurization. Dairy Ind. **5**, 13–14 (2009)
3. Infrared pasteurizers [electronic resource]. http://ecomash.ru/pasterizatory/elektropaste-rizatory-s-infrakrasnym-nagrevom-serii-a1-ope2
4. Kuzmichev, A.V., Malyshev, V.V., Tikhomirov, D.A.: Efficiency of combined pasteurization of milk by UV and IR irradiation. Light. Eng. **5**, 6–9 (2010)
5. Kuzmichev, A., Lamonov, N., Tikhomirov, D.: Modular installation for liquid treatment by infrared radiation of a thin layer: Pat. 2389397 Grew. Federation: IPC A23L 3/005 (2006/01). Applicant and patentee of GNU VIESH. No. 2008143729/02; statement. 06.11.2008; publ. 20.05.2010, bul. No. 14, 4 p
6. Fomina, E.: Optimization of the scheme and parameters of IR pasteurizer. Bull. Saratov State Tech. Univ. **1**(10), 127–131 (2006)
7. Spalart, P.R., Rumsey, C.L.: Effective inflow conditions for turbulence models in aerodynamic calculations. AIAA J. **45**(10), 2544–2553 (2007)
8. Tikhomirov, D., Kuzmichev, A.: Calculation of infrared emitter electrical installation for pasteurization of liquid products. Mech. Electr. Agric. **5**, 14–17 (2013)
9. Tikhomirov, D., Kuzmichev, A.: Calculation of pasteurization chamber of electric installation for heat treatment of liquid products. Mach. Equip. Village **7**, 21–24 (2014)
10. Smith, W.L., Lagunas-Solar, M.C., Cullor, J.S.: Use of pulsed ultraviolet laser light for the cold pasteurization of bovine milk. J. Food Prot. **65**(9), 1480–1482 (2002). 77. U.S
11. Cho, Y.S., Song, K.B., Yamda, K.: Effect of ultraviolet irradiation on molecular properties and immunoglobulin production-regulating activity of β-lactoglobulin. Food Sci. Biotechnol. **19**(3), 595–602 (2010)
12. Kuzmichyov, A.V., Malyshev, Vladimir V., Tikhomirov, Dmitry A.: Efficiency of the combined pasteurisation of milk using UV and IR irradiation. Light Eng. **19**(1), 74–78 (2011)
13. Letaev, S.: Justification of parameters of the decontamination of milk on the farm ultraviolet and infrared radiation: the dissertation … candidate of technical Sciences: 05.20.02, 145 p (2012)
14. Chernykh, E.: Influence of ultraviolet processing of cow milk on its biochemical, technological and hygienic properties: the dissertation … candidate of biological Sciences: 03.00.04 (2006)
15. Klimenko, S., Tikhomirov, D.: Application of UV and IR radiation for pasteurization of milk in a thin-film installation. Trends in science and education. Part 4. Ed. SIC "L-Magazine", pp. 14–17 (2018)
16. Schneider, S., Warthesei, J.: Stability of Vitamin D in fluid milk. University of Minnesota, St. Paul. 298G, Teaming Up for Animal Agriculture, July 31–August 4 1989

CAIAS Simulator: Self-driving Vehicle Simulator for AI Research

Sabir Hossain, Abdur R. Fayjie, Oualid Doukhi, and Deok-jin Lee[✉]

Department of Mechanical and Automotive Engineering, Kunsan National
University, 558, Daehak-ro, Gunsan, Jeonbuk 54150, Republic of Korea
{sabir,abdurrfayjie,doukhioualid,
deokjlee}@kunsan.ac.kr

Abstract. This paper presents a simulation environment which includes virtual
structures of a low-cost embedded designed car for the autonomous driving test,
tracks, obstacles, and environments. A cross-platform game engine, Unity 3D,
empowers the embedded designed car to check and trial new tracks, parameters
and calculations in the 3D environment before the real-time test. The virtual
environment fabricates the domain such like that it is the mimics of the activity
of a genuine car and Unity 3D are utilized to incorporate the embedded designed
car into the test situation while the car's movements and steering angle can serve
as an examination premise. Distinctive driving situations were utilized to analyze how the sensors respond when they are connected to genuine circumstances
and are also utilized to confirm the impacts of other parameters on the scenes.
Options are available to choose flexible sensors, monitor the output and
implement any autonomous driving, steering prediction, deep learning and end-
to-end learning algorithm.

Keywords: Simulator · Autonomous vehicle · AI research · Sensor fusion
Virtual environment

1 Introduction

Sensorimotor control in three-dimensional environments remains a major challenge in
machine learning and robotics [1, 2]. Autonomous vehicle development is one the
example among them. Because, the setup is particularly challenging due to complex
vehicle dynamics, distinctive tracks with the different angular path, curved road
markings; and the response to the motion of various actions that may be in view at any
given time; the necessity to quickly accommodate with the conflicting objectives, such
as obstacles and grass. The infrastructure costs and the logistical difficulties of training
and testing systems in the real physical world are main impediments for research in
autonomous driving. A significant amount of manpower and funds are involved in case
of instrumenting and operating even in a driverless robotic car. And a single vehicle is
far from sufficient for collecting the requisite data that cover the multitude of corner
cases that must be processed for both training and validation [3]. Also, a single vehicle
is a long way from adequate for gathering the imperative information that covers the
huge number of corner cases that must be prepared for both training and validation. For

© Springer Nature Switzerland AG 2019
P. Vasant et al. (Eds.): ICO 2018, AISC 866, pp. 187–195, 2019.
https://doi.org/10.1007/978-3-030-00979-3_19

this, a replaceable plan like simulation is needed to continue autonomous driving research. Simulation can provide safe experiment since in real world there are possibilities of different casualties. More significantly, researchers found simulator more feasible for trialing the new approaches of self-driving algorithms [4]. Moreover, manually setting up the environment every time in real-time is costlier and time-consuming than the manual setup in a simulator. For a simple change, the experiment could cost more money and time in the case of reality whereas in simulation it's less. There are many commercial companies who already producing simulation environment for various purpose. But, due to the limitation in open-source products and lacking in sensors mode, a simulator is required maintaining all the necessary tools for autonomous driving research. In this paper, we present CAIAS simulator for embedded self-driving car research. Using this simulator, it is possible to train, test and verify the driving models. Feasible options are given to choose sensor and study different self-driving algorithms like deep reinforcement learning [5], end-to-end learning [6], etc.

2 Development of Simulation Engine

This simulator is built on Unity 3D software. This Game engine is the core of this simulator development. Unity 3D is a multi-platform game development tool as well as a fully integrated game engine, which provides functions such as rendering engine, physics engine, scripting engine, lightmap and scene management and supports these three programming languages, JavaScript, C# and Boo [7]. The main technology characters of Unity 3D are the component model, event-driven model, and class relationships [8]. Unity 3D is very popular in recent years. The efforts and the organizations that work on the development of virtual simulators in Europe and all over the world are numerous. Unity 3D as it is perfect for developing independent small-scale game apps and multiplatform game engine for the creation of interactive 3D content. Unity 3D has a good user interface and powerful interactive design module. So, this integrated platform is best for creating 3D simulator or other interactive contents such as virtual reconstructions or 3D animations in real time [9].

2.1 Interactive

First, the simulator is made compatible for Windows, Linux and Mac OS with both 32-bit and 64-bit. In the simulator, the interaction is between the agent car and the environment. The agent car model is designed in CATIA DS as a prototype of the real embedded car and then imported in the Unity environment (Fig. 1). To bolster the connection, a server-client is established to render the simulation using Socket. This Socket is responsible for the communication between the agent which works as a server and the model algorithm which works as a client.

2.2 Kinematic Bicycle Model

The nonlinear continuous time equations that describe a kinematic bicycle model [10] which is considered to build the car model of the simulator in an inertial frame are

$$\dot{x} = v \, \cos(\psi + \beta) \tag{1}$$

$$\dot{y} = v \, \sin(\psi + \beta) \tag{2}$$

$$\dot{\psi} = \frac{v}{l_r} \sin(\beta) \tag{3}$$

where x and y are the coordinates of the center of mass in an inertial frame (X, Y). ψ is the inertial heading and v is the speed of the vehicle. l_f and l_r represent the distance from the center of the mass to the front and rear axles, respectively. β is the angle of the current velocity of the center of mass with respect to the longitudinal axis of the car [11]. So, this 2-DOF car model can be controlled yaw and longitudinal motions.

2.3 Number of Environments

In the simulation environment, two different worlds with two different tracks are provided. The user can choose preferable tracks to train and test the model. One of the environment track is based on the Kunsan National University main stadium athletic field (Fig. 3). Basically, this approach is taken since every university has a similar racing track which has 8 lanes athletic and anyone can check the model from simulation to real life.

The reason for converting one real-life scenario to convert to a virtual environment is to check and test some algorithm which can imitate the behavior from the image. So, having one scenario as like real life will give the advantage to evaluate those algorithms as well. Another track is randomly designed just to check all the steering angles and speeds of the agent car produced during the simulation. This environment contains the countryside road, trees, random obstacles, grasses and driving lanes (Fig. 4).

Inside the environment, different types of obstacles are positioned to make the model training more robust. Skybox is used to make the environment look more realistic. Skyboxes are a wrapper around your entire scene that shows what the world looks like beyond your geometry [12].

2.4 Types and Construction of Sensors

Three types of sensors configuration are provided in the agent vehicle. The user can choose one or multiple sensors to generate results according to their model.

RGB Camera. Unity 3D has a built-in script for the camera. A Camera is a device through which the player views the world like a simple camera. Here specifying pixel value is one of the important things. Single camera and multiple cameras are used as a sensor input (Fig. 5). The idea of three multiple cameras is from end-to-end learning using convolutional network [13].

LiDAR Sensor. To imitate the LiDAR sensor and acquire the LiDAR data as a form of an image, we simulate each individual laser in the physic engine using raycasting. This implementation is intuitive and accurate. For each laser in the simulated LiDAR, a raycast is used to detect the distance. In the update loop, if the timer exceeds the limit, a list of raycast will be a trigger to gather distance information. [14]. The result of all raycast is stored in a depth map image which is shown in the picture. The green image below the picture of the environment is the distance matrix correspond to the positions of all the obstacle in the environment (Fig. 6).

Depth Camera Sensor. The camera function can generate a depth or motion vector texture in unity. This is a minimalistic G-buffer Texture that can be used for post-processing effects or to implement custom lighting models. It can be used as an input of depth camera. It produces the drawing of the depth sensor completely based on the experiment layout (Fig. 7).

2.5 Auxiliary Setting and Other Appliance

The physics of car used for the driving simulation system is similar to normal car dynamics. The functions and particle effects in the imported Unity were used to create trees, grass, object, climate and lighting effects to simulate the real environment (as shown in Figs. 2 and 5). C# was used to program the control code that would be imported to the scenes to meet the requirement of establishing simulation environments in this study. Graphical Screen resolution and display quality can be changed at the beginning. The user should consider changing the display quality according to their CPU/GPU processing power. The user can choose the track and environment regarding the sensor options. There are also settings for both training and testing in the simulator display. From the car user interference display, car speed and the angle are visible. It is possible to get the steering angle and speed during the training mode.

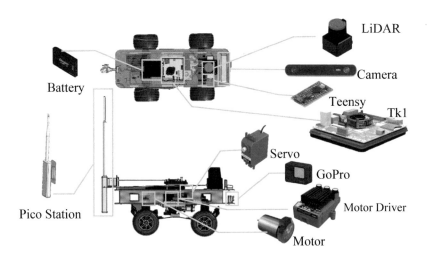

Fig. 1. This figure shows the detailed design of embedded system inside the vehicle. CATIA DS software is used to design the prototype vehicle.

Fig. 2. Embedded designed vehicle for self-driving in a test environment. The cad model is converted to fbx file and later imported with real-life physical parameter to unity engine.

Fig. 3. Real life image of the 8-lane athletic track of Kunsan National University (Left Image) which is the very feasible place to perform the experiment. It is also the easiest place to perform experiment since almost all university has an athletic track. Eight-lane athletic track in the simulator (Right Image)

Fig. 4. The second virtual environment on the simulator (Left Image). The figure on the right side represents the map of the second environment.

3   Result

Finally, this simulation platform provides a flexible specification of sensor suites and environmental conditions. The simulator is made compatible for all common operating systems along with different CPU bit version.

Fig. 5. Single camera to take image input (Figure on top) and Multiple Camera (Left-Right-Center) monitoring and receiving data from the environment. (Figure on bottom)

Fig. 6. LiDAR sensor raycasting over different objects and below that depth map of the distance matrix corresponding to the positions of all the obstacle in the environment.

Fig. 7. These figures show the output of RGB camera and depth camera simultaneously

3.1 Debugging and Performance Analysis in Unity

Application troubleshooting involves getting knowledge into how the application is structured and executed, assembling information to assess genuine performance, assessing it against desire, at that point deliberately disengaging and eliminating the issues. While debugging or troubleshooting, it is urgent to continue controlled way with the goal that to know what change particularly brings about an alternate result. Unity comes with a built-in profiler which provides per-frame performance metrics, which can be used to help identify bottlenecks [15].

3.2 Test Result Evaluation from the Performance

We implemented two different approaches to the algorithm. Table 1 reports the number of times car effectively finished test condition under two distinct scene. The Track scene is eight-lane athletic track. In this track, speed and steering angle is perfect since the model is only trained in this scenario. The other track: all the possible turn and obstacles are almost efficiently tackled by the car. Results introduced in Table 1 recommend a few general conclusions. In general, the execution of all strategies are working fine and the achievement rate is good. Hence, the results verify the efficacy of the simulator.

Table 1. Qualitative evaluation of performance on both tracks. The higher value indicates the better results.

Scenarios	Athletic track		Country side track	
	End-to-end learning	DRL	End-to-end learning	DRL
Straight	99	95	99	92
Left turn	76	69	96	90
Right turn	88	78	98	87
Brake	87	74	89	70
Obstacle avoid	83	72	79	66

4 Conclusion

The simulator provides the compatibility of Windows, Linux, and Mac system. The binary files can be used to trial, monitor and train. In this study, the simulator provides two tracks among them one scene is produced based on the real images of the Kunsan National University main stadium athletic track. Sketchup is used to generate the drawing and blender were used to create the 3D model to enhance the accuracy of the environment of the athletic stadium track. All the objects, roads, etc., were imported into Unity project file to create the simulation environment. Different input data like RGB camera data, depth image data, and LiDAR data can be used to check the feasibility of any learning algorithm through socket.IO. Inside simulator, the client can create and prepare the frameworks and after that assess them in controlled situations. The feedback provided by the simulator enables detailed analyses that highlight particular failure modes and opportunities for future work. We trust that this test system will help all the personal in self- driving research.

4.1 Future Modification for More Robust Result

- Three types of sensors configuration are provided in the agent vehicle. The user can choose including different climate conditions, for example, radiant, stormy, sunny, rainy, and snowy alongside automatic volatile incident, for example, sudden intersections by walkers, barricades, vehicle, and bike blind spot, and other crisis that can happen while driving.
- Adding weather effects on driving (wet roads = more slippery, low sunset = blinding glare, fog = reduced visibility)
- Day/night cycles.
- Generating the analogous complexity driving events by using realistic artificial intelligence (AI) in the traffic vehicles.
- Constructing good city geometry (like road modules that can be plugged together, multilane roads, intersections, parking lots, etc.

Acknowledgments. This research was supported by Unmanned Vehicles Advanced Core Technology Research and Development Program through the National Research Foundation of Korea (NRF), Unmanned Vehicle Advanced Research Center (UVARC) funded by the Ministry of Science, ICT & Future Planning, the Republic Of Korea (No. 2016M1B3A1A01937245) and by the Ministry of Trade, Industry & Energy (MOTIE) under the R&D program (Educating Future-Car R&D Expert). (N0002428). It was also supported by Development Program through the National Research Foundation of Korea (NRF) (No. 2016R1D1A1B03935238).

References

1. Tresilian, J.: Sensorimotor control and learning: an introduction to the behavioral neuroscience of action. In: Behavioral Neuroscience. Palgrave Macmillan (1805) (2012)
2. Dosovitskiy, A., Ros, G., Codevilla, F., Lopez, A., Koltun, V.: CARLA: an open urban driving simulator. In: CORL (2017)

3. Dosovitskiy, A., Ros, G., Codevilla, F., Lopez, A., Koltun, V.: CARLA: an open urban driving simulator. In: Proceedings of the 1st Conference on Robot Learning, California, USA (2017)
4. Tang, W., Wan, T.R.: Synthetic vision for road traffic simulation in a virtual environment. In: Intelligent Agents for Mobile and Virtual Media, pp. 176–185. Springer, London (2002)
5. Sallab, A.E., Abdou, M., Perot, E., Yogamani, S.: Deep reinforcement learning framework for autonomous driving. Electron. Imaging **19**, 70–76 (2017)
6. Bojarski, M., Testa, D.D., Dworakowski, D., Firner, B., Flepp, B., Goyal, P.: End to end learning for self-driving cars. In: Computer Vision and Pattern Recognition, CoRR (2016). arXiv preprint arXiv:1604.07316
7. Liao, H., Qu, Z.: Virtual experiment system for electrician training based on kinect and Unity 3D. In: Proceedings of the 2013 International Conference on Mechatronic Sciences, Electric Engineering and Computer (MEC), pp. 2659–2662. IEEE (2013)
8. Xie, J.: Research on key technologies base Unity 3D game engine. In: 7th International Conference on Computer Science & Education (ICCSE), pp. 695–699. IEEE (2012) (July Edn.)
9. Luca, V.D., Meo, A., Mongelli, A., Vecchio, P., Paolis, L.T.D.: Development of a virtual simulator for microanastomosis: new opportunities and challenges international. In: International Conference on Augmented Reality, Virtual Reality and Computer Graphics, pp. 65–81. Springer, Cham (2016)
10. Rajamani, R.: Vehicle Dynamics and Control. Science & Business Media, Springer (2011)
11. Kong, J., Pfeiffer, M., Schildbach, G. and Borrelli, F.: Kinematic and dynamic vehicle models for autonomous driving control design. In: IEEE Intelligent Vehicles Symposium (IV), pp. 1094–1099. (2015) (June Edn.)
12. Wu1, J., Li1, Y., Liu1, Q., Su, G., Liu, K.: Research on application of Unity 3D in virtual battlefield environment. In: 2nd International Conference on Control, Automation, and Artificial Intelligence (CAAI), Advances in Intelligent Systems Research, vol. 134 (2017)
13. Chi, L., Mu, Y.: Deep Steering: Learning end-to-end driving model from spatial and temporal visual cues (2017). arXiv preprint arXiv:1708.03798
14. Wang, Y.: web article about self-driving car simulation. From: wangyangevan.weebly. com/lidar-simulation. Archived from the original on July 2017. Accessed 16 Feb 2018
15. Luo, M., Claypool, M.: Uniquitous: implementation and evaluation of a cloud-based game system. In: Unity in Computer Science and Interactive Media & Game Development (GEM), Worcester, MA 01609, USA, pp. 1–6. IEEE (2015)

Vision-Based Driver's Attention Monitoring System for Smart Vehicles

Lamia Alam and Mohammed Moshiul Hoque[(✉)]

Department of Computer Science & Engineering, Chittagong University
of Engineering & Technology, Chittagong 4349, Bangladesh
lamiacse09@gmail.com, mmoshiulh@gmail.com

Abstract. Recent studies revealed that the driver's inattention is one of the most prominent reasons for car accidents. Intelligent driving assistant system with real time monitoring of the driver's attentional status may reduce the accident rate that mostly occurred due to lack of attention. In this paper, we presents a vision-based driver's attention monitoring system that estimates the driver's attentional status in terms of four categories: attentive, distracted, drowsy, and fatigue respectively. The attentional status is classified with a variety of parameters such as, percentage of eyelid closure over time (PERCLOS), yawn frequency and gaze direction. Experimental results with different subjects show that the system can classify the driver's attentional status with a reasonable accuracy.

Keywords: Computer vision · Human computer interaction
Attentional status · Yawn frequency · Gaze direction

1 Introduction

Inattention and distraction of drivers are the most obvious reasons for car accidents. According to world health organization (WHO), every year the lives of more than 1.25 million people are cut short as a result of a road traffic crash [1]. Since a fraction of second distraction may cause a severe mishap of lives and wealth and hence active attention of driver is mandatory while driving a car. Intelligent driving assistance system with real time monitoring of driver's attention may reduce accident rate that are mostly occur due to inattentiveness and in turn improve the efficiency in driving. It is quite challenging task for computer vision to monitor the driver's level of attention in real time and aware him/her while level of attention is not adequate for safe driving.

Level of attention may be low during driving for some reasons. For example, while drivers are involved in texting or talking over the cell phone, their eyes off the road due to mind wandering, sleepiness or tiredness. In case of a driver, it is an important issue to keep his/her attention level high while s/he is driving for the sake of his/her as well as the passenger's life. Meanwhile the level of attention of a driver is the measure of concentration while driving in terms of

© Springer Nature Switzerland AG 2019
P. Vasant et al. (Eds.): ICO 2018, AISC 866, pp. 196–209, 2019.
https://doi.org/10.1007/978-3-030-00979-3_20

his/her physical, physiological and behavioral parameters. In case of driver, if s/he loses her/his attention while driving due to some physiological behaviors or cognitive engagement, a serious accident may happen within a second, which may cause serious injuries to the driver as well as passengers [2]. Therefore, driving with adequate level of attention plays a significant role in reducing the accident rate as well as assures safe journey. In this paper, we propose an attention monitoring system that may be introduce in smart vehicle to monitor the level of attention of the driver in real time and it helps to create awareness while s/he is in inattentive or inadequate level of attentional status for safe driving.

Driver's attention monitoring system may be designed in two approaches: sensor-based and vision-based. In the sensor-based system, several sensor need to be embedded in the driver's body to estimate status of attention. Embedding sensors in human body is very complex, uncomfortable and sometimes provides noisy data. In this work we propose computer vision-based approach due to its less complexity and low cost. The propose system seek real-time record to obtain some physical indicators: PERCLOS, yawn frequency and gaze direction found to be the most reliable and valid determination of a driver's attention.

2   Related Work

Although several famous auto companies are conducting researches on driver inattention monitoring systems there is still a quite challenging task to develop a reliable, fully functional and cost-efficient methods in a real driving context. Four main approaches have been developed to detect driver inattention, such as subjective, physiological, driving-behavior-based, and visual-feature-based approaches [3]. Subjective approach identifies the drowsiness of driver through some ratings on their level of drowsiness-verbally or through questionnaires and suggests countermeasure to drowsiness different doses of caffeine depending on level of drowsiness [4].

Physiological approaches involve analysis of vital signals such as brain activity, heart rate, and pulse rate. A recent system is developed to detect vigilance level using not only a driver's electroencephalogram (EEG) signals but also driving contexts as inputs [5]. However, as physiological approaches often require electrodes that are attached to the driver's body, which are intrusive in nature and, therefore, may cause annoyance to the driver, some wireless system has been introduced based on EEG signals [6,7]. Driving-behavior-information-based approaches evaluate the driver's performance over time. Schoiack patent a method that determines and verifies a state of driver alertness by receiving a response at a steering wheel [8]. The feature-based approach analyzes visual features from the driver's facial images. Shibli et al. proposed a driving assistance framework that can estimate the driver's attention and determine his/her level of attention while driving using a simple webcam by estimating his/her face direction, gaze direction, mouth movement, and head pose [9]. Few vision based systems were developed to estimate the driver's attentional status based on distraction [10], PERCLOS [11], facial angel with lip motion [12] and head/gaze direction [13] respectively. Most of these works detected inattentiveness using only one or two visual parameters. In contrast to these, we propose a system that can use three visual cues (i.e., eyes movement, pupil movement, change in mouth region and head orientation) simultaneously, and estimates the driver's level of attention in real time. All computed parameters are combined to estimate the attentionional status of the driver.

3   Proposed Framework

The main purpose of this work is to develop a framework that can estimate status of driver's during driving in terms of different level of attention. Although there are lots of factors involved in determining the level of attention in this work we considered only four factors: eye movements, pupil movements, change in mouth region, and head orientation. Figure 1 illustrates the schematic diagram of our proposed system. The proposed system consists of four major modules. Details description is given in following subsections.

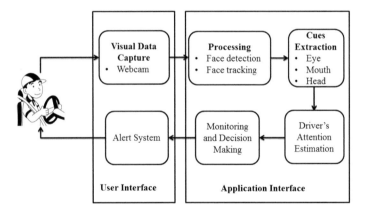

Fig. 1: Schematic diagram of driver's attention monitoring system

Driver's attentionional focus is captured in terms of frontal face by a simple USB webcam, Logitech C170. The video data is captured by the system and sent to the application interface.

3.1   Processing

Detecting the driver's face is the primary step of processing module. This module takes a frame of a video sequence and performs some preprocessing on frame such as gray scale conversion. For face detection, Haar-like features and cascade AdaBoost [14] is used. Face detection algorithm returns a rectangular representation of each of the faces in the frame. For now, we are only interested in the

"Closest" face, and we determine this based on the largest area of the found rectangle using integral image. As extracting cues to characterize attentional status face image analysis and recognition in each frame, performing the face detection algorithm for all frames is computationally complex. Therefore, after detecting face in the first frame, face tracking algorithms is used to track driver's face in the next frames unless the face is lost. In order to track the detected face correlation tracker tool from dlib library [15] is used to just keep track of the relevant region (detected face) from frame to frame. This tool implemented a method that learns discriminative correlation filters for estimating translation and scale independently. Compared to an exhaustive scale space search scheme, this tracker provides improved performance while being computationally efficient [16]. For each frame, the module checks if the correlation tracker is actively tracking a face. If the tracker is actively tracking a face in the image, it updates the tracker. Depending on the quality of the update, a rectangle around the face is drawn indicated by the tracker and cue extraction starts. Figure 2 shows the output from processing step.

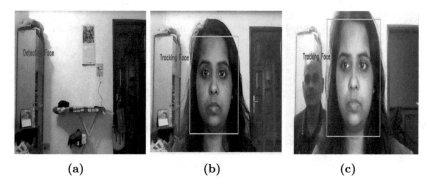

Fig. 2: Results of processing step: (a) detecting, (b) tracking and (c) more than one face appears

3.2   Cues Extraction

Given the face region, we have used facial landmark detector to localize key points of interest along the facial structure of the face region. The pre-trained facial landmark detector inside the dlib library is used to estimate the location of 68 (x, y)-coordinates that map to facial structures on the face. Given these facial landmarks we extracted some visual cues from the face part to characterize driver's attentional status. The indexes of the 68 facial landmark coordinates are the 68 points mark-up used by the iBUG 300-W dataset [17]. In this work we divided cues from driver's visual behaviors for inattentiveness detection into three general categories:

Cues Related to the Eye Region. Eye is the most important area of the face where the cues of drowsiness, fatigue and distraction appear. The cues related to eye region include calculation of Eye Aspect Ratio (EAR) and eye center localization to estimate percentage of eyelid closure over time (PERCLOS) and eye gaze respectively to characterize the attentional status. Six (x, y)-coordinates related to each eye (right or left) are extracted. Figure 3 shows the points detected by facial land mark predictor. After the detection of eyes, eye areas are individually passed to estimate EAR and to detect the center of eye. EAR describes the proportional relationship between width and height of eye. For each eye, i (right or left) EAR_i is calculated using Eq. (1) from [18].

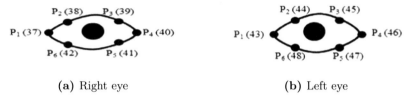

(a) Right eye   (b) Left eye

Fig. 3: Point used to calculate EAR and eye center localization

$$EAR_i = \frac{|P_2 - P_6| + |P_3 - P_5|}{2|P_1 - P_4|}, \tag{1}$$

where, $P_1, \ldots, P_6$ are 2-D coordinates depicted in Fig. 3. *EAR* of both eyes is averaged then using Eq. (2).

$$EAR = \frac{EAR_{Right} + EAR_{Left}}{2}. \tag{2}$$

The *EyeState* (open and closed) for each frame (f) as in Eq. (3).

$$EyeState = \begin{cases} Closed, & \text{if } EAR_f = th_{EAR} \\ Open, & \text{otherwise} \end{cases}. \tag{3}$$

Eyes state was determined to detect eye blink and PERCLOS. Figure 4 shows the opening state and closing state detected by the system.

   In order to estimate the gaze direction center of eyes are detected first. To facilitate the detection, region of interest (ROI) i.e. eye images for both eyes are extracted and re-sized using the points P_1, P_3, P_4 and P_6. Then the histogram is calculated for distribution analysis. Taking the histogram into account, a threshold is set according to the maximum count and the image size. Afterwards, the pixels with higher value above the threshold is eliminated as skin pixels. Dilation and erosion operations are performed to remove small noises and to decrease the size of the white region respectively. Figure 5 shows the step in our system for right eye.

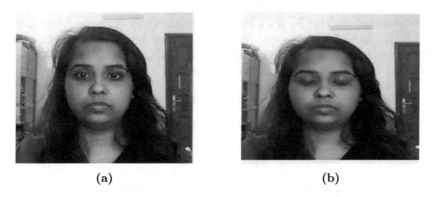

(a)   **(b)**

Fig. 4: Detection of eye state: (a) open eyes and (b) closed eyes

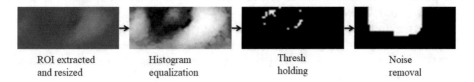

ROI extracted   Histogram   Thresh   Noise
and resized   equalization   holding   removal

Fig. 5: Pre-processing steps for detecting center of eye (right eye)

Visible eyeball area is considered an ellipse. Instead of complex ellipse fitting algorithms we have implemented the border following algorithm proposed by Suzuki and Abe [19] on the binary image for boundary detection. The ellipse center indicates the exact position of the eye center. To do so, moments are applied [20] on the detected ellipse. The centroid $(\bar{x}, \bar{y})$ of the ellipse is calculated using Eq. (4).

$$\bar{x} = \frac{m_{10}}{m_{00}}, \bar{y} = \frac{m_{01}}{m_{00}}, \tag{4}$$

where, for the ellipse with pixel intensities $I(x, y)$, moments m_{ij} are computed as,

$$m_{ij} = \sum_{x,y} I(x, y) x_j y_i. \tag{5}$$

After the center $(\bar{x}, \bar{y})$ of eye for both left and right eye have been determined, Eye Gaze Direction (EGD) is classified as left, front and right using the Eq. (6).

$$EGD = \begin{cases} Left, & \text{if } \theta > 8° \\ Right, & \text{else if } -8° < \theta, \\ Front, & \text{otherwise} \end{cases} \tag{6}$$

where, $\theta = \tan^{-1}(\frac{\triangle x}{\triangle y})$ and $\triangle x = \bar{x}_R - \bar{x}_L, \triangle y = \bar{y}_R - \bar{y}_L$, where, $(\bar{x}_L, \bar{y}_L)$ and $(\bar{x}_R, \bar{y}_R)$ corresponds to the center of left and right eye.

Cues Related to the Mouth Region. Yawn is one of the key symptoms of fatigue [21]. Mouth is wide open is larger in yawning compared to speaking. After detecting the face, we at first isolated the Mouth Region (MR). The coordinates of the *MR* (Fig. 6) are empirically defined as:

$$\begin{bmatrix} x_1 \\ y_1 \end{bmatrix} = \begin{bmatrix} x + \frac{w}{4} \\ y + \frac{11h}{16} \end{bmatrix} \tag{7}$$

$$\begin{bmatrix} x_2 \\ y_2 \end{bmatrix} = \begin{bmatrix} x + \frac{3w}{4} \\ y + h \end{bmatrix}, \tag{8}$$

where, (x, y) is the coordinate of the starting point of detected face with a width of w and height of h. After the region of interest has been identified, operations are performed on this region to measure the changes in mouth area due to wide opening of mouth while yawning.

In order to measure the area of the mouth, we implemented the algorithm [19] mentioned previously to capture the contour of the mouth. To do so, using threshold value (τ), we get an irregular segmentation, S_{MR} of the dark area inside MR as bellow:

$$S_{MR}(x, y) = \begin{cases} 255, & \text{if } MR(x, y) \geq \tau \\ 0, & \text{otherwise} \end{cases}. \tag{9}$$

Due to the dependence of the segmentation on the intensity value of the region, at first the MR is converted in to gray scale image and in order to contrast enhance the darkest and the brightest mouth region histogram equalization is applied. After the segmentation, some noise such as, the shade area under lower lip is eliminated by applying the contour finding algorithm. The yawn is assumed to be modelled with a large vertical mouth opening. When the mouth starts to open, the mouth contour area starts to increase in subsequent frames. Figure 6 shows MR detected by our system.

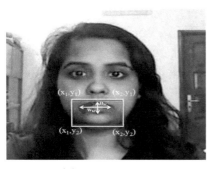

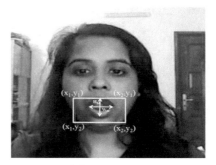

(a) Mouth closed   (b) Mouth open

Fig. 6: Mouth region detected by our system

The rate of such increase is calculated using the ratio (R_M) of width (W_M) to height (H_M) of mouth (Eq. 10) and used as an indication of yawning:

$$R_M = \frac{H_M}{W_M}. \tag{10}$$

If mouth is closed, this ratio is low and it is higher when mouth is open. The number of states to the point where the mouth is wide open is larger in yawning compared to speaking and can be used in differentiating these two states. We marked the mouth state wide opened due to yawning when, $R_M > Th_Y$; where Th_Y is the threshold value calculated empirically. In this paper, we consider that the driver is yawning if we find a significant number of consecutive frames (3s) where the mouth is wide open. The number of yawning is denoted as Y_N. Initially $Y_N = 0$, as the monitoring begins Y_N is updated using following equation:

$$Y_N = \begin{cases} Y_N + 1, & \text{if } R_M > Th_Y \\ Y_N, & \text{otherwise} \end{cases}. \tag{11}$$

Cues Related to the Head. The head pose can be used for detecting distraction behaviors among the categories defined for inattentive states. The normal face orientation while driving is frontal. If the driver's face orientation is in other directions for an extended period of time, it is assumed to be distractions. In order to estimate the head pose at first we need to localize n facial landmarks points corresponding to the selected head model points $(n = 15)$ (Fig. 7a) [17] and then perform head pose calculation. Once the facial landmarks for head models are detected using dlib's facial landmark detector (Fig. 7b), we calculated algebraically the head position and rotation. 3-D Head Pose, h can be expressed using 6 Degree of Freedom (DOF) (Eq. 12) [22] i.e. three for rotations, $r = (r_x, r_y, r_z)^T$, and three for translations, $t = (t_x, t_y, t_z)^T$, where (r_x, r_y, r_z) are represented as Euler angles pitch (β), yaw (α), and roll (γ).

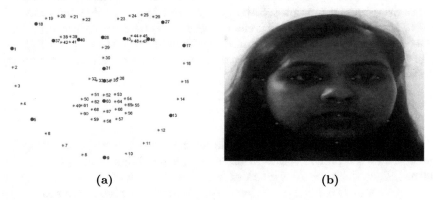

(a)   (b)

Fig. 7: Points (a) used to estimate head pose (marked with blue dots) and (b) detected by our system

$$h = (r, t)^T. \tag{12}$$

Given n 2-D facial landmark points on an input image, $p_{(2 \times n)}$, and their corresponding, reference 3-D coordinates, $P_{(3 \times n)}$—selected on a fixed, generic 3-D face model, head pose is determined by solving the following pinhole model that obtains the 3-D to 2-D projection of the 3-D landmarks onto the 2-D image using a perspective transformation:

$$s[p, 1]^T = A[R|t]P^T, \tag{13}$$

where, s is scaling factor, A is a camera matrix and $[R|t]$ is joint rotation-translation matrix. Then vector $r = (r_x, r_y, r_z)$ is obtained from matrix R using Rodrigues rotation formula:

$$R = \cos\theta I + (1 - \cos\theta)rr^T + \sin\theta \begin{bmatrix} 0 & -r_z & r_y \\ r_z & 0 & -r_x \\ -r_y & r_x & 0 \end{bmatrix}, \tag{14}$$

where, I is vector in $\mathbb{R}^3$ and $\theta = \|r\|_2$.

From the obtained vector r, we got the Euler angle yaw(α), which is used to model the user's attention by estimating the Face Direction (FD) by using Eq. (15).

$$FD = \begin{cases} Left, & \text{if } -90° \leq \alpha < -30° \\ Right, & \text{else if } 30° < \alpha \leq 90° \\ Front, & \text{otherwise} \end{cases} \tag{15}$$

3.3   Attention Estimation of Driver

In this paper, we have proposed following criterion to determine inattention: Drowsiness is one of the major reasons for driver to be inattentive. Here, to detect drowsiness we monitored the eye behavior by estimating the **PERCLOS**. **PERCLOS** is one of the most reliable parameter used to detect drowsiness. **PERCLOS** is the percentage of duration of closed-eye state in a specific time interval T_1 (1 min or 60 s), excluding the time spent on normal closure (eye blinks) and can be defined as follows:

$$PERCLOS = \frac{t}{T_1} \times 100\%, \tag{16}$$

where, t is the duration that eye were closed. Eye blink is a reflex that closes and opens the eyes rapidly. Studies revealed that, a real blink of an eye takes 300–400 ms. Since there are 1000 ms in each second, a blink of an eye takes around 1/3 of a second. The duration of closed-eyes state in driver's blink increases when drowsiness occurs. So in order to avoid confusion of taking an eye blink as a state of driver's inattentiveness **PERCLOS** is estimated. A higher value of **PERCLOS** indicates higher drowsiness level and vice versa. In order to detect driving fatigue, along with **PERCLOS** another criterion used is yawning. Excessive yawning (1–4 yawns per minute) is associated with fatigue. Whenever, a yawning is detected, corresponding counter is incremented to keep track of Yawning Frequency (YF) using Eq. (17).

$$YF = \frac{Y_N}{T_2},\tag{17}$$

where, T_2 is the time window. To monitor the distraction state of a driver, we focused on estimating driver Gaze Direction (GD). To obtain the gaze direction of the driver, we need to take into account both **FD** and **EGD**. For a driver, the nominal **GD** is frontal. Looking at other directions for an extended period of time (T_3) may indicate distraction. **GD** is computed as the composition of face direction **FD** and **EGD** for over T_3 time and can be defined as,

$$GD = \{FD, EGD\}.\tag{18}$$

If head rotation is wide enough, or driver is wearing sun glass eyes might not be visible, so the **GD** is calculated using only the known **FD**.

3.4 Monitoring and Decision Making

The visual behavior of driver is continuously captured, processed, extracted and updated accordingly in the declarative memory. The information provided by previous component is used to determine if and what decisions must be made based on the Status of Attention (SoA) of the driver estimated for past few frames. The most common decision making might be whether to alert the driver by generating warning message if *SoA* value indicating inattentive status is detected using the Eq. (19).

$$SoA = f(PERCLOS, YF, GD) = \begin{cases} Drowsy, & \text{if } PERCLOS \geq 0.125 \\ Fatigue, & \text{elseif } YF > 1 || 0.048 \leq PERCLOS < 0.125 \\ Distracted, & \text{elseif } GD == Right || GD == Left \\ Attentive, & \text{otherwise} \end{cases}\tag{19}$$

4 Experimental Results

The system is developed and tested on a Windows 10 PC with an Intel Core i5 1.60 GHz processor and 4 GB RAM. The system is developed using Python, OpenCV library.

In order to evaluate the system performance the accuracy of the system was measured using the Eq. (20) for video sequence.

$$Accuracy = \frac{D_F}{T_F} \times 100\%,\tag{20}$$

where, D_F is the number of frames in which attentional status was correctly recognized and T_F is the total number of frames in the test sequence.

4.1   Accuracy of Estimating Attentional Status

We tested the proposed system for four types of attentional status: attentiveness, drowsiness, fatigue and distraction. We asked 3 drivers [average age = 38 years] to interact with the system, each of them spent 4 min in front of the system posing different expressions. Figure 8 shows sample attentional status classification snapshots. Table 1 shows the accuracy of different attentional status using the Eq. (20).

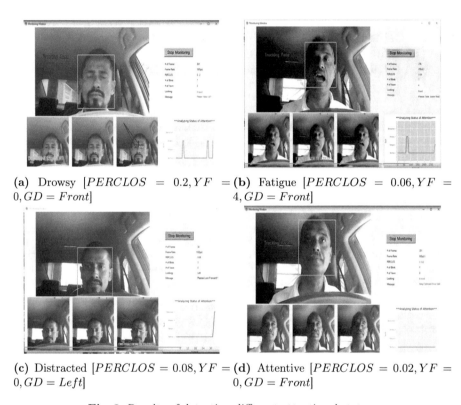

(a) Drowsy [$PERCLOS = 0.2, YF = 0, GD = Front$]

(b) Fatigue [$PERCLOS = 0.06, YF = 4, GD = Front$]

(c) Distracted [$PERCLOS = 0.08, YF = 0, GD = Left$]

(d) Attentive [$PERCLOS = 0.02, YF = 0, GD = Front$]

Fig. 8: Results of detecting different attentional status

The results shows that the system works quite satisfactory for measuring attentional status. Its accuracy in determining drowsiness is very good, whereas while estimating gaze direction some issues such as distance from camera, wearing glasses plays crucial role in decreasing the accuracy level.

Table 1: Accuracy of different attentional status

Participant	No. of frames	Accuracy (%)			
		Drowsiness	Fatigue	Distraction	Attentiveness
1	1280	100	98	92	93
2	1360	100	89	88	89
3	1090	93	92	87	86
	Average	97.67	93	89	89.34

4.2   Overall System Accuracy

The overall accuracy for the system was determined using Eq. (20). Proposed system was tested in four sequences that contain more than 25000 frames. Sequences were captured using a video camera placed on the car dashboard at a distance (0.6 m–0.9 m) from the driver under different daylight conditions ranging from broad daylight to parking garages with subject wearing glasses and not wearing glasses, talking in the phone in order to test the robustness of the system. Figure 9 shows the experimental setup. Talking in the phone didn't affect much in this case. Table 2 shows the overall accuracy of the proposed system in detecting attentional status.

Fig. 9: Experimental setup

Table 2: Overall accuracy of the system in determining attentional status

Sequence	D_F	N_F	Accuracy(%)
S_1	5714	6279	91
S_2	5328	7200	74
S_3	5740	6450	89
S_4	4088	5679	72
		Average	81.5

S_1: Broad day light and subject not wearing glass;
S_2: Broad day light and subject wearing clear glass;
S_3: Parking garage and subject not wearing glass;
S_4: Parking garage and subject wearing glass

Results indicate that detection accuracy is high in Sequence-1 and 3 than Sequence-2 and 4, which suggests that our system works well in different lighting conditions. The detection performance falls when wearing clear glasses due to reflection, which is the area of our system that needs more work considering the current state of art.

5   Conclusion

Driver's inattention is the most prominent factor for car clashes. A system that monitors the driver's attention and alert him/her when level of attention is inadequate may reduce car accident rate. The paper proposed a vision-based driver's attention monitoring system by classifying the attentional status into attentive and inattentive (fatigue, drowsy and distracted) state which will be helpful in preventing accident. Attentional status is classified using three parameters-PERCLOS, yawn frequency and gaze direction. Experimental result reveals that the system can detect the attentional status with reasonable accuracy. Future research will involve detection of yawning using facial landmarks to establish stronger facial geometric constraints. We also plan to include alarming system to create the awareness of the driver while level of attention is low and car tracking system.

References

1. Global Status Report on Road Safety 2015, World Health Organization, WHO Press, Switzerland. http://www.who.int/entity/violence_injury_prevention/road_safety_status/2015/en/index.html
2. Klauer, S.G., Dingus, T.A., Neale, V.L., Sudweeks, J.D., Ramsey, D.J.: The impact of driver inattention on near-crash/crash risk: an analysis using the 100-car naturalistic driving study data. Technical report, National Highway Traffic Safety Administration, Washington, DC, USA (2006)
3. Arun, S., Sundaraj, K., Murugappan, M.: Driver inattention detection methods: a review. In: IEEE Conference on Sustainable Utilization and Development in Engineering and Technology (STUDENT), pp. 1–6, Kuala Lumpur (2012)
4. De Valck, E., Cluydts, R.: Slow-release caffeine as a countermeasure to driver sleepiness induced by partial sleep deprivation. J. Sleep Res. **10**, 203–209 (2001)
5. Guo, Z., Pan, Y., Zhao, G., Cao, S., Zhang, J.: Detection of driver vigilance level using EEG signals and driving contexts. IEEE Trans. Reliab. **67**(1), 370–380 (2018)
6. Wang, H., Dragomir, A., Abbasi, N.I., Li, J., Thakor, N.V., Bezerianos, A.: A novel real-time driving fatigue detection system based on wireless dry EEG. Cogn. Neurodynamics, 1–12. Springer, Netherlands (2018)
7. Li, G., Chung, W.Y.: Combined EEG-Gyroscope-tDCS brain machine interface system for early management of driver drowsiness. IEEE Trans. Hum.-Mach. Syst. **48**(1), 50–62 (2018)
8. Schoiack, M. M. V.: Driver drowsiness detection and verification system and method. U.S. Patent 8,631,893 B2 (2014)
9. Shibli, A. M., Hoque, M. M., Alam, L.: Developing a vision-based driving assistance system. In: 2018 International Conference on Emerging Technologies in Data Mining and Information Security (IEMIS), Kolkata, India (2018)
10. Chien,J.-C., Chen, Y.-S., Lee, J.-D.: Improving night time driving safety using vision-based classification techniques. Sensors **17**(10), 2199 (2017)
11. Mandal, B., Li, L., Wang, G.S., Lin, J.: Towards detection of bus driver fatigue based on robust visual analysis of eye state. IEEE Trans. Intell. Transp. Syst. **18**(3), 545–557 (2017)

12. Chowdhury, P., Alam, L., Hoque, M. M.: Designing an empirical framework to estimate the driver's attention. In: 5th International Conference on Informatics, Electronics & Vision (ICIEV), pp. 513–518, Dhaka, Bangladesh (2016)
13. Vicente, F., Huang, Z., Xiong, X., Torre, F.D.I., Zhang, W., Levi, D.: Driver gaze tracking and eyes off the road detection system. IEEE Trans. Intell. Transp. Syst. **16**(4), 2014–2027 (2015)
14. Freund, Y., Schapire, R.E.: A decision-theoretic generalization of on-line learning and an application to boosting. In: Computational Learning Theory, pp. 23–37. Springer, Heidelberg (1995)
15. De, K.: Dlib-ml: a machine learning toolkit. J. Mach. Learn. Res. **10**, 1755–1758 (2009)
16. Martin, D., Häger, G., Khan, F.H., Felsberg, M.: Accurate scale estimation for robust visual tracking. In: Proceedings of the British Machine Vision Conference. BMVA Press (2014)
17. Facial Point Annotations. https://ibug.doc.ic.ac.uk/resources/facial-point-annotations/
18. Soukupova, T., Cech, J.: Real-time eye blink detection using facial landmarks. In: Cehovin, L., Mandeljc, R., Struc, V. (eds.) 21st Computer Vision Winter Workshop. Rimske Toplice, Slovenia (2016)
19. Suzuki, S., Abe, K.: Topological structural analysis of digitized binary images by border following. Comput. Vis. Graph. Image Process. **30**(1), 32–46 (1985)
20. Hu, M.K.: Visual pattern recognition by moment invariants. IRE Trans. Inf. Theory **8**, 179–187 (1962)
21. Arbuck, D.: Is yawning a tool for wakefulness or for sleep? Open J. Psychiatry **3**(1), 5–11 (2013)
22. Chang, F.-J., Tran, A.T., Hassner, T., Masi, I., Nevatia, R., Medioni, G.: Face-PoseNet: making a case for landmark-free face alignment. In: 2017 IEEE International Conference on Computer Vision Workshops (ICCVW), pp. 1599–1608, Venice, Italy (2017)

Characterizing Current Features of Malicious Threats on Websites

Wan Nurulsafawati Wan Manan[✉], Abdul Ghani Ali Ahmed,
and Mohd Nizam Mohmad Kahar

Faculty of Computer System and Software Engineering, University Malaysia
Pahang, Pekan, Malaysia
{safawati,abdulghani,mnizam}@ump.edu.my

Abstract. The advance growth of cybercrime in recent years especially in high critical networks becomes an urgent issue to the security authorities. They compromised computer system, targeting especially to government sector, ecommerce and banking networks rigorously and made it difficult to detect the perpetrators. Attackers used a powerful technique, by embedding a malicious code in a normal webpage that resulted harder detection. Early detection and act on such threats in a timely manners is vital in order to reduce the losses which have caused billions of dollars every year. Previously, the detection of malicious is done through the use of blacklisting repository. The repository or database was compiled over time through crowd sourcing solution (e.g.: PishTank, Zeus Tracker Blacklist, StopBadWare.. etc.). However, such technique cannot be exhaustive and unable to detect newly generated malicious URL or zero-day exploit. Therefore, this paper aims to provide a comprehensive survey and detailed understanding of malicious code and URL features which have been extracted from the web content and structures of the websites. We studied the characteristic of malicious webpage systematically and syntactically and present the most important features of malicious threats in web pages. Each category will be presented along with different dimensions (features representation, algorithm design, etc.).

Keywords: Cybercrime · Malicious website · Malicious features

1   Introduction

Nowadays, the Internet has grown tremendously in communication technologies. The internet provides an essential infrastructure to the online business users especially promoting of businesses across online platform, with many applications including online-banking, e-commerce, and social networking.

Please note that the AISC Editorial assumes that all authors have used the western naming convention, with given names preceding surnames. This determines the structure of the names in the running heads and the author index.

© Springer Nature Switzerland AG 2019
P. Vasant et al. (Eds.): ICO 2018, AISC 866, pp. 210–218, 2019.
https://doi.org/10.1007/978-3-030-00979-3_21

In fact, in today's digital age, it plays an important role in various domains including digital economy and IoT devices. As a result, it becomes a demanding trend in transforming business practices towards the web, and continuously increasing. However, with the advancement of this technology, crime rate over the Internet has risen massively, with various sophisticated techniques to attack and stealing users' personal data.

With this in mind, cybercrime becomes the main threat of this technology. Several of attacks include rogue websites that sell counterfeit goods, financial fraud by tricking users into revealing confidential information which finally leads to identity breach or stealing users' money, or even installing malware in the users' system [1].

As reported for the past two years, 2016 will be remembered with numerous breaches of personal data by largest companies that offer online services. One of the most remarkable is, Ransomware attack which caused billions of dollar lost. Besides, variety of techniques are used to implement malicious attack through websites. Figure 1 below illustrates some examples of attack classification.

Fig. 1. Website attack classification

These malicious web attacks can be classified into several types such as, cross sites scripting (XSS), code injection, URL redirection, Drive-by – download, malvertising, phishing sites, hacking attempts and many more.

With millions of users population having the internet access, malicious content on the web has become predominant attacking technique. Furthermore, the distribution of compromised URL is one of the common attack operations by the perpetrator to compromise any user's account, using the internet to access and steal personal information from other users.

Most of this attacks are related to JavaScript, which is one of the most important programming language for client-side scripting. This code embedded in HTML pages runs into webpage locally or can be fetched from the remote server. Besides, it is generally used in web pages to improve the interactivity and functionality of their websites [20]. Correspondingly, in this paper we proposed and validate a set of features extracted from website in order to distinguish between malicious and benign webpage. In particular, the features considered in the paper aim at characterizing classes of web pages/content components that are frequently used by malware writers to launch an attack trough a webpage. The webpage which could contain pieces of malicious applications or could be a channel to perform redirection to their malicious pages. The features have been chosen based on various comparisons from different researchers' point of view.

The chosen features can be used to describe about feature characteristics and categories which have been used in malicious web detection procedure. The proposed set of features is expected to correctly classify malicious websites with a high level of precision rate. The remainder of this paper is organized as follows. In Sect. 2, we present the related works on previous research in detecting malicious websites. Section 3 discusses on characteristic representation in identifying malicious website. The last section concludes the proposed ideas and future work.

2 Related Works

A large number of techniques [2, 3] have been proposed for the purpose of confronting malicious website. In the last few years, blacklisted URL method has become a preferred technique in detecting a malicious web. It recorded the list of a web address which had been declared earlier as a malicious web. The advantages of blacklist-based technique include easy implementation and low false positive rate [4]. However, this technique was unable to detect malicious URLs that are not listed in the database and it was difficult to maintain the URL list, since new URL expands and is generated every day.

The main focus in this review is to group a set of features from the webpages and use them to deciding whether a website contains a threat, or not. Meanwhile, most of these techniques are depending on features extraction. Some methods execute the page in order to extract the features and due to the time and resource consuming of this method. As a result, detecting such malicious websites quickly with high precisions is necessitated.

Garera [5] has proposed the extraction of the features from URLs is only possible to tell whether or not a URL belongs to a phishing attack without requiring any knowledge of the corresponding page data.

McGrath and Gupta have studied Phishing URLs anatomy and showed that phishing URLs properties significantly differ from normal distribution and contained a target brand name. Besides, the length of the domain name also became a strong indicator of Phishing URL.

While [6, 7] have proposed a usage of several features for host-based category, such as, IP Address Properties, WHOIS is information, location, Connection speed and DNS-related properties.

In HTML, [8] has proposed several features categories: (1) length of document, (2) average length of words, (3) word count, (4) distinct word count, (5) word count in a line, (6) the number of NULL characters, (7) usage of string concatenation, (8) unsymmetric HTML tags, (9) the link to remote source of scripts, (10) invisible objects.

While in JavaScript, [9] also used several similar features with other researcher [10, 11, 14] such as; escape(), eval(), link(), unescape(), exec(), link(), and search() functions.

Another important criteria in detecting malicious JavaScript is identifying obfuscated JavaScript code. Kim et al. [12] has suggested that, the increasing length of string compared with the normal string is another major criteria in defining Obfuscated JavaScript.

3   Characteristic Representation

In this survey, we focus primarily on the static analysis techniques where a web page, such as its textual content, features of its HTML and JavaScript code, and characteristics of the associated URL will be considered. We have identified several categories that we gather for malicious website features based on most related research from several researches. Figure 2 below depicts our four category of malicious websites based feature/characteristic analysis which are, URL Based features, Content Based features, Network protocol and HTTP Response features [2, 3, 7, 13, 14].

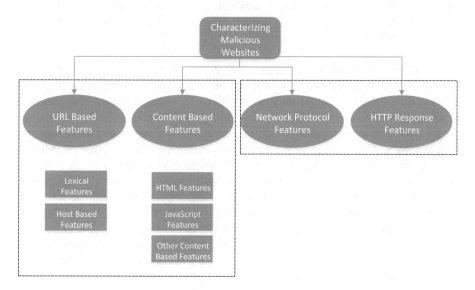

Fig. 2. General classification of malicious web page detection based

3.1   URL Based Features and Classification

A URL is the abbreviation of Uniform Resource Locator, specifically the address of a resource on the Internet. A URL indicates the location of data and resource on Internet as well as the protocol which has been used to access the World Wide Web. A URL has several main components namely: (i) protocol, (ii) subdomain, (iii) domain name, (iv) top level domain, and (v) path domain. Protocol indicates what protocol to use between communication devices; HTTP, HTTPS, FTP, etc. Whilst, the primary domain is the most vital part in URL system, where it provides name given to the Internet Protocol (IP) through Domain Name System (DNS).

As stated earlier, a URL based features, can be divided into two sub categories in order to define their characteristic that exist in URL links: Lexical Features and Host based Features.

A. *Lexical Features*

Lexical features are feature elements of the URL name or string. It is defined based on how the URL "looks" or "differently look" in the eyes of the users. It is also based on textual properties itself in order to define the URL links of which may look like benign or malicious URL. Based on [2], this feature classification is used together with several other features (e.g. host-based features) to improve model performance.

B. *Host Based Features*

Host based features are associated with the hostname properties of the URL [6]. Malicious URL is normally hosted in trusted hosting providers as it looks like legitimate URL. Besides, it is very common that malicious URL comes with no DNS name and hosted by well-known Top Level Domain (TLD) [9].

3.1.1   Features/Attributes of URLs

For malicious URL based detection method, we have suggested several features which can be used to define useful information in deciding whether the website are malicious or benign.

Features below are associated with suspicious URL patterns and characters. We have compared from several related researchers who have done thorough research in characterizing malicious web pages using URL features. They listed characters such as: (1) length of URL string, (2) length of the host name, (3) number of subdomain, (4) number of dots (.), (5) number of specific symbol (hypens(-), underscores(_),forwarded slashes(/) equal sign(=)), (6) Suspicious Port Number, (7) Number of TLD and out of TLD position, (8) Suspicious words, (9) Domain features, (10) Number of backslash (/), (11) Country matching, (12) Length of path, (13) DNS record features, (14) WHOIS record, and (15) Value of TTL. Table 1 below depicts the summary of the suggested URL properties which can be used as a reference for future proposed malicious web detection.

3.2   Content Based Features and Classification

Content based features are content which are obtained upon opening or downloading the website. It can be divided into three sub categories in order to define their malicious characteristic that exist in website: HTML features, JavaScript Features and other content categories based features.

A. *HTML Features*

HTML features are based on information about the whole page content and on structural information derived from HTML code. It was analyzed based on website page length, the white spaces percentages and the location of elements in the web pages [9]. This HTML features, as proposed by [8, 9] used lexical features from HTML of the websites.

Table 1. Summary of proposed URL features

	Features name	Description
1	Length of URL	Length of the URL
2	Hostname	Length of hostname
3	Suspicious strings	Whether URL has "@", "//', "?", "=","-","_"
4	Number of subdomain	Number of dots in domain
6	Suspicious Port Number	Whether Well-known port numbers for HTTP or HTTPs
7	Number of TLD and out of TLD position	More than one TLD in URL, and out of TLD position
8	Suspicious words	Whether URL has suspicious terms or words used
9	Domain features	Whether primary domain is similar to whitelisted domains, the length of domain, or IP is used as a domain
10	Number of "/"	Number of backslash in URL
11	Country matching	TLD country, and country are equal or not
12	Path	Length of the file path in the URL
13	DNS record	Whether URL has DNS record
14	WHOIS record	Domain age in WHOIS record
15	Value of TTL	TTL value of domain

B. *JavaScript Feature*

JavaScript is the most popular scripting language which used similar concept based as HTML (lexical and statistical). The JavaScript analysis features based content type of text or script in the webpage an inline <script> element.

Other Content Based Features

Other Content based features are content which is obtained from other source such as based on visual features, action or behavioral analysis of the users. For example, features based on wall posts, shared linked, multimedia data, chat and message logs etc. [15, 16].

3.2.1   Features/Attributes of JavaScript and HTML in Webpages

For malicious HTML, and JavaScript features, we have suggested several items which can be used to define useful information in deciding whether the website are malicious or benign. As previously defined, HTML are based both on statistical information about the raw content of structural information. Based on some previous researches which have been conducted by several researchers [2, 8, 17], agreed that common features below been used in detecting malicious activity. Table 2 below depicted proposed usage of features for HTML.

While JavaScript features result from the static analysis of either a JavaScript file content type of text/javascript of each script included in a web page <script> element. Common features that similarly defined by other researcher [3, 8, 13, 18] can be summarized in Table 3 below.

Table 2. HTML based features

No	HTML-based features	
	Features name	Description
1	Document length	The length of document
2	Number of words	Words count in a line
3	number of lines	Average length with words
4	Blank spaces	Number of blank spaces in a line
5	Number of NULL Char	Number of e escape sequence \0
6	Number of hidden elements	Number of hidden attribute in webpages
7	Number of iframes	Number of <iframe> tag inline frame, used to embed another document within the current HTML document
8	Unequal HTML tags	HTML tags appear in an inline frame unsequence
9	String sequence	Sequence of string exist in document

Table 3. JavaScript based features

No	JavaScript- based features	
	Features name	Description
1	eval()	Built-in functions. Function evaluates or executes an argument
2	escape()	Function which encode a string, makes a string portable
3	unescape()	The dual of escape() which function decodes an encoded string.
4	window.open ()	Method opens a new browser window
5	CharCodeAt()	Combination of method and function
6	fromCharCode ()	
7	parseInt()	
8	Replace()	Type of method which replace specific subsequence of characters
9	link()	Some of the suspicious Javascript function which could occur frequently in many different attack
10	exec()	
11	search()	

3.3   Network Protocol Features/HTTP Response Features of Suspected Malicious Performed in Webpages

The Network protocol and HTTP response also can be considered as an important element in detecting malicious website. This features could indicate the existing of malicious website. As mentioned by Li Xu [19], these network layer features had contributed important indication as malicious URLs that caused crawler to send multiple Domain Name Service (DNS) queries and connect multiple web servers which cause high volume communication. Proposed features which have been suggested are: *Duration, Source_app_byte, Avg_remote_pkt_rate, Dist_remote_TCP_port.*

4 Conclusion

Malicious website has become a critical issue to the web users nowadays. In recent years, not only the malicious website is on the rise, but also the skill of perpetrator or attacker has become more sophisticated. There are many existing techniques with various characteristic of features suggested have been proposed in detecting malicious website. However, some of the features proposed only focus on certain detection web threats issues, such as, malicious URL or malicious JavaScript in web pages only. Therefore, we have proposed several elements in detecting the malicious websites with combination various features hopefully able to detect the malicious activity or threat in website accurately and precisely with minimum duration.

Acknowledgement. This study was fully funded by the Ministry of Higher Education in Malaysia (RDU 160106).

References

1. Ahmed, A.A., Li, C.X.: Locating and collecting cybercrime evidences on cloud storage: review. In: 2016 International Conference on Information Science and Security, ICISS 2016 (2017)
2. Sahoo, D., Liu, C., Hoi, S.C.H.: Malicious URL detection using machine learning: a survey, pp. 1–21 (2017)
3. Awathe, A.: Malicious web page detection through classification technique : a survey, vol. 8491, pp. 74–79 (2017)
4. Akiyama, M., Yagi, T., Itoh, M.: Searching structural neighborhood of malicious URLs to improve blacklisting. In: Proceedings of 11th IEEE/IPSJ International Symposium on Applications and Internet, SAINT 2011, pp. 1–10 (2011)
5. Garera, S., Provos, N., Chew, M., Rubin, A.D.: A framework for detection and measurement of phishing attacks. In: Proceedings of 2007 ACM Workshop on Recurring Malcode - WORM 2007, p. 1 (2007)
6. Ma, J., Saul, L., Savage, S., Voelker, G.: Identifying suspicious URLs: an application of large-scale online learning. In: Proceedings of the 26th Annual International Conference on Machine Learning, pp. 681–688 (2009)
7. Ma, J., Saul, L.K., Savage, S., Voelker, G.M.: Beyond blacklists : learning to detect malicious web sites from suspicious URLs. In: World Wide Web Internet Web Information System, pp. 1245–1253 (2009)
8. Hou, Y.T., Chang, Y., Chen, T., Laih, C.S., Chen, C.M.: Malicious web content detection by machine learning. Expert Syst. Appl. **37**(1), 55–60 (2010)
9. Canali, D., Cova, M., Vigna, G., Kruegel, C.: Prophiler : a fast filter for the large-scale detection of malicious web pages categories and subject descriptors. In: Proceedings of International World Wide Web Conference, pp. 197–206 (2011)
10. Choi, H., Zhu, B.B., Lee, H.: Detecting malicious web links and identifying their attack types. WebApps **11**, 11 (2011)

11. Eshete, B.: Effective analysis, characterization, and detection of malicious web pages. In: Proceedings of 22nd International Conference on World Wide Web companion, pp. 355–360 (2013)
12. Kim, B., Im, C., Jung, H.: Suspicious malicious web site detection with strength analysis of a javascript obfuscation. Int. J. Adv. Sci. Technol. **26**, 19–32 (2011)
13. Canfora, G., Visaggio, C.A.: A set of features to detect web security threats. J. Comput. Virol. Hacking Tech. **12**(4), 243–261 (2016)
14. Seshagiri, P., Vazhayil, A., Sriram, P.: AMA: static code analysis of web page for the detection of malicious scripts. Proc. Comput. Sci. **93**, 768–773 (2016)
15. Saquib, S., Ali, R.: Malicious Behavior in Online Social Network
16. Neeraja, M., Prakash, J.: Detecting Malicious Posts in Social Networks Using Text Analysis, vol. 5, no. 6, pp. 2015–2017 (2016)
17. Eshete, B.: Security and Privacy in Communication Networks, vol. 106, p. 2015 (2013)
18. Fraiwan, M., Al-Salman, R., Khasawneh, N., Conrad, S.: Analysis and identification of malicious javascript code. Inf. Secur. J. **21**(1), 1–11 (2012)
19. Xu, S., Bylander, T., Maynard, H.B., Sandhu, R., Xu, M.: Detecting and characterizing malicious websites (2014)
20. Bielova, N.: Survey on JavaScript security policies and their enforcement mechanisms in a web browser. J. Log. Algebr. Program. **82**(8), 243–262 (2013)

A Deep Learning Framework on Generation of Image Descriptions with Bidirectional Recurrent Neural Networks

J. Joshua Thomas$^{(\boxtimes)}$ and Naris Pillai$^{(\boxtimes)}$

Department of Computing, School of Engineering, Computing, and Built Environment, KDU Penang University College, 32, Anson Road, George Town, Penang, Malaysia
joshopever@yahoo.com, narispillai@gmail.com

Abstract. The aim of the paper is to develop a deep learning framework for a model that generates natural descriptions of pictures (data) and their sections so as to search out a lot of insights. Image recognition is one in all the promising applications of visual objects. In this study, a small-scale food image data set consisting of 5115 pictures of fourteen classes and an eight-layer CNN was made to acknowledge these pictures. CNN performed far better with associate degree overall accuracy 54%. The approach influences information sets of images and their patterns bi directional recurrent neural network (BRNN) will concerning the intern- model correspondences between prediction and visual information for calorie estimation. Data expansion techniques were applied to extend the dimensions of trained images, that achieved a considerably improved accuracy of 74% stop the over fitting issue that occurred to the CNN while not misclassified.

Keywords: Deep learning · Convolutional Neural Network
Data augmentation · Malaysian food chain

1 Introduction

Digital data, in all shapes and sizes, is growing exponentially. The internet is processing 1826 petabytes of data per day [1]. In 2012, digital information grew nine times in volume in just five years [2]; and by 2020 its amount in the world is expected to reach 35 trillion gigabytes [3]. The high demand of exploring and analyzing big data has encouraged the use of data-hungry machine learning algorithms like deep leaning (DL). DL has gained huge success in a wide range of applications such as computer games, speech recognition, computer vision, natural language processing, self-driving cars, among others [4]. It is safe to say now DL is changing our everyday life. In Year 2018, DL represented AI technologies were ranked at the top position [5]. Deep learning allows computational models that are composed of multiple processing layers to learn representations of data with multiple levels of abstraction. These approaches have intensely improved the state-of-the-art in speech recognition, visual object recognition, object detection and many other domains such as drug discovery and genomics.

© Springer Nature Switzerland AG 2019
P. Vasant et al. (Eds.): ICO 2018, AISC 866, pp. 219–230, 2019.
https://doi.org/10.1007/978-3-030-00979-3_22

Conventional machine-learning techniques were limited in their ability to process natural data in their raw form. For decades, constructing a pattern-recognition or machine-learning system required careful engineering and considerable domain expertise to design a feature extractor that transformed the raw data (such as the pixel values of an image) into a suitable internal representation or feature vector from which the learning subsystem, often a classifier, could detect or classify patterns in the input. Representation learning is a set of methods that allows a machine to be fed with raw data and to automatically discover the representations needed for detection or classification. For classification tasks, higher layers of representation amplify aspects of the input that are important for discrimination and suppress irrelevant variations. An image, for example, comes in the form of an array of pixel values, and the learned features in the first layer of representation typically represent the presence or absence of edges at particular orientations and locations in the image. The second layer typically detects objects by spotting particular arrangements of edges, regardless of small variations in the edge positions. The third to eighth layer may assemble treatments into larger combinations that correspond to parts of familiar objects, and subsequent layers would detect objects as combinations of these parts. To compute an objective function that measures the error (or distance) between the output scores and the desired pattern of scores. The machine then modifies its internal adjustable parameters to reduce this error. These adjustable parameters, often called weights, are real numbers that can be seen as 'buttons' that define the input–output function of the machine. In a typical deep-learning system, there may be hundreds of millions of these adjustable weights, and hundreds of millions of labelled examples with which to train the machine. The rest of the article is organised as sub-sections reflecting the literature review, proposed algorithmic model, dataset annotation, methodology, Implementation, evaluation and conclusion.

2 Related Work

2.1 Convolutional Neural Network (CNN)

This project was proposed by [7] under the title "Food Image Recognition by Using Convolutional Neural Networks (CNN)". As in the title, the researcher chose Convolutional Neural Networks (CNN) deep learning architecture to achieve image recognition [8]. CNN is a type of deep learning approach that consists of multilayer neural network [9]. The layers are used to extract information and combine features from a given two-dimensional data which in this case would be the food image. The research also stated the reason behind the decision which was due to CNN ability in learning optimal feature form pictures adaptively. For the image recognition, the system utilizes 5822 pictures for ten classifications of food with five layers of Convolutional Neural Networks (CNN) layer in the framework's machine learning design [8]. Each layer consists of two components which are convolution and max-pooling and each of them extracts a different number of features maps such as 32, 64 and 128 respectively.

At the very end of the CNN architecture, a fully connected layer to connect all the neuron from the first CNN layer is implemented before the actual output stage. In the output layer, 10 softmax neurons presented which correspond to the ten categories of food chosen for the image recognition. The initial three layer is aimed to perform a different number of highlight mapping procedure to pull and pass useful data to the following layer. For the testing and implementation of CNN, the researcher used "keras" package, which is a neural network library [10], on top of "theano" python library using Spyder virtual environment. Another project which uses similar architecture was from. This system was aimed to achieve fruit classification and grading. CNN plays its role here by extracting the features for the fruit image after trained on hundreds of images with associated labels. The CNN architecture here only contains 4 layers which are Convolution, Pooling, Flattening and Full Connection. The CNN architecture used for this system was created based on Keras on top of the TensorFlow framework. The convolution layer feature detector in terms of image sharpening, edge detection and enhancement, median, and emboss were applied to an image to create feature map. Relu activation function is used here to crack the linearity on the input image. In the pooling layer, it helps in mapping the correct highlights in a given picture, which can be utilized to outline the same object in different pictures. In flattening layer, the entire pooled map is converted into a single vector as it will be the input component for the neural network. Finally, all the neurons are joint in the last layer. For the CNN model training, the researchers use Central Processing Unit (CPU) instead of Graphical Processing Unit (GPU) as GPU may require more power to train the model compared to CPU [11].

2.2 Recurrent Neural Network (RNN)

The first RNN based project considered for this field was from Karol Gregor, Ivo Danihelka Alex Graves and Daan Wierstra who used RNN for image generation. The RNN is used in a network called Deep Recurrent Attentive Writer (DRAW). DRAW is an encoder link that packs the actual pictures given in the training, and a decoder that rebuilt pictures subsequent to getting codes [12]. The next project which utilizes RNN was from Bolan Su and Shijian Lu for an "Accurate Scene Text Recognition" system. In their proposed system the researchers implemented LSTM architecture as well due to RNN long preparing process as the error course integral rots exponentially coupled with the sequence [13]. Besides, Histogram of Oriented Gradient (HOG) was also integrated into to the system for feature extraction process. RNN nodes were replaced with LSTM internal memory structure to determine the output activation of the network with the input help at a time to store it in the internal memory at t-1. LSTM also helped RNN in drawing out errors in the training process as well as recalling contextual data [13]. RNNLIB was used for the multilayer RNN implementation and testing. In the RNN output layer, Connectionist temporal classification (CTC) were adapted for unsegmented data labeling. The RNN architecture for this system was trained using the back-propagating algorithm. In this work shares the high-level goal of densely annotating the contents of images with many works before us. Barnard et al. [14] and Socher et al. [15] studied the multimodal correspondence between words and images to annotate segments of images. Several works [16–19] studied the problem of holistic scene understanding in which the food type, objects and their spatial support in the image is inferred. However, the focus of these works is on correctly labeling food types, objects and calorie information with a fixed set of categories, while our focus is on richer and higher-level descriptions of the food types. The task of describing images with sentences has also been explored. A number of approaches pose the task as a retrieval problem, where the most compatible annotation in the training set is transferred to a test image [20] or where training annotations are broken up and stitched together [21]. Several approaches generate image captions based on fixed templates that are filled based on the content of the image [22] or generative grammars [23, 24], but this approach limits the variety of possible outputs. Most closely related to us, Kiros et al. [25, 26] developed a log bilinear model that can generate full sentence descriptions for images, but their model uses a fixed window context while our Recurrent Neural Network (RNN) model conditions the probability distribution over the next word in a sentence on all previously generated words. Convolutional Neural Networks (CNNs) [27] have recently emerged as a powerful class of models for image classification and object detection [28]. On the sentence side, our work takes advantage of pertained word vectors [28] to obtain low-dimensional representations of words. Finally, Recurrent Neural Networks have been previously used in language modeling [29], but additionally condition these models on images.

2.3   Proposed Model

Image classification is the task of assigning a single label to an image (or rather an array of pixels that represents an image) from a fixed set of categories. A complete pipeline for this task is as follows:

- **Input:** A set of N images, each labeled with one of K different classes. This data is referred to as the training set.
- **Learning** (aka Training): Use the training set to learn the characteristics of each class. The output of this step is a model which will be used for making predictions.
- **Evaluation:** Evaluate the quality of the model by asking it to make predictions on a new set of images that it has not seen before (also referred to as the test set). This evaluation is done by comparing the true labels (aka ground truth) of the test set with the predicted labels output by the learned model.

The aim of our model is to generate training of image regions. During training, the input to our model is a set of images and their corresponding sentence descriptions (Fig. 1). The propose model that aligns sentence snippets to the visual regions that they describe through a multimodal embedding. Then treat these correspondences as training data for a second, multimodal Recurrent Neural Network model that learns to generate the snippets. The formal approach for solving the problem of image classification can be broken down into several key components which discuss next subsections.

Score Function
The score function maps the raw data to class scores. For a linear classifier, the score function can be defined as:

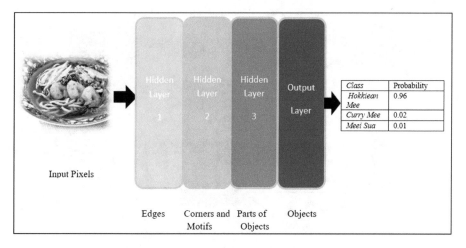

Fig. 1. Insertion model of multilayer perceptron

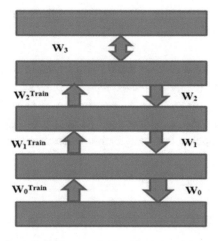

Fig. 2. A hybrid network algorithm generative model

$$f(x_i, W, b) = Wx_i + b.$$

Where xi represents the input image. The matrix W, and the vector b are the parameters of the function, and represent the weights and bias respectively.

Loss Function

The loss function quantifies the match between the predicted scores and the ground truth labels in the training data.

The loss function (also referred to as the cost function or objective) can be viewed as the unhappiness of the predicted scores output by the score function. Intuitively, the loss would be low if the predicted scores match the training data labels closely.

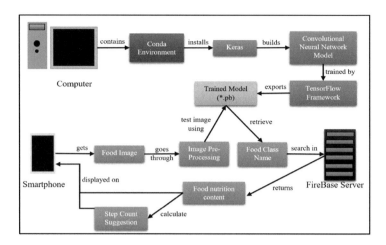

Fig. 3. Deep leaning neural network tensor flow model of the proposed prototype

Otherwise the loss would be high. Next section explains the classifiers which is the most important function.

Classifiers

There exists two common classifiers that are often used in image classification tasks: the **SVM** Classifier and the **Softmax** Classifier.

SVM Classifier

The SVM classifier uses the hinge loss (also referred to as max-margin loss, or SVM loss). For the i-th example in our data, the hinge loss is given as:

$$L_i = \sum_{j \neq y_i} \max\left(0.s_j - s_{y_i} + \Delta\right).$$

where delta is a hyperparameter which represents that the SVM loss function in equation

Softmax Classifier

The *Softmax* classifier uses the cross entropy loss (also referred to as softmax loss). For the i-th example in our data, the cross entropy loss is given as: where fj means the j-th element of the vector of class scores f. Note that the softmax classifier uses the softmax function to squash the raw class scores s into normalized positive values that sum to one, so that the cross entropy loss can be applied.

$$L_i = -\log \frac{e^{f_{y_i}}}{\sum_j e^{f_j}}.$$

The softmax classifier can be represented as

$$f_j(z) = \frac{e^{z_j}}{\sum_k e^{z_k}}.$$

It takes a vector of real-valued scores (in z) and squashes it to a vector of values between zero and one, which sum to one.

2.4   Annotation of Dataset

Data Collection
The collected our dataset (5115 images as dated 30/4/2018) using the Google Image Search [30], Bing Image Search API and on the spot image collection at the restaurants, Hawkers (Street food) stalls and street food. The work has explored the use of ImageNet [31] and Flickr [3] for collecting images. However, the images from Google and Bing to be much more representative of the classes they belonged to, compared to the images from ImageNet and Flickr. ImageNet and Flickr seem to have a lot of false images (images which clearly do not belong to the class). Hence decided to use the images could collect from Google and Bing.

Preprocessing Techniques
The re-sized all of our images to have height, width and channel dimensions of 32, 32 and 3 ($32 \times 32 \times 3$) respectively. This was done primarily for computational efficiency in performing the experiments. The filtered out images which are unable to resize to our specified height, width and channel requirements. Unfortunately, this meant losing approximately 10% of the data from our original dataset. Figure 4 shows an experiment that has been invoked over the 5115 images as part of the image dataset. In Table 1 has illustrate the class distribution from the dataset which is growing in number of images. For example, nyonya, pre-packed food (ready-to-eat), street food are yet to populate the dataset. It is estimated, the dataset will reach the maximum of 5,000 images.

Proposed Greedy Algorithm for Transfer Learning
The idea behind the greedy algorithm is to allow each model in the sequence to receive a different representation of the data. The model performs a non-linear transformation on its input vectors and produces as output the vectors that will be used as input for the next model in the sequence. Figure 2 shows a multilayer generative model in which the top two layers interact via undirected connections and all of the other connections are directed. The undirected connections at the top are equivalent to having infinitely many higher layers with tied weights. There are no intra-layer connections and, to simplify the analysis, all layers have the same number of units. It is possible to learn sensible (though not optimal) values for the parameters W_0 by assuming that the parameters between higher layers will be used to construct a complementary prior for W_0. This is equivalent to assuming that all of the weight matrices are constrained to be equal. The task of learning W_0 under this assumption reduces to the task of learning a Restricted Boltzmann machine (RBM) and although this is still difficult, good approximate solutions can be found rapidly by minimizing contrastive divergence. Once W_0 has been learned, the data can be mapped through W_{T0} to create higher-level "data" at the

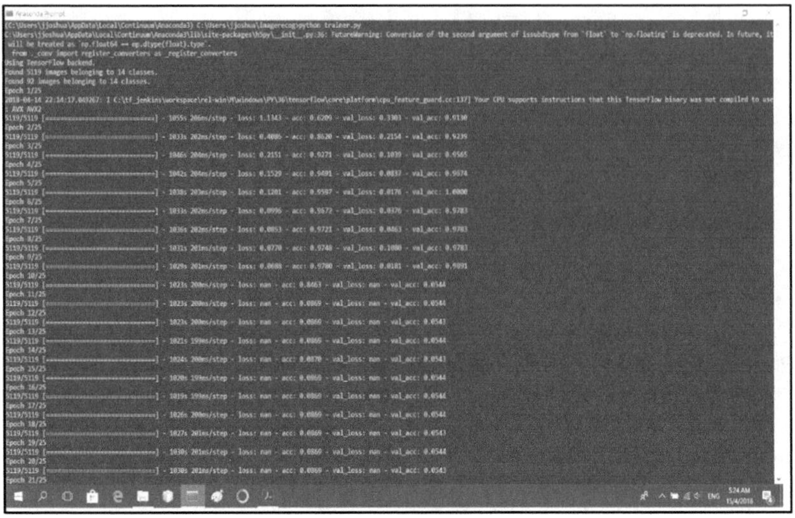

Fig. 4. Initial training model with Convolutional Neural Network with tensor flow

Table 1.

Food items	Number of images
Ais Kacang	445
Ayam Percik	266
Bak Kut The	216
Banana Fritters	196
Char Kuey Teow	330
Chicken Rice	364
Hokkien Mee	372
Idly	471
Ketupat	297
Laksa	514
Lemang	274
Mee Goreng	589
Nasi Lemak	400
Roti Canai	385

first hidden layer. There is a fast, greedy learning algorithm that can find a fairly good set of parameters quickly, even in deep networks with millions of parameters and many hidden layers. The learning algorithm is unsupervised but can be applied to labeled data by learning a model that generates both the label and the data.

3   Methodology

A hybrid generative model in Fig. 2 has explained the core implementation methodology of the work. The main idea is that there's much stuff you do every time you start
your tensor flow project, so wrapping all this shared stuff will help you to change just
the core idea every time you start a new tensorflow work. In Fig. 3 the CNN model
framework has invoked with trained model here the image size is of 32 × 32 × 3 input
for with the layer for multiclass classification. In Table 1 has described the 14 classes.
Each layer trends with height and width go down channels up to eight layers. A couple
of convolutional/pool layers followed by the fully connected with approximate
parameters.

The images are trained with three levels of epochs 100, 400, and 600 respectively.
Figure 5(a) has shown the execution of the training models for the above epochs.
Figure 6 has illustrated the SVM classifier with the two layer fully connected classification accuracy of the food types the learning rate has used the SGM to measure rate
of the score. The multiclass classifiers has run the preliminary epoch 1 and the loss in
entropy Fig. 5(b). TensorFlow framework was chosen to train the food recognition
model due to the available support and documentation for the framework [33]. With the
huge amount of resource, the chances of failing to achieve the feature are very low
compared to a framework that has fewer projects or resources for developers. Besides,
Keras will be used on top of Tensor Flow as an interface since existing food recognition project researchers as well as developers confirm that Keras able to help in
creating layers in CNN architecture with least amount of code and clear-cut implementation that is easy to understand [34]. Table 1 is the input labelled image data
folders with the number of images to be trained by the CNN.

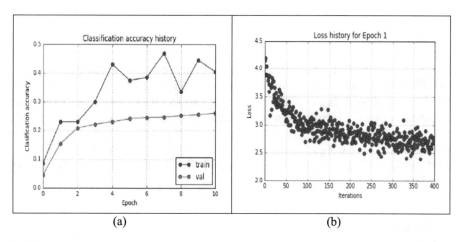

(a)   (b)

Fig. 5. (a) Classification accuracy history of fully connected two layers. (b) Reduction in loss in
accuracy first epoch of CNN

In Fig. 4, the implementation of the framework has invoked with NVIDIA GTX 960 M CPU with i7-6800 GHz with 16 GB of RAM. The initial training has to run for 24 h for 100 epochs. We could improve this by initiating GPU with at least GTX 1070 for better performance on prediction and accuracy of the greedy algorithm performance.

4   Evaluation

The results of the proposed work will analyse the CNN architectures on how accurate in solving the complete calorie, labelling of images problem which has written in research statement and objectives section. In this article we have used CNN to classified and recognize the food images. The initial work of Bidirectional recurrent neural work (BRNN) are advice in this article. The working nature of hybrid algorithm as compared with normal execution of training model between the results are evaluated, the generation of image descriptions are verified with the probability distribution functions and the accuracy of the training model values are examined.

Using the images features with a linear SVM classifier we were able to get a validation accuracy of 0.21, using SGD with a learning rate of 1e-03 and regularization strength of 1e+00. Using the images features with a Two Layer Fully Connected Neural Network gave much better performance. We got the best validation accuracy of 0.26 while using SGD as our update rule with a learning rate of 0.9, learning rate decay of 0.8 and regularization strength 0. The corresponding test set accuracy was 0.27. After the five conv layers, we added two fully connected layers with 1024 and 20 neurons respectively. For the last layer we use softmax with cross entropy Loss. The best validation accuracy of 0.40 was achieved using the Adam [16] update rule with a learning rate of 1e-04. The test set accuracy was 0.40.

5   Conclusion

The results observe that convolutional neural networks are quite suitable for the task of classifying food dishes, and improve traditional machine learning approaches at this task. The initial work has used CNN and uses transfer learning approach looks most promising, especially because both the training and validation accuracy are improving with the number of epochs (i.e. we have not over fit our model). This suggests that more data (and/or running it for more epochs) could improve the accuracy metric with multiclass classification using recurrent and bi directional neural networks will be considered.

References

1. National Security Agency Statement. https://www.nsa.gov/news-features/press-room/statements/2013-08-09-the-nsa-story.shtml
2. Gantz, J., Reinsel, D.: Extracting value from chaos (2011). https://www.emc.com/collateral/analyst-reports/idc-extracting-value-from-chaos-ar.pdf

3. Gantz, J., Reinsel, D.: The digital universe decade-are you ready? (2010). https://www.emc. com/collateral/analyst-reports/
4. Howard, J.: The business impact of deep learning. In: Proceedings of the 19th ACM SIGKDD International Conference on Knowledge Discovery and Data Mining, p. 1135 (2013)
5. Top Strategic Technology Trends for 2018. http://www.gartner.com/technology/research/ top-10-technology-trends/
6. De Sousa Ribeiro, F., Caliva, F., Swainson, M., Gudmundsson, K., Leontidis, G., & Kollias, S. (2018, May). An adaptable deep learning system for optical character verification in retail food packaging. In: IEEE Conference on Evolving and Adaptive Intelligent Systems
7. Yunzhou, Z., et al.: Remote mobile health monitoring system based. J. Healthc. Eng. 6(3), 717–738 (2015)
8. Lu, Y.: Food Image Recognition by Using Convolutional Neural Networks (CNNs), Michigan (2016). eprint arXiv:1612.00983
9. Lecun, Y., Bottou, L., Bengio, Y., Haffner, P.: Gradient-based learning applied to document recognition. Proc. IEEE 86(11), 2278–2324 (1998)
10. Ketkar, N.: Introduction to Keras. In: Deep Learning with Python. Apress, Berkeley (2017)
11. Naik, S., Patel, B.: Machine vision based fruit classification and grading. Int. J. Comput. Appl. (0975–8887) 170(9), 22–34 (2017)
12. Karol, G., Ivo, D., Alex, G., Daan, W.: DRAW: A Recurrent Neural Network For Image Generation (2015). https://arxiv.org/abs/1502.04623. Last accessed 14 Apr 2018
13. Su, B., Lu, S: Accurate scene text recognition based on recurrent neural network. In: Cremers, D., Reid, I., Saito, H., Yang, MH. (eds.) Computer Vision – ACCV 2014. ACCV 2014. Lecture Notes in Computer Science, vol. 9003. Springer, Cham (2015)
14. Barnard, K., Duygulu, P., Forsyth, D., De Freitas, N., Blei, D.M., Jordan, M.I.: Matching words and pictures. JMLR (2003)
15. Socher, R., Fei-Fei, L.: Connecting modalities: semisupervised segmentation and annotation of images using unaligned text corpora. In: CVPR (2010)
16. Li, L.-J., Socher, R., Fei-Fei, L.: Towards total scene understanding: classification, annotation and segmentation in an automatic framework. In: IEEE Conference on Computer Vision and Pattern Recognition, 2009. CVPR 2009, pp. 2036–2043. IEEE (2009)
17. Gould, S., Fulton, R., Koller, D.: Decomposing a scene into geometric and semantically consistent regions. In: 2009 IEEE 12th International Conference on Computer Vision, pp. 1–8. IEEE (2009)
18. Fidler, S., Sharma, A., Urtasun, R.: A sentence is worth a thousand pixels. In: CVPR (2013)
19. Li, L.-J., Fei-Fei, L.: What, where and who? Classifying events by scene and object recognition. In: ICCV (2007)
20. Socher, R., Karpathy, A., Le, Q.V., Manning, C.D., Ng, A.Y.: Grounded compositional semantics for finding and describing images with sentences. TACL (2014)
21. Kuznetsova, P., Ordonez, V., Berg, T.L., Hill, U.C., Choi, Y.: Treetalk: composition and compression of trees for image descriptions. Trans. Assoc. Comput. Linguist. 2(10), 351–362 (2014)
22. Yao, B.Z., Yang, X., Lin, L., Lee, M.W., Zhu, S.-C.: I2t: image parsing to text description. Proc. IEEE 98(8), 1485–1508 (2010)
23. Yatskar, M., Vanderwende, L., Zettlemoyer, L.: See no evil, say no evil: description generation from densely labelled images. Lex. Comput. Semant. (2016)
24. Chen, X., Fang, H., Lin, T.-Y., Vedantam, R., Gupta, S., Dollar, P., Zitnick, C.L.: Microsoft coco captions: data collection and evaluation server. arXiv preprint arXiv:1504.00325 (2015)

25. Frome, A., Corrado, G.S., Shlens, J., Bengio, S., Dean, J., Mikolov, T., et al.: Devise: a deep visual-semantic embedding model. In: NIPS (2013)
26. Karpathy, A., Joulin, A., Fei-Fei, L.: Deep fragment embeddings for bidirectional image sentence mapping. arXiv preprint arXiv:1406.5679 (2014)
27. LeCun, Y., Bottou, L., Bengio, Y., Haffner, P.: Gradientbased learning applied to document recognition. Proc. IEEE **86**(11), 2278–2324 (1998)
28. Mikolov, T., Sutskever, I., Chen, K., Corrado, G.S., Dean, J.: Distributed representations of words and phrases and their compositionality. In: NIPS (2013)
29. Szegedy, C., Liu, W., Jia, Y., Sermanet, P., Reed, S., Anguelov, D., Erhan, D., Vanhoucke, V., Rabinovich, A.: Going deeper with convolutions. arXiv preprint arXiv:1409.4842 (2014)
30. Google image search. https://images.google.com/
31. Image-net. http://www.image-net.org/
32. Flickr. https://www.flickr.com/
33. Abadi, M., Barham, P., Chen, J., Chen, Z., Davis, A., Dean, J., Devin, M., Ghemawat, S., Irving, G., Isard, M., Kudlur, M., Levenberg, J., Monga, R., Moore, S., Murray, D.G., Steiner, B., Tucker, P., et al.: TensorFlow: a system for large-scale machine learning. In: 12th USENIX Symposium on Operating Systems Design and Implementation (OSDI 2016), 28 January, Issue 12, pp. 265–267 (2016)
34. Zhang, X.: Deep Learning - Michael Hahsler (2017). http://michael.hahsler.net/SMU/EMIS8331/tutorials/Deep_Learning_Zhang.pdf. Last accessed 25 Apr 2018

Feature of Operation PV Installations with Parallel and Mixed Commutation Photocells

Pavel Kuznetsov[1] and Leonid Yuferev[2(✉)]

[1] FSAEI HE, Sevastopol State University, Sevastopol
Russian Federation (Ukraine)
Pavelnik2@gmail.com
[2] FSBSI, Federal Scientific Agro Engineering Center VIM, Moscow,
Russian Federation
leouf@ya.ru

Abstract. The article deals with practical application of photovoltaic cells in a variety of connecting options. For maximum efficiency of photovoltaic installations is necessary to take account of all the peculiarities of their work. As the title implies the article describes the peculiarities of photovoltaic cells in parallel and mixed commutation in uniform and non-uniform solar radiation, arising as a result of partial shading or soiling. It is given the theoretical explanation of the causes of the decrease of energy efficiency photovoltaic cells with the above commutation. It is represented experimental confirmation of this fact by the example of industrial plants using photovoltaic panels PS-250 and the single-crystal transformers produced by company "KVAZAR". Results of the research will be useful in the developing and design of photovoltaic installations, especially those with probable partial shading.

Keywords: Photovoltaic · PV · Parallel · Mixed · Connection
Shading

1 Introduction

Currently, the world's becoming more common power plants, running on renewable energy sources, among which, one of the most promising plants using solar energy. The advantages of this energy source are environmental friendliness, which makes it possible to use it practically at any scale without causing damage to the environment, as well as availability in almost every point of our planet, differing by radiation density by no more than two times [1]. In addition, the modular design of photovoltaic systems allows them to be designed for almost any power, which makes these installations a universal and reliable solution that is used both in industrial production of electric power and in small power supply systems [2].

Despite all the advantages, in solar photovoltaic installations there are several features of the work that must be considered when designing them for maximum energy production. One such feature is the non-linear internal resistance of the photoelectric converters (PEC) [3]. This phenomenon manifests itself especially negatively

© Springer Nature Switzerland AG 2019
P. Vasant et al. (Eds.): ICO 2018, AISC 866, pp. 231–238, 2019.
https://doi.org/10.1007/978-3-030-00979-3_23

in conditions of uneven illumination, shading, or pollution of the PEC. To achieve the required electrical power in solar power plants are used in series, parallel and mixed switching PEC. With each version of switching, the decrease in efficiency occurs in different ways [4]. In this paper, we consider some features of the operation for parallel and mixed switching of a photomultiplier in conditions of uniform and not uniform illumination. Investigation of this issue is very important, because at present many existing solar installations are using such types of switching.

2   Research

Parallel switching of the photovoltaic cell into a solar photoelectric device shown in Fig. 1.

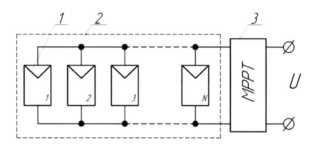

Fig. 1. Installation with parallel solar cells in the battery (1 - PEC; 2 - solar photovoltaic installation; 3 - the device for finding the maximum power point and optimizing the output power (MPPT)).

The current-voltage characteristic of the uniformly illuminated battery in the first quad-welt, with an accuracy sufficient for engineering calculations described analytically by the Eq. (1). This equation is derived according to the Kirchhoff law and the Shockley equation written for a "simplified" equivalent circuit for replacing the PEC, taking into account that the current at the output of the unit will be equal to the sum of the currents of each PEC $\left(I = \sum_{i=1}^{n} I_i\right)$ and the voltage will be equal to the voltage at the output of each PEC $(U = U_1 = U_2 = \cdots = U_n = const)$:

$$I = I_{ph} - I_d = n_p\left[I_{ph} - I_0\left[\exp\left(\frac{q(U + IR_s)}{AkT}\right) - 1\right]\right], \tag{1}$$

where n_p is the number of parallel connected PEC in the installation; I - load current, (A); I_{ph}- photocurrent, (A); I_d - the current flowing through the diode, (A); I_0- reverse saturation current, (A); q - is the electron charge, $(1.602 \cdot 10^{-19}$ C); U - output voltage, (V); k - the Boltzmann constant, $(1.381 \cdot 10{-23}$ J/K); T - the absolute temperature of the solar cell, (K); R_s- series resistance of the solar cell, (Ohm); A - coefficient of ideality, depending on the thickness of the p-n-junction and the material, takes values for silicon PEC from 1.2 to 5 (Fig. 2).

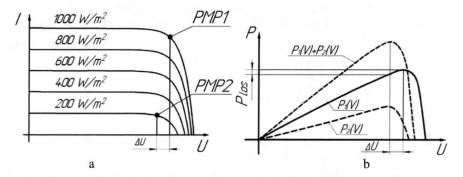

Fig. 2. Current-voltage (a) and power (b) characteristics at different illumination PEC

The maximum power of such an installation determined by the Eq. (2):

$$P_{max} = U_{max}I_{max} = U_{max}\sum_{i=1}^{n} I_i = U_{max}n_p\left[I_{ph} - I_0\left[\exp\left(\frac{q(U_{max} + IR_s)}{AkT}\right) - 1\right]\right], \quad (2)$$

where P_{max} - maximum power of the photovoltaic installation, (W); U_{max} - the installation voltage at the maximum power point, (V); I_{max} - installation current at maximum power point (A); I_i - the current of the i-th PEC, (A).

It can be assumed that in case of uneven illumination of the PEC with the same temperature, the maximum power of the installation will decrease by the amount of decrease in the maximum power of the individual PECs, due to their shading. However, this happens somewhat differently because the voltage of the PEC at the points of

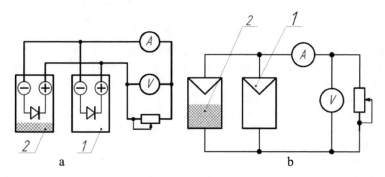

Fig. 3. Functional circuit experimental installation: a - with PEC; b - with photoelectric modules (1 - PEC (photoelectric panel); 2 – shading)

maximum power (PMP1 and PMP2) is not the same for different values of the illumination, as shown in Fig. 3.

Due to the different voltage, it is not possible to obtain electrical energy from the installation at the maximum power point of each PEC. The value of this difference can be found from Eq. (3), solving it with respect to the voltage for each of the points:

$$
\begin{aligned}
\Delta U = U_{max1} - U_{max2} = \\
= \frac{AkT}{q} \ln\left(\frac{I_{ph1}}{I_{01}} - \frac{I_{max1}}{I_{01}} + 1\right) - I_{max1}R_s - \frac{AkT}{q} \ln\left(\frac{I_{ph2}}{I_{02}} - \frac{I_{max2}}{I_{02}} + 1\right) - I_{max2}R_s \approx \\
\approx \frac{AkT}{q} \ln\left(\frac{(I_{ph1}-I_{max1})I_{02}}{(I_{ph2}-I_{max2})I_{01}}\right) - R_s(I_{max1} + I_{max2}),
\end{aligned}
\tag{3}
$$

where U_{max1}, U_{max2} - the voltage at the point of maximum power of PEC1 and PEC2, (B); I_{ph1}, I_{ph2} - photocurrent of PEC1 and PEC2, (A); I_{max1}, I_{max2} - load currents PEC1 and PEC2 at the points of maximum power, (A); I_{01}, I_{02} - reverse saturation currents of PEC1 and PEC2, (A).

An analysis of the characteristic of such a facility shows that the voltage at the absolute maximum point (AMP) will with some assumption correspond to the voltage at the point of maximum power of the shaded PEC in the range of the radiation intensity divergences from 10 to 100%. From this it follows that in such a case, PEC having a capacity greater brightness value will be limited to less illuminated PEC (Fig. 5b). Then the power losses will be:

$$
\begin{aligned}
P_n = P_{max1} - P_1 = I_{max1}U_{max1} - I_1U_{max2} = \\
= \left[I_{ph1} - I_{01}\left[\exp\left(\frac{q(U_{max1} + I_{max1}R_s)}{AkT}\right) - 1\right]\right]U_{max1} - \\
- \left[I_{ph1} - I_{01}\left[\exp\left(\frac{q((U_{max1}-\Delta U) + I_{max1}R_s)}{AkT}\right) - 1\right]\right]U_{max2}
\end{aligned}
\tag{4}
$$

For experimental confirmation of these events was assembled, functional diagram is shown in Fig. 3.

As a PEC in the experiment, two silicon single crystal plates produced by the company "KVAZAR" with the dimensions of 100×100 mm, pseudo square filters and two commercially produced PS-250 photovoltaic panels from Progeny Solar, USA, were used in the experiment. Their use presupposes a mixed connection of the PEC, but since in the experiment the panels are evenly illuminated, then their use in the experiment is permissible and gives an opportunity to obtain results that are more reliable. The intensity of radiation was PEC $IR = 1000\,BT/M^2$, and the other - $(0, 1\ldots1)IR$. The illumination was measured with two luxmeters UT381 and U116. The experimental results are presented in Fig. 4.

It can be seen from the graphs that in such a case the PEC having normal illumination loses up to 7.4% of the power, depending on the illumination of the shaded PEC, which can be significant in photovoltaic systems with mixed commutation, as will be shown below.

With mixed switching of the PEC, the installation current is composed of the currents of each parallel PEC array $\left(I = \sum_{i=1}^{n} I_i\right)$, and the voltage is equal to the voltage at the output of each such array, which in turn will be composed of voltages each PEC

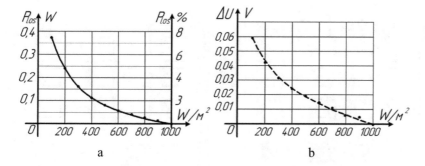

Fig. 4. Plots of power loss (a) and voltage difference (b) PEC of radiation intensity

included in this array $\left(U = \sum_{i=1}^{n} U_i\right)$. Thus, the current-voltage characteristic, with a uniform illumination will be described by Eq. (5). The maximum output power of the plant will be determined according to the first part of Eq. (6).

$$I = n_p \left[I_{ph} - I_0 \left[\exp\left(\frac{q(U + IR_s)}{n_s(AkT)}\right) - 1\right]\right], \tag{5}$$

where n_s - the number of sequentially connected PEC in a parallel array.

In case of uneven illumination of a photovoltaic plant with a mixed PEC connection, the reduction in energy production occurs for the reasons that arise during serial and parallel commutation, but with some peculiarities. These features are related to the fact that when changing the PMP of any of the parallel connected arrays, it becomes impossible to select the maximum power from each group due to the voltage difference (ΔU), which leads to a decrease in the battery power by an amount exceeding the reduction in the maximum power of individual groups, due to their shading. At the same time, power reduction occurs in different ways depending on the nature of the shading - uniform and non-uniform (Fig. 5).

With uniform shading and setting of the PMP to the point of absolute maximum, the total power loss of the solar installation is characterized by the system of Eq. (8):

$$\begin{cases} P_{los} = U_{1max} n_p \left[I_{ph\,o} - I_0 \left[\exp\left(\frac{q(U_{1max} + I_{1max}R_s)}{n_s(AkT)}\right) - 1\right]\right] - U_{2max}\left[n_p\,_{sh}\left[I_{ph\,sh} - \right. \right. \\ \left. -I_0 \left[\exp\left(\frac{q(U_{2max} + I_{2max}R_s)}{n_s(AkT)}\right) - 1\right] + n_{po}\left[I_{ph\,il} - I_0\left[\exp\left(\frac{q(U_{1max} + I_{1max}R_s)}{n_s(AkT)}\right) - 1\right]\right]\right], \\ \frac{q(U_{1max} + I_{1max}R_s)}{n_s(AkT)} - \ln\dfrac{I_0 + I_{ph}}{I_0\left(1 + \frac{q(U_{1max} + I_{1max}R_s)}{n_s(AkT)}\right)} = 0, \\ n_{p\,sh}I_{ph\,3} + n_{p\,il}I_{0\,il} - \exp\left(\frac{q(U_{2max} + I_{1max}R_s)}{n_s(AkT)}\right) \times \\ \times \left(\frac{q}{n_d(AkT)}\left(n_{psh}I_{0sh}(U_{2max} + I_{2max}R_d) + n_{pil}I_{0il}(U_{2max} + I_{1max}R_d)\right)\right) = 0, \end{cases} \tag{6}$$

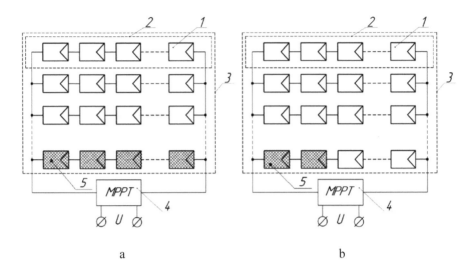

a   b

Fig. 5. Photovoltaic installation mixed switched PEC: a - uniform shading; b - non-uniform shading (1 - normally illuminated PEC; 2 - array of PEC; 3 - solar installation; 4 - MPPT; 5 - shaded PEC).

where P_{los}- loss of the solar installation, (W); $n_{p\,sh}$ и $n_{p\,il}$ - the quantity of arrays of PEC that have shading and normal illumination; n_p- number of parallel connected arrays PEC; $I_{ph\,sh}$ и $I_{ph\,il}$ - photocurrent of the shaded and illuminated array PEC, (A); U_{1max}, U_{2max} - the values voltage in the PMP of parallel arrays of PEC without shading and with shading, (V); I_{1max} и I_{2max} - values of currents in PMP without shading and with shading, (A).

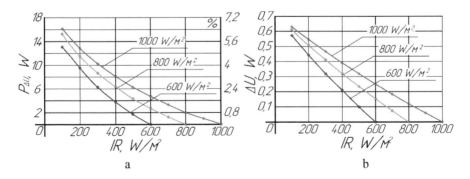

a   b

Fig. 6. Graphs of the dependence of losses of a photoelectric installation with mixed PEC connection PEC and uniform partial shading on the radiation intensity in the shadow and setting the PMP to the absolute maximum point for different illumination: a - loss from inconsistency in voltage, b - value of mismatch voltage

The results of modeling the losses of a solar installation consisting of three PV-250 photovoltaic modules, with a uniformly shaded one module, are shown in Fig. 6:

When MPPT is installed at the local maximum point, the value of the total losses of the solar installation will also be characterized by the system of Eq. (10), except that the value of the output voltage of the solar installation will be determined from the equation:

$$\frac{q(U_{1max}+I_{1max}R_s)}{n_s(AkT)} - \ln\frac{I_0+I_{ph}}{I_0\left(1+\frac{q(U_{1max}+I_{1max}R_s)}{n_s(AkT)}\right)} = 0. \tag{7}$$

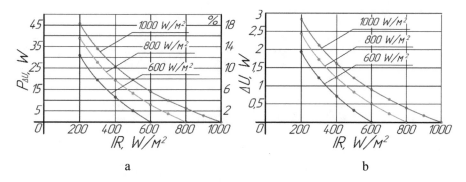

a   b

Fig. 7. Graphs of the dependence of losses of a photoelectric installation with mixed PEC connection PEC and uniform partial shading on the radiation intensity in the shadow and setting the PMP to the local maximum point for different illumination: a - loss from inconsistency in voltage, b - value of mismatch voltage

The results of modeling the same SFU for such a case are shown in Fig. 7.

3   Conclusion

It can be seen from the graphs that due to uneven illumination, the solar photoelectric installation significantly reduces its energy efficiency due to the inconsistency in the voltage of parallel PEC arrays, even with a uniform shading character. Energy losses when setting the mode of selection of maximum power to the point of absolute maximum are up to 6.4%, and when installed at the point of local maximum - up to 18%. This problem can be solved using the method invented by P.N. Kuznetsov and A. A. Borisov, which allows selecting power from the installation, equal to the sum of the maximum powers of separate parallel-connected PEC arrays [5]. Experiments carried out at the Sevastopol solar power plant, at the Sevastopol State University and other operating facilities confirmed the efficiency of its use, even with this type of shading - the introduction of a device that implements the method, avoids this component of the loss of electricity.

References

1. Badescu, V.: Modeling Solar Radiation at the Earth's Surface. Springer, Berlin (2014)
2. Rauschenbach, H.S.: Solar Cell Array Design Handbook: The Principles and Technology of Photovoltaic Energy Conversion. Springer, New York (2008)
3. Liu, T., Liu, Z., Ren, J., et al.: Operating temperature and temperature gradient effects on the photovoltaic properties of dye sensitized solar cells assembled with thermoelectric–photoelectric coaxial nanofibers. Electrochim. Acta **279**, 177–185 (2018)
4. Balato, M., Costanzo, L., Vitelli, M.: Series-parallel PV array re-configuration: maximization of the extraction of energy and much more. Appl. Energy **159**, 145–160 (2015)
5. Kuznetsov, P., Borisov, A.: The method of selection of electrical energy from photo-voltaic panels. Patent RF 2634590 (2017)

The Functional Dependencies of the Drying Coefficient for the Use in Modeling of Heat and Moisture-Exchange Processes

Alexey N. Vasilyev$^{(\boxtimes)}$ ⓘ, Alexey A. Vasilyev ⓘ,
and Dmitriy A. Budnikov ⓘ

Federal State Budgetary Scientific Institution, "Federal Scientific
Agroengineering Center VIM" (FSAC VIM), Moscow, Russia
{vasilev-viesh, dimml3}@inbox. ru, lex. of@mail. ru

Abstract. When modeling and performing calculations on heat and moisture-exchange in a grain layer, it is considered that the drying coefficient remains constant. However, our observations allowed to put forward the hypothesis that the value of the drying coefficient varies depending on the parameters of the drying agent. With the use of variety of electrical effects for intensification of grain drying, the data was obtained which reveals the change in the rate of drying. This allows suggesting that drying coefficient depends not only on the parameters of utilized air but also on parameters of electrophysical effects. A number of experiments were conducted to get the regression equations reflecting how drying coefficient depends on these factors. Experimental data was processes with the use of application software. Obtained regression equations turned out to adequately describe the relation between drying coefficient and parameters of the drying agent and electrophysical effects. These dependencies can be used to simulate the process of drying more accurately and to look for the optimal modes of drying.

Keywords: Grain drying · Drying coefficient · Electrically activated air
Microwave field · Process modelling · Parameters optimization

1 Existing Approaches to Modeling of Grain Drying

Authors of the work use the classical system of equations [1–5] when modelling the processes of heat and moisture exchange in a grain layer during grain drying. In this system of equations, grain moisture content at a particular moment of time is determined with the following equation:

$$\partial W/\partial\tau = -K\left(W - W_p\right), \tag{1}$$

where W is current grain moisture content, %; K is the drying coefficient, 1/h; W_p is equilibrium grain moisture content, %; τ is time, h.

In some cases [6, 7], equation derived by Okun' [8, 9] is used to find the value of the drying coefficient:

© Springer Nature Switzerland AG 2019
P. Vasant et al. (Eds.): ICO 2018, AISC 866, pp. 239–245, 2019.
https://doi.org/10.1007/978-3-030-00979-3_24

$$K = 7,110^{-2}e^{0,05\,T}, \tag{2}$$

where

T is the temperature of the drying agent, °C.

In this relation, temperature of the dryings agent is the only variable. As in the existing technological processes of high temperature drying the velocity of the drying agent remains unchanged, it is understandable why only air temperature was used in the equation to calculate the value of the drying coefficient. However, in active ventilation bunkers with radial air distribution, velocity of the drying agent changes when passing through the layer [10–12]. Which is more, it has been recently studied with computer simulations how the change of air supply at certain moments during grain drying can influence intensity and power consumption of the process [5]. Wide range of research is also conducted on how to apply electrophysical effects to increase efficiency and speed of grain drying [13–16]. Therefore, the task was set to obtain the regression equations of how the drying coefficient depends on the parameters of used air and settings of used electrophysical effects. For this purpose, special experimental setup was developed and constructed, allowing carrying out studies of the drying process with use of various electrophysical effects [6].

2 Brief Description of Experimental Studies

Experimental studies were carried out for convective drying, for the temperature range from 20 to 50 °C. This range is chosen because overall goal of the research is to study the influence of electrophysical effects on the process of drying carried out with active ventilation and low-temperature. Grain portions with the initial moisture content of 16, 20 and 24% were used for the experiment. Electrically activated air (air saturated with negatively charged ions) and electromagnetic field of microwave frequency of 2,45 GHz were used as electrophysical effects. Density of negative air ions at the entrance to the grain layer was maintained constant by means of the air ions source at the level of $3,5 \times 10^{10}$ m^{-3}. Such concentration was set with regard to the required threshold limit value (TLV) of air ions. Speed of the drying agent during the experiment was maintained in the range between 0.2 and 2 m/s. Drying agent speed was measured with the use of thermo-anemometer installed at the place where air goes out from the grain layer. Initial grain moisture content was determined with the moisture meter Fauna-M, portions of grain were weighted before being placed to the setup. Grain moisture content in percents was determined during drying with the following dependency:

$$W = \frac{m_1 - m_2}{m_1} 100, \tag{3}$$

where m_1 is weight of the portion of grain before drying, kg; m_2 – weight of the portion of grain after drying, kg.

Volume and thickness of grain layer were kept the same in the experiments. For this, grain was poured into a tube of dielectric material with the height of 10 cm. In the study of microwave-convective drying, the microwave field was affecting the grain until temperature at the center of the layer did not reach 55 °C. After that, the magnetrons were turned off and then turned on again when the temperature dropped to 50 °C. By changing the distance from the tube with grain to the antenna of the waveguide, energy of the microwave field was adjusted to be in the range between 1000 and 4000 $\frac{kJ}{m^3}$.

For the research, multi-factorial experiments was planned. During the experiments, graphs of grain drying were obtained for different parameters of air and various power densities of the microwave field. With all kinds of drying, value of the drying coefficient was calculated with the known equation [17, 18]:

$$K = \frac{2,3\left(lg\left(W_1 - W_p\right) - lg\left(W_2 - W_p\right)\right)}{\tau_2 - \tau_1},$$ (4)

where W_1 and W_2 is moisture content of the drying grain at moments of time τ_1 and τ_2 accordingly.

3   Obtaining Regression Equations

With the use of MATLAB, regression analysis was performed on the experimental data, which allowed to obtain regression equations. All obtained regressions dependencies were checked for adequacy, their usability was confirmed. Thus, for convective drying the following dependency was obtained:

$$K = \frac{2,3\left(lg\left(W_1 - W_p\right) - lg\left(W_2 - W_p\right)\right)}{\tau_2 - \tau_1}.$$ (5)

Obtained three-dimensional dependency graph $K = f(T, V)$ for initial grain moisture content of 20% is shown in the Fig. 1.

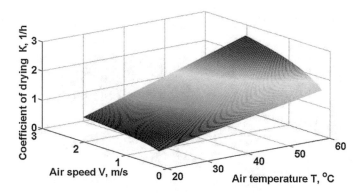

Fig. 1. Graphical illustration of dependency $K = f(T, V)$ of the drying coefficient on the speed and temperature of drying agent in the process of convective grain drying

When drying grain with electrically activated air, concentration of air ions was not used in the regression equation because it remained unchanged throughout the drying process. As the result, the following regression equation was received:

$$K = 9,48T^{1,31} \cdot V^{0.029} \cdot W^{-2,229}. \tag{6}$$

The graphs show that the effect of change in the speed of the drying agent on value of the drying coefficient manifests itself in a greater degree when the air is heated. Such situation makes it possible to optimize the process of convective drying on duration of drying and energy costs. Three-dimensional graph which illustrates the mentioned dependency for the initial grain moisture content of 20% and concentration of negative ions in the air entering the grain layer at the level of 3.5×10^{10} m^{-3} is shown in in the Fig. 2.

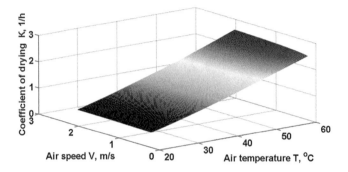

Fig. 2. Graphical representation of dependency of the drying coefficient on speed and temperature of the drying agent in the process of convective grain drying with electrically activated air

In experiments with microwave convective grain drying, power density of the microwave field was changed by means of placing the experimental batch of grain at a certain different distances from the magnetron [19]. The following equation was obtained as a result of experimental data processing for the grain drying:

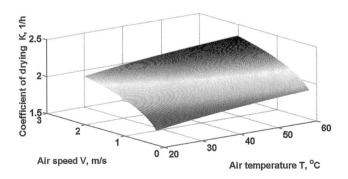

Fig. 3. Graphical representation of dependency $K = f(T, V)$ of the drying coefficient on the speed and temperature of the drying agent during microwave-convective grain drying

$$K = 0,606T^{0,162} + 1,155V^{0,183} - 7,429W^{-0,493} + 1550Q_v^{-1}. \qquad (7)$$

Three-dimensional graph of this dependency for the initial grain moisture content of 20% and the microwave field power density of 1000 $\frac{kJ}{m^3}$ is shown in Fig. 3.

Graphs presented in the figures allow to perform visual evaluation of how applying electrical technologies can impact the speed of the process of grain drying.

4   Conclusions

Comparison of obtained regression Eqs. (5), (6) and (7) shows that the use of electrophysical effects changes how grain moisture content affects the drying coefficient. So, for example, coefficient W illustrating degree of grain moisture content is positive in Eq. (5) while having negative values in Eqs. (6) and (7). Previous studies [20] allow to conclude that overall coefficient of moisture diffusion increases when electrophysical effects are applied. This in turn contributes to improvement in drying efficiency when speed of the drying agent is increased. However, it should also be considered that the increase in grain moisture content contributes to greater "shielding" of microwave field and prevents the field from penetrating into the depth of the grain layer. Air ions are also more active absorbed by wet grain, which results in the rise of unevenness of grain layer drying.

Obtained regression dependencies allow to carry out modeling of the processes of convective drying, microwave – convective drying, drying with electrically activate air.

Consideration of the changes in the value of the drying coefficient will allow, with the help of modeling of the process, to determine optimal operating modes of electrical equipment of drying setups.

References

1. Nejadi, J., Nikbakht, A.M.: Numerical simulation of corn drying in a hybrid fluidized Bedâlnfrared dryer. J. Food Process. Eng., 0145–8876 (2017)
2. Dimitrov, N.: Effect of the storage duration and moisture on the temperature pattern of respiration of soft wheat grain, University of Food Technology. Аграрни науки (България), pp. 1313–6577 (2014)
3. Ospanov, A.B., Vasilyev, A.N., Budnikov, D.A., Karmanov, D.K., Vasilyev, A.A., Baymuratov, D.Sh., Toksanbaeva, B.O., Shalginbaev, D.B.: Sovershenstvovanie protsessov sushki i obezzarazhivaniya zerna v SVCh pole [Improvement of drying and disinfection of grain in the microwave field]. Almaty`: Nur-Print (2017). (in Russian)
4. Rudobashta, S.P., Zueva, G.A., Kartashov, E.M.: Heat and mass transfer when drying a spherical particle in an oscillating electromagnetic field. Theor. Found. Chem. Eng. **50**(5), pp. 718–729 (2016)
5. Vasilyev, A.N., Budnikov, D.A., Gracheva, N.N., Severinov, O.V.: Sovershenstvovanie tehnologii sushki zerna v plotnom sloe s ispol`zovaniem elektrotehnologiy, ASU i modelirovaniya protsessa [Improvement of the technology of grain drying in a dense layer with the use of electrotechnologies, automated control systems and process modeling]. M.: FGBNU FNACz VIM (2016). (in Russian)

6. Lauva, A., Aboltins, A., Palabinskis, J., Karpova-Sadigova, N.: Grain drying experiments with unheated air. In: International Journal of Agricultural and Environmental Information Systems, vol. 7, Issue 2 Editors-in-Chief: Petraq Papajorgji and François Pinet (2006). https://doi.org/10.4018/ijaeis

7. Viesx, M.: Metodicheskie rekomendacii po optimizacii sushki plotnogo sloya [Methodological recommendations for optimizing the drying of a dense layer] (1978). (in Russian)

8. Okun`, G.S.: Metody raschyota prodolzhitel`nosti sushki otdel`nogo zerna pshenitsy i zernovogo sloya. V kn. Mashiny` dlya posleuborochnoj potochnoj obrabotki semyan. Teoriya i raschyot mashin, tehnologiya i avtomatizatsiya processov [Methods for calculating the duration of drying of an individual grain of wheat and a grain layer. In the book. Machines for post-harvest flow processing of seeds. Theory and calculation of machines, technology and automation of processes]/Pod red. Z.L. Ticza. M.: Mashinostroenie, pp. 290–308 (1967). (in Russian)

9. Okun`, G.S., Verczman, I.I., Esakov, Yu.V.: Raschet prodolzhitel`nosti i energoemkosti processa sushki zerna v sloe s pomoshch`yu EVM [Calculation of the duration and energy intensity of the process of grain drying in a layer with the help of a computer]. Sb. nauch. tr. VIM. M.: VIM, T. 100, pp. 73–80 (1984). (in Russian)

10. Vasilyev, A.N., Severinov, O.V.: K raschyotu teplo- i vlagoobmena v plotnom sloe zerna [To the calculation of heat and moisture exchange in a dense layer of grain]. Innovatsii v sel`skom hozyaystve [Innovations in agriculture]. №1 (11), pp. 4–8 (2015). (in Russian)

11. Garyaev, A.B., Sorochinskij, V.F., Goryacheva, E.M.: Matematicheskaya model` processa perenosa vlagi pri aktivnom ventilirovanii zerna v elevatorah [Mathematical model of the process of moisture transfer for active grain ventilation in elevators]. V sbornike: Povyshenie effektivnosti protsessov i apparatov v himicheskoj i smezhnyh otraslyah promyshlennosti sbornik nauchnyh trudov Mezhdunarodnoy nauchno-tehnicheskoy konferencii, posvyashchyonnoy 105-letiyu so dnya rozhdeniya A. N. Planovskogo, pp. 320–324 (2016). (in Russian)

12. Vasilyev, A.N., Severinov, O.V.: Planirovanie i metodika eksperimental`ny`h issledovaniy po opredeleniyu vliyaniya parametrov vozduha i zerna na koefficient sushki [Planning and methods of experimental studies to determine the effect of air and grain parameters on the drying coefficient]. Innovatsii v sel`skom hozyajstve [Innovations in agriculture]. № 1 (16), pp. 118–123 (2016). (in Russian)

13. Kharchenko, V., Vasant, P.: Handbook of Research on Renewable Energy and Electric Resources for Sustainable Rural Development (2018). https://doi.org/10.4018/978-1-5225-3867-7

14. Vasilyev, A.N., Gracheva, N.N., Budnikov, D.A.: Kriterial`noe uravnenie sushki zerna aktivnym ventilirovaniem elektroaktivirovannym vozduhom [Criterion equation of grain drying with active ventilation with electrically activated air], Nauchnyy zhurnal KubGAU [Elektronnyy resurs]. Krasnodar: KubGAU. №73(09). Rezhim dostupa (2011) http://ej.kubagro.ru/2011/09/pdf/35.pdf. (in Russian)

15. Vasilyev, A.N., Budnikov, D.A., Vasilyev, A.A.: Variant resheniya uravneniy teplo- i vlagoobmena v zernovom sloe pri SVCh-konvektivnom vozdeystvii [Variant of the solution of the equations of heat and moisture exchange in the grain layer under microwave convection influence]. Mezhdunarodnyy nauchno-issledovatel`skiy zhurnal [International Research Journal]. № 11–6 (42), pp. 53–56 (2015). (in Russian)

16. Vasant P., Alparslan-Gok, S.Z., Weber, G.-W.: Handbook of Research on Emergent Applications of Optimization Algorithms (2018). https://doi.org/10.4018/978-1-5225-2990-3

17. Lykov, A.V, Energiya, M.: Teoriya sushki [The theory of drying] (1968). (in Russian)

18. Norton, C.I.: Grain drying rates and zone depths at the steady state. Dissertation Submitted to the Graduate Faculty in Partial Fulfillment of The Requirements for the Degree of doctor of philosophy Major Subjects: Agricultural Engineering Mechanical Engineering, Iowa State University of Science and Technology Ames, Iowa (1959)
19. Vasilyev, A.N., Budnikov, D.A., Krausp, V.R., Dubrovin,V.A. i dr.: Metody energeticheskogo vozdeystviya na semena prioritetnyh zernovyh i ovoshchnyh kul'tur razlichnyh sortov, rasteniya i sel`skohozyaystvennye materialy. Kontseptsiya ispol`zovaniya elektrotehnologiy dlya obrabotki kormov, udobreniy, othodov rastenievodstva. Nauchno obosnovannye parametry energosberegayushchih kombinirovannyh ustanovok dlya obezzarazhivaniya vozduha i poverhnostey [Methods of energy impact on seeds of priority grain and vegetable crops of various varieties, plants and agricultural materials. the concept of the use of electrical technologies for processing feed, fertilizer, crop waste. scientifically grounded parameters of energy-saving combined installations for air and surface disinfection]. Otchet o NIR. Federal`nyy nauchnyy agroinzhenernyy center VIM [Report on research. Federal Scientific Agro-Engineering Center VIM] (2017). (in Russian)
20. Gracheva N.N.: Ispol`zovanie elektroaktivirovannogo vozduha dlya intensifikatsii sushki zerna aktivny`m ventilirovaniem [Use of electrically activated air for intensification of drying of grain by active ventilation]. Dissertatsiya na soiskanie uchenoy stepeni kandidata tehnicheskih nauk/Azovo-Chernomorskaya gosudarstvennaya agroinzhenernaya akademiya. Zernograd (2012). (in Russian)

Investigation Model for Locating Data Remnants on Cloud Storage

Khalid Abdulrahman[1], Abdulghani Ali Ahmed[1(✉)],
and Muamer N. Mohammed[2]

[1] Faculty of Computer Systems & Software Engineering,
Universiti Malaysia Pahang, Kuantan, Malaysia
kkkhalid@yahoo.com, abdulghani@ump.edu.my
[2] State Company for Internet Services, The Ministry of Communications of Iraq,
Baghdad, Iraq
muamer.scis@gmail.com

Abstract. Cloud storage services allow users to store their data online and remotely access, maintain, manage, and back up their data from anywhere through the Internet. Although this storage is helpful, it challenges digital forensic investigators and practitioners in collecting, identifying, acquiring, and preserving evidential data. This research proposes an investigation scheme for analyzing data remnants and determining probative artefacts in a cloud environment. Using the Box cloud as a case study, we collect the data remnants available on end-user device storage following the accessing, uploading, and storing of data in the cloud storage. The data remnants are collected from several sources, such as client software files, Prefetch, directory listings, registries, browsers, network PCAP, and memory and link files. Results indicate that the collected data remnants are helpful in determining a sufficient number of artefacts about investigated cybercrimes.

Keywords: Forensic science · Digital forensic · Cloud storage
Cybercrime investigation · Box cloud · Evidence collection · Data remnants
Artefacts

1 Introduction

Cloud storage can be considered a component of cloud computing. This storage can also be a model of data storage in which the digital data are stored in logical pools, the physical storage spans multiple servers (and often locations), and the physical environment is typically owned and managed by a hosting company. The providers of cloud storage are responsible for keeping the data accessible and available and the physical environment protected and running smoothly.

World Networks [1] stated that "there are questions should be asked from any business that anticipates using cloud based on services, the question is: What can my cloud provider do for me and for my data in terms of digital forensics data in the event of any legal dispute, criminal or civil cases, or data breaches?" Other studies have compared actual providers. Cloud service providers vary, and this difference

© Springer Nature Switzerland AG 2019
P. Vasant et al. (Eds.): ICO 2018, AISC 866, pp. 246–256, 2019.
https://doi.org/10.1007/978-3-030-00979-3_25

complicates cloud storage-based forensics because each provider has distinct rules, requirements, and guidelines. Dropbox, Google Drive, SkyDrive, and Live Drive are examples of cloud storage services that need to be investigated further.

This research attempts to focus on providing help for investigators in terms of maintaining the integrity and confidentiality of users' data as these data are deleted or moved from one device to another using cloud storage during their investigations. By increasing the number of crime involving cloud storage, process of acquiring the digital evidences from cloud storage becomes harder and more difficult. The strength of cloud storage is allowing users to upload data to the web and to share these data with others anywhere and anytime as long as they are connected to the Internet. Data can be uploaded or shared from one computer to another without leaving traceable evidence; thus, cloud technology is creating considerable challenges for forensic investigation and the possibility of cybercrime detection and prevention [2–7].

2 Related Work

This Section reviews articles related to this research and discusses the findings, methodologies, limitations, and conclusions of each article. Quick and Choo identified means of acquiring files uploaded and accessed using Dropbox [8, 9]. A standard PC with a virtual machine (VM) was installed with the Windows 7 operating system; various PCs were used to examine different cases, particularly within the forensic analysis of the client software Dropbox in numerous browsers, which included Microsoft Internet Explorer, Google ChromeTM, and Mozilla Firefox. This research determined the data remnants and artefacts left on a Windows 7 hard drive after using Dropbox. These data remnants included usernames, passwords, browsers, software access, data stored in accounts, and time frames found on the file metadata. Data from Enron Corpus were used to test accounts created through three service suppliers. MD5 values were produced. VMs were created using VMware Player 4.0.1. A VM for every service supplier was used for testing, and Base-VM files were utilized as the control media to discover newly created files after each scenario. The contributions and limitations are summarized in Table 1.

Table 1 Contributions and limitations of Quick and Choo's research

Contributions	Limitations
Verified how to obtain files uploaded to and stored in Dropbox	Timestamps were manipulated by each service
Revealed the data that remains on a Windows 7 PC that uses a Dropbox application	Only Dropbox was used in the case study, and the results cannot be proven applicable to different service providers
These data include usernames, passwords, browsers, software access, remaining data stored in the accounts, and the time frame found on file metadata	

Chung et al. [10], who focused on data stored on servers, stated that finding user activities upon service subscription is the most difficult aspect of investigating a cloud storage service [11]. Most cloud companies are unwilling to release the information of user activities, which may be found in the log files of a cloud server, to protect the personal data and privacy of their clients. The study aimed to find traces left on PCs that accessed Amazon S3, Dropbox, Google Docs, and Evernote for cloud storage. The study also proposed a process for the forensic investigation of cloud storage services and described important elements of the investigation process. Internet Explorer and Firefox were used to access cloud services to locate the data left in temporary log files. Traces of system installation log and database files are left in the registry when an application is installed on a Windows system. These files are vital because they contain traces of cloud storage service usage. Forensic investigators can then obtain original documents and related metadata from client devices, given that certain cloud services sync the files stored on the cloud server to the clients' hard drives. The contributions and limitations are summarized in Table 2.

Table 2 Contributions and limitations of Chung et al.'s research

Contributions	Limitations
Discovered files retained in PCs and smartphones after access to cloud services from Amazon S3, Dropbox, Google Docs, and Evernote	The research did not compare MD5 values at each phase to show file integrity
	The proposed model should be further investigated using providers of cloud storage service other than those used in the research
Proposed a process model for the forensic investigation of cloud storage services and described important elements of such investigation	

3   Cybercrime Evidence Collection Model

This section explains the method of collecting cybercrime evidence that may be found in computer cloud storage. The following statements show the particular software and hardware that we used in conducting this research. The process that we adopted for recovering the evidence material involved different methodologies, such as keyword search across digital media, recovery of deleted files, extraction of registry and log file information, and tools such as the SQLite Viewer, HxD, AccessData FTK Imager, AccessData FTK, Wireshark, and Event Viewer.

Implementation, which was the final process of our project, required all the detailed specifications that we gathered during the system and software design phase. This stage was performed in the form of a real-time experiment. In the implementation step, the experiment used 27 VMs to produce results. Three to four VMs were deployed accordingly to six computers that were used for testing purposes to allow the experiment to be performed within the shortest time possible to avoid resource waste. To ensure that all the processes could be conducted smoothly, the specific requirements and design specifications were studied briefly before the real-time experiment. All the necessary software and hardware devices were then prepared before the integration and system testing phase to avoid any unnecessary problem that would disrupt the succeeding testing process.

After the requirements were determined, all the data analyses were illustrated in the form of a flowchart. All the analyzed requirements and information that we gathered during the requirement definition and analysis phase were then implemented in the system design phase. This stage allowed us to design the overall flow of the process to be approved in the implementation phase. The flowchart was also used for guidance in the conduct of the experiments to enable the acquisition of results within the planned time frame. It provided an overview of the accomplishment of the succeeding research experiment, thereby preventing waste of time. The collected information enabled the unification of the specifications that were crucial in implementing our experiment. The flowchart of the experiment process and the testing parts is shown in Fig. 1.

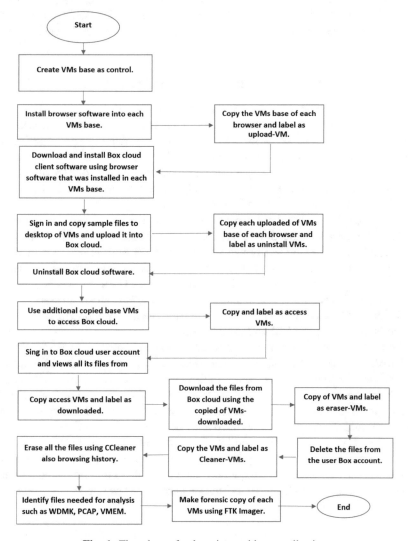

Fig. 1 Flowchart of cybercrime evidence collection

4   Experiment Results and Analysis

In this research, forensic image files retrieved from each scenario, memory dump files, and network PCAP files were copied from hard drives. We examined and analyzed all the forensic images that were copied using different tools, such as AccessData FTK Version 5.5, Wireshark Network Analyzer 2.0.5, AccessData FTK Imager Version 3.4.2.6., and HxD Version 1.7.7.1. We conducted analyses to locate the content of data remnants that were created and left on the PCs' hard drives. These data remnants that we found will help forensic investigators visualize the activities and other necessary information of clients. Users who access https://my.box.com/ through the specified web browsers will have their accounts' usernames appear at the top right hand corner of their browsers. Devices synced to mobiles and PCs with the Box account will be shown, together with the amount of spaces used and available for each user. Furthermore, the homepages of Box websites will show the details of the stored files and folders, the last modification date, and the sizes of the files included in these folders. The information icon provided in each file allows users to view the metadata of their files.

4.1   Box Cloud Base Images Created by Firefox Browser

For this case, base images were created before each scenario to serve as control. We completed this step to ensure that no files or related artefacts were present before browser installation and scenario implementation. We ran keyword searches with different terms to conduct the image analysis using Firefox. The Box software was installed on a desktop where the pictures and files were uploaded. The analysis indicated that no data and artefacts were originally present in relation to the browser and Box client software files before each installation.

Our test displayed that the downloaded Box client software was from www.box.com for Windows and the downloaded files were stored in "C:\Users\ [username] \Downloads\" with the name BoxSyncSetup. When we ran the software, a window appeared that requested the user to agree with the conditions required in installing Microsoft.NET 4.5.2 to continue the installation process. The same scenario happened in all of the VMs used for the installation process of the Box client software. After the installation of Microsoft.NET 4.5.2, the executable file box_Windows.exe was copied into the hard drive of the VM created previously.

Our analysis showed that the installed Box client software was in the "C:\Program Files\BoxSync" folder in the six uploaded VMs. The Box sample files and folders also appeared in the default Box Sync folder, which was at "C:\Users\[username] \Box Sync". Meanwhile, a drive called "Box Sync (P:\)" was in My Computer. The default box drive contained all of the files that the user had saved and synced to their website account. It also showed all the details of each file in the folder.

The use of Access Data FTK Toolkit to analyze files contained in Box sync folders allows forensic examiners to easily identify all the files necessary for the investigation process. Timestamps attached to these files, such as the dates of modification, creation, and last access, may be valuable to the investigation. Moreover, the synchronization of files in BoxSync Drive folders with those in web browsers allows examiners to obtain crucial data and important evidence regarding their investigations.

4.2   Evidence Directories

The evidence that we collected are explained briefly in the following subsections.

A. Directory Listing of VM Hard Drives

An analysis of link files showed that extension filenames with.lnk associated with the sample files of the Box cloud, such as picturetesting2.lnk and picturetesting3.lnk, were placed in "C:\Users\[username]\ AppData\Roaming\Microsoft\Windows\Recent", where Upload-VMs, Download-VMs, Copy&Move-VMs, CCleaner-VMs, and Eraser-VMS were located. However, other link files were not found within the five Access-VMs. Thus, link files were created only after the files were downloaded and opened. Forensic investigators can benefit from the Prefetch files of applications running on a system. Windows normally creates a Prefetch file when an application is run for the first time from a certain location to speed up the loading time of this application. Relevant data are shown in these files in the history of users' applications. Prefetch files remain on a system (in "C:\Windows\Prefetch") even after the Box client software and its files are uninstalled and deleted, respectively.

B. Event Logs Files

Event logs were part of our analysis. These logs were observed and analyzed using the built-in Windows Event Viewer. The files presented in Event Viewer were shown in the VMs throughout testing for all the events that occurred. The Windows Firewall exception list was provided with the rule when the Box client software was installed. Another rule, called "Windows Communication Foundation Net.TCP Listener Adapter (TCP-In)," was provided after Microsoft's installation. NET was necessary for the Box client software installation. Each software installation process was stored in the event log file "Microsoft-Windows-Windows Firewall with Advanced Security% 4Firewall.evtx", which was in the "%SystemRoot%\system32\winevt\logs" folder.

C. VM Cache Files

The test revealed that the VMware is suitable for locating information storages regarding guest applications in a directory (called "caches"). The .vmdk file of each VM resided under the directory. Through analyzing cache directories, forensic examiners can recover useful information, such as shortcuts present in users' PC, filenames, timestamps of shortcut creation, shortcut icons, and timestamps of the initial runs of applications. The ubiquity of virtualization presents opportunities for examiners to locate artefacts in the cache directories of users' hard drives. During the installation of VMware Tools on a guest, <VMHome>s\caches was created on the host. PC \Documents\VirtualMachines\caches\<VMHome>\Caches\GuestAppsCache\appdata contained 241 files, each of which was named a long hex value and the extension appinfo .appicon.

The 241 files contained the path to the Box client application and other information. The sister file, which opened 7e866a683c41ec902f0fafea6e4f.appinfo in a hex editor, showed the installation process of the Box client software into the hard drive. The presence of this file specified that the user had installed Box previously and important evidence may be present in the Box account. No changes were observed in any of the . appinfo or .appicon pair of files. The contents of <VMHome>\caches\GuestAppsCache remained untouched as a user opened an application within the context of the guest VM. Content files remain in their location as long as the VMs are in use.

D. RAM Analysis

In this research, memory captured the VMEM files in their locations in all scenarios before the VMs were shut down. This analysis result was conducted by using AccessData FTK Imager. The analysis process showed that the terms of Box appeared in the VMEM files in the seven implemented scenarios, except the first scenario of base VMs. Thus, the website URL (www.box.com) appeared in all VMEM files except Base VMs.

Box account information, such as the username, was in the Upload-VMs, Access-VMs, Copy&Move-VMs, and Download-VMs. The usernames were often around several texts, including "usernam", "username[]L", "username", and "username&&$". The passwords of Box accounts were shown in plaintext around "%s&password", "Passwo.rd", "PASSWRxD_Dc[]@>>&", and "*eà [password]". This text can be used as guide by forensics investigators in running keyword searches to find potential Box account information of criminals. Furthermore, data carving was retrieved from thumbnail pictures and the partial picture files recovered from the Box sample files in the VMEM files for Access-VMs, Upload-VMs, Copy&Move-VMs, and Download-VMs. A VMEM file is a backup of a virtual machine's paging file. It appears only when the virtual machine is running or has crashed.

E. Thumbcache Files

For the seven Base-VMs, a thumbcache file analysis illustrated that no pictures of the Box sample existed prior to Box client software installation and access. Furthermore, the stored Box sample pictures were not found in the Uninstall-VMs and CCleaner-VMs or Access-VMs. The databases that keep various content thumbnail images in systems are called the thumbcache. However, in our study, thumbnail samples related to the files of the stored Box sample were found in the Upload-VMs, Download-VMs, Eraser-VMs, Uninstall-VM, and in the other VMs, including Move&Copy-VMs and Upload-VMs. Therefore, after certain files are accessed, uploaded, or downloaded into an account, only the thumbnail images will be kept in the thumbcache files.

F. Network PCAP Files

With the use of Network Miner 2.0 and Wireshark, PCAP files (also known as Packet Capture data) were created during the live network capture during each Box access through different browsers. Usually, PCAP files are created to assist forensic investigators in analyzing the packet sniffing and the characteristics of a data network. Examiners can extract information regarding file remnants associated to Box and network activities by analyzing network capture files.

Port 443 (https) and then Port 80 (http) were used to observe network traffic. Our analysis illustrated that a session with IP addresses ranging from 107.152.26.0 to 107.152.25.255 was registered under www.box.com when a Box account was accessed using the client software or a web browser from Fort Lauderdale, Florida, United States. A different session with IP addresses ranging from 54.192.72.0 to 54.192.72. 255 was then found on Port 80 and then Port 443, which was registered for cloudfront. net at Woodbridge, New Jersey, United States. In addition, sessions with IP addresses ranging from 216.58.221.0 to 216.58.221.225 and registered under www.google-anaytics.com were found from source port TCP 443. We also found evidence indicating the use of certain anti-forensic tools by criminals, as explained in the following.

4.3   Anti-forensic Techniques and Uninstallation of Box Client Software

To delete the downloaded files from the Box account and to uninstall the Box client software (BoxSyn) as part of the test, Eraser and CCleaner were ran within the computer hard drive. However, Box-related data residue, such as cache files, sync files, and web browsing history, were still found within the hard drive although the client software had been uninstalled. The anti-forensic application toolkit failed to delete all the files related to Box. Table 3 shows a summary for the findings of Box cloud installation analysis.

Table 3 Summary of Box cloud installation analysis

Name of files	Location of files
Box sample files deleted	C:\Program Files (x86)\
Box Sync folder	C:\ProgramFiles\OpenSSH\home\[username]\AppData\local \Box Sync
Box Sync folder	C:\Program Files\OpenSSH\home\[username]\Box Sync
Boxsync folder	C:\Program Files\OpenSSH\home\[username]\AppData\Boxsync
Boxsync	C:\Users\[username]\AppData\local\temp\Box Sync
Box Sync folder	C:\Users\[username]\AppData\local\Box Sync
data.db	C:\Users\ [username]\AppData\Local\box folder
BoxSyncSetup.exe	C:\Program Data\Package caches{2812d567-60c1-45ad-a7b0-9de32e4d94df}
BoxSyncSetup.exe	C:\ProgramData\User\[username]\Appdata\Local\Temp\{2812d567-60c1-45ad-a7b0-9de32e4d94df}\.be
BoxSyncSetup.exe	C:\ProgramData\User\[username]\Appdata\Local\Temp\{3012d567-60c1-45ad-a7b0-9de32e4d94df}\.cr

4.4   Comparative Evaluation

As an objective of this research, tests and the implementation process for the determination of the location and the data remaining on users' computers prior to the usage of a cloud storage application for new cloud storage were accomplished. The research

Table 4 Summary of analysis findings

Type of VMs	Password	Username	Software	Sample files	Keyword search term
Located data remnants					
Base-VMs	Nil	Nil	Nil	Nil	Nil
Upload-VMs	Found in RAM	Found in RAM	BoxSync_Windows.exe was found after downloading. The location of client software installation and the BoxSync sample files uploaded were found	Prefetch and link files were found	Multiple matches of keyword search obtained
Access-VMs	Found in RAM	Found in RAM	Nil	The information of BoxSync software access was found in cookies, browsing history, page files, and unallocated spaces	Multiple matches of keyword search obtained
Move&Copy-VMs	Found in RAM	Found in RAM	Nil	The file that we moved and copied were found	Multiple matches of keyword search obtained
Download-VMs	Found in RAM	Found in RAM	Nil	The downloaded files were stored in the hard drives of the VMs	Multiple matches of keyword search obtained
Eraser-VMs	Nil	Nil	Nil	The information of BoxSync software access was found in cookies, browsing history, page files, and unallocated spaces. The deleted files were still available in the unallocated spaces	Multiple matches of keyword search obtained
CCleaner-VMs	Nil	Nil	Nil	The information of BoxSync software access was found in	Multiple matches of

(*continued*)

Table 4 (*continued*)

Located data remnants

Type of VMs	Password	Username	Software	Sample files	Keyword search term
				cookies, browsing history, page files, and unallocated spaces. The deleted files were still available in the unallocated spaces	keyword search obtained
Uninstall-VM	Nil	Nil	Nil	The information of BoxSync software access was found in cookies, browsing history, page files, and unallocated spaces. The deleted files were still available in the unallocated spaces	Multiple matches of keyword search obtained

conducted by [8, 9], who used Dropbox as their case study, did not include information regarding the analysis of VM Cache files, which is beneficial for obtaining information regarding the history of files that were previously ran on a user's computer.

Box's cloud directory listing showed that certain synced files were the same as those found in Box accounts, thereby becoming beneficial for forensics investigators in obtaining evidence offline. Moreover, the analysis of client software in this research indicated that unlike "filecache.dbx" and "host.db" files (created by Dropbox), "data. db" files (created by Box) contained three pieces of information. These files from Box contained information regarding Box sync, web browsing information used to access Box, and the number of ports used in accessing Box. By contrast, "filecache.db" and "host.db" contained only information regarding the history of filenames synchronized with Dropbox. Table 4 summarizes the findings of Box cloud analysis.

5 Conclusion and Future Works

Throughout this research, the hard drives of the VMs were the main location for data remnants that were left after we accessed or downloaded and stored files in a Box account. Users leave sufficient evidence that can be used by forensic examiners for support and guidance through the investigation of related cases. This research helps forensic investigators find the particular file locations of necessary evidence and retrieve these data while reducing the time consumed in solving such cases.

The data obtained by the analysis of the Box client software showed that all information, such as Box usernames, passwords, network traffic, session IDs, Prefetch file listings, link files, and browsing history, were regarded as definitive clues in specifying file content, which is critical in solving such cases. The analysis of volatile data is efficient and sufficient in identifying Box account access and the networking devices used in accessing Box accounts. This process results from Box account synchronization and display of all browsers used for client access. Every device used to access and synchronize with the Box account saves, holds, and identifies all evidence potentially related to investigated cases.

This research was limited in that only Box cloud was used as a case study. Future research adopting the same methodology as in the current study must be conducted to examine other newly developed cloud storage services on the market to test the suitability of this methodology to other cloud service providers.

Acknowledgements. Funding support provided by the Ministry of Higher Education in Malaysia (No. RDU160106).

References

1. Messmer, E.: Cloud forensics: in a lawsuit, can your cloud provider get key evidence you need? Netw. World (2013). Last Accessed from http://www.networkworld.com/news/2013/030613-cloud-forensics-267447.html
2. Montelbano, M.K.: Cloud Forensics. Champlain College (2013)
3. Ahmed, A.A.: Investigation approach for network attack intention recognition. Int. J. Digit. Crime Forensics (IJDCF) **9**(1), 17–38 (2017)
4. Ahmed, A.A., Khay, L.M.: Securing user credentials in web browser: review and suggestion. In: 2017 IEEE Conference on Big Data and Analytics (ICBDA). IEEE, pp. 67–71 (2017)
5. Ahmed, A.A., Kit, Y.W.: MICIE: a model for identifying and collecting intrusion evidences. In: 2016 12th International Conference on Signal-Image Technology and Internet-Based Systems (SITIS). IEEE, pp. 288–294
6. Ahmed, A.A., Li, C.X.: Locating and collecting cybercrime evidences on cloud storage. In: 2016 International Conference on Information Science and Security (ICISS). IEEE, pp. 1–5 (2016)
7. Ahmed, A.A., Li, C.X.: Analyzing data remnant remains on user devices to determine probative artifacts in cloud environment. J. Forensic Sci. **63**(1), 112–121 (2018)
8. Quick, D., Choo, K.K.R.: Dropbox analysis: data remnants on user machines. Digit. Investig. **10**(1), 3–18 (2013)
9. Quick, D., Choo, K.K.R.: Google drive: forensic analysis of data remnants. J. Netw. Comput. Appl. **40**, 179–193 (2014)
10. Chung, H., Park, J., Lee, S., Kang, C.: Digital forensic investigation of cloud storage services. Digit. Investig. **9**(2), 81–95 (2012)
11. Taylor, M., Haggerty, J., Gresty, D., Hegarty, R.: Digital evidence in cloud computing systems. Comput. Law Secur. Rev. **26**(3), 304–308 (2010)

An Intelligent Expert System for Management Information System Failure Diagnosis

Kamal Mohammed Alhendawi[1] and Ala Aldeen Al-Janabi[2(✉)]

[1] Faculty of Management Sciences, Al-Quds Open University, Ramallah,
Palestine, Israel
Hindawi.kamal@yahoo.com
[2] Ahmed Bin Mohammed, Military College, Doha, Qatar
alaaljanabi@abmmc.edu.qa

Abstract. The purpose of this study is to develop and validate a new expert system for detecting failure in the web system interaction design. This system aims at helping the top management and IS developers to justify IT investment through diagnosing the interaction capabilities of user interface and online communication tools. The research methodology consists of a five-step process which facilitates the development of the expert system and, consequently, the diagnosis of the interactivity features of the Web information system. The five process components are as follows: reviewing related empirical studies; extracting the core diagnosis factors; designing and implementing the expert system; testing the expert system; and deploying the expert system. The validation achieved by a sample of IS developers and professionals demonstrates the effectiveness of the proposed detection framework. Based on the feedback collected from the IS developers who tested the developed expert system, the proposed framework seems promising, and can be even be applied in other related area such as human resources management.

Keywords: Artificial intelligence (AI) · Expert system · Information system
Knowledge-based system · Web system diagnosis

1 Introduction

In comparison with traditional systems, the expert system is considered to be one of the most advanced systems. It is defined as an interactive computer solution, which documents experience, intuition, judgment and other information in order to provide knowledgeable advice [1, 2]. The majority of expert systems applications can be categorized into the following: diagnosis and/or advisory, design, planning or selection, configuration, data interpretation, scheduling, and training and support [3]. An expert system is also known as a knowledge-based system. The knowledge based system consists of four main parts: user interface, inference engine, knowledge base and knowledge engineering tool [4]. Moreover, professionals in the field state that the expert system provides valuable feedback and advice to non expert users in different fields [5].

© Springer Nature Switzerland AG 2019
P. Vasant et al. (Eds.): ICO 2018, AISC 866, pp. 257–266, 2019.
https://doi.org/10.1007/978-3-030-00979-3_26

In the context of information systems, diagnosing the failure of system in terms of its features and effectiveness is very important as there is no attempt to use the expert systems in detecting the system failure. Therefore, this study represents an initiative to detect the failure of Web information system based on the standards of software engineering. Based on a literature review of software engineering and human computer interaction, the interface design and the available communication tools are considered as the main factors for detecting the problem of users'interactivity [6–9].

To diagnose the interactivity of the system, this study aims at developing a diagnostic expert system in order to detect the failure in the Web system interactivitiy.

2   Theoretical Foundations

As a study that takes the initiative to utilize the expert systems in the field information system management, it seems appropriate to divide this section into two sub-section, including key factors of IS evaluation, as well as the background regarding the concept of expert systems.

2.1   Key Factors of Information System Assessment

At present, the interaction design is not only focused on system development, but also on product design and development [10]. This is also confirmed by [11] who indicated that interaction design is considered as one of the main key user-centered design disciplines.

[12] mentioned that the interaction term was suggested by Bill Moggridge and Bill Verplank in the mid-1980s. In this respect, Verplank highlighted that the interaction design is derived from the user interface design which is a computer science.

Moreover, [12] indicted that interaction design focuses on form design as well as behavioral actions. [13] also pointed out that interaction design is associated with the industrial design of software products, while the interactivity on the internet is defined as the extent to which the organizations share the online exchange with others, regardless of the restrictions relating to distance as well as time [14]. Another important point in this regard is that communication and collaboration can be considered as the key dimensions of interaction [15].

Within the context of the current research study, the interaction design is defined as: the degree to which WPS enable the organization employees to engage in online exchange with others through the user interface facilities and the quality of the available communication tools [14, 16], such as profiling, e-mail links, discussion form, feedback form, FAQ page, group subscription, web layout, and web site structure [17–19]. Thus, interaction design quality will be measured through two dimensions: user interface quality and communication tools quality. The measures of these two dimensions are adapted from several standard scales [17–19]. The following table expresses the concept and dimensions of the interaction design quality (See Table 1).

Consequently, these two factors are considered as standard factors in the software engineering and human computer interaction fields for assessing or diagnosing the software interactivity level [6–9]. Therefore, they can be considered as a base for

Table 1 The definition of interaction design quality dimensions

User satisfaction dimensions	Definition	Adapted from
D1: User interface quality.	The extent to which user interface layout such as profiling and links enables the employees to properly interact with the system.	[17–19]
D2: Communication quality.	The degree to which the WPS provides online communication tools such as feedback, discussion forum, and FAQ in order to allow knowledge sharing among employees.	[17]

building the expert system rules, and thus, detecting the weaknesses of the software interactivity.

2.2   Expert System Concepts

In this context, it is important to mention that there are many AI tools that have numerous applications in different fields. Neural networks, genetic algorithm, Bayesian network, pattern recognition, and expert system are examples on the AI tools that could be applied in order to efficiently generate the knowledge need for decision making. Figure 1 illustrates a sample of AI tools which are developed by an AI-community.

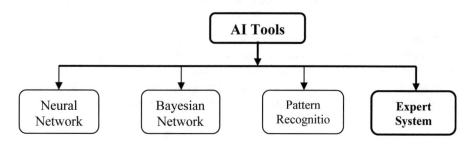

Fig. 1 Artificial intelligence tools

Based on Fig. 1, it can be inferred that expert system is a field of artificial intelligence. Expert System is considered as one AI discipline that has applicability in several domains and sciences, such as information system, medicine, chemistry, finance and management. In practical terms, the expert system is a computerized program that utilizes the knowledge and inference procedures in order to find the solution for the problems which formerly required a human expertise [20]. An expert system is a knowledge based systems consisting of a set of components which includes a user interface, an explanation facility, a knowledge base, an inference engine and a working memory [21]. Explanation facility is responsible for providing the reasoning behind a particular conclusion. The Knowledge base stores the knowledge in terms of rules. Working memory is the database that stores the facts used by the rules [22].

While the inference engine is responsible for selecting which rule to fire and with what priority. With regard to Agenda, it is a prioritized list of rules whose conditions are fulfilled by facts. Pattern matcher makes comparison between rules and facts [5]. Figure 2 demonstrates the components of an expert system.

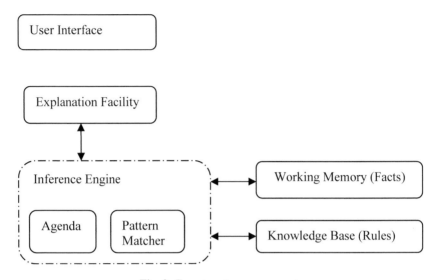

Fig. 2 Expert system components

As the expert system is a computer program used for extracting knowledge for non-expert users, it is essential to use some specialized software such as CLIPS in order to represent and extract the knowledge based solution. Practically, CLIPS stands for "C Language Integrated Production System". Due to its low cost and flexibility in knowledge representation, it is used to implement the proposed expert system model including facts, rules and advice [2, 5].

3 Methodology

To achieve the main objective of this study, the researchers followed a set of methodological steps: First, identifying the problem of IS failure which encounters the interactivity level of the Web system users. Second, collecting the knowledge of the system from the related literature, as well as articles in the field of information system evaluation and management. Third, using CLIPS language to implement the rule-based expert system, which stores the knowledge in terms of rules.

Based on Fig. 3, the methodology of this study mainly consists of three stages. In the first stage, the problems of interactivity level of information system are divided into two main types based on literature review. Therefore, the interaction failures can be classified into user interface and online communication failures.

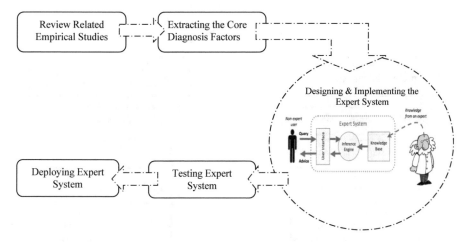

Fig. 3 The research methodology

4   Design and Implementation of Knowledge

Building the expert system requires five steps: knowledge acquisition, knowledge representation, design interfacing, knowledge updating, and system evaluation [23].

In the knowledge acquisition stage, the researchers collected the knowledge or experience in the field of Web information system evaluation (as much as possible) where the knowledge was surveyed based on the related literature review in the interaction design area [6, 7, 9, 17].

In the knowledge representation stage, the most suitable approach for representing data was selected, and a detailed expert layout was designed in order to logically organize the data and to identify the needed rules. The relationship between the system rules is illustrated in Fig. 4:

Figure 5 explains the sample code for some of the rules that were implemeneted using CLIPS. More details regarding the expert system implementation can be found in Appendix A.

In the design interfacing stage, the user interface was designed in its preliminary form. It was then modified in response to user feedback. Figure 6 shows the start-up screen of the program, while Fig. 7 illustrates a sample result of the expert system for the first choice.

In the knowledge updating stage, the researchers updated the knowledge based on feedback collected from users (i.e. ten system developers). This was done in order to improve the accuracy of the expert system results as much as possible.

Finally, in the system evaluation stage, the collected recommendations were summarized and taken into consideration to produce or deploy the final version of the expert system. With regard to the implementation, CLIPS was used a programing language in order to design the diagnostic expert system.

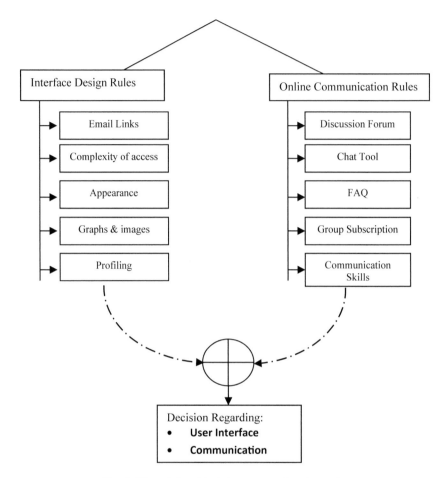

Fig. 4 The relationship between expert system rules

```
(defrule IQ-1
(ifYesNoSelect yes)  ?retractOpt1 <- (ifYesNoSelect? yes)
 (not (ifYesNoSelect1 ?))
=>
(retract ? retractOpt1) (printout t crlf crlf  "Does the employee can check
general information of profile via Web system." crlf crlf "Answer :  ")

(assert (ifYesNoSelect1 (read))))
```

Fig. 5 CLIPS code for implementing rule-based expert (Rule 1)

```
Expert System for Web Interaction Diagnosis

Select Your Choice

    1- User Interface Design - Check

    2- Online Communication Tools - Check

    3- Exit The Program

    Your Selection is: 1
```

Fig. 6 The start up menu of the expert system

```
Are system's services are designed to be easily accessed (y/N)?

Your Answer:      Y

Does employee can check general information of profile via Web system (y/N)?

Your Answer:      Y
```

Fig. 7 Sample questions for detecting user interface design

5   Result and Conclusion

It is vital to address the problem of failure diagnosis of the web IS interaction because senior management and IS developers need to justify their costly IT investment in terms of its interactivity level and, ultimately, its ongoing value to the organization. In light of this, the current study was primarily concerned with the development and validation of a new expert system for detecting the failure in the web system inter-action. The interaction quality was assessed in terms of the software design standards of user interface design and online communication tools. The expert knowledge base system was implemented using CLIPS expert system language. In the test phase, the accuracy of system detection was evaluated by ten specialized system developers and found that this system has a reasonable efficiency to be adopted for diagnosing the Web systems interaction. As a future work, the authors feel that the existing expert system should be developed in order to cover all other IS aspects, including information output and service features.

Appendix A: CLIPS Code of the Developed Expert System

```
;;---------------------------------- Main Menu ----------
--------------------
(reset)
(clear)
 (defrule start-up   (not (F ?))
=>   (printout t   "     Enter the ID of the Information
System Quality Factor For Check" crlf
"1- Interface Design  Quality  (IDQ) Check" crlf crlf
"2- Communication Quality (ComQ) Check " crlf crlf
"3- Exit The Program …. " crlf crlf crlf " ID of
Information System Quality is:   " )

(assert ( F(read))))
;;-------------------------------- Rule IDQ 0 ------------
------------------
 (defrule  IQ-0
(not (ifYesNoSelect ?)) (F 1)
?retractOpt1 <- (F 1)
 =>
 (retract ?retractOpt1)
(printout t crlf crlf  " Do you want to check the quality
of interface design" crlf crlf "Answer:   ")

(assert (ifYesNoSelect (read))))
;;------------ ----------------- Rule IDQ 1 -------------
--------------------
(defrule IQ-1
(ifYesNoSelect yes)
 ?retractOpt1 <- (ifYesNoSelect? yes)
 (not (ifYesNoSelect1 ?))
 =>
(retract ? retractOpt1) (printout t crlf crlf  "Is the
employee can check general information of profile and
organization via the system."crlf crlf "Answer :   ")

(assert (ifYesNoSelect1 (read)))  )
;;----------- ------------------- Rule IDQ 2 -------------
-------------------
(defrule IQ-2
(ifYesNoSelect1 yes)
?retractOpt1 <- (ifYesNoSelect1? yes)
(not (ifYesNoSelect2 ?))
 =>
```

```
(retract ?retractOpt1) (printout t crlf crlf  "Is the
system presents an organized list of specific e-mail link
to each employee contact.." crlf crlf "Answer:   ")

(assert (ifYesNoSelect2 (read))))
;;------------ ----------------- Rule IDQ 3 -------------
-------------------
(defrule IQ-3    (ifYesNoSelect2 yes)
?retractOpt1 <- (ifYesNoSelect2? yes)
(not (ifYesNoSelect3 ?))
=>
(retract ?retractOpt1) (printout t crlf crlf  "Is The
system's services are designed to be easily accessed."
crlf crlf "Answer:   ")

(assert (ifYesNoSelect3 (read))) )
;;------------ ----------------- Rule IDQ 4 -------------
-------------------
(defrule IQ-4
(ifYesNoSelect3 yes)
 ?retractOpt1 <- (ifYesNoSelect3? yes)
 (not (ifYesNoSelect4 ?))
=>
(retract ?retractOpt1) (printout t crlf crlf  "Is The
system's look/appearance is unambiguous.  " crlf crlf
"Answer:   ")

(assert (ifYesNoSelect4 (read))) )
;;------------ ------------- Rule IDQ 5 ------------------
-------------------
(defrule IQ-5
(ifYesNoSelect4 yes)
?retractOpt1 <- (ifYesNoSelect4? yes)
 (not (ifYesNoSelect5 ?))
 =>
(retract ?retractOpt1) (printout t crlf crlf  "Is There is
compatibility between graphics (colors, graphs, images)
and content " crlf crlf "Answer:   ")

(assert (ifYesNoSelect5 (read))) )
;;------------ ------------- Rule IDQ 6 ------------------
-------------------
(defrule IQ-6
(ifYesNoSelect5 yes)
?retractOpt1 <- (ifYesNoSelect5? yes)
=>
(retract ?retractOpt1) (printout t crlf crlf  "The
Information System suffering has a problem encountered the
interface design" crlf crlf " Thank you for using Info.
Sys. Expert System  …. " crlf crlf))
```

References

1. Harvey, J.J.: Expert systems: an Introduction. Int. J. Compu. Appl. Technol. **1**, 53–60 (1988)
2. Dalkir, K.: Knowledge Management in Theory and Practice. Taylor & Francis (2013)
3. Allwood, R.J.: Techniques and Applications of Expert System in The Construction Industry. Ellis Horwood Series in Civil Engineering, 1st edn. England (1989)
4. Lin, H.-C. K., Chen, N.-S., Sun, R.-T., Tsai, I.-H.: Usability of affective interfaces for a digital arts tutoring system. Behav. Inf. Technol. (ahead-of-print), 1–12 (2012)
5. Hopgood, A. A.: Intelligent Systems for Engineers and Scientists, 3rd edn. Taylor & Francis (2011)
6. Benbya, H., Passiante, G., Belbaly, N.A.: Corporate portal: a tool for knowledge management synchronization. Int. J. Inf. Manage. **24**(3), 201–220 (2004)
7. Puntambekar, A.: Software Engineering And Quality Assurance. Technical Publications (2010)
8. Law, C.C., Ngai, E.W.: ERP systems adoption: an exploratory study of the organizational factors and impacts of ERP success. Inf. Manage. **44**(4), 418–432 (2007)
9. Zhang, P., Nah, F.F.-H., Preece, J.: Guest editorial: HCI studies in management information systems. Behav. Inf. Technol. **23**(3), 147–151 (2004)
10. Edeholt, H., Löwgren, J.: Industrial design in a post-industrial society: a framework for understanding the relationship between industrial design and interaction design. In: 5th Conference European Academy of Design (2003)
11. Holmlid: Interaction design and service design: expanding a comparison of design disciplines. In: Nordic conference on service design and service Innovation (2009)
12. Cooper, A., Reimann, R., Cronin, D.: The Essentials of Interaction Design. Wiley Press, Indianapolis (2007)
13. Moggridge, B.: Designing Interactions. MIT Press (2007)
14. Albrecht, C.C., Dean, D.L., Hansen, J.V.: Marketplace and technology standards for B2B e-commerce: progress, challenges, and the state of the art. Inf. Manage. **42**, 865–875 (2005)
15. Lawson-Body, A., Limayem, M.: The impact of customer relationship management on customer loyalty: the moderating role of web site characteristics. J. Comput. Mediated Commun. **9**, 4 (2004)
16. Julier, G.: From visual culture to design culture. Des. Issues **22**, 1 (2006)
17. Lawson-Body, A., Willoughby, L., Logossah, K.: Developing an instrument for measuring e-commerce dimensions. J. Compu. Inf. Syst. **51**(2), 213 (2010)
18. Muylle, S., Moenert, R., Despontin, M.: The conceptualization and empirical validation of website user satisfaction. Inf. Manage. **41**, 213–226 (2004)
19. Yoo, B., Donthu, N.: Developing a scale to measure the perceived quality of an internet shopping site SiteQual. Quart. J. Electron. Commer. **2**(1), 31–45 (2001)
20. Tyler, A.R.: Expert Systems Research Trends: Nova Science Publishers (2007)
21. Shu-Hsien, L.: Expert system methodologies and applications - a decade review from 1995–2004. Expert Syst. Appl. **28**, 93–103 (2005)
22. Chung, P., Hinde, C., Moonis, A.: Developments in applied artificial intelligence. In: Proceedings of 16th International Conference on Industrial and Engineering Applications of Artificial Intelligence and Expert Systems, IEA/AIE 2003, Laughborough, UK, 23–26 June. Springer (2003)
23. Ismail, N., Ismail, A., Atiq, R.: An overview of expert systems in pavement management. Eur. J. Sci. Res. **30**, 99–1111 (2009)

Fuzz Test Case Generation for Penetration Testing in Mobile Cloud Computing Applications

Ahmad Salah Al-Ahmad[1] and Hasan Kahtan[2(✉)]

[1] Department of Computer Science, Universiti Teknologi, Mara, Malaysia
ahmad.salah.85@gmail.com
[2] Department of Software Engineering, Faculty of Computer Systems
and Software Engineering, Universiti Malaysia Pahang (UMP), Lebuhraya Tun
Razak, Gambang, 26300 Kuantan, Pahang, Malaysia
hasankahtan@ump.edu.my

Abstract. Security testing for applications is a critical practice used to protect data and users. Penetration testing is particularly important, and test case generation is one of its critical phases. In test case generation, the testers need to ensure that as many execution paths as possible are covered by using a set of test cases. Multiple models and techniques have been proposed to generate test cases for software penetration testing. These techniques include fuzz test case generation, which has been implemented in multiple forms. This work critically reviews different models and techniques used for fuzz test case generation and identifies strengths and limitations associated with each implementation and proposal. Reviewing results showed that previous test case generation methods disregard offloading parameters when generating test case sets. This paper proposes a test case generation technique that uses offloading as a generation parameter to overcome the lack of such techniques in previous studies. The proposed technique improves the coverage path on applications that use offloading, thereby improving the effectiveness and efficiency of penetration testing.

Keywords: Penetration testing · Software testing · Security testing
Test case generation

1 Introduction

Information technology security is a dominant issue. Most organisations are attempting to improve their security levels. To achieve this goal, firms attempt to uncover hidden security vulnerabilities in applications, networks and other devices that they use. Penetration testing can be used to discover such vulnerabilities. This practice was defined in a previous work as 'the art of finding an open door' [1]. Researchers are attempting to redefine and improve penetration testing by considering it a post-deployment vulnerability assessment task that is conducted as an isolated test process in a manual and even ad-hoc fashion [2–5].

© Springer Nature Switzerland AG 2019
P. Vasant et al. (Eds.): ICO 2018, AISC 866, pp. 267–276, 2019.
https://doi.org/10.1007/978-3-030-00979-3_27

Penetration testing is flexible and scalable in testing for security vulnerabilities. Furthermore, this practice can be automated [6, 7]. Therefore, numerous researchers have proposed penetration testing models for the web and service-oriented applications [2, 7, 8]. Many have attempted to use penetration testing to test web applications, web service mobile applications and databases with different models and frameworks to find hidden vulnerabilities that may be exploited by attackers to harm applications, interrupt systems and steal data [6, 8–12]. Penetration testing includes certain phases, a main one of which is test case generation [13]. This phase is an effective means of enhancing the quality and security of software and systems. Test case generation primarily aims to find security-relevant weaknesses in implementation that may result in crashes of systems-under-test (SUTs) or anomalous behaviour [14] or vulnerabilities that can be exploited to hack the system [15]. Penetration test case generation generates invalid, random or unexpected input to a program in testing for unseen vulnerabilities [16–18]. The generation can be considered a black box testing technique used to find flaws in software by feeding random input into applications and monitoring for crashes [16]. According to these definitions, multiple models and techniques are available in the literature and in practice for penetration test case generation. For example, test case generation was used in a previous study to generate random input into a command line that was used to leverage security vulnerabilities [18].

Fuzzing is a negative testing technique that feeds malformed and unexpected input data to a program to reveal security vulnerabilities [19]. This method is effective in finding security vulnerabilities in software [20]. Fuzz testing is an interface robustness testing method that is used to stress the interfaces of SUTs with invalid input data [14]. Therefore, fuzz test case generation techniques need to be studied to ensure support for new technologies. Offloading is a new method used to provide scalability for over-coming the limitations of mobile devices, such as limited resources, by delegating certain tasks to the cloud [21]. The implementation of offloading is complex and affects the execution path, depending on the device and environmental parameters. Currently, demand for the implementation of offloading is exponentially growing because most mobile implementations have shifted to it [22]. Together, the complexity and the widespread usage of offloading have led to the necessity of the study and comparison of this technology against the available penetration test case generation techniques.

This work critically studies and analyses the existing models and techniques used for fuzz test case generation with respect to offloading technology. Furthermore, this work proposes a test case generation technique for penetration testing that considers the use of offloading when generating test case sets. This approach improves the efficiency and effectiveness of penetration testing of applications that use offloading, such as mobile cloud computing applications.

2   Fuzz Test Case Generation for Penetration Testing: Related Work

Fuzzing is an important technique for penetration test case generation. This practice is a form of negative testing that feeds malformed and unexpected input data to a program to reveal security vulnerabilities [19]. Likewise, fuzz testing is an effective technique for finding vulnerabilities in applications [20] and serves as an interface testing of robustness by stressing the interface of an application with data input [14, 15]. The main advantages of fuzz testing are its simplicity and capability to explore a finite state space [23]. Fuzzing is an effective means of improving the quality and security of software and systems. However, randomly generated input, which is used for penetration test case generation, is often redundant and frequently misses certain program behaviour entirely [24].

The adoption of fuzzing has the disadvantages of considerable time and processing power consumption, thereby making its implementation expensive. This large resource consumption is related to the massive number of generated test cases, which exponentially increases with application size. In expressing the relationship between the size of the application and number of generated test cases, Nageswaran [25] found that the number of generated test cases is equal to the number of function points raised to the power of 1.2. Therefore, multiple models, techniques and tools have been proposed to decrease the resource consumption of fuzz testing. In the literature, processes that use fuzzing for test case generation are called fuzzers [14, 20, 26]. According to published works, different categories of fuzzers are used [15], including random-based, template-based, block-based, dynamic-based and smart fuzzers. The only fuzzers that use full knowledge about the application-under–test are smart fuzzers; this knowledge is applied to fuzz data only in certain situations that can be reached by the model [14, 15].

The main point of this study is offloading technology, which allows application tasks to be delegated to the cloud. Offloading is implemented in different ways. For example, some models suggest offloading all processes to the cloud [27, 28], whereas other models suggest elastically splitting applications between the cloud and mobile devices in a dynamic or static manner [29–31]. Therefore, offloading changes the execution path on the basis of certain parameters related to the cloud, client and network.

Generating test cases that do not consider offloading may prevent the tester from testing certain paths and functionalities that may be open to attackers because of hidden variabilities. For instance, a test case used to test the function 'A' may use two implementations, namely, one in the cloud and another in the client. Testing both needs to consider offloading to generate test cases for both scenarios and thus run both function implementations on the cloud and on the client.

The models and techniques for test case generation were selected based on the methodology described in Fig. 1, which adopts the Systematic Literature Review (SLR) described in [32, 33]. The SLR began by identifying the nominated records from the selected databases using the following search strings: (i) 'penetration testing' and 'test case generation'; (ii) 'test case generation' and 'penetration testing'; (iii) 'penetration testing' and 'case generation' and (iv) 'penetration testing' or 'test case generation'. Google Scholar (https://scholar.google.com), ACM Digital Library (http://portal.acm.org), IEEE Xplore (http://ieeexplore.ieee.org), ScienceDirect (http://www.sciencedirect.com) and SpringerLink (http://linkspringer.com) have been selected as these are currently the most relevant sources in software and security engineering [34].

Subsequently, these records were screened to remove non-fuzz test case generation techniques and other records that do not describe such methods. A total of 72 articles were assessed for eligibility. During the full-text eligibility assessment, the articles that

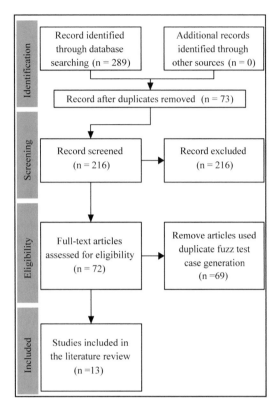

Fig. 1. Literature review methodology

did not propose new test case generation methods and used others' models for penetration testing have been removed. Finally, 13 articles were selected for the study. Table 1 summarises the selected articles about the models and techniques used in fuzz test case generation. These models disregard offloading and the device state; moreover, they generate a large number of test cases, thereby making testing all models expensive. Test cases are randomly generated without targeting the important aspects of the applications under testing.

The chart shows that the reviewed fuzz test case generation used in penetration testing can mainly be categorised into code [24, 39, 40], modelling [19, 20, 38], domain values [9, 18], pattern bases [35, 36] and session-based models and techniques [17, 23, 37].

3   Discussion and Proposed Technique

Table 1 analyses previous test case generations model and techniques on the basis of the parameter category used when generating test cases. The table also shows that previous test case generation models and techniques disregard offloading, which is one

Table 1. Models and techniques for penetration test case generation

Ref.	Method	Strength	Limitations
[24]	Uses constraint language to generate test cases, explores code with an interpreter and defines paths	Uses runtime values to analyse dynamic code	Disregards offloading
[9]	Determines the values for the other input vectors using an alphanumeric value for the password and setting	Choosing a value on the basis of the domain.	Disregards offloading
[20]	Generates UML state machine model on the basis of the manual analysis, static analysis or dynamic instrumentation	Used with complex structured inputs	Disregards offloading
[35]	Uses protocol builder, diagnoses and debugs facilities and compliments a compatibility layer with a GUI	Allows user to choose the type of protocol to use	Disregards offloading
[19]	Uses architecture and call graph model, generates test cases for each scenario with input that cover domain values, executes test cases and monitors results	Automates and uses domain values to generate test cases	Disregards offloading
[17]	Uses crawled forms on regular expressions as a validator	Generates test cases related to web applications	Can fuzz web forms only and disregards offloading
[18]	Generates sequences of characters using fuzzer on the basis of target component input type	Automatically interacts with the UI of a target application	Generates fuzz test case with one variable Disregards offloading
[23]	Uses GUI exploration and determines the context of the input by searching for keywords in the hints and the widget IDs associated with editable text boxes and in the visible text labels next to them through context determination	Used for mobile GUI exploration	Disregards offloading
[36]	Builds patterns that include test type, steps and leak type and uses these patterns to generate test cases	Reflects real memory leak issues and causes	Disregards offloading
[37]	Uses session tokens to send SQL injection from admin to client	Update log status immediately	Disregards offloading
[38]			

(*continued*)

Table 1. (*continued*)

Ref.	Method	Strength	Limitations
	Create the source code of the RESTful API and creates test cases to test the implementation	Generates test cases to test the source code and implementation	Resource-intensive and disregards offloading
[39]	Uses entire source codes files to locate interface to be used in generating all test cases	Can reach pages that are not linked with URL at the front-end pages	Requires all code files to be examined and disregards offloading
[40]	Uses inferential metamorphic testing to reduce false positives in SQL injection penetration tests	Reduces number of false positive	Disregards offloading

of the most important characteristics of the new software development model. Disregarding offloading reduces the path coverage, thereby compromising the efficiency of the penetration testing when testing applications that use offloading, such as those in mobile cloud computing [41, 42].

The proposed technique, which is shown in Fig. 2, overcomes this lack of fuzz test case generation techniques that use offloading parameters for penetration testing. This method proposes to use offloading parameters collected from the application documentation and developers or obtained from the application source code used in generating the test cases. In this instance, if the list-sorting function is implemented in the cloud and on the local device and selecting the function to execute is based on client status data, then should generate test cases that cover all the applicable client statuses.

This research gap opens a new research direction in the domain of fuzz test case generation in penetration testing. Future research will attempt to propose a new fuzz test case generation technique that considers offloading for penetration testing. The approach proposed in the current study improves the penetration testing coverage path, thereby enhancing security for applications used in mobile cloud computing and others that use offloading (Table 2).

4   Conclusion

Penetration testing is an important defence tactic in the domain of digital security because it is used to reveal hidden vulnerabilities. This work has shown the importance of test case generation in penetration testing and reviewed the state of art in conducting fuzz test case generation techniques and models. Certain models for fuzz test case generation used in penetration testing have been reviewed considering offloading parameters. Reviewing results showed that previous penetration fuzz test case generation models and techniques disregard offloading, which is used in advanced technologies to augment tasks. Additionally, this study has proposed a new fuzz test case generation technique that uses offloading as a parameter for generation. The proposed technique improves path coverage when conducting penetration testing over

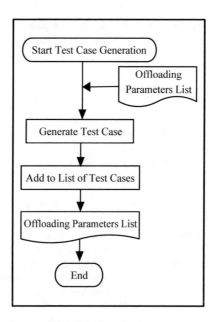

Fig. 2. Proposed technique for test case generation

Table 2. Analysis of models and techniques for penetration test case generation

Reference	Code-based	Model-based	Domain values	Pattern bases	Session-based	Offloading
[24]	✓	×	×	×	×	×
[9]	×	×	✓	×	×	×
[20]	×	✓	×	×	×	×
[35]	×	×	×	✓	×	×
[19]	×	✓	×	×	×	×
[17]	×	×	×	×	✓	×
[18]	×	×	✓	×	×	×
[23]	×	×	×	×	✓	×
[36]	×	×	×	✓	×	×
[37]	×	×	×	×	✓	×
[38]	×	✓	×	×	×	×
[39]	✓	×	×	×	×	×
[40]	✓	×	×	×	×	×

applications that use offloading, thereby improving the efficacy and effectiveness of penetration testing. In future, this proposed technique needs to be formulated in detailed steps for improved ease of use.

Acknowledgments. This research is supported by the Department of Research and Innovation of University Malaysia Pahang under Fundamental Research Grant Scheme (FRGS) RDU170102.

References

1. Geer, D., Harthorne, J.: Penetration testing: a duet. In: 18th Annual Computer Security Applications Conference. IEEE, Las Vegas (2002)
2. Xiong, P., Peyton, L.: A model-driven penetration test framework for Web applications. In: Eighth Annual International Conference on Privacy, Security and Trust (PST). IEEE, Ottawa (2010)
3. Xu, W., Groves, B., Kwok, W.: Penetration testing on cloud—case study with owncloud. Glob. J. Inf. Technol. **5**(2), 87–94 (2016)
4. Goel, J.N., et al.: Ensemble based approach to increase vulnerability assessment and penetration testing accuracy. In: 2016 International Conference on Innovation and Challenges in Cyber Security (ICICCS-INBUSH). IEEE (2016)
5. Zhao, J., et al.: Penetration testing automation assessment method based on rule tree. In: IEEE International Conference on Cyber Technology in Automation, Control, and Intelligent Systems (CYBER), Shenyang, China (2015)
6. Deptula, K.: Automation of cyber penetration testing using the detect, identify, predict, react intelligence automation model. In: Postgraduate School. Naval Postgraduate School, Monterey, p. 139 (2013)
7. Xing, B., et al.: Design and implementation of an XML-based penetration testing system. In: International Symposium on Intelligence Information Processing and Trusted Computing (IPTC). IEEE, Huanggang (2010)
8. Mainka, C., Somorovsky, J., Schwenk, J.: Penetration testing tool for web services security. In: 8th IEEE World Congress on Servicess. IEEE, Honolulu (2012)
9. Halfond, W.G., Choudhary, S.R., Orso, A.: Penetration testing with improved input vector identification. In: International Conference on Software Testing Verification and Validation. IEEE, Lillehammer (2009)
10. Jones, G.: Penetrating the cloud. Netw. Secur. **2013**(2), 5–7 (2013)
11. Kang, B.-H.: About effective penetration testing methodology. J. Secur. Eng. **5**(5), 10 (2008)
12. LaBarge, R., McGuire, T.: Cloud penetration testing. Int. J. Cloud Comput. Serv. Arch. (IJCCSA) **2**(6), 43–62 (2013)
13. Al-Ahmad, A., Abu Ata, B., Wahbeh, A.: Pen testing for web applications. Int. J. Inf. Technol. Web Eng. (IJITWE), **7**(3), 1–13 (2012)
14. Schneider, M., et al.: Online Model-based behavioral fuzzing. In: IEEE Sixth International Conference on Software Testing, Verification and Validation Workshops (ICSTW). IEEE, Luxembourg (2013)
15. Schneider, M., et al.: Behavioral fuzzing operators for UML sequence diagrams. Springer (2013)
16. de Graaf, M.: Intelligent fuzzing of web applications. In: Software Engineering. Universiteit van Amsterdam: Digital Academic Repository – UBA, University of Amsterdam (2009)
17. Färnlycke, I.: An approach to automating mobile application testing on Symbian smartphones: functional testing through log file analysis of test cases developed from use cases. In: School of Information and Communication Technology (ICT), KTH Royal Institute Of Technology (2013). Opent Access in DiVA

18. Karami, M., et al.: Behavioral analysis of android applications using automated instrumentation. In: IEEE 7th International Conference on Software Security and Reliability-Companion (SERE-C), Gaithersburg, Maryland, USA (2013)
19. Mahmood, R., et al.: A whitebox approach for automated security testing of Android applications on the cloud. In: Proceedings of the 7th International Workshop on Automation of Software Test (AST). IEEE, Zurich (2012)
20. Yang, Y., et al.: A model-based fuzz framework to the security testing of TCG software stack implementations. In: International Conference on Multimedia Information Networking and Security (MINES). IEEE, Hubei (2009)
21. Kaur, S. Sohal, H.S.: Hybrid application partitioning and process offloading method for the mobile cloud computing. In: Proceedings of the First International Conference on Intelligent Computing and Communication. Springer, Singapore (2017)
22. Shiraz, M., et al.: A study on the critical analysis of computational offloading frameworks for mobile cloud computing. J. Netw. Comput. Appl. **47**(1), 47–60 (2015)
23. Rastogi, V., Chen, Y., Enck, W.: Appsplayground: automatic security analysis of smartphone applications. In: Proceedings of the Third ACM Conference on Data and Application Security and Privacy. ACM, New Orleans (2013)
24. Wassermann, G., et al.: Dynamic test input generation for web applications. In: Proceedings of the International Symposium on Software Testing and Analysis. ACM, Seattle (2008)
25. Nageswaran, S.: Test effort estimation using use case points. In: Quality Week, San Francisco, California, USA (2001)
26. Mendez, X.: SQL injection fuzz strings (from wfuzz tool) (2014). https://wfuzz.googlecode. com/svn/trunk/wordlist/Injections/SQL.txt. Accessed 11 Sept 2015
27. Huang, D., et al.: MobiCloud: building secure cloud framework for mobile computing and communication. In: Fifth IEEE International Symposium on Service Oriented System Engineering (SOSE). IEEE, Nanjing (2010)
28. Kumar, K., Lu, Y.-H.: Cloud computing for mobile users: can offloading computation save energy? IEEE Comput. Soc. **43**(4), 51–56 (2010)
29. Zhang, J., Sun, D., Zhai, D.: A research on the indicator system of cloud computing security risk assessment. In: International Conference on Quality, Reliability, Risk, Maintenance, and Safety Engineering (ICQR2MSE). IEEE, Chengdu (2012)
30. Giurgiu, I., et al.: Calling the cloud: enabling mobile phones as interfaces to cloud applications. In: Middleware, pp. 83–102. Springer, New York (2009)
31. Kovachev, D. Klamma, R.: Beyond the client-server architectures: a survey of mobile cloud techniques. In: 1st IEEE International Conference on Communications in China Workshops (ICCC). IEEE, Beijing (2012)
32. Singh, Y., et al.: Systematic literature review on regression test prioritization techniques. Inform. (Slov.) **36**(4), 379–408 (2012)
33. Budgen, D., Brereton, P.: Performing systematic literature reviews in software engineering. In: Proceedings of the 28th International Conference on Software Engineering. ACM, Shanghai (2006)
34. Brereton, P., et al.: Lessons from applying the systematic literature review process within the software engineering domain. J. Syst. Softw. **80**(4), 571–583 (2007)
35. Zeisberger, S., Irwin, B.: A fuzz testing framework for evaluating and securing network applications. In: the Annual Southern Africa Telecommunication Networks and Applications Conference (SATNAC), London, UK (2011)
36. Shahriar, H., North, S., Mawangi, E.: Testing of memory leak in android applications. In: IEEE 15th International Symposium on High-Assurance Systems Engineering (HASE). IEEE, Miami (2014)

37. Kaushik, M., Ojha, G.: Attack penetration system for SQL injection. Int. J. Adv. Comput. Res. **4**(2), 724 (2014)

38. Fertig, T., Braun, P.: Model-driven testing of restful apis. In: Proceedings of the 24th International Conference on World Wide Web. ACM (2015)

39. Zhu, Z.: Automated penetration testing for PHP web applications, Georgia Institute of Technology, pp. 48, November 2016

40. Liu, L., et al.: An inferential metamorphic testing approach to reduce false positives in SQLIV penetration test. In: IEEE 41st Annual Computer Software and Applications Conference (COMPSAC) (2017)

41. Al-Ahmad, A.S., Aljunid, S.A., Sani, A.S.A.: Mobile cloud computing testing review. In: International Conference on Advanced Computer Science Applications and Technologies (ACSAT). IEEE, Kuala Lumpur (2013)

42. Paranjothi, A., Khan, M.S., Nijim, M.: Survey on three components of mobile cloud computing: offloading, distribution and privacy. J. Comput. Commun. **5**(06), 1 (2017)

Application of Travelling Salesman Problem for Minimizing Travel Distance of a Two-Day Trip in Kuala Lumpur via Go KL City Bus

Wan Nor Ashikin Wan Ahmad Fatthi[1(✉)], Mea Haslina Mohd Haris[1], and Hasan Kahtan[2]

[1] Department of Mathematics, Universiti Teknologi MARA Cawangan Selangor, Kampus Dengkil, 43800 Dengkil, Selangor, Malaysia
ashikin7463@uitm.edu.my, mealina@tmsk.uitm.edu.my
[2] Faculty of Computer Systems and Software Engineering, Universiti Malaysia Pahang (UMP), Lebuhraya Tun Razak, 26300 Gambang, Kuantan Pahang, Malaysia
hasankahtan@ump.edu.my

Abstract. Kuala Lumpur is a cosmopolitan urban centre of Malaysia and has received more than 11 million tourists per year. Tourists usually spend a few days in Kuala Lumpur to visit as many attractions as possible. However, planning such trips can be challenging for tourists who are unfamiliar with the city. Moreover, they are restricted by time and budget constraints. One of the free charter public transports in Kuala Lumpur is the Go KL City Bus. This study aims to assist tourists or travellers (domestic or international) in optimizing their trip around Kuala Lumpur via the Go KL City Bus. A mathematical approach called travelling salesman problem is used to identify the shortest distance between the places of interest. The study proposes a solution on the basis of a two-day tour route for selected tourist attractions in Kuala Lumpur. Results show that the shortest distance of routes for the first and second days are 48.64 km and 46.96 km, respectively. This study aims to promote tourism in Malaysia, thereby contributing to the country's economic and tourism growth.

Keywords: Tourism route · Shortest distance · Travelling salesman problem Optimization

1 Introduction

Tourism is an important service industry in Malaysia. A total of 25.9 million visitor arrivals and RM 82.2 billion tourist receipts were recorded in 2017 [1]. Tourism brings sustainable impact to Malaysia's development, economic growth and other related service industries such as transport, hotels, food and beverages, shopping mall and entertainment [2]. In 2017, the total shopping receipts increased, with foreign tourists recording an average stay of six nights. The top ten countries of origin of these tourists are Singapore, Indonesia, China, Thailand, Brunei, India, South Korea, Japan, the Philippines and the United Kingdom [1].

© Springer Nature Switzerland AG 2019
P. Vasant et al. (Eds.): ICO 2018, AISC 866, pp. 277–284, 2019.
https://doi.org/10.1007/978-3-030-00979-3_28

Kuala Lumpur, the capital city of Malaysia, can be overwhelming for first-time tourists. It is a self-contained city that offers safety, modern amenities, beautiful landscapes and numerous attractions [3–5]. Such amenities include affordable accommodations, public transportation options (i.e. KTM, LRT, Monorail, Rapid KL bus), shopping centres and many dining spots. Among the top tourist attractions in Kuala Lumpur are the Petronas Twin Tower, Aquaria KLCC, Chinatown, Jamek Mosque, Central Market, Merdeka Square, Petaling Street, Bintang Walk and Batu Caves [1].

Tourists now consider various reasons when visiting a destination, such as recreation, adventure, nature, food and heritage [6–11]. Visiting an unfamiliar city like Kuala Lumpur requires thorough planning. The following are common questions tourists ask when planning their trip: (1) What are the places of interest (POIs) to visit during the trip? (2) Which of these POIs should be visited first? (3) Is there any travel route suggestion for a two or three-day trip? (4) How to get to each POI?

Tourists wish to visit every attraction during their holiday trip. However, they also have to deal with limited time and budget [12]. Thus, they need to choose among the POIs. Accordingly, tourists gather information from different sources about the selected POIs, identify connecting routes and prepare an itinerary [13]. Preparing an itinerary requires tourists to consider the opening and closing hours of each POI, duration of visit to each POI, starting and ending place to visit and available budget.

The Go KL City Bus, which is owned by the Land Public Transport Commissions of Malaysia, is one of the public transportation options in the central business district (CBD) of Kuala Lumpur. Go KL offers a free bus service for an all-day ride and stops at various main attractions and business centres in the city. The buses are disabled- and eco-friendly. They run every 5 min during peak hours and 15 min during normal hours. Go KL City Bus operates daily from 6:00 AM to 11:00 PM and offers four routes, namely, the red, blue, green and purple routes, with a total of 65 stops between them [14].

This study aims to propose a two-day trip route in Kuala Lumpur via Go KL City Bus. The travelling salesman problem (TSP) method is used to identify the shortest distance between ten select tourist attractions. TSP minimises the distance between places, thereby reducing the travel time of tourists. This paper is organised as follows. Section 2 describes the background and theory applied in this study. Section 3 presents the problem description. Section 4 highlights the model formulation. Section 5 describes the findings of this study, and Sect. 6 concludes.

2   Background

The shortest path problem is one of the most fundamental problems in graph theory. It involves finding a path between two vertices in a given graph such that the sum of the weights of the constituent's edges is minimised [15]. The shortest path problem can be divided into two categories, namely, static shortest path and dynamic shortest path. In dynamic shortest path network, the travel time (cost) of the link varies with starting time on the link. On the contrary, static shortest path network has fixed topology and link costs [15]. The effectiveness of the shortest path problem is measured via total distance travelled, cost and time.

TSP is related to the shortest path theory [16]. TSP is a nondeterministic polynomial-time hard problem in combinatorial optimization and widely studied in operational research and computer science. TSP is defined in an undirected graph $G = \{V, A, W\}$, where $V = \{1, \cdots, N\}$, is a set of nodes, $A = \{(i,j)|i \neq j\}$ is a set of edges and W is a symmetrical weight matrix. The objective of a TSP is to find the shortest Hamilton route on G that visits every node exactly once and then returns to the starting point [16, 17].

TSP applications have been extended to route planning, logistics, maritime, medical services, manufacturing of microchips, DNA sequencing, vehicle routing, robotics, airport flight scheduling, time and job scheduling of machines [18–22]. Various approaches using exact and heuristics techniques have been proposed for solving TSP, such as branch and bound, branch and cut, dynamic programming and genetic algorithm [16, 17].

In route planning, TSP can be described as finding the route for the shortest tour between cities. Each city is visited only once, and the returning and starting point of the tour is the same. This finding is possible only if N number of cities and M distance path between the cities are given to perform the shortest path search between the cities [15–17].

Whilst solving TSP, two approaches can be considered. The first approach is finding the exact solution, the optimal Hamilton cycle in the graph. The exact solution can be derived by finding all the Hamilton cycles and choosing which cycle with the least weight. However, problem arises when the number of vertices is high. The second approach is identifying the optimal value via an approximate solution, such as the branch and bound method (BnB) [23]. BnB can be employed when the number of vertices in the graphs does not exceed 60.

3   Problem Description

Go KL is a public transportation service that offers free charters for commuters within the CBD of Kuala Lumpur. This bus service first operated in 2012 with two main routes, the green and purple Lines. In 2014, two additional routes called the red and blue Lines were introduced. Currently, Go KL City Bus comprises four routes: the green, purple, red and blue routes which have 14, 15, 19 and 17 stops, respectively. Go KL City Bus takes a total of 65 stops in between the four routes.

As depicted in Fig. 1, most of the stops of Go KL City Bus are in close proximity to various tourist attractions. For instance, the National Museum, National Mosque and Chow Kit Market are among the stops in the red route. Pavilion, Pasar Seni and KL Tower are the drop points in the purple route. However, passengers are required change buses from red line to blue line to reach Pavilion. Similarly, an exchange Go KL City Bus in Bukit Bintang brings passengers to the Kuala Lumpur City Centre (KLCC) in the green route.

This study proposes a route and visiting places for tourists via the Go KL City Bus. Ten top attractions were selected for planning a two-day trip, visiting five attractions

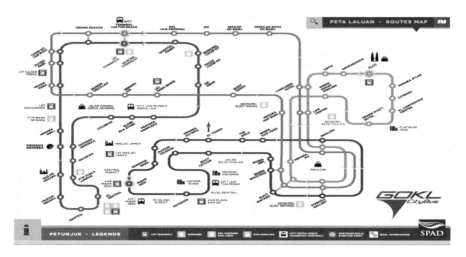

Fig. 1. GO KL City Bus route map

per day. The five attractions for the first day are Central Market, Pavilion, National Museum, KL Tower and National Mosque. The five attractions for the second day are Dataran Merdeka, KLCC, Bukit Bintang, Pasar Chow Kit and KL Convention Centre. TSP is utilised to find the shortest distance between attraction places, thereby reducing travel time. KL Central has been selected as the starting and ending point for the two-day trip.

Primary data collection was performed manually. It was conducted by riding a Go KL City Bus in Kuala Lumpur. It also involved travelling through all 10 places to identify the distance and travel time in between stations. Details of the time (min) and distance (km) were listed down. In addition, tourists were interviewed to obtain their feedback on the service of Go KL City Bus. Secondary data collection was performed to obtain the exact distance between the attractions. Such information was collected from Land Public Transport Commission.

4   Model Formulation

We refer to the mathematical model proposed by Jiang and Yang [24] to formulate an asymmetric TSP. Suppose that the coordinates of vertex set V are known and that a distance matrix $D = (d_{ij})$ is defined by edge set E. Let y_{ij} be a decision variable associated with each edge (i, j), where the decision variable $y_{ij} = 1$. The route from city indices i to j, represents the path selected by a salesman, whereas $y_{ij} = 0$ represents the path the salesman did not select. In addition, suppose that S is the proper subset of V, and $|S|$ denotes the number of vertices included in the set S. These notations and indices can be used to formulate the mixed integer programming formulation of TSP as follows:

$$\text{Min } Z = \sum_{j=1}^{n} y_{ij} = 1 \tag{1}$$

Subject to:

$$\sum_{j=1}^{n} y_{ij} = 1, \quad i \epsilon V, \tag{2}$$

$$\sum_{i=1}^{n} y_{ij} = 1, \quad j \epsilon V, \tag{3}$$

$$\sum_{i \epsilon S} \sum_{j \epsilon S} y_{ij} \leq |S| - 1 \quad \forall S \subset V, |S| \neq \emptyset$$
$$y_{ij} \in \{0, 1\} \tag{4}$$

The objective function (1) is used to minimise the total tour distance. Constraints (2) indicate that the tourist can only travel to the city (attraction place) j once. The departure constraints (3) indicate that the tourist can only depart from city i once. Constraints (2) and (3) ensure that the tourist travels to each city (attraction) only once, without eliminating the possibility of any subtour. However, the elimination constants (4) prevent the formation of any subtour.

5   Results and Discussion

To obtain the result, TSP Solver was utilised in the model computation to find the shortest route for the visiting places. The distance between the places was inputted into the TSP Solver, as shown in Fig. 2. The output shows an order of visiting places with minimum total distance on the first day of the trip. In addition, the validity and accuracy of the TSP result were tested via manual calculation using an Excel Spreadsheet. Five selected places must be visited per day. Thus, the calculation was performed based on the mathematical logic of five factorial (5!) with 120 possible routes to be traversed by tourists. Accordingly, the result shows that the minimum distances calculated by the TSP solver and Excel Spreadsheet are similar in value, as depicted in Table 1.

Table 1 shows the comparison of results obtained using Excel Spreadsheet and TSP Solver. The calculation was performed for a two-day trip in Kuala Lumpur. Table 1 shows the two possible routes attained from Excel Spreadsheet. The first route starts at point A to D, D to F, F to B, B to E, E to C and C to A, whereas the second route begins at point A to D, D to F, F to C, C to B, B to E and E to A. The total distance of both routes is 48.64 km.

By contrast, the route provided by TSP Solver is from A to D, D to F, F to C, C to B, B to E and E to A. Nevertheless, the total distance of both calculations is similar. The suggested visiting order for the first-day trip via Go KL City Bus is from KL Central to National Museum, followed by National Mosque to Pavilion, Pavilion to Central Market and finally back to KL Central as the starting point.

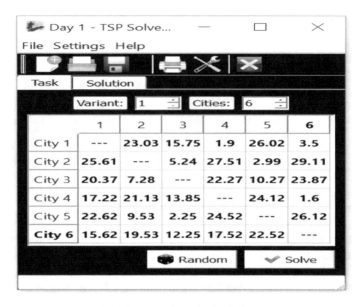

Fig. 2. Input data of TSP Solver

Table 1. Results of TSP Solver and Excel Spreadsheet

Day	First day	Second day
Shortest route (TSP Solver)	A – D – F– C – B – E – A	A – K – H – I – J – G – A
Shortest path (Excel Spreadsheet)	A – D – F – B– E– C– A A– D – F– C– B– E– A	A – G – I – K – H – J – A A – G – K –H – I – J – A A – I – K– H – J – G – A A – K – H – I – J – G – A
Total distance	48.64 km	46.96 km

Note: A = KL Central, B = Central Market, C = Pavilion, D = National Museum, E = KL Tower, F = National Mosque, G = Dataran Merdeka, H = KLCC, I = Bukit Bintang, J = Pasar Chow Kit, K = KL Convention Centre, km = kilometer

For the second-day trip, the result of the Excel Spreadsheet shows four possible routes with minimum distance. The output of TSP Solver provides a route with similar distance of 48.64 km to the route derived via the Excel Spreadsheet . Therefore, an identical route from both the Excel Spreadsheet and the TSP Solver was selected as the suggested route for the second-day trip which is from A to K, K to H, H to I, I to J, J to G and G to A. The suggested visiting order for the second-day trip by riding Go KL City Bus is from KL Central to KL Convention Centre, followed by KLCC to Bintang Walk, Bintang Walk to Jalan Chow Kit, Jalan Chow Kit to Dataran Merdeka and finally back to KL Central.

6   Conclusion and Recommendations

TSP provides an effective and efficient shortest route approach. It enables tourists to optimise cost, time and distance of travelling with limited time. In this study, TSP was utilised to find the shortest distance between tourist attractions in Kuala Lumpur. A two-day trip that follows an order of visiting places is possible via the free charter bus service Go KL City Bus. Such a strategy can assist tourists in planning their visit to Malaysia. Hence, it helps promote Malaysia's tourism industry.

Go KL is a public transport service in Kuala Lumpur. This service is convenient because it offers free fare for an all-day ride. The drop points en route consist of several tourist attractions places. However, this study considered only ten attraction places in Kuala Lumpur. Such limitation is due to the fact that one day is insufficient for tourists (local/international) to explore Kuala Lumpur. The selected attraction places are KLCC, KL Tower, Merdeka Square, Bintang Walk, Merdeka Square, Central Market and others.

Several aspects can be considered to achieve better results. For instance, the variables, nodes, alternative methods and tourist opinions can be included for future study. The number of variables, such as the time, cost and time windows, can be incorporated as well. The number of nodes which refer to tourist attractions can be extended to more than 10. Apart from BnB, other optimization methods such as the Chinese postman problem, Prim's algorithm and Kruskal's algorithm can be considered to solve the problem. In addition, various software, such as Python, MATLAB and TORA, can be employed for fast computation.

Acknowledgement. This study is funded by Research Management Centre, Universiti Teknologi MARA (UiTM) Selangor under the grant number 600-IRMI/MYRA 5/3/LESTARI (063/2017). Also, the authors would like to thanks Norshahirah Md Salim, Nurul Syamila Mat Sukri and Nur Hidayatul Nadzirah Amirudin for their contribution.

References

1. Tourism-Malaysia: Tourism Malaysia 2016 Annual Report (2016)
2. Sangaran, G., Jeetesh, K.: The effects of job satisfaction towards employee turnover in the hotel industry: a case study of hotels in Kuala Lumpur City Center. J. Tour. Hosp. **4**(142), 2167–0269.1000142 (2015)
3. Sarkar, S.K.: Urban ecotourism destinations and the role of social networking sites; a case of Kuala Lumpur. Ecotourism Pap. Ser. **39** (2016)
4. Wong, B.K.M., Musa, G., Taha, A.Z.: Malaysia my second home: the influence of push and pull motivations on satisfaction. Tour. Manag. **61**, 394–410 (2017)
5. Amir, A.F., et al.: Sustainable tourism development: a study on community resilience for rural tourism in Malaysia. Procedia-Soc. Behav. Sci. **168**, 116–122 (2015)
6. Chandran, S.D., et al.: Medical tourism: why Malaysia is a preferred destination? Adv. Sci. Lett. **23**(8), 7861–7864 (2017)
7. Ordóñez de Pablos, P., Aung, Z.: Tourism and Opportunities for Economic Development in Asia (2017)
8. Mason, P.: Tourism Impacts, Planning and Management. Routledge (2015)

9. Horner, S., Swarbrooke, J.: Consumer Behaviour in Tourism. Routledge (2016)
10. Veal, A.J.: Research Methods for Leisure and Tourism, Pearson UK (2017)
11. Ahmad, H., Jusoh, H.: Family-friendly beach tourism in Malaysia: whither physical or man-made determinants. Rev. Soc. Econ. Perspect. RSEP, 141 (2018)
12. Kamisan, B.P.: Generation Y's Perception and Involvement in Homestay Programmes: A Case Study of Kampung Sarang Buaya, Muar, Johor, Malaysia. School of Hotel and Tourism Management, The Hong Kong Polytechnic University (2015)
13. Xiang, Z., Magnini, V.P., Fesenmaier, D.R.: Information technology and consumer behavior in travel and tourism: insights from travel planning using the internet. J. Retail. Consum. Serv. **22**, 244–249 (2015)
14. Suruhanjaya Pengangkutan Awam Darat, S. A Smarter Way to Travel (2018). Available from https://www.gokl.com.my/
15. Broumi, S., et al.: Applying Dijkstra algorithm for solving neutrosophic shortest path problem. In International Conference on Advanced Mechatronic Systems (ICAMechS 2016). IEEE (2016)
16. Taccari, L.: Integer programming formulations for the elementary shortest path problem. Eur. J. Oper. Res. **252**(1), 122–130 (2016)
17. Bakar, S.A., Ibrahim, M.: Optimal solution for travelling salesman problem using heuristic shortest path algorithm with imprecise arc length. In: AIP Conference Proceedings. AIP Publishing (2017)
18. Anderson, R., et al.: Finding long chains in kidney exchange using the traveling salesman problem. Proc. Natl. Acad. Sci. **112**(3), 663–668 (2015)
19. Braekers, K., Ramaekers, K., Van Nieuwenhuyse, I.: The vehicle routing problem: state of the art classification and review. Comput. Ind. Eng. **99**, 300–313 (2016)
20. Ganganath, N., et al.: Trajectory planning for 3D printing: a revisit to traveling salesman problem. In: 2016 2nd International Conference on Control, Automation and Robotics (ICCAR). IEEE (2016)
21. Deo, N.: Graph Theory with Applications to Engineering and Computer Science. Courier Dover Publications (2017)
22. Arnesen, M.J., et al.: A traveling salesman problem with pickups and deliveries, time windows and draft limits: case study from chemical shipping. Comput. Oper. Res. **77**, 20–31 (2017)
23. Subramanyam, A., Gounaris, C.E.: A branch-and-cut framework for the consistent traveling salesman problem. Eur. J. Oper. Res. **248**(2), 384–395 (2016)
24. Jiang, Z.-B., Yang, Q.: A discrete fruit fly optimization algorithm for the traveling salesman problem. PLoS One **11**(11), e0165804 (2016)

An Analysis of Structure Heterogeneity of Lithium Silicate Melts

Vu Tri Vien[1(✉)], Mai Van Dung[2,3], Nguyen Manh Tuan[2],
Tran Thanh Nam[1], and Le The Vinh[1]

[1] Faculty of Electrical and Electronics Engineering, Ton Duc Thang University,
No. 19 Nguyen Huu Tho Street, Tan Phong Ward, District 7, Ho Chi Minh City,
Vietnam
{vutrivien, tranthanhnam, lethevinh}@tdtu.edu.vn
[2] Institute of Applied Materials Science, Vietnam Academy of Science and
Technology, No. 01 TL29 Str., Thanh Loc Ward, District 12, Ho Chi Minh City,
Vietnam
[3] Thu Dau Mot University, No. 6, Tran van on Street, Phu Hoa Ward, Thu Dau
Mot City, Binh Duong Province, Vietnam

Abstract. In this paper, the structure of the lithium -silicate melt under different pressure have been investigated. The structure is analyzed via SC-particle and SC-cluster. Our simulation reveals the structural heterogeneity in local environment and chemical composition. The densification of the melt is accompanied with decreasing the radius of core of SC-particle and number of large SC-particles.

Keywords: Dynamics · Structure heterogeneity · Lithium silicate melts

1 Introduction

Understanding the structure and dynamics of silicate liquids is important from both fundamental and practical point of view [1–7]. According to experimental studies, the silicate liquids suffer from peculiar properties. For instance, silica-rich liquids have negative pressure dependence of shear viscosity although silica-poor liquids in contrast have positive pressure dependence [8–11]. The increase in number of high-coordinated atom [12–14] or decrease in Si–O–Si angle [15–17] are thought to be a cause of negative pressure dependence. Further, silicate liquids exhibit the dynamics heterogeneity (DH). It means that the liquid comprises separate mobile and immobile regions where the atom mobility is extremely low or high. The heterogeneity in the structure seems to be responsible for DH. To clarify the diffusion mechanism, the molecular dynamics (MD) simulation is an appropriate tool. There are many MD simulations on the negative pressure dependence for silicate liquid [18–21]. The numerical techniques

© Springer Nature Switzerland AG 2019
P. Vasant et al. (Eds.): ICO 2018, AISC 866, pp. 285–292, 2019.
https://doi.org/10.1007/978-3-030-00979-3_29

such as multi-correlation function, visualization and cluster analysis [22–33] are widely used to investigate DH. Recently, the simulation of silicates focuses mainly on the organization of TOx units [34–38]. Here T is the network former or network modifier element. The TOx are connected to each other through bridging oxygen. The polymerization degree calculated by the ratio of non-bridging oxygen per tetrahedron is used to establish the relationship between dynamics and structure. In this study, an SC-cluster model will be presented. The proposed model does not require the definition of connectivity and gives more detail about local environment of ions in network-forming liquid.

2 Computational Method

MD simulation is carried out for a model consisting of 500 Si, 1250 O and 500 Li. The pressure varies in the range up to 30 GPa. We employ the Born–Mayer potential. Initial configuration is generated by randomly placing all atoms in a simulation box and heating up to 6000 K. Then the obtained sample is cooled down to 2500 K and relaxed until reach the equilibrium. Firstly, we prepare a sample at ambient pressure and 2500 K. Secondly, the sample at higher pressure is produced by compressing the equilibrated sample and relaxing for long time. More detail about preparing the LS2 model can be found in [14]. The obtained model reproduces well the structure of the melt.

A simplex is defined as a sphere passing four oxygen atoms. We find simplexes which contain one or more cation atoms inside. Then we determine largest sphere which has the same center as the simplex. In addition, the area between surfaces of those spheres contains only oxygen atoms. Such micro-region is called SC-particle. It consists of core and peripheral shell. The core contains cation, while the shell contains oxygen atoms (see Fig. 1a). The SC-particle is characterized by the radius of the core RC, length of the shell DS and the (s, c) number. In which, s is the number of oxygen in the shell and c is the number of the cation in the core, respectively. Two adjacent SC-particles may have common cation or oxygen (see Fig. 1b, c). The SC-cluster is defined as a set of SC-particle where each SC-particle and its adjacent SC-particle have at least one common cation atom (see Fig. 1c, d). The SC-cluster has the following properties. Firstly, it consists of two separate space regions called the oxygen and cation part. The cation part contains cation atoms, while the oxygen part contains oxygen atoms. Secondary, two adjacent SC-clusters do not have any common cation, but they may have common oxygen. A SC-cluster may comprise one or more SC-particles.

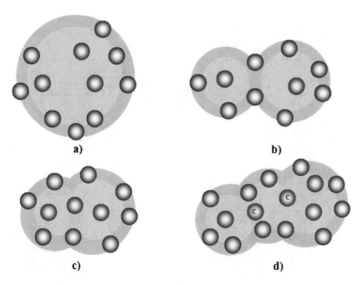

Fig. 1. The schematic illustration of SC-cluster. (a) SC-particle; here the black, blue sphere represents oxygen and cation, respectively; the grey and brown region represents shell and core of SC-particle, respectively; (b) Two adjacent SC-particles having two common oxygen atoms, here the sphere with 'c' letter represents the common atom for adjacent SC-particles; (c) SC-cluster with two SC-particles; (d) SC-cluster with three SC-particles, here the grey and brown region represents cation-part and oxygen-part of SC-cluster, respectively. (Color figure online)

3 Results and Discussion

In Table 1 we show characteristics of SC-particle for two configurations. For a low-pressure configuration, the number s varies from 1 to 4. Meanwhile the number c ranges from 4 to 9. Moreover, the majority of SC-particle is the types with c = 2–4. The core radius of the type with c = 2, 3, 4 increases with the number s. The length of shell in contrast is bigger except [4, 7] type. In general, the length of shell complexly depends on the pressure.

A [s,c] SC-particle in fact is a sphere which has radius of $R_c + D_c$ and contains s + c atoms inside. Therefore, the average atomic density of SC-particle can be estimated as $\rho SC = (s + c)/[4\pi(<R_c> + <D_s>)^3/3]$. Because the melt contains a number of SC-particle types, the low-density and high-density regions must have different concentrations of high-density and low-density SC-particles. As shown from Table 1, the parameter ρSC varies in a wide range depending on the pressure. In particular, at ambient pressure the high-density SC-particles ($\rho SC > 0.115$) except [2, 4] and [4, c] types. However, in the case of 20 GPa the high-density SC-particles ($\rho SC < 0.810$) are [4, 4], [4, 6] and [4, 7]. The majority of low-density SC-particle are [2, 4] and [2, 5] for the low-pressure configuration, while those in high-pressure configuration are [2, 5] and [2, 6].

Figure 2 shows the average radius of core as a function of pressure. One can see that the radius of core significantly decreases with increasing pressure. In addition, it

Table 1. The characteristics of SC-particle in the configuration at ambient pressure and 20 GPa; msc, <Rc>, <Ds> is the number of SC-particle, the average radius of core and length of shell, respectively; ρSC is the average atomic density of SC-particle.

$[s, c]$	0 GPa				20 GPa			
	m_{sc}	$<R_c>$, Å	$<D_s>$, Å	ρ_{SC}, Å^{-3}	m_{sc}	$<R_c>$,Å	$<D_s>$,Å	ρ_{SC}, Å^{-3}
[1, 4]	47	1.92	0.33	0.105	58	1.72	0.26	0.154
[1, 5]	16	1.89	0.59	0.094	39	1.76	0.59	0.110
[1, 6]	**3**	**1.97**	**0.95**	**0.067**	**61**	**1.73**	**0.78**	**0.106**
[2, 4]	552	2.11	0.32	0.100	2	1.83	0.83	0.076
[2, 5]	279	2.19	0.48	0.088	461	1.87	0.32	0.159
[2, 6]	76	2.25	0.52	0.090	214	1.95	0.44	0.140
[2, 7]	26	2.39	0.56	0.084	85	2.02	0.53	0.130
[2, 8]	6	2.32	0.76	0.082	11	2.06	0.56	0.133
[2, 9]	**1**	**2.32**	**0.91**	**0.078**	**3**	**2.10**	**0.56**	**0.140**
[3, 4]	76	2.30	0.32	0.093	61	1.99	0.30	0.139
[3, 5]	67	2.30	0.52	0.085	56	2.02	0.49	0.121
[3, 6]	40	2.36	0.66	0.078	29	2.10	0.55	0.115
[3, 7]	12	2.33	0.76	0.081	10	2.07	0.65	0.119
[3, 8]	**1**	**2.27**	**0.79**	**0.092**	**1**	**1.94**	**0.72**	**0.140**
[4, 4]	6	2.42	0.37	0.088	1	2.01	1.03	0.068
[4, 5]	9	2.46	0.47	0.085	2	2.17	0.81	0.081
[4, 6]	9	2.46	0.57	0.086	58	1.72	0.26	0.308
[4, 7]	**1**	**2.42**	**1.08**	**0.061**	**39**	**1.76**	**0.59**	**0.202**

strongly depends on the number of oxygen in the shell. Thus the densification of the melt is mainly realized via decreasing the radius of core and number of SC-particles which have large number s or c (large SC-particle).

In Fig. 3, we show the pressure dependence of number of SC-cluster. Here mCk is the number of SC-cluster comprising k SC-particle. It can be seen that mCk decreases with increasing k. The chemical composition of SC-cluster is determined by $C_{Li}/(C_{Li} + C_{Si})$ and $C_O/(C_{Li} + C_{Si})$, where C_{Li}, C_{Si} is the number of Li and Si in the cation-part, respectively; CO is the number of O in the oxygen-part. Figure 4 shows the pressure dependence of those ratios. For convenience we denote that the small SC-cluster consists of a SC-particle, while the large SC-cluster comprises more than one SC-particle. For small SC-clusters the $C_{Li}/(C_{Li} + C_{Si})$ and $C_O/(C_{Li} + C_{Si})$ is in average equal to 0.3 and 4.75, respectively. In Table 2 we show characteristics of several large SC-clusters. From Table 2 one can see that $C_{Li}/(C_{Li} + C_{Si})$ and $C_O/(C_{Li} + C_{Si})$ for large SC-cluster is much larger and smaller respectively than one for small SC-cluster. Therefore, Li atoms tend to reside in large SC-clusters, while Si atoms mainly locate in small SC-clusters.

SC-clusters can be divided into three groups. First and second group include the SC-cluster which cation-part contains only Si or Li. The cation-part of SC-cluster of third group contains both Li and Si. Table 3 shows characteristics of three groups for

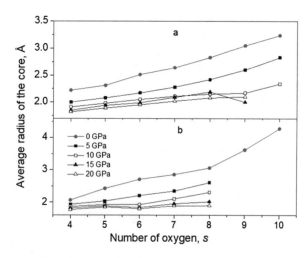

Fig. 2. The average radius of core as a function of number of oxygen in the shell. (a) SC-particle with $c = 1$; (b) SC-particle with $c = 2$.

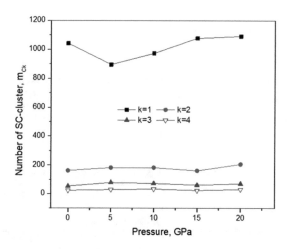

Fig. 3. The pressure dependence of number of SC-cluster; here k is the number of SC-particles.

the configuration at ambient pressure. One can see that about 80% of total Si belongs to first group, while third group contains 41% of total Li. Because the diffusion of cation is impeded by oxygen due to chemical T–O bond, hence cation diffuses fast in the region with lower density of oxygen. This follows that the regions just mentioned represent the diffusion pathway where Li atoms move fast. Overall the heterogeneity in local structure includes following issues:

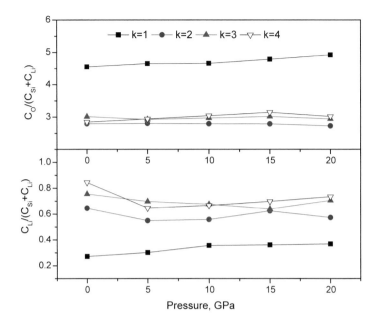

Fig. 4. The pressure dependence of $C_O/(C_{Li} + C_{Si})$ and $C_{Li}/(C_{Li} + C_{Si})$ for SC-cluster; here k is the number of SC-particles.

Table 2. The characteristics of three groups for SC-cluster at ambient pressure.

The number of SC-particles	11	12	13	14	18	19
$C_{Li}/(C_{Li} + C_{Si})$	0.829	0.838	0.851	0.793	0.896	0.894
$C_O/(C_{Li} + C_{Si})$	3.006	4.177	3.237	3.579	2.569	3.162

Table 3. The characteristics of three groups for SC-cluster at ambient pressure.

Group	Atom in the cation-part	Number of SC-cluster	Number of SC-particles	Number of cations	Averaged number of cations per SC-cluster
1	Si	774	775	791	1.02
2	Li	389	479	588	1.51
3	Si and Li	167	452	621	3.72

(i) There are two types of SC-cluster. The first type has a spherical form and comprises one SC-particle. Second type (large SC-cluster) in contrast comprises more than one SC-particle. Most Si atoms belong to first type, while second type contains the majority of Li atoms. The ratio $C_O/(C_{Li} + C_{Si})$ for first type is much smaller than one for second type.

(ii) Large SC-clusters tend to locate nearby forming large space regions where the majority of atoms are Li and the oxygen atoms surrounding those regions have the low density. Such regions represent the diffusion pathway for Al.

4   Conclusions

An analysis on structure and dynamics of LS2 melt is carried out. It is shown that the low-pressure configuration of the melt exhibits the DH. Moreover the liquid comprises separate regions where the mobility of atom is extremely low or high. The mobile and immobile O atoms tend to reside in regions which have high and low density of Al, respectively. The similar trend is observed for Li atoms. The mobile Si atoms in contrast reside in regions with low density of Si. Our simulation also reveals the heterogeneity in local environment and chemical composition for the melt.

The structure of the melt is analyzed through SC-particle and SC-cluster. SC-cluster consists of oxygen-part and cation-part. Adjacent SC-clusters do not have common cation, but may have common oxygen. It is shown that the densification of the melt is accompanied with decreasing the radius of core of SC-particle and number of large SC-particle. Further, we show that the liquid comprises two types of SC-cluster. Most Si atoms belong to first type, while second type contains the majority of Li atoms. Large SC-clusters tend to locate nearby forming space regions which represent the diffusion pathway for Li.

Acknowledgment. This research is funded by Vietnam National Foundation for Science and Technology Development (NAFOSTED) under grant number 103.05-2017.345.

References

1. Smedskjaer, M.M.: Front. Mater **23**, 1 (2014)
2. Hehlen, B., Neuville, D.R.: J. Phys. Chem. B **119**, 4093 (2015)
3. Du, J.: J. Am. Ceram. Soc. **92**, 87 (2009)
4. Neuville, D.R., et al.: Am. Mineral. **93**, 228 (2008)
5. Durrani, S.K., et al.: Mat. Sc. Pol. **28**, 459 (2010)
6. Poe, B.T., Romano, C., Zotov, N., Cibin, G., Marcelli, A.: Chem. Geol. **174**, 21 (2001)
7. Allwardt, J.R., et al.: Am. Mineral. **90**, 1218 (2005)
8. Kushiro, I.: J. Geophys. Res. **81**, 1955 (1976)
9. Scarfe, C.M., Mysen, B.O., Virgo, D.: Carnegie Inst. Wash. Yearbook **78**, 574 (1979)
10. Suzuki, A., Ohtani, E., Funakoshi, K., Terasaki, H., Kubo, T.: Phys. Chem. Miner. **29**, 159 (2002)
11. Rubie, D.C., Ross, C.R., Carroll, M.R., Elphick, S.C.: Am. Miner. **78**, 574 (1993)
12. Xue, X., Stebbins, J.F., Kanzaki, M., Poe, B.T., McMillan, P.F.: Am. Miner. **76**, 8 (1991)
13. Poe, B.T., et al.: Nature **276**, 1245 (1997)
14. Banhatti, R.D., Heuer, A.: Phys. Chem. Chem. Phys. **3**, 5104 (2001)
15. Navrotsky, A., Geisinger, A.L., McMillan, P., Gibbs, G.V.: Phys. Chem. Miner. **11**, 78 (1985)
16. Wakabayashi, D., Funamori, N., Sato, T.: Phys. Rev. B **91**, 014106 (2015)
17. Noritake, F., Kawamura, K.: J. Comp. Chem. Jpn. **14**, 124 (2015)
18. Karki, B.B., Bhattarai, D., Stixrude, L.: Phys. Rev. B **76**, 104205 (2007)
19. Baggain, S.K., Ghosh, D.B., Karki, B.B.: Phys. Chem. Miner. **42**, 393 (2015)
20. Y. Wang el al., Nat. Comm. **5**, 3241 (2014)
21. Bauchy, M., Guillot, B., Micoulaut, K., Sator, N.: Chem. Geol. **346**, 47 (2013)
22. Mizuno, H., Yamamoto, R.: Phys. Rev. E **84**, 011506 (2011)

23. Garrahan, J., Chandler, D.: Phys. Rev. Lett. **89**, 035704 (2002)
24. Hung, P.K., Hong, N.V.: Eur. Phys. J. B **71**, 105 (2009)
25. Horbach, J., Kob, W.: Phys. Rev. B **60**, 3169 (1999)
26. Ediger, M.: Annu. Rev. Phys. Chem. **51**, 99 (2000)
27. Weeks, E.R., et al.: Science **287**, 627 (2000)
28. Kegel, W.K., Blaaderen, A.V.: Science **287**, 290 (2000)
29. Tata, B.V.R., Mohanty, P.S., Valsakumar, M.C.: Phys. Rev. Lett. **88**, 018302 (2002)
30. Glotzer, S.C.: J. Non-Cryst. Solids **274**, 342 (2000)
31. Cates, M.E., Puertas, A.M., Fuchs, M.: J. Phys. Chem. B **109**, 6666 (2005)
32. Cooper, A.W., Harrowell, P., Fynewever, H.: Phys. Rev. Lett. **93**, 135701 (2004)
33. Mauro, J.C., Vargheese, K.D., Tandia, A.: J. Chem. Phys **132**, 194501 (2010)
34. Narayanan, B., Reimanis, I.E., Ciobanu, C.V.: J. Appl. Phys. **114**, 083520 (2013)
35. Okuno, M., Zotov, N., Schmucker, M., Schneider, H.: J. Non-Cryst. Solids **351**, 1032 (2005)
36. Poe, B.T., Mcmillan, P.T., Angell, C.A., Sato, R.K.: Chem. Geol. **96**, 333 (1992)
37. Zhang, Z., Zheng, K., Yang, F., Sridhar, S.: ISIJ Int. **52**, 342 (2012)
38. Takei, T., Kameshima, Y., Yasumori, A., Okada, K.: J. Mater. Res. **15**, 186 (2000)

Mathematical Modeling of the Work of the Flow-Meter Flowmeter-Doser

Alexey N. Vasilyev[1], Alexey A. Vasilyev[1], Dmitry A. Shestov[2],
Denis V. Shilin[2(✉)], and Pavel E. Ganin[2]

[1] Federal State Budgetary Scientific Institution "Federal Scientific
Agroengeneering Center VIM" (FSAC VIM), Moscow, Russia
lex. of@mail. ru
[2] National Research University "Moscow Power Engineering Institute"
(NRU MPEI), Moscow, Russia
deninfo@mail. ru

Abstract. In this paper, a mathematical simulation of the closed-loop control of a flow-through flow meter-metering device with two installed load cells on the loading side of the material is presented. The presented model was tested on the boundary conditions and all the dependencies obtained were checked for adequacy. The obtained data show that the determination of the optimal speed of conveyor belt allows to increase the accuracy of dosage of bulk material.

Keywords: Loose material · Fuzzy regulator · Batching error
Force-measuring sensor · Simulation

1 Mathematical Description Mass Feeding Feeder Formation and Weighting Algorithm

For accurate dosing of bulk material, weight batchers are used, which can be continuous or batch-acting [12]. One of the most common continuous weighers is belt conveyors [1–3]. There are different methods for weight continuous dosing of bulk material, which, among themselves, have structural differences [4–6]. In this paper, we will present the results of mathematical modeling of the operation of closed flow control by a flow meter-metering device with two installed load-measuring sensors on the bulk material loading side [15, 16], the structural diagram, which is shown in Fig. 1.

Figure 1 shows the supports (1), the load cells x_T and y_T (2), the belt conveyor (3), the feeder (4). To determine the analytical dependence of the readings of load cells from the incoming mass on the conveyor belt, the characteristics of the readings of the sensors were measured from the displacement of the mass of a different standard along the axis of the installed load cells and along the entire length of the conveyor belt. The results of experiments in a graphic form are shown in Fig. 2.

As can be seen from Fig. 2, the experimental results for each mass standard are the plane passing through the origin. This indicates that the arithmetic mean of the readings of the force-measuring sensors for the same mass lying on the axis parallel to the axis

© Springer Nature Switzerland AG 2019
P. Vasant et al. (Eds.): ICO 2018, AISC 866, pp. 293–299, 2019.
https://doi.org/10.1007/978-3-030-00979-3_30

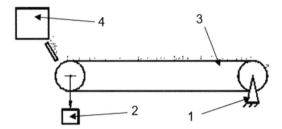

Fig. 1. Structural diagram of the flow meter-metering device with two installed load cells on the loading side of bulk material.

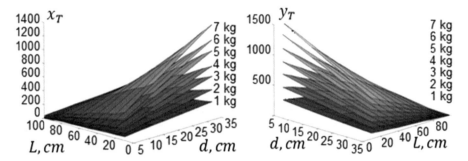

Fig. 2. Structural diagram of the flow meter-metering device with two installed load cells on the loading side of bulk material.

of the sensors will always be equal. Then, to calculate the newly arrived mass, we use the formula presented below:

$$F = \frac{x_T + y_T}{2} - \sum_{i=0}^{k}\left[\left(\frac{x_i + y_i}{2}\right) \cdot \left(L - \sum_{j=0}^{n}\frac{v_{cj}}{f_j}\right)\right]. \tag{1}$$

x_T, y_T – current readings of load cells; x_i, y_i – previous indications of load cells for each time interval; f_j – frequency of interrogation of load cells for each time interval, Hz; L – length of the conveyor belt, m; v_{cj} – speed of conveyor belt movement for each time interval, m/s; F – value of newly received mass, kg. The above equation calculates the value of the newly received mass, minus the mass that fell off the conveyor for a given period of time.

To minimize the errors in measuring the incoming mass, it is necessary to automatically determine the optimal speed of the conveyor belt in automatic mode [13]. This is due to the fact that the measurement error depends on the speed of the conveyor belt and the feeder flow, which organizes the flow of loose mixtures onto the conveyor. To minimize the weighing error, it is necessary to ensure the operation of the unit in the most loaded mode, in order to avoid the accumulation of an error while idling, while avoiding the load of the device above its capacity by increasing the speed of the conveyor belt.

In order to avoid the inertia of the system, as well as the influence of dynamic effects, it is necessary to determine the average value of the suspended bulk material, in the analysis of which the dynamic disturbances of the structure and all sorts of short-term stops (increase) of the flow of loose mixtures will not affect the speed correction thereby will release the operative space of the computer system.

$$F_s = \left(\sum_{i=0}^{\tau} \frac{F}{f_j} \right), \tau = \sum_{j=0}^{n} \frac{1}{f_j}. \tag{2}$$

F_s – mean value of the received mass for a period of time τ, kg; τ – time interval, s. Then the average volume of V_S will be:

$$V_S = \frac{F_S}{\rho}. \tag{3}$$

ρ – bulk density, kg/m^3. Based on the design data of the mechanism for feeding bulk material to the conveyor belt, the maximum possible volume of material V_m, that the installation is capable of adopting is presented below:

$$V_m = \sum_{i=0}^{n} \left(a \cdot b \cdot \frac{v_{ci}}{f_i} \right). \tag{4}$$

$a \cdot b$ – flow area feeder of the flow meter-metering device, which has rectangular shapes, width a and height b, kg. Then the deviation from the given loading of the belt of the flowmeter-dispenser by loose material will be:

$$\Delta V = K \cdot V_m - V_s. \tag{5}$$

K – fill factor $(0 \leq K \leq 1)$. As a result, ΔV takes the form:

$$\Delta V = K \cdot \sum_{i=0}^{n} \left(a \cdot b \cdot \frac{v_{ci}}{f_i} \right) - \frac{F_s}{\rho}. \tag{6}$$

To regulate this indicator, we use an intelligent system, based on fuzzy logic [7, 8].

$$v_T = \textbf{Fuzzy}(v_c, \Delta V). \tag{7}$$

v_T – current conveyor belt speed, m/s.

2 Approbation of the Presented Mathematical Model

The presented mathematical model, designed to determine the optimum speed of the conveyor belt of a flow meter-metering device, was tested on the boundary conditions and all the dependences obtained were checked for adequacy [9–11, 14]. Figure 3 presents the results of mathematical modeling with a constant flow of bulk material of 5 kg/s, a constant speed of the conveyor belt of 0.3 m/s, where 1 – transient characteristic of the dependence of the quantity imitating the flow of bulk material from time; 2 – transient characteristic of the dependence of conveyor belt speed on time; 3 – transient characteristic of the dependence of the arithmetic mean value of load cells x_T and y_T on time; 4 – the transient characteristic of the dependence of the amount of mass, descended from the conveyor on time.

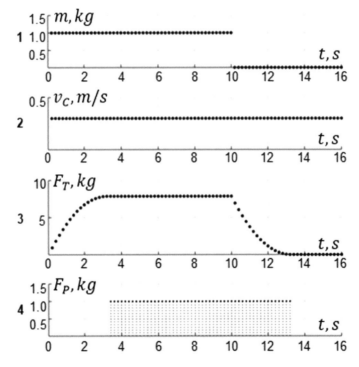

Fig. 3. Results of mathematical modeling with a constant supply of the feeder and the speed of the conveyor belt.

The Fig. 3 shows that the indication of the arithmetic mean of the load cells x_T and y_T assumes a constant value at the moment of the appearance of the value of the mass different from zero, descended from the conveyor, which indicates the adequacy of the results obtained, provided the feeder is continuously fed and the conveyor belt runs.

3 The Results of Mathematical Modeling in Determining the Optimal Speed of Conveyor Belt

Figure 4 presents the results of mathematical modeling with variable feed of bulk material and adjustable conveyor belt speed, where 1 – transient characteristic of the quantity dependence simulating the flow of bulk material from time; 2 – transient characteristic of the dependence of conveyor belt speed on time; 3 – transient characteristic of the dependence of the arithmetic mean value of load cells x_T and y_T on time; 4 – the transient characteristic of the dependence of the amount of mass, descended from the conveyor on time.

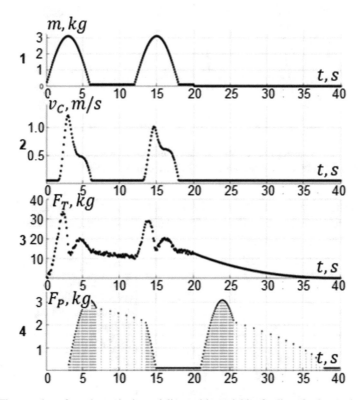

Fig. 4. The results of mathematical modeling with variable feeding feeder and adjustable conveyor belt speed.

In the Fig. 4, it is seen that when the feeder of the bulk material decreases, the speed of the conveyor belt automatically decreases, which leads to the accumulation of bulk material on the tape and a reduction in the batching error. Figure 5 shows the transient characteristics of the calculated amount of weighted weight from the time at the maximum speed of the conveyor belt ($v_C = 3\,\text{m/s}$), the minimum speed of the conveyor belt (based on their geometric design features $v_C = 0,45\,\mathrm{m/s}$) and during operation closed loop on the basis of a fuzzy controller.

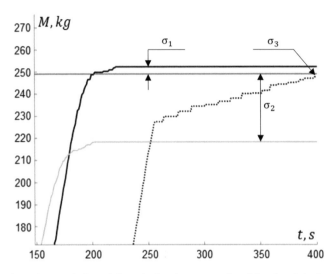

Fig. 5. Transient characteristics of the calculated amount of weighted weight from the time at the maximum, minimum and adjustable speed of the conveyor belt.

The dosing error in determining the optimal speed of the conveyor belt by the Fuzzy regulator was less than 0.5% of the total sum of the weighed mass (Table 1).

Table 1. Dosing error at different conveyor belt speed

Frequency of information processing from sensors, f [Гц]	Weight of material to be weighed, $\sum m$ [kg]	Conveyor belt speed, v_C [m/s]	Weighted mass according to the developed algorithm, M [kg]	Weighing error, σ [%]
50	249,1308	0,45	252,3624	−1,30%
50	249,1308	3,00	218,3818	12,34%
50	249,1308	Fuzzy controller	248,1342	0,40%

4   Conclusions

In conclusion, it should be noted that the authors of the paper proposed a mathematical model that describes a closed loop control flow meter, built on the basis of a fuzzy controller. The presented model was tested on the boundary conditions and all the obtained dependences were checked for adequacy and can be used to find the optimal control laws for such installations.

References

1. Shestov, D.A., Krausp, V.R.: Energy-saving, flowmeter-doser of dry feed in livestock and poultry. In: Scientific Methods and Experience of Computerization of Management of Innovative Projects of the Agroindustrial Complex Till 2020, GPU VIESH, 336c (2010, in Russian)
2. Shestov, D.A.: ACS with a universal feed mill with a flowmeter-doser of dry forages. In: Collected Papers, XI International Scientific and Practical Conference "Automation and Information Support of Production Processes in Agriculture", Part 2, FSUE Izvestiya UDP RF, 732c (2010, in Russian)
3. Shestov, D.A., Krausp, V.R.: Accounting-dosing control system for the electrified process of preparing fodder mixtures. In: Proceedings of the 8th International Scientific and Technical Conference "Energy Supply and Energy Saving in Agriculture", Part 5, VSU VYESH, 240s. (2012, in Russian)
4. Shestov, D.A.: Automatic control system for dosing and weighing of bulk materials. - M: Agricultural machines and technologies, VNIM. Agricultural machinery and technologies. Sci. Prod. Inf. Mag. (1), 52c (2013, in Russian)
5. Shestov, D.A.: Accounting and dosing of loose materials. The newest directions of development of agrarian science in the work of young scientists. V MNPK SMU. SB Rosselkhozaka-demi. Novosibirsk, 448s (2012, in Russian)
6. Shestov, D.A.: Analysis of the processes of dosing and weighing of loose materials. Innovations in agriculture Issue number 1. Theor. Sci. Pract. J. GNU VIESH (2012, in Russian). Access mode: http://smu.xn-b1abqswjr6d.xn-p1ai/liter.html, free
7. Kang, I.-J., Kwon, J.H., Moon, S.-M., Hong, D.: A control system using butterworth filter for loss-in-weight feeders (2014). https://doi.org/10.7736/kspe.2014.31.10.905
8. Siva Vardhan, D.S.V., Shivraj Narayan, Y.: Development of an automatic monitoring and control system for the objects on the conveyor belt (2015). https://doi.org/10.1109/mami.2015.7456594
9. Shestov, D.A.: The method of calculating the productivity of continuous belt feeder. Innovations in agriculture. Issue number 2. Theor. Sci. Pract. J. (2012, in Russian). Access mode: http://xn-b1abqswjr6d.xn-p1ai/insel2.pdf, free
10. Shestov, D.A.: Random errors of a random function of a load cell. Innovations in agriculture. Theor. Sci. Pract. J. (1) (2014, in Russian). Access mode: http://smu.xn-b1abqswjr6d.xn-p1ai/liter.html, free
11. Shestov, D.A., Kravov, M.R., Shilin, D.V.: Dynamic weighing of bulk materials on a flow-through flowmeter. In: Conference of Students and Graduate Students, Moscow (2017, in Russian)
12. Aleksandrović, S., Damnjanović, V.: Volume flow measurement of bulk solids on conveyor belts. Int. J. Transp. Logist. ISSN 1451-107X
13. Sick, B.: Volume flow measurement of bulk material on conveyor belts, Operation manual (2001)
14. Haußecker, H., Geißler, P.: Handbook of Computer Vision and Applications. Academic Press, San Diego (1999)
15. Siemens, Weighing and feeding guide, Operation manual. www.siemens.com
16. Agg-net, Load-Out And Dosing Control Systems, Operation manual. www.agg-net.com

Framework for Faction of Data in Social Network Using Link Based Mining Process

B. Bazeer Ahamed[1(✉)] and D. Yuvaraj[2]

[1] Department of Computer Science and Engineering,
Balaji Institute of Technology and Science, Warangal, Telangana, India
bazeerahamed@gmail.com
[2] Department of Communication and Computer Engineering,
Cihan University, Erbil, Kurdistan Region, Iraq
yuva.r.d@gmail.com

Abstract. Recent online social networks such as Twitter, Facebook, and LinkedIn have hurriedly grown in reputation. The resulting accessibility of a social network data supplies an unparalleled occasion for data analysis and mining researchers to resolve useful and semantic information in a broad range of fields such as social sciences, marketing, management, and security. Still, unprocessed social network data are enormous, noisy, scattered, and susceptible in nature, in which some challenges is faced when applying data mining tools and analyzing tasks in storage, efficiency, accuracy, etc. In addition to that there are many problems related to the data collection and data conversion steps in social network data preparation. We focused on the endeavor for privacy preserving social network conversion which provides method for better protection and identification of privacy for social network users and to maintain the convenience of social network data.

Keywords: Social network · Social network analysis · Link mining
Learning model

1 Introduction

The number of social network users around the world rises from 1.47 billion in 2012 to 1.73 billion in 2013, an 18% increase. By 2017, the global social network audience will reach 2.55 billion [5]. Another study in April 2013 reveals that social networking has been ranked as the most popular content category in worldwide engagement, accounting for 27% of all time spent online [6]. The high penetration of Internet-enabled devices such as personal computers, smart phones, and tablets, online social networks have become easily accessible platforms for users to communicate and share information. The primary objectives of social network analysis are to handle large-scale social network data, extract actionable patterns, and gain insightful knowledge about dynamic and multifaceted social networks. Therefore, social network analysis is of significant value for many applications domains such as policy making, advertising, and homeland security. Social network data available for analysis are usually voluminous, structurally complex, heterogeneous, and dynamic in nature, and can be

© Springer Nature Switzerland AG 2019
P. Vasant et al. (Eds.): ICO 2018, AISC 866, pp. 300–309, 2019.
https://doi.org/10.1007/978-3-030-00979-3_31

broadly classified as content and linkage data [2]. Content data contain texts, images, and other multimedia data, which are explicitly generated by social network users. Linkage data are essentially graphs where individual users are represented by vertices, and relationships or interactions between individuals are represented by edges. On each type of social network data, a wide range of analysis tasks can be performed to reveal valuable information.

1.1 Data Preparation: Steps and Challenges

The main objectives of data preparation are to process raw datasets, reduce time and space costs, enhance data quality with better interpretability and accuracy, and limit disclosure of sensitive information. The data preparation process can be detailed into four main steps namely data collection, data cleaning, data reduction, and data conversion [1]. In the following subsections, we discuss the challenges and issues in each data preparation step, with a focus on the factors related to social network data. Online social network data are collected from social network service providers who possess the overall social network data of their service users. The collection process is normally performed automatically through programs or scripts [10]. Some social media websites such as Facebook and Twitter provide APIs for data crawling. The main challenges of social network data collection are limited processing power and storage space. Online social networks are usually huge as measured by user population size, user-generated content volume, and update velocity. Take Twitter, one of the most popular micro blogging social networks, as an example. The number of registered users on Twitter reached one billion in 2013, and collectively, Twitter users now send over 500 million posts every day [3].

Density refers to the "connections" between participants. Density is defined as the number of connections a participant has, divided by the total possible connections a participant could have. For example, if there are 20 people participating, each person could potentially connect to 19 other people. A density of 100% (19/19) is the greatest density in the system. A density of 5% indicates there is only 1 of 19 possible connections. Centrality focuses on the behavior of individual participants within a network. It measures the extent to which an individual interacts with other individuals in the network. The more an individual connects to others in a network, the greater their centrality in the network.

In-degree and out-degree variables are related to centrality. In-degree centrality concentrates on a specific individual as the point of focus; centrality of all other individuals is based on their relation to the focal point of the "in-degree" individual. Out-degree is a measure of centrality that still focuses on a single individual, but the analytic is concerned with the out-going interactions of the individual; the measure of out-degree centrality is how many times the focus point individual interacts with others.

A sociogram is visualization with defined boundaries of connections in the network [11]. For example, a sociogram which shows out-degree centrality points for Participant A would illustrate all outgoing connections Participant A made in the studied network. Social network data contain a lot of informal user-generated content, which is inevitably accompanied by noise, spam, and inconsistency. The purpose of data cleaning is to remove noisy and irrelevant data from useful information [4]. It improves data quality and then improves the accuracy of analysis results. The typical issues that data cleaning deals with are listed as follows:

Noise and spam, Data inconsistency and Data incompleteness: The goals of data reduction are to represent data in a reduced form (with or without information loss) which has much smaller volume, and to make sure that analysis on the reduced data produces the same or almost the same outputs as on the original data. The classic technologies of social network data reduction include feature extraction and data compression.

2   Related Work

The current development of social network privacy protection techniques and discuss the weaknesses of existing approaches which our work can improve.

2.1   Organization of Social Networks Integration Using Link Types Method

The characteristic that is frequently overlooked when classifying in social networks is the fortune of data in the link set, Consider, for instance, the variety of link types within only Facebook. An individual can have "regular" friends, "best" friends, siblings, cousins, co-workers, classmates from high school and college. Previous work has either ignored these link types or, worse, used them solely on the basis of whether to include or exclude them from consideration [7]. For instance, Facebook now has over 150 million users. Facebook is only one example of a social network that is for general connectivity [9].

Some prior work compares the structural properties of samples obtained using different methods with those of the original graph and discusses the sampling bias of different methods. For example, Leskovec et al. [8] study sampling methods of the three strategies and find some sampling by exploration methods (i.e., random walk and forest fire) outperform vertex selection and edge selection methods in accurately representing both the static and evolutionary patterns of the original graph, Gjoka et al. [13] find the Metropolis-Hashing random walk algorithm and a re-weighted random walk sampling method can produce approximately uniform user samples on Facebook. Maiya et al. [10] investigate the bias of different sampling strategies and show that certain types of bias are beneficial for many applications as they "push" the sampling process towards inclusion of specific properties of interest (e.g., high-expansion or high-degree vertices) (Fig. 1).

A lot of differential privacy algorithms on social networks implement edge differential privacy. This is because under this definition, many network statistics have low and bounded sensitivity, and thus the algorithms can generate relatively accurate outputs with low levels of noise added. For example, Mir et al. [15] propose a method to generate differentially private Kronecker graph models of an input network, based on which synthetic networks that mimic important properties of the input network can be generated.

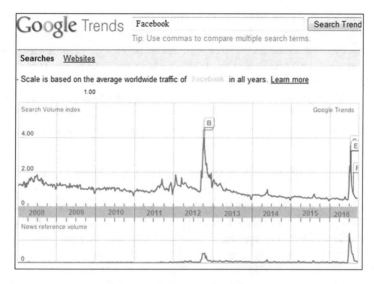

Fig. 1. Average traffic scale for face book

The proposition of classification in social network data extends far beyond the simple case of targeted advertising. This could be used for addressing classification problems in radical networks. By using the link structure and link types among nodes in a social network with known radical nodes, we can attempt to classify unknown nodes as radical or non-radical nodes.

To deal with the above problem, we choose the network-only Naive Bayes classier method since Naive Bayes classification combined with collective Inference techniques which provide an efficient solution with acceptable accuracy in practice. We adapt this relational Naive Bayes classier to incorporate the type of the link between nodes into the possibility calculations (Fig. 2).

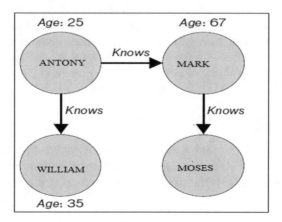

Fig. 2. Unidirectional attributed graph Knows relationship

Facebook's privacy issues are more complex than Twitter's, meaning that a lot of status messages are harder to obtain than Tweets, requiring 'open authorization' status from users. Facebook currently stores all data as objects and has a series of APIs, ranging from the Graph and Public Feed APIs to Keyword Insight API [14]. The results of searching will contain the unique ID for each object. When returning the individual ID for a particular search result, one can use https://graph.facebook.com/ID to obtain further page details such as number of 'likes'. This kind of information is of interest to companies when it comes to brand awareness and competition monitoring.

3 Relational Methods that Incorporate Link Based Types

The classification algorithms model deals with the social network data as a graph, where the set of nodes are a specific, homogeneous entity in the representative network. Edges are added based on specific constraints from the original data set [8]. For instance, in Facebook data, the edges are added based on the existence of a friendship link between the nodes. A graph structure is created, a classification algorithm is then applied that uses the labels of a nodes neighbors to probabilistically apply a label to that node. One of the problems with even relational classifiers is that if the labels of a large portion of the network are unknown, then there may be nodes for which we cannot determine a classification.

3.1 Directed and Undirected Links

Links can be undirected (e.g., "shares information with") or directed (e.g., "seeks advice from"). Directed links can be one-way or two-way. Social network analysis addresses both undirected and directed networks.

3.2 Density and Links Per Node

Density is the number of links that exist in a network divided by the maximum possible number of links that could exist in the network [12]. All of the social network analysis metrics assume that the numbers of nodes and links that exist in a network are known; we use N to refer to the number of nodes and M to refer to the number of links. The maximum possible number of links in a network depends on N and on whether the network is undirected or directed. For an undirected network, the maximum possible number of links is $N(N-1)/2$; for a directed network it is $N(N-1)$.

3.3 Hubs and in Degree Centrality

Hubs are individuals in a network with the most influence. Whether hubs bridge across clusters or bond within a cluster (or some combination), they are highly sought-after by other network members [13]. Hubs of influence in a network are best measured using directed links.

$$Cb_i = \sum_j^N \sum_k^N \frac{gjik}{gjk}, i \neq k \neq j \qquad (1)$$

Where gjik denotes a number of shortest paths linking nodes j and k passing through the node i, gjk a number of paths not including the node I, N denotes a number of nodes (Fig. 3).

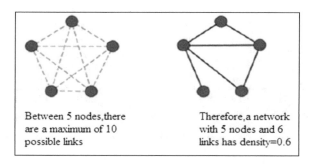

Between 5 nodes,there are a maximum of 10 possible links

Therefore,a network with 5 nodes and 6 links has density=0.6

Fig. 3. An example of density in an undirected network

The classifier to determine the probability that a particular node, xi, is of a particular class, ci, given the entire set of its character, by the formula.

$$\Pr(xi = ci|T) = \frac{\Pr(T|xi = ci)\Pr(xi = ci)}{\Pr(T)} = \Pr(xi = ci) \times \prod_{tict} \frac{\Pr(ti|xi = ci)}{\Pr(ti)} \qquad (2)$$

The general nBC assumes that all link types are the same, and that the probability of a particular nodes class is influenced by the class of its neighbors. Since the details of a particular node are factored in when we establish these priors, we do not duplicate this in the nBC calculations (Fig. 4).

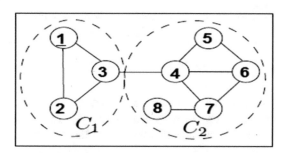

Fig. 4. Edge distribution sequences

$$Pr(xi = ci\,|\,N) = \frac{Pr(N|xi = ci)}{Pr(N)} = Pr(xi = ci) \times \prod_{ni\in N} \frac{Pr(ni|xi = ci)}{Pr(ni)} \qquad (3)$$

Our first experiment was to use only the local Bayes Classifier with Relaxation Labeling to establish a baseline accuracy of a non-relational classifier on our particular dataset. We perform analysis on the datasets collected over the one-month period and also present results on a daily basis (Table 1).

Table 1. Daily sampling ratios for users

Daily statistic	Complete user #		Sample #1		Sample #2	
	No. of user	Sample ratio	No of user	Sample ratio	No of user	Sample ratio
Daily avg.	45,416	44.55%	25,625	29.01%	19791	10.58%
Standard. Dev.	67,776	39.97%	1,21,028	26,05%	17451	9.28%

However, after 40% of the nodes are labeled, the gains from additional nodes in the training set are minimal. This does show that even though we do not consider any relationships at all, by simply using a method of supervised learning, we can improve on the naive method of guessing the most populous group. This improvement is evident even in a situation where most of the class values for the nodes are unlabeled (Fig. 5).

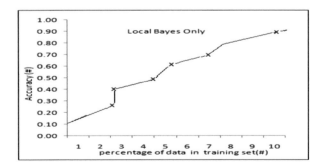

Fig. 5. Local Bayes method

Our second experiment was conducted to establish a performance baseline of an existing relational classifier on our extended dataset. The experiment was conducted using only one attribute the production company as the determinant of relationships between movies, whereas we consider all attributes to be indicative of relationships. This large number of relationships appears to inject a higher degree of error into our trials as opposed to simply using a single attribute. These observations tell us that the sample datasets tend to underestimate the amount of reciprocal relationships (Table 2).

Table 2. Average daily sampling ratios in relational

Daily statistic	Complete user #		Sample user #1		Sample user #2	
	Local user	Relational user	Local user	Relational user	Local user	Relational user
Daily avg	84.14%	15.59%	83.23%	15.76%	84.21%	15.79%
Std. dev.	0.56%	0.49%	0.53%	0.51%	0.76%	0.71%

Other factors can influence results also. While relational user could not reach an acceptable level of reidentification in these measurements (resulting recall rates at most around 20%), the variant produced better rates, though it was also incapable of reaching recall rate significantly higher than 70%, Again, we find that extending the sampling period improves the estimation of the proportions of reciprocal and directed relationships. Next, we study how many of the relationships in the complete dataset are captured by the sample datasets. We calculate the recall of reciprocal, directed, and all relationships and list the results (Fig. 6).

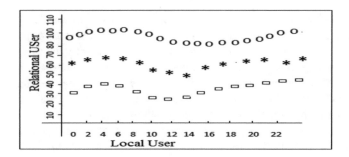

Fig. 6. Average comparison of local vs relational user

However, many applications that utilize user interaction information need a complete view of user relationships, for example, analyzing user network properties and studying network-based information diffusion. For these applications, the sample datasets do not provide sufficient information. We also notice that the recall of reciprocal relationships is generally smaller than that of directed relationships, which indicates that reciprocal relationships are harder to capture from sample data (Fig. 7).

In the final set of analysis, we study the intensity of users being mentioned. This piece of information is important for studying user roles However, many of the users mentioned frequently in the complete dataset are not observed in a one-day sample, Extending the sampling period to one month greatly improves the results.

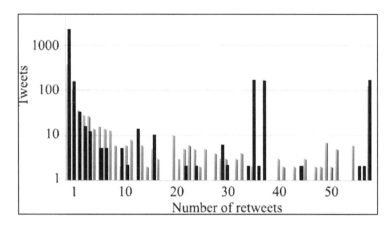

Fig. 7. Comparison of tweet and Retweet distribution

4 Conclusion

We have provide an expressive study of Local data with relational data, few samples are obtained from the ontology's stream. These two data streams are data sources for a variety of research and commercial applications. By comparing the sample data with the corresponding complete data in different aspects, we explore the nature of the sample data, their partiality, and how well they represent the complete data stream. Our results provide insights about the sample data obtained from the data stream. We find that the data streams with the user point and relational user point access priorities which provide samples of the entire public tweets with actual sampling ratios of around 0.96% and 9.6% respectively. The sample datasets truthfully reflect the daily and hourly activity patterns of the Twitter users in the complete dataset. We find that extending the sampling period or increasing the sampling rate both help to improve user coverage. By carefully examining the users that are difficult to sample, we find that the majority of them are extremely inactive with very low average local accessing data frequency. Although our results provide new information about the quality of relational data streams, they are limited by the scope of the datasets, which were collected based on a set of global users.

References

1. Gupta, P., Bhatnagar, V.: Data preprocessing for dynamic social network analysis. In: Data Mining in Dynamic Social Networks and Fuzzy Systems. IGI Global (2013)
2. Charu, C.: Aggarwal. Social Network Data Analytics. Springer Publishing Company, Incorporated, USA (2011)
3. Smith, C.: By the numbers: 220 amazing twitter statistics. WWW page (2014)
4. Sengstock, C., Gertz, M.: Latent geographic feature extraction from social media. In: Proceedings of the 20th International Conference on Advances in Geographic Information Systems, SIGSPATIAL 2012, pp. 149–158 (2012)

5. Emarketer. Worldwide social network users: 2013 forecast and comparative estimates. WWW page, June 2013
6. Tatham, M.: Experian marketing services reveals 27 percent of time spent online is on social networking. WWW page, April 2013
7. Sofean, M., Smith, M.: A real-time architecture for detection of diseases using social networks: design, implementation and evaluation. In: Proceedings of the 23rd ACM Conference on Hypertext and Social Media, HT 2012, pp. 309–310 (2012)
8. Sun, X., Wang, H., Li, J., Pei, J.: Publishing anonymous survey rating data. Data Min. Knowl. Discov. **23**(3), 379–406 (2011)
9. Tan, E., Guo, L., Chen, S., Zhang, X., Zhao, Y.: Spammer behavior analysis and detection in user generated content on social networks. In: Proceedings of the IEEE 32nd International Conference on Distributed Computing Systems, ICDCS 2012, pp. 305–314 (2012)
10. Wang, Y., Wu, X., Wu, L.: Differential privacy preserving spectral graph analysis. In: Pei, Jian, Tseng, Vincent S., Cao, Longbing, Motoda, Hiroshi, Xu, Guandong (eds.) PAKDD 2013. LNCS (LNAI), vol. 7819, pp. 329–340. Springer, Heidelberg (2013). https://doi.org/10.1007/978-3-642-37456-2_28
11. Wondracek, G., Holz, T., Kirda, E., Kruegel, C.: A practical attack to de-anonymize social network users. In: Proceedings of the 2010 IEEE Symposium on Security and Privacy, SP 2010, pp. 223–238 (2010)
12. Xiao, Q., Chen, R., Tan, K.L.: Differentially private network data release via structural inference. In: Proceedings of the 20th ACM SIGKDD International Conference on Knowledge Discovery and Data Mining, KDD 2014, pp. 911–920 (2014)
13. Yue, Wang, Xintao, Wu: Preserving differential privacy in degree correlation based graph generation. ACM Trans. Data Privacy **6**, 127–145 (2013)
14. Zhao, W.X., Jiang, J., Weng, J., He, J., Lim, E.P., Yan, H., Li, X.: Comparing twitter and traditional media using topic models. In: Proceedings of the 33rd European Conference on Advances in Information Retrieval, ECIR 2011, pp. 338–349 (2011)
15. Chen, S., Zhou, S.: Recursive mechanism: towards node differential privacy and unrestricted joins. In: Proceedings of the 2013 ACM SIGMOD International Conference on Management of Data, SIGMOD 2013, pp. 653–664 (2013)

Application of Various Computer Tools for the Optimization of the Heat Pump Heating Systems with Extraction of Low-Grade Heat from Surface Watercourses

A. Sychov[1(✉)], V. Kharchenko[1], P. Vasant[2], and G. Uzakov[3]

[1] FSBSI Federal Scientific Agroengineering Center VIM, 1st Institutsky proezd, 5, 109456 Moscow, Russian Federation
arsenikus@yandex.ru
[2] Department of Fundamental and Applied Sciences, Universiti Teknologi Petronas, Seri Iskandar, Malaysia
[3] Karshi Engineering Economic Institute, Mustakillik Str, 225, 180100 Karshi, Uzbekistan

Abstract. A number of computer tools have been applied to solve optimization issues of the heat pump heating systems with extraction of low grade heat from surface watercourses are described. Among them: 3D modeling in CAD KOMPAS-3D, calculations using numerical methods and programming elements in MathCAD, measurements using a specially designed computer monitoring system. The example of a novel design of the submersible water-brine heat exchanger is presented.

Keywords: Water-source heat pump · Low-grade heat · Watercourse Monitoring system

1 Introduction

In many countries heat pump installations (HPI) implemented and actively used in the heating sector for a long time. Especially they are suitable for heating of private houses with no connection to the main gas pipeline where the electric grid or deliverable fuel can be the only power sources. The main obstacles for a more widespread adoption of such plants are the high cost and, consequently, long payback period of installations using heat of soil or aquatic environments. On the other hand, more affordable air source heat pumps are characterized by low efficiency at low temperatures of outdoor air. The last type is suitable mostly only for regions with relatively mild climates.

A significant share of the total capital expenditure in the construction of HPI is the cost of the heat-transfer system for the selection of low-grade heat. If you use ground or groundwater as a heat source significantly reduce installation costs in most cases not possible due to the need for a huge excavation. However, in the case of the existence of the open water heat source suitable in parameters there is an opportunity to reduce installation costs.

P. Vasant et al. (Eds.): ICO 2018, AISC 866, pp. 310–319, 2019.
https://doi.org/10.1007/978-3-030-00979-3_32

Particularly promising in this respect is the use of the heat of the watercourse. Analysis of the situation with the practice of creating such HPI shows that, due to the low experience and lack of research in this area, in many cases not optimal solutions are used. That leads to increased costs, and sometimes to situations where the characteristics of the installation are much worse than expected. In some cases when you try to design HPI with the use of classical methods of selection of heat from the water mass the results of preliminary calculations make this project unattractive to the customer, and this idea was rejected. In the same time more in-depth approach to the issue and the use of other technical solutions could make installation much more cost-effective.

It should also be noted that in addition to natural aquatic environments, there are many relatively warm water bodies and watercourses, bearing the bargain anthropogenic or geothermal heat, which are at HPI is possible to significantly reduce the cost of heating.

There are several possible methods of selection of heat from a reservoir or watercourse [1]. The most simple, inexpensive and effective at first glance seems to be open loop without intermediate heat carrier, that is, with the extraction and subsequent discharge of water. But this scheme is not possible in all cases and due to significant drawbacks usually not recommended for use. Thus, in practice mostly applied passive methods for the selection of heat with circuit of an intermediate heat carrier. The most widely used method is based on the laying on the bottom of the so-called mats made of polyethylene pipes. That can be called an analog of the horizontal ground collectors at HPI, which use the heat of the soil. However, despite the simplicity of design and low cost of polyethylene pipes, such a scheme of selection of heat from the aquatic environment is not always the most rational.

We reviewed ways of enhancing technical and economic characteristics of heat pump systems using the heat of the water environment, especially the watercourse.

A water stream in full can be called a medium with high heat transfer, and to achieve the best technical-economic performance of heat pumps, which use the heat of the watercourse, it is advisable to use this feature. In Fig. 1 is a block-diagram showing the principal steps to be taken in the design of the heat exchanger and the overall contour of the selection of low-grade heat from the watercourse to achieve estimated goals.

As can be seen from the diagram, this way requires a more careful approach to the calculation and optimization of parameters, but designed according to this manner HPI with metal immersion heat exchangers in many cases should be more profitable than HPI with bottom mats, and in some cases this is the only acceptable solution.

2 Practical Solutions

One can offer several different designs of water-brine heat exchangers, designed for selection heat from the stream corresponding to the above criteria of maximum efficiency. A variant of such heat exchanger basing on use of a flat coil of metal tubing of circular cross-section is modeled in CAD KOMPAS-3D and shown in Fig. 2 [2].

In this design, enhancing the heat transfer is primarily achieved through the use of the natural movement of water in the direction of flow in the flow core for heat transfer

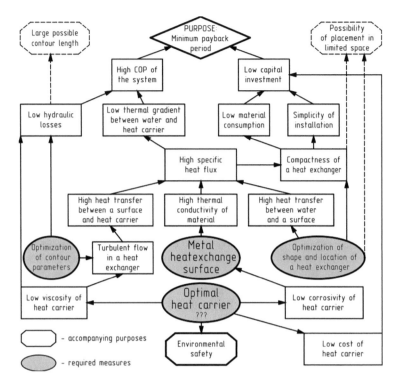

Fig. 1. Possible ways to improve technical and economic indicators of heat pump based systems in case of warmth selection from a watercourse.

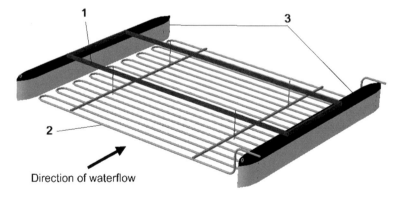

Fig. 2. Submersible floating water-brine heat exchanger: 1 – frame, 2 – coil-pipe, 3 – floats.

processes intensification. It is known that the rate of flow of water in an open channel takes the highest values near the surface, and in the case of the ice-cover area of greatest flow velocity is shifted inland, closer to the middle of the stream. For installing the heat exchanger in the zone of greatest velocity it is equipped with floats, which gives it buoyancy and ropes and anchors so that the heat exchanger can be positioned and retained in the area of best heat transfer (Figs. 3 and 4).

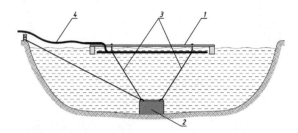

Fig. 3. Arrangement of the heat exchanger in a watercourse: 1 – floating heat exchanger, 2 – anchors, 3 – ropes, 4 – flexible hose.

The improving heat transfer characteristics also occur due to the fact that this construction and arrangement of the heat exchanger permit to direct the flow of water in the direction of straight segments of pipe of the coil that intensify the process of heat transfer.

When using the heat exchanger in the freezing conditions of the watercourse for the period the ice-cover it is advisable to pull the cables closer to the bottom (Fig. 4b, d), and the rest of the time to keep the surface of the watercourse (Fig. 4a, b), thus the coil will be in the areas of highest velocity and will not be frozen in the ice. In the same time, even a significant decrease in the level of water in the canal will not result in drying up pipes of the coil, as the heat exchanger will start to drop after the water level.

To test the described method of extraction of heat from the watercourse, as well as for testing other technical solutions aimed at improving technical and economic indicators of HPI, there was collected the experimental setup, which was a heat pump heating and air conditioning system of residential house water–to-air type with capacity up to 7 kW (Fig. 5). The system of selection of low-potential heat based on the floating heat exchanger was mounted on a specially selected ice-free watercourse (Fig. 6).

3   Calculation Program

For this scheme, the efficiency of the entire system depends on parameters such as the size and configuration of the submersible heat exchanger, the composition and specific consumption of heat carrier and others. The total coefficient of performance (COP) of the whole installation is also affected by the power required for circulation. To determine the best configuration and optimization of all parameters for the specific initial conditions previously a special calculation program in MathCAD was compiled [3].

The process of heat carrier heating in a coil-pipe, not covered by ice, is described by the differential equation as below:

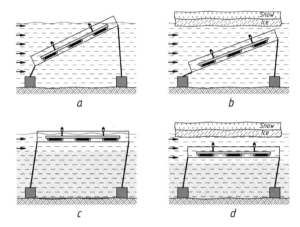

Fig. 4. Examples of arrangement of the heat exchanger in a watercourse depending on conditions: a – clear channel; b – existence of an ice cover; c – silted channel; d – the presence of bottom sludge and ice cover.

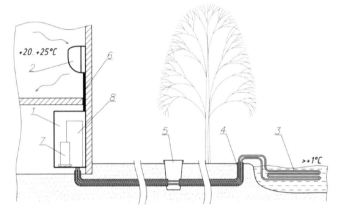

Fig. 5. Schematic diagram of the experimental installation 1 – outdoor unit; 2 – indoor unit; 3 – water-brine heat exchanger; 4 – heat-insulated underground pipeline; 5 – caisson; 6 – freon line; 7 – compressor; 8 – brine-freon heat exchanger.

$$\frac{dT(x)}{dx} = \frac{\pi \cdot d \cdot K \cdot (T_R - T(x))}{G \cdot C}, \tag{1}$$

where $T(x)$ – temperature of heat carrier along the path through the heat exchanger; d – average pipe diameter; K – coefficient of heat transfer from water to heat carrier; T_R – temperature of river water; G – flow rate of heat carrier; C – specific heat of heat carrier.

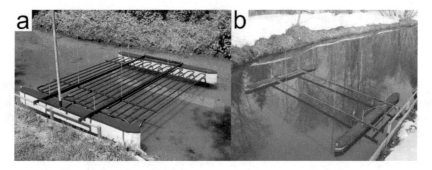

Fig. 6. Experimental sample of the floating heat exchanger: a – lifted over water in the summer; b – in working position in the winter

In cold countries such as Russia, Finland, Sweden, etc. the operation of the designed system can be associated with a possibility of icing, that is with a formation of ice layer of different thicknesses on walls of the heat exchanger, which is located in water [4].

To design heat exchangers taking into account the possibility of forming an ice layer on the coil-pipe surface, another differential equation was derived:

$$\frac{dT(x)}{dx} = \frac{\pi \cdot (d_O + 2 \cdot \Delta_I(T(x))) \cdot \alpha_I(\Delta_I(T(x))) \cdot (T_R - 273.15)}{G \cdot C}, \quad (2)$$

where $T(x)$ – temperature of heat carrier depending on the path traveled through the heat exchanger; d_O – outside pipe diameter; $\Delta_I(T(x))$ – steady-state thickness of the ice layer on the surface of the pipe, depending on the temperature of heat carrier at a given point of the coil-pipe; $\alpha_I(\Delta_I(T(x)))$ – coefficient of heat transfer from water to the ice-covered pipe, depending on the thickness of the ice layer; Δ_I at a given point of the coil-pipe; T_R – temperature of river water; G – flow rate of heat carrier; C – specific heat of heat carrier.

This equation is obtained at the condition that the temperature of the outer surface of the ice layer is 0 °C (273.15 K), which means a constant temperature gradient between river water and the ice surface at a variable coefficient of heat transfer, which depends on the outer diameter of the ice covered pipe:

$$\alpha_I(\Delta_I) = \frac{1}{2} \cdot \lambda_W \cdot \mathrm{Pr}_W^{0.38} \cdot \sqrt{\frac{V_W}{v_W \cdot (d_O + 2 \cdot \Delta_I)}}, \quad (3)$$

where λ_W, Pr_W, v_W – thermal conductivity, Prandtl number and kinematical viscosity of river water, V_W – the speed of water in the river, d_O – outside pipe diameter.

The dependence of ice layer thickness on heat carrier temperature at a given point of the coil-pipe $\Delta_I(T(x))$ is in turn calculated out of the constancy of the linear density of the heat flux through the pipe wall in this section:

$$
\begin{aligned}
&\left(\frac{d_I + d_O + 2 \cdot \Delta_I(T(x))}{2}\right) \cdot K \cdot (T_R - T(x)) = \\
&= (d_O + 2 \cdot \Delta_I(T(x))) \cdot \alpha_I(\Delta_I(T(x))) \cdot (T_R - 273.15)
\end{aligned}
\tag{4}
$$

where d_I, d_O – inside and outside pipe diameter, K – coefficient of heat transfer from water to heat carrier, $\alpha_I(\Delta_I(T(x)))$ – coefficient of heat transfer from water to ice-covered pipe, depending on the thickness of the ice layer at a given point of the coil-pipe.

The differential equation (1) has an analytic solution, which simplifies the calculations:

$$
T(x) = T_R - \exp\left(\ln(T_R - T_0) - \frac{\pi \cdot d \cdot K \cdot x}{G \cdot C}\right),
\tag{5}
$$

where T_0 – temperature of heat carrier at the inlet to the coil-pipe.

The differential equation (2) does not have a simple analytical solution, therefore, to calculate heat carrier temperature in this case numerical methods for solving differential equations available in the MathCAD environment, such as the "rkadapt" and "rkfixed" commands, are used.

The algorithm of calculation and optimization of the river heat exchanger includes a lot of subroutines, conditional operators, cycles and iterations, and the performed works, thus, demonstrates the wide possibilities of the MathCAD package, which proved to be indispensable for the solution of the task.

4   Monitoring System

When conducting research of the experimental heat pump system operation, it is necessary to monitor simultaneously a multitude of operating parameters. One of the main tasks is to optimize the installation characteristics. Therefore, it is necessary to monitor in real time the result parameters such as performance and COP of the installation that can't be measured directly, but can only be calculated from other data.

In this regard the special system for monitoring the operating parameters of the installation with the possibility of receiving in nearly real-time mode (with an insignificant delay compared to the characteristic reaction time of the installation) the values of performance and efficiency indicators has been developed.

The monitoring system is based on a simple 8-channel analog-to-digital converter (ADC), however, due to the non-standardity of the task, it is performed according to the developed novel scheme and assumes a special algorithm for data processing by means of a specially written plugin for a personal computer. Besides, a method has been developed that allows one ADC channel to feed a signal from two or more sensors, which greatly expands the capabilities of ADC.

In system it is used the following sensors: many temperature sensors based on NTC thermal resistors, electric power meter, heat carrier flow meter, compressor and fan motor speed sensors. Other indicators are calculated on the basis of data from these sensors.

Another feature of the system (to be exact the feature of the plugin) is the function of automatic detection the steady-state operating modes of the installation with the calculation and output of their duration and averages for the steady-state indicators. Heat pump system with a long contour of the intermediate heat carrier and large volume of liquid in it has significant inertia and any purposeful change during the experiment of any initial parameter or essential change of conditions, independent of the operator (for example, changes in air or watercourse temperature) is accompanied by a prolonged (from several minutes to tens of minutes) transition process which in most cases isn't of interest for analysis. To study the influence of various parameters on the efficiency of the installation, it is required to determine the average performance and efficiency indicators over long periods of steady operation, accompanied by only a small fluctuation in values. Due to these it was important to provide that the monitoring system automatically according to defined criteria detects such established operating modes and collects data on them in the separate table (file).

The Fig. 7 shows the program window while the system is running, which displays the current system parameters and the graphs of their changes over time.

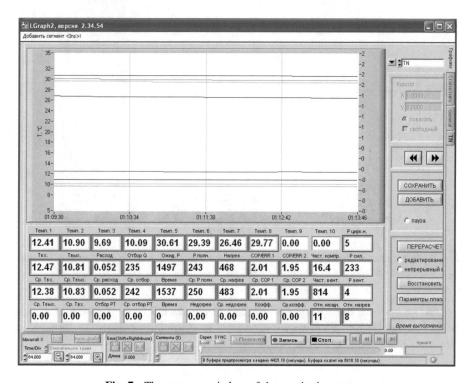

Fig. 7. The program window of the monitoring system

Thus, the monitoring system provides the receipt and processing of a large amount of data in an automatic mode, what is almost impossible to achieve by manual measurement of individual parameters due to the large number of simultaneously changing indicators.

5   Results

Mainly, the mounted experimental setup with the monitoring system was necessary for check and specification of the calculation program, which further allows to carry out studies on the mathematical model and to calculate an optimal configuration for any case. But some visual comparative results can be obtained directly from the experimental setup. For example, Fig. 8 shows dependence between linear heat transfer coefficient of the coil-pipe and zone of the arrangement of the floating heat exchanger.

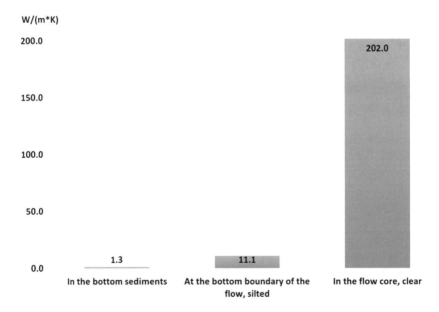

Fig. 8. Experimental results of the comparison of the linear heat transfer coefficient of the coil-pipe for various conditions of an arrangement

As can be seen, the arrangement of the coil-pipe in a flow core greatly intensifies heat exchange. That explains why it is possible to refuse huge bottom mats in favor of compact metal heat exchangers.

6   Conclusion

The use of surface water, especially channels, small rivers and other watercourses as sources of low-grade heat for heat pump systems allows reducing the cost of such systems. To achieve high technical and economic indicators such HPI new, most optimal in each case technical solutions are required and nowadays computer tools are very useful in solving these problems.

As one of these solutions can serve the proposed submersible floating water-brine heat exchanger. Introduction to the practice of this and other solutions that can reduce the cost of heat pump installations and payback period would promote wider dissemination of such systems. A particularly promising application of systems such as described above appear to be in areas where widely used irrigation system for watering and irrigation of agricultural structures. Such areas include some territories of southern Russia, southern Kazakhstan, almost all territory of Uzbekistan, and so on.

References

1. Spitler, J.D., Mitchell, M.S.: Surface water heat pump systems. In: Rees, S.J. (ed.) Advances in Ground-Source Heat Pump Systems, pp. 225–246. Woodhead Publishing (2016). https://doi.org/10.1016/b978-0-08-100311-4.00008-x
2. Sychov, A.O., Kharchenko, V.V.: Heat supply of a rural house using low-potential heat of open water currents. Mech. Electrif. Agric. **1**, 14–17 (2015, Russian language)
3. Kharchenko, V.V., Sychev, A.O.: Optimization of the low-temperature circuit of a heat pump system based on the heat of surface water. Altern. Energy Ecol. **7**, 31–36 (2013, Russian language)
4. Kharchenko, V.V., Sychev, A.O.: Calculation of influence of icing on effectiveness of low-potential heat selection from water environment. Energy Autom. **4**, 21–29 (2017, Russian language)

Epilepsy Detection from EEG Signals Using Artificial Neural Network

Amer A. Sallam[1], Muhammad Nomani Kabir[2]([envelope]),
Abdulghani Ali Ahmed[2], Khalid Farhan[2], and Ethar Tarek[1]

[1] Faculty of Engineering and Information Technology,
Taiz University, Taiz, Yemen
amer.sallam@gmail.com
[2] Faculty of Computer Systems and Software Engineering,
University Malaysia Pahang, 26300 Gambang, Pahang, Malaysia
{nomanikabir, abdulghani}@ump.edu.my,
kkkhalid@yahoo.com

Abstract. In the field of medical science, one of the major recent researches is the diagnosis of the abnormalities in brain. Electroencephalogram (EEG) is a record of neuro signals that occur due the different electrical activities in the brain. These signals can be captured and processed to get the useful information that can be used in early detection of some mental and brain diseases. Suitable analysis is essential for EEG to differentiate between normal and abnormal signals in order to detect epilepsy which is one of the most common neurological disorders. Epilepsy is a recurrent seizure disorder caused by abnormal electrical discharges from the brain cells, often in the cerebral cortex. This research focuses on the usefulness of EGG signal in detecting seizure activities in brainwaves. Artificial Neural Network (ANN) is used to train the data set. Then tests are conducted on the test data of EEG signals to identify normal (non-seizure) and abnormal (seizure) states of the brain. Finally, accuracy is computed to evaluate the performance of ANN. The experiments are carried out on CHB-MIT Scalp EEG Database. The experiments show plausible results from the proposed approach in terms of accuracy.

Keywords: Electroencephalogram · Artificial neural networks
Discrete wavelet transform

1 Introduction

A disease causes an abnormal condition to the body, often affecting organs. Any change from the normal condition of a body or an organ is exhibited by a characteristic set of symptoms and signs. Epilepsy is one of the world's most common neurological diseases and approximately 50 million people of the world's population currently suffers from epilepsy according to the World Health Organization factsheet (2018) [1]. Temporary electrical disturbance of the brain causes epileptic seizures. Sometimes seizures may go unnoticed depending on their occurrence, and sometimes may be confused with other events. Antiepileptic drugs have helped treat millions of patients.

© Springer Nature Switzerland AG 2019
P. Vasant et al. (Eds.): ICO 2018, AISC 866, pp. 320–327, 2019.
https://doi.org/10.1007/978-3-030-00979-3_33

Analysis of brain signals using Electroencephalogram (EEG) is used for detecting the brain diseases. EEG is the recording of electrical activity of the brain from scalp. It measures the voltage fluctuations resulting from ionic current flows within the neurons of the brain. EEG procedure is non-invasive, painless and harmless. EEG recording of patients suffering from epilepsy shows two categories of abnormal activity: interictal-abnormal signal recorded between epileptic seizures; and ictal-the activity recorded during an epileptic seizure. EEG signals can be decomposed into five EEG sub-bands: *delta, theta, alpha, beta and gamma* [2].

Delta waves are slow and its frequency range is less than 4 Hz. Theta frequency ranges from 4 to 8 Hz. The amplitude of the delta wave is high. The frequency of alpha waves is from 8 to 12 Hz while beta waves are of the range from 12 to 30 Hz. Gamma waves are the highest brainwave frequency and its frequency range is from 30 to 60 [3].

Several works have been done on identification of epilepsy from EEG signals using different techniques. Shoeb [2] used a Machine-Learning approach to detect epileptic seizure. Mirowski et al. [4] classified different patterns of EEG synchronization to predict a seizure. Orhan et al. [5] developed a model based on multilayer neural network to classify normal and epileptic behavior from EEG signals. Salem et al. [6] used Discrete Wavelet Transform (DWT) and Ant Colony algorithm to detect epileptic seizure from EEG Signals. Li e al. [7] proposed a method for detecting normal, interictal and epileptic signals using wavelet-based envelope analysis (EA) combined with DWT, and neural network ensemble (NNE). Gupta et al. [8] proposed a technique based on an autoregressive moving average (ARVA), a self-similar Gaussian random process and a 10-fold support vector machine to classify EEG signals as pre-ictal, interictal, and ictal states using the features: parameters involved in ARVA and Hurst values in Gaussian random process. Acharya et al. [9] used a thirteen-layer deep convolutional neural network for detecting the epileptic abnormalities from EEG signals.

In the proposed work, the dataset is divided into two groups – training and testing sets. First, EEG signals are decomposed into five sub-band signals using Discrete Wavelet Transformation (DWT). Then, artificial neural network (ANN) is used to train the data. Finally, the tests are conducted on the testing sets to identify the given EEG signal as normal or abnormal (epileptic). We used EEG signals obtained from CHB-MIT Scalp EEG Database [10] to conduct the experiments.

The rest of the paper is organized as follows. Section 2 provides the methodology of the work. Section 3 presents the test results and finally, Sect. 4 concludes the paper.

2   Methodology

EEG signal classification with ANN is illustrated by Figs. 1 and 2. Figure 1 shows how the signal segments are trained with the known output or target values, in which ANN neuron weights are computed. At the beginning, signal segments go through DWT filterbank process to be transformed into the five frequency ranges. Then the frequency spectrum in Decibel at different frequency ranges are calculated and averaged.

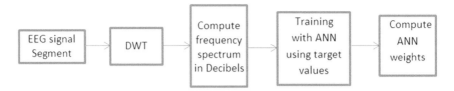

Fig. 1. Training process of EEG signal segments

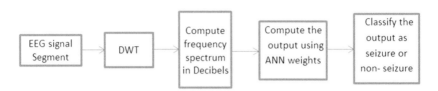

Fig. 2. Testing process of EEG signal segments

In Fig. 2, testing procedure of EEG signal segments is presented. Similar to training process, signals first undergo DWT filterbank process to split different frequency ranges, followed by computation of frequency spectrum which is then averaged over each range. This results in five values of frequency spectrum in Decibel, providing the inputs of ANN. Next, the results are computed using the same weights as obtained in training stage. Finally, the output is classified as seizure or non-seizure.

Figure 3 demonstrates how DWT filterbank splits the EEG signals into five frequency ranges - gamma, beta, alpha, theta and delta. Discrete Cosine Transform, Discrete Wavelet Transform are widely used in frequency-domain transformation [11, 12]. Figure 4 presents the ANN design performed in Matlab. As mentioned before, the five averaged frequency-spectrum values in Decibel are used as inputs for ANN. Ten hidden neurons are used in the hidden layer of the design and the number of outputs is kept as five. There are two possible outputs - seizure represented as [1, 1, 1, 1, 1] and non-seizure as [−1, −1, −1, −1, −1].

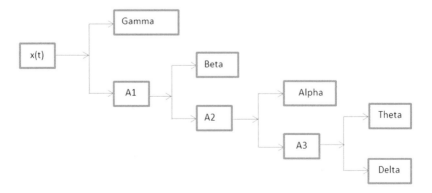

Fig. 3. DWT Filterbank to split EEG signal

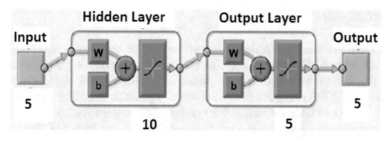

Fig. 4. ANN diagram for EEG signal segments in Matlab.

During the training phase, the neural weights in ANN are calculated and they are used in testing phase. The output from the testing phase is obtained as T = [T_1, T_2, T_3, T_4, T_5]. Using the output, we find T$^-$ using

$$T^- = \sum_{i=1}^{5} \min(0, T_i) \tag{1}$$

and T$^+$ by the form:

$$T^+ = \sum_{i=1}^{5} \max(0, T_i) \tag{2}$$

Finally, the decision parameter D is computed by

$$D = \begin{cases} -1, & \text{if} \quad T^- > T^+ \\ 1, & \text{otherwise.} \end{cases} \tag{3}$$

The decision parameter determines whether the EEG signal segment is with seizure and non-seizure. Specifically, if $D = -1$, the signal segment is non-epileptic, and if $D = 1$ indicates the segment is epileptic. After testing with different sample sets, the accuracy A of the proposed method is computed using the formula:

$$A = \frac{TP + TN}{TP + TN + FP + FN} \times 100\% \tag{4}$$

where TP is the number of sets which are true-positive (i.e., seizure identified as seizure); TN, true-negative (i.e., non-seizure identified as non-seizure); FP, false-positive (i.e., seizure identified as non-seizure) and FN, false-negative (i.e., non-seizure identified as seizure).

3 Test Results

In the tests, we use an EEG epilepsy dataset from (CHB-MIT Scalp EEG Database) to verify our method. The dataset contains normal and abnormal (epileptic) cases. We trained the dataset with two or more seizures per patient and tested on 916 h of continuous scalp EEG sampled at 256 Hz from 24 patients. The EEG samples are split into sets of 100-seconds recording. We conducted three experiments namely, E1, E2 and E3 and the experimental setup and the overall results are shown in Table 1. In the columns two and three, the training sets for normal case and abnormal case are given. Finally, the accuracy obtained from each experiment is presented.

Table 1. Accuracy of the test results.

Experiment	Number of training sets		Number of testing sets		Accuracy	
	Normal cases	Abnormal cases	Normal cases	Abnormal cases	Normal case	Abnormal case
E1	100	100	1000	1000	83%	78%
E2	200	200	1000	1000	91%	88%
E3	1000	1000	1000	1000	94%	93%

The detailed results are given in Table 2. The first column provides the sample name where the first two characters imply the experiment number; the second two characters denote the sample identification number. For example, E2S2 implies the sample is from experiment E2 with the sample identification number S2. Columns two to six present the values of the output T. Then values of T^+ and T^- are computed and the decision D is made. The last column provides the original result O from the database for a comparison of our method. Note that the value of D or O as -1 indicates a normal case and 1 denotes an abnormal case.

In experiment E1 where the network was trained with 100 normal and 100 abnormal cases, the result from the training phase is illustrated in Fig. 5. The Performance is shown over epochs (iterations) by computing Mean Square Error (MSE) that represents the deviation between network output and the target. The percentage of similarity between network output and network target is computed by the value of Regression R.

Experiment E2 that considered 200 normal and 200 abnormal cases as training sets exhibits better accuracy; however, the network performance is lower with more regression value R as shown in Fig. 6. Training results in experiment E3 with 1000 normal and 1000 abnormal sets provide the highest accuracy as demonstrated in Table 1 are the worst in terms of performance and regression as presented in Fig. 7 since more data are involved during training phase.

Table 2. Decision from experimental results and the actual results from some samples.

Sample (Feature)	T_1	T_2	T_3	T_4	T_5	T^+	T^-	D	O
E1S1	-0.872	-0.981	-0.873	**-0.989**	-0.976	0	-4.692	-1	
E1S2	**-1**	-0.923	-0.927	-0.919	-0.889	0	-4.658	-1	-1
E1S3	**-0.995**	-0.981	-0.952		-0.93	-0.920	0	**-4.779**	-1	
E2S1	**-0.894**	-0.774	-0.772	-0.816	-0.646	0	-3.902	-1	
E2S2	**-0.677**	-0.286	-0.287	-0.209	-0.38	0	-1.839	-1	-1
E2S3	-0.998	-0.994	**-0.999**	-0.952	-0.953	0	**-4.896**	-1	
E3S1	-0.214	-0.196	-0.289	-0.215	**-0.300**	0	-1.214	-1	
E3S2	-0.244	-0.202	**-0.245**	-0.133	-0.203	0	-1.027	-1	-1
E3S3	-0.130	-0.240	**-0.441**	-0.308	**-0.461**	0	**-1.580**	-1	
E1S4	0.905	0.925	**0.930**	0.909	0.899	4.570	0	1	
E1S5	-0.999	0.930	0.890	**0.935**	0.921	3.767	-0.999	1	1
E1S6	0.913	0.925	0.922	0.929	**0.987**	4.676	0	1	
E2S4	**0.902**	0.783	0.711	0.748	0.812	3.956	0	1	
E2S5	0.707	**0.899**	0.890	0.895	0.859	4.250	0	1	1
E2S6	0.940	0.928	**0.947**	0.879	0.933	4.627	0	1	
E3S4	0.501	0.320	**0.538**	0.500	0.534	2.393	0	1	
E3S5	0.250	0.288	0.473	**0.521**	0.390	1.922	0	1	1
E3S6	0.236	0.261	0.383	**0.535**	0.426	1.841	0	1	

Highest values for both normal and abnormal   Second highest values for normal cases

Lowest values for both normal and abnormal   Second highest values for abnormal cases

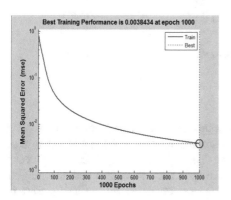
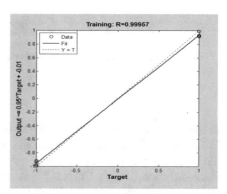

Fig. 5. Training performance and regression results for experiment E1.

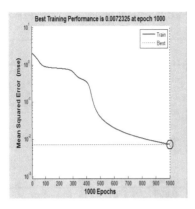
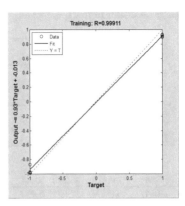

Fig. 6. Training performance and regression results for experiment E2.

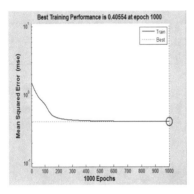
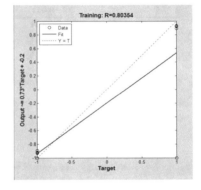

Fig. 7. Training performance and regression results for experiment E3.

4 Conclusion

In this research, we propose an ANN based approach to detect epileptic seizures using EEG signals. The signals are divided into two groups, namely, training and testing sets. EEG signals are decomposed into five sub-band signals using Discrete Wavelet Transformation (DWT). Then, ANN is used to train the data to compute the weights associated with the neurons. Finally, the tests are conducted on the testing sets to identify the given EEG signal as seizure or seizure-free. We used CHB-MIT Scalp EEG Database [10] to conduct the experiments. The experiments show plausible results from the proposed approach. Thus, the approach can be used to detect the epileptic state using EEG signals, facilitating early measures and precautions during a seizure state. However, more study is needed for better understanding of the mechanisms of the epileptic disorders. Further research can be conducted on improving the accuracy of seizure detection from EEG records with some other technique that can provide better insights of this widespread brain disorder.

Acknowledgments. This work was supported by RDU project number 1603102 from University Malaysia Pahang.

References

1. World Health Organization Factsheet: On Epilepsy (2018). http://www.who.int/news-room/fact-sheets/detail/epilepsy
2. Shoeb, A.H.: Application of machine learning to epileptic seizure onset detection and treatment. Doctoral dissertation, Massachusetts Institute of Technology (2009)
3. Moran, L.V., Hong, L.E.: High vs low frequency neural oscillations in schizophrenia. Schizophr. Bull. **37**(4), 659–663 (2011)
4. Mirowski, P., Madhavan, D., LeCun, Y., Kuzniecky, R.: Classification of patterns of EEG synchronization for seizure prediction. Clin. Neurophysiol. **120**(11), 1927–1940 (2009)
5. Orhan, U., Hekim, M., Ozer, M.: EEG signals classification using the K-Means clustering and a multilayer perceptron neural network model. Expert Syst. Appl. **38**(10), 13475–13481 (2011)
6. Salem, O., Naseem, A., Mehaoua, A.: Epileptic seizure detection from EEG signal using discrete wavelet transform and ant colony classifier. In: IEEE International Conference on Communications (ICC), pp. 3529–3534. IEEE (2014)
7. Li, M., Chen, W., Zhang, T.: Classification of epilepsy EEG signals using DWT-based envelope analysis and neural network ensemble. Biomed. Signal Process. Control **31**, 357–365 (2017)
8. Gupta, A., Singh, P., Karlekar, M.: A novel signal modeling approach for classification of seizure and seizure-free EEG signals. IEEE Trans. Neural Syst. Rehabil. Eng. **26**(5), 925–935 (2018)
9. Acharya, U.R., Oh, S.L., Hagiwara, Y., Tan, J.H., Adeli, H.: Deep convolutional neural network for the automated detection and diagnosis of seizure using EEG signals. Comput. Biol. Med. (in press)
10. CHB-MIT Scalp EEG Database. https://www.physionet.org/pn6/chbmit/
11. Ernawan, F., Kabir, M.N.: A robust image watermarking technique with an optimal DCT-psychovisual threshold. IEEE Access. **6**, 20464–20480 (2018)
12. Ernawan, F., Kabir, M.N.: A blind watermarking technique using redundant wavelet transform for copyright protection. In: 14th International Colloquium on Signal Processing & Its Applications (CSPA), pp. 221–226. IEEE (2018)

Evaluation of the Silicon Solar Cell Parameters

Valeriy Kharchenko[1(✉)], Boris Nikitin[1], Pavel Tikhonov[1],
Vladimir Panchenko[1,2], and Pandian Vasant[3]

[1] FSBSI "Federal Scientific Agroengineering Center VIM",
1st Institutskij Proezd 5, 109428 Moscow, Russia
{kharval,pancheska}@mail.ru
[2] Russian University of Transport, Obraztsova Street 9, 127994 Moscow, Russia
[3] Universiti Teknologi Petronas, 31750 Tronoh, Ipoh, Perak, Malaysia
pvasant@gmail.com

Abstract. For estimation and prediction of PV solar cell parameters there was
suggested and developed the methodology based on acceptance for consideration the important peculiarity internal photoeffect which mind-set in the fact that
photon of sunlight can form only one electron-hole pair despite of its energy
level and that the solar radiation is characterized by spectral distribution of
photons of various length of a wave. The main point of the methodology is
consideration the process of electron-hole pairs formation under influence of
photons of given site of solar spectrum with semiconductor (silicon) used as
solar cell substrate. Some results of this kind calculation are represented in the
paper.

Keywords: Semiconductor · Solar cells · Sunlight · Photons · Solar irradiation

1 Introduction

Solar energy is increasingly used in all spheres of human activity. With her help it is
possible to obtain electricity in photovoltaic solar cells or heat in the form of hot water
in solar water heaters. More and more widely devices combining both types of above
mentioned instruments, so-called PV Thermal modules are using Kharchenko et al.
(2018). In solar cells and in combined PV Thermal modules, the efficiency of solar
cells, as well as a number of other parameters, is of great importance. In some cases it is
highly desirable to be able to determine these parameters well in advance of the
application.

Possibility to predict and evaluate parameters of PV solar cells and modules with
high degree of accuracy on the all steps of their fabrication and practical use is the
essential factor for development of works on perfection of parameters of solar power
systems. In connection with this it is necessary to develop new technical means. Well
proved methodology, based on theoretical calculations can be very useful for this. The
approach based on the analysis of interaction of an solar cells initial material with
photons, amount and energy of which is determined by their position in the spectrum of
solar radiation looks to be perspective. The amount of photons and their energies for
each line of spectrum can be obtained from the standard solar radiation of 1000 W/m^2,

suggested and accepted by the International Electrotechnical Committee and corresponding table of the standard solar radiation spectral structure (Bird et al. 1983). The sunlight spectrum especially at the earth surface is complicated enough and depends on a number of factors, such as a thickness of a layer of air at sunlight passage to a surface of the atmosphere mass (AM), content of gaseous impurity etc. In Poulek and Libra (2006) a number of the factors influencing a spectrum of sunlight are consistently considered. These results were be used when considering processes of sunlight photons and semiconductor interaction.

2   Results of Investigations

At the beginning the above mentioned standard table was supplemented with obtained by calculations values connected to photons energy of respective wavelengths such as spectral photon density of a standard solar radiation, a derivative of energy spectral density and photons energy, and also the density of photon flows in each wavelengths sub-diapason of the considered table. Photons energy was represented as in joule as well in electron-volt equivalent that is very convenient for further investigations. Tables obtained required a lot of space. That's why they were not included in the paper in original form and represented as graphs. Figure 1 shows graph of the energy density distribution for the above mentioned standard solar radiation vs wavelength.

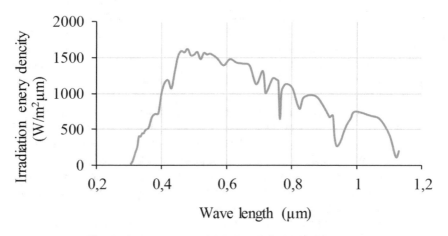

Fig. 1. Energy density of solar irradiation vs wave length

Figure 2 illustrates the results of calculation for spectral photon density distribution of the solar radiation flow 1000 W/m^2. Some solar cell parameters calculated in view of solar radiation spectral structure are described below.

Studies in which purposefully processes in solar cells are considered in a context of the above mentioned factors especially concerning to the given photons position in the solar irradiation spectrum, acting in a working zone, and the mechanism of their interaction with component of solar cells (especially directly in the p-n junction area, in

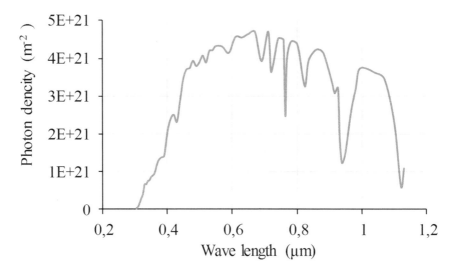

Fig. 2. Photon density for different wave length of standard solar radiation, μm.

the base, doped layer and on contacts) were described earlier (Kharchenko et al. 2009). Unlike mechanism of the interaction of the high energy quantum of the electromagnetic irradiation with electron (Kompton-effect), when energy of the photon could be transmitted to electron partly, at the photo effect photon is absorbed completely. The part of the photon energy spends for breakup chemical link of valence electron in semiconductor (for silicon $E_{cl} = 1,1$ eV), the remain part of energy disperses in the volume and transfer to increase of electrons kinetic energy. Even photons with highest energy (part of spectrum with wavelength about 0,3 μm) are not capable to form more than one electron-hole pair since specific consumption of energy for creation one electron-hole pair in silicon makes the value 3,55 eV (Poulek and Libra 2006).

On the basis of these data there were obtained a number of interesting results. Particularly the diagram of the spectral dependence of a silicon layer thickness in which the radiation flow of the given wavelength diminishes in e time was constructed. There was shown that the share of absorbed photons, for instance, in silicon layer are defined by the layer thickness and absorption coefficient corresponding to length of the waves of the standard spectrum of the solar radiation.

There was calculated solar irradiation absorption coefficient α in silicon vs wavelength (Fig. 3). It was shown that the absorption coefficient in silicon with wavelength increasing falls and becomes lower ($\alpha = 10$ μm^{-1}).

In the range of wave's lengths about 1 μm silicon becomes more transparent for long wave photon. However, for lengths of the waves about 0,3 μm coefficient of the absorption is enough high ($\alpha = 104$ μm^{-1}), that explains high absorbing ability of heavy doped layer of the solar cell in this area of the spectrum.

Physically this parameter shows on what thickness silicon layer weakens light flow of given wavelength in e times. Analysis of this curve shows that for left part of the standard solar spectrum i.e. for photon with wavelength λ about 0,3 μm this value is evaluated as 0,2 μm, but for lengths of the waves around 1 μm this value makes 100 μ.

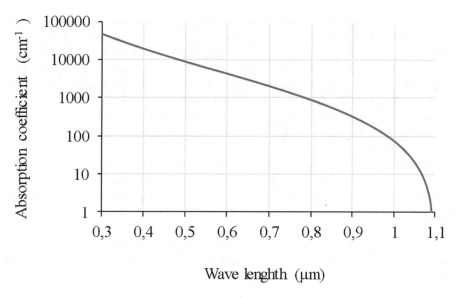

Fig. 3. Absorption coefficient of a solar irradiation α in silicon vs wave length.

Analytical expression for absorption coefficient in silicon according (Kharchenko et al. 2009) looks like below.

$$\alpha = 0,526367 - 1,4425\lambda^{-1} + 0,585368\lambda^{-2} + 0,0399, \tag{1}$$

where: λ – wavelength (μm).

Value of inverse absorption coefficient (namely layer thickness) vs wavelength is presented at Fig. 4.

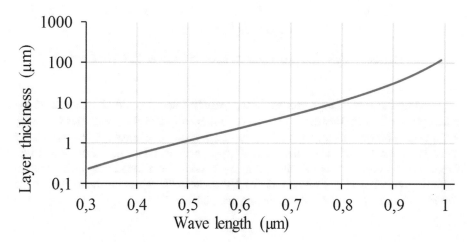

Fig. 4. Layer thickness decreasing in e time vs wave length.

This result enables to value the transmission factor for light flows of different wave's lengths and silicon layers of different thicknesses. The graphs illustrating results of these kind calculations have shown that transmission of the irradiation through silicon layers increases with increase of wavelength and approaches to 1 at the wave length around 1 μm. The transmission factor also increases at reduction of the layers thickness. Theoretical estimations show that at the thickness of silicon layer equal zero dependency is transformed in vertical direct line that is in good correspondence with requirements of the optimum silicon photoelectric converters operation (Vasiliev and Landsman 1971).

These results give an opportunity to realize method of nondestructive control of the "dead" layer thickness in already fabricated wafers after diffusion, i.e. during technologic processes of solar cells fabrication. For this objective it is intended to use results of short circuit current measurements under laser irradiation. The integrated dependence of the predicted value of the solar cell photocurrent on the thickness of highly doped ("dead") layer was constructed on the basis of the analysis of spectral dependence of the solar radiation transmission through silicon layers of different thickness (Fig. 5).

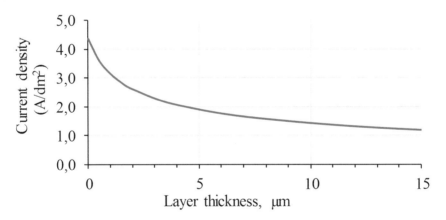

Fig. 5. Short circuit current density vs doped layer thickness at 1000 W/m² and AM 1.5.

Among the parameters representing the greatest interest could be emphasized the efficiency coefficient. However, at work on improvement of this parameter it is very important to imagine clearly and take into consideration those restrictions, which are caused by the nature of an initial semi-conductor material and the nature of the sunlight itself. The approach specified above has been used for a theoretical estimation of extremely possible effectiveness ratio of PV solar cells depending on width of the forbidden gap of an initial semiconductor material.

Expression for efficiency coefficient of a solar cell looks like below (Vasiliev and Landsman 1971):

$$\eta = \frac{i_{sc} \cdot U_{oc} \cdot FF}{R}, \qquad (2)$$

where i_{sc} – density of a short circuit current, U_{oc} – open circuit voltage of solar cell, FF – fill-factor of I-U curve as a factor of filling of area $U_{oc} \cdot i_{sc}$ by $U_{opt} \cdot i_{opt}$, R – level of light exposure of the photo converter, including standard level of solar radiation AM 1,5 (1000 W/m^2).

The parity i_{sc} and R is any constant k and thus expression (2) takes a form:

$$\eta = k \cdot U_{oc} \cdot FF \qquad (3)$$

The estimation of efficiency of solar cells is made on the basis of theoretical (idealized) I-V curve under which it is stipulated such interrelation of a current and voltage at which consecutive resistance of a solar cell is equal zero, and its shunting resistance is equal to infinity. The algorithm of an estimation of the efficiency coefficient of a solar cell with given value of forbidden gap width of the semiconductor is reduced to sequence of calculations in each range $\Delta\lambda$ of the solar irradiation spectrum.

Figure 6 represent results of calculations of theoretical (utmost) efficiency of solar cell vs. width of the forbidden gap of an initial semiconductor (Arbusov and Evdokimov 2007).

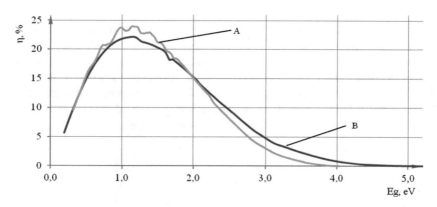

Fig. 6. Theoretical (utmost) efficiency of solar cell vs width of the forbidden gap of an initial semiconductor for standard terrestrial (A) and space (B) spectrum of solar radiation

In addition the similar curves were obtained for different temperatures of solar cells operation. These results, submitted at Fig. 7 (Kharchenko et al. 2010), show that temperature of operation is a very important parameter which should be taken under strong control to provide more efficient mode of operation of solar power plants.

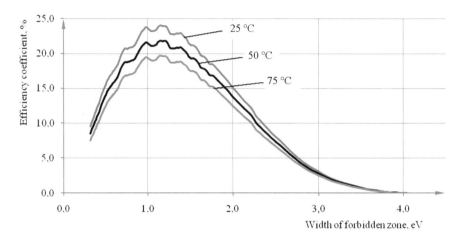

Fig. 7. Solar cell efficiency coefficient vs width of forbidden zone of semiconductor for different temperature of operation.

This approach gives an opportunity to estimate an influence of concentrated solar radiation on the efficiency of solar cells. The *FF* value for this case could be expressed as below (Arbusov and Evdokimov 2007):

$$FF = \frac{U_{xx(0)}}{U_{xx(R_{max})}} = \frac{1}{2 - e^{-\frac{R_{max}}{k_0 R_0}}} \tag{4}$$

Figure 8 represents results of fill-factor (FF) calculations for low levels of solar radiation (Calculations have been performed for $k_0 = 6$). As follows from Fig. 8, for such converters FF parameter with solar radiation growth smoothly decreases from limiting meaning 1 (corresponding to zero level of illumination) to the minimum value 0,616 at 6000 W/m². In this range of radiation for photo converters of similar quality the theoretical efficiency should not change, as with growth of level of solar radiation the increase of U_{oc} will be compensated by adequate decrease in FF factor.

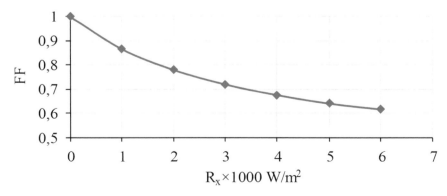

Fig. 8. Calculated FF values for low solar radiation levels, $R_x \times 10^3$ W/m²

Results of FF calculations for more wide range of solar irradiance could be seen in Fig. 9.

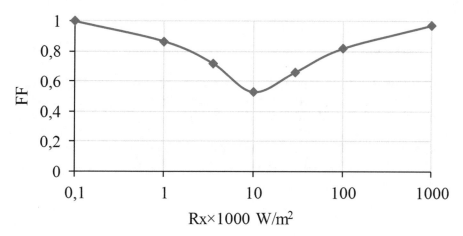

Fig. 9. FF values for wide range of solar radiation levels, $R_x \times 10^3$ W/m²

At higher levels of solar radiations such converters should have FF meanings like it is shown in Fig. 9. After passage of some minimum (around radiation level about 10 Suns) FF value starts to increase and at ultrahigh levels of incident radiation come nearer to the limit, namely to 1.

Last years a wide circulation has received so-called PVThermal systems. PVT system is a device, transforming a solar energy in electricity by means of PV cells and in thermal by means of a thermal absorbing element (absorber).

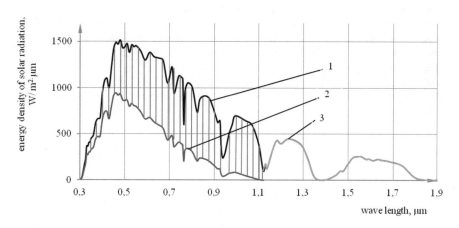

Fig. 10. Distributions of solar radiation energy in PVT system on heat and electricity: 1 – share of energy absorbed in PV cell; 2 – share of absorbed energy transformed into heat in the solar cell volume; 3 – long wave part of spectrum passed through silicon and transformed into heat behind cell structure.

It is important to realize what part of solar radiation could be used for heat and electricity production in such kind of devices. Suggested approach gives an opportunity to identify these shares, which can be found from result of calculations represented at Fig. 10.

3   Conclusion

The presented results are interesting both from a practical and scientific point of view, since they not only help to solve practical problems in the design of solar modules and batteries, but also allow for a deeper review of the processes occurring in the solar cell during the conversion of solar radiation into electrical energy. In spite of the fact that some of the results obtained are based on both theoretical calculations and experimental data, in general, it is desirable to conduct special experiments to verify the theoretical predicted data.

References

Arbusov, Yu.D., Evdokimov, V.M.: PV fundamentals. UNESCO-BRESCE 2007, Moscow, VIESH, p. 292 (2007)

Bird, R.E., Hulstrom, R.L., Lewis, L.J.: Terrestrial solar spectral, data sets. Solar Energy **30**(6), 563–573 (1983)

Kharchenko, V., Nikitin, B., Sherban, D., Simashkevich, A., Bruk, L., Usatiy, I.: Estimation of solar cell parameters in view of solar radiation spectral structure. Mold. J. Phys. Sci. **8**, N3–4, 387–391 (2009).

Kharchenko, V., Nikitin, B., Tikhonov, P., Adomavicius, V.: Utmost efficiency coefficient of solar cells versus forbidden gap of used semiconductor. In: Proceedings of the 5th International Conference on Electrical and Control Technologies ECT-2010, Kaunas, Lithuania, pp. 289–294, 6–7 May 2010. ISSN 1822-5934

Kharchenko, V., Panchenko, V., Tikhonov, P., Vasant, P.: Cogenerative PV thermal modules of different design for autonomous heat and electricity supply. In: Handbook of Research on Renewable Energy and Electric Resources for Sustainable Rural Development, pp. 86–119 (2018). https://doi.org/10.4018/978-1-5225-3867-7.ch004

Poulek, V., Libra, M.: Solar energy, CUA Prague, pp. 25–31 (2006). ISBN 80-213-1489-3

Vasiliev, A.M., Landsman, A.P.: Poluprovodnikovie fotopreobrazovateli. Sov. Radio, Moskva, 148 (1971)

Optimization of Microclimate Parameters Inside Livestock Buildings

Gennady N. Samarin[1]([⊠]) (ID), Alexey N. Vasilyev[2] (ID),
Alexander A. Zhukov[1], and Sergey V. Soloviev[1]

[1] Federal State Budgetary Educational Institution of Higher Education,
State Agricultural Academy of Velikie Luki, 2, Lenina Prospect, Velikie Luki,
Pskov region 182112, Russia
samaringn@yandex.ru
[2] Federal State Budgetary Scientific Institution "Federal Scientific
Agroengeneering Center VIM" (FSAC VIM), Moscow, Russia
vasilev-viesh@inbox.ru

Abstract. Concentration of a large number of animals in the same building, as foreseen by standard projects of large livestock complexes. These places high demands on micro-climate parameters, even a short-term change. It can lead to large economic and economic losses.

Investigating energy-saving methods of forming microclimate parameters, we summarized the disparate data of scientific research of various scientists in the field of veterinary hygiene and veterinary medicine. According to these information, we see to obtain the minimum cost of livestock products, we must control and manage the parameters of the microclimate and maintain them at the optimum level. One of the main directions in energy saving is the development of models of microclimate systems. It saves time and money when we have chose effective solutions of microclimate systems at the stage of their justification and development. Summarizing energy-saving measures in one project is not a guarantee of the most effective solution. Each of the solutions can be economically profitable in itself, but their combination in one project can give the opposite result. We have been selecting the optimal set of microclimate systems from a number of possible, it is necessary to use the scientific method of system analysis, which you allow to estimate the consequences of each solution in advance. The theoretical principles of the formation of heat and mass exchange processes are described in the article: in machines with animals that determine the analytical dependences of the influence of various microclimate parameters on the productivity of animals, the feed consumption; in the microclimate system of livestock buildings. Account the results of the theoretical and experimental studies have carried out, a mathematical model of the microclimate system of the cattle-breeding premises was developed, which it is determined by a system of three equations. In the basis of this mathematical model, an algorithm and a computer program for calculation have been developing that it allows to optimize the main design, technological and energy parameters of the system.

Keywords: Animal · Animal husbandry · Microclimate · Optimization
Energy · Productivity · Feed consumption

© Springer Nature Switzerland AG 2019
P. Vasant et al. (Eds.): ICO 2018, AISC 866, pp. 337–345, 2019.
https://doi.org/10.1007/978-3-030-00979-3_35

1 Introduction

Intensification of livestock, as a basis for the implementation of Russia's food program, requires the development and introduction of new technologies for keeping livestock, are ensuring the creation of healthy herds and increasing their productivity. It is the realization of hereditary qualities of animals [1, 2].

The formation of a regulatory microclimate on farms requires a large amount of energy. The cost of the microclimate is close to the cost of feeding animals [3–5].

Nowadays a significant number of studies have been carried out in our country and abroad. They related to the development, comparative evaluation of efficiency and the introduction of various systems and facilities for normalizing the microclimate on farms [6–12].

Accordingly, the research related to the development of energy-efficient microclimate systems in livestock buildings. Especially at a high cost of energy resources are topical. The solution of this problem is associated with a great economic effect.

Therefore, the aim of our work optimizes the parameters of the microclimate of the cattle-breeding premises, in which the creation and maintenance will require the least amount of energy in obtaining the greatest productivity of animals.

2 Materials and Methods

Taking into account the results of the theoretical and experimental studies performed, a mathematical model of the microclimate system was developed, which is determined by a system of three Eqs. (1, 2, 3).

Equations of motion of air through the installation

$$N = f_1 (L_{29}, \Delta P\{L_{29}, \rho_{29}, F, \xi, h_{18}\}) \tag{1}$$

Heat transfer equation

$$t_{K(29)} = f_2 (G_{29}, G_{18}, t_{H(29)}, t_{H(18)}, E) \qquad \rightarrow \qquad \begin{array}{c} L_{29} \rightarrow max \\[2mm] Q \rightarrow min \end{array} \tag{2}$$

The equation of mass transfer

$$G_{K(29)} = f_3 (G_{29}, G_{18}, h_{18}, \mu_{Hi}, \eta_i) \tag{3}$$

where N – driving power, W; f_1, f_2, f_3 – functional; L_{29} – volume air flow through the installation, m^3/s; ΔP – differential pressure at the inlet and outlet of the instalation, Pa; ρ_{29} – the air density, kg/m^3; F –cross-sectional area of the installation, m^2; ξ – local coefficient of resistance; h_{18} – depth immersion of hoses in the water, m; $t_{K(29)}$ – final temperature of air at the outlet of the installation, $^\circ$C; G_{29} – mass air flow through the installation, kg/s; G_{18} – mass flow of water through the installation, kg/s; $t_{H(29)}$ – initial air temperature at the inlet to the installation, $^\circ$C; $t_{H(18)}$ – initial water temperature at the input to the installation, $^{\circ}$C; E – is the coefficient of heat exchange efficiency; $G_{K(29)}$ – the final mass air flow at the output of the installation, kg/s; μ_{Hi} – is initial concentration of ammonia, oxygen, hydrogen sulphide, carbon dioxide and dust in the air, respectively, kg/m^3; η_i – is coefficient of efficiency of air purification from harmful gases and dust; Q – the energy consumption, J.

In order to optimize the microclimate parameters in the cattle-breeding premises, in which the least amount of energy is required to produce and maintain them. The mathematical model of the microclimate system is developed, which for clarity is presented in a general graphical form in the Fig. 1.

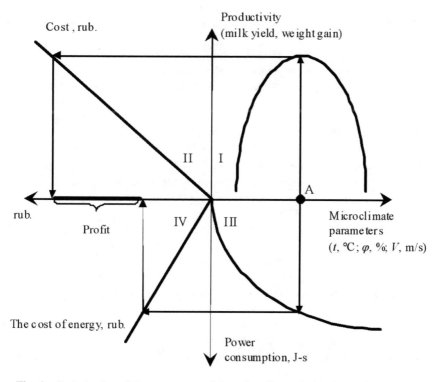

Fig. 1. Optimization of the parameters of the microclimate inside livestock buildings.

I quadrant. It shows the dependence of the productivity of animals (milk yield, average daily weight gain, egg production of a bird, etc.) on various parameters of the microclimate of the cattle-breeding premises. The parabolic dependence (see Fig. 1) is observed at such parameters of microclimate, such as temperature and relative humidity, velocity of air inside the livestock building. To achieve this goal, we study, generalize and establish the patterns of action on animals of the above factors, propose and develop low energy-consuming methods and ways of eliminating negative actions, seek positive technological influences of factors on productivity, product quality and the ethology (behavior) of animals [5].

Based on these dependencies, regression equations (Table 1) were obtained for the influence of microclimate parameters on the weight of pigs and feed consumption.

Table 1. The influence of microclimate parameters on the body weight gain of pigs and feed consumption.

Type of animals, indicator	The regression equation	Limits
1	2	3
Taking into account the ambient temperature, t_B, °C:		
weight gain of pigs K_P, %	$K_P = -0,19329 \cdot (t_B)^2 + 5,8148 \cdot (t_B) + 57,1175$	$-20 \leq t_B \leq +40$
feed consumption K_K, %	$K_K = 0,063808 \cdot (t_B)^2 - 4,7716 \cdot (t_B) + 170,58$	$-20 \leq t_B \leq +40$
Taking into account the speed of air movement V in terms of temperature t_B, °C:		
weight gain of pigs K_P, %	$K_P = -0,15985 \cdot (t_B)^2 + 3,9766 \cdot (t_B) + 79,8362$	$+5 \leq t_B \leq +50$ winter– 0,2 м/s summer– 0,6 м/s
Taking into account the relative humidity of the ambient air, φ_B, %:		
weight gain of pig, K_P, %	$K_P = -0,011 \cdot (\varphi_B)^2 + 1,22504 \cdot (\varphi_B) + 66,404$	$50 \leq \varphi_B \leq 100$
feed consumption, K_K, %	$K_K = 0,021492 \cdot (\varphi_B)^2 - 2,3674 \cdot (\varphi_B) + 164,2$	$50 \leq \varphi_B \leq 100$
Taking into account the concentration of carbon dioxide in the indoor air, μ_{44} (CO_2), %:		
weight gain of pig, K_P, %	$K_P = 1,9268 \cdot (\mu_{44})^2 - 27,4072 \cdot (\mu_{44}) + 97,3947$	$0 \leq \mu_{44} \leq 6$
Taking into account the concentration of ammonia in the indoor air, μ_{17} (NH_3), µkg/m³:		
weight gain of pig, K_P, %	$K_P = -9,7 \cdot 10^{-6} \cdot (\mu_{17})^3 + 0,00284 \cdot (\mu_{17})^2 - 0,88675 \cdot (\mu_{17}) + 101,4292$	$0 \leq \mu_{17} \leq 150$
feed consumption, K_K, %	$K_K = 0,56 (\mu_{17}) + 100,2$	$0 \leq \mu_{17} \leq 20$
Taking into account the concentration of hydrogen sulphide in the internal air, μ_{34} (H_2S), µkg/m³:		
weight gain of pig K_P, %	$K_P = -0,1 \cdot (\mu_{34}) + 100$	$0 \leq \mu_{34} \leq 1000$
Taking into account the illumination of the room, E_O, lx:		
weight gain of pig, K_P, %	$K_P = 0,000101 \cdot (E_O)^3 - 0,01304 \cdot (E_O)^2 + 0,60949 \cdot (E_O) + 90,214$	$0 \leq E_O \leq 80$
feed consumption, K_K, %	$K_K = 0,001753 \cdot (E_O)^2 - 0,2836 \cdot (E_O) + 111,0387$	$0 \leq E_O \leq 80$
Taking into account the production of noise (sound pressure), Z_d, Pa:		
weight gain of pig, K_P, %	$K_P = -0,00254 \cdot (Z_d)^2 + 0,051 \cdot (Z_d) + 99,994$	$0 \leq Z_d \leq 80$

II quadrant. This sector of the coordinate system reflects the value of production per time unit. (s, min).

III quadrant. The energy requirements for creation and maintenance of the specified value of the microclimate parameter are displayed.

IV quadrant. In this sector of the coordinate system displays the cost of energy.

After the formation of all four sectors in the coordinate system, you can determine: by setting the value of the microclimate parameter (point A) and moving along the arrows - the energy costs, production efficiency.

After the formation of all four sectors of the coordinate system, you can determine: by setting the value of the microclimate parameter (point A) and moving in the directions indicated by the arrows - the energy expenditure, profit.

As a criterion for optimization, the objective function is adopted-the energy value. The product obtained from the animal (productive energy), taking into account the energy balance of the organism [13–15]

$$Q_P = \Sigma Q_K + \Sigma Q_B \pm \Sigma Q_{OBC} - \Sigma Q_G \rightarrow max, \tag{4}$$

where $\sum Q_K$ – energy received by animals with food, J;

$$\Sigma Q_K = Q_{JPK} \cdot \prod_{i=1}^{n} \hat{E}_{\hat{E}i} \cdot K_H, \quad at \quad K_H \geq [K_H], \tag{5}$$

where Q_{JPK} – energy feed nutrition в, J/kg;

$\prod_{i=1}^{n} \hat{E}_{\hat{E}i}$ – the product of the coefficients, taking into account the effect of micro-climate parameters on the fuel feed (Table 1); K_H – normative balanced feed consumption for the period of operation of the microclimate system, J; $[K_H]$ – minimum permissible normative balanced feed consumption for the period of operation in the microclimate system, J [4]; $\sum Q_B$ – the energyis received by animals with water, J; $\sum Q_{OBC}$ – the energy is received from the environment/given to the environment, J; $\sum Q_G$ – energy released by animals, J;

The task of optimizing the economic parameters of the microclimate system in the mathematical plan is reduced to finding the minimum value of the adopted objective function [16–23]. The specific reduced costs for creating and maintaining the optimal microclimate CS'

$$CS' = DC + E_H \cdot KB \rightarrow min, \tag{6}$$

where DC – specific direct costs of creation and maintenance of an optimal micro-climate, rub./kg;

$$DC = \sum_{i=1}^{n} DC_i = \frac{\tilde{N}_{\hat{E}} + \tilde{N}_{ZP} + \tilde{N}_{\hat{A}} + \tilde{N}_{\hat{O}\hat{\imath}} + \tilde{N}_{\hat{O}\hat{A}} + \tilde{N}_{\hat{O}\hat{A}}}{\tilde{N}\hat{I}}, \tag{7}$$

where C_K – costs for animal feed for the period of receipt of gross production (CO), rub.; C_{ZP} – salaries of maintenance personnel of the microclimate system for the period of receipt of CO, rub.; C_A – deductions for depreciation of technological equipment of the microclimate system for the period of obtaining CO, rub.; C_{TO} – deductions for maintenance of technological equipment of the microclimate system for the period of obtaining *CO*, rub.; C_{TE} – the cost of fuel and electricity during the operation of the microclimate system for the period of obtaining *CO*, rub.; C_{XB} – costs for chemicals in the operation of the microclimate system for the period of obtaining *CO*, rub.; *CO* – he gross production obtained during the period of operation of the system of microclimate, kg;

$$CO = \prod_{i=1}^{n} \hat{E}_{ji} \cdot M_M, \tag{8}$$

where $\prod_{i=1}^{n} \hat{E}_{ji}$ – product of coefficients, taking into account the effect of microclimate parameters planned productivity of the animal (Table 1); M_M – the planned productivity of the animal, taking into account the genetic potential for the period of operation of the microclimate system, kg; E_H – normative coefficient of capital investments $E_H = 0,15$ [16]; $KB = \sum_{i=1}^{n} \hat{E}\hat{A}_i$ – total specific investment in the technological process, rub./kg.

On the basis of the mathematical model, a calculation algorithm and a computer program for calculation have been developed, which it makes possible to optimize the technological, energy and economic indices of various microclimate formation technologies in the livestock buildings by the method of sequential analysis of options.

The application program is implemented in the form of an imitation system that it allows specialists in the mode of direct dialogue with the computer. It calculates the possible consequences of the decisions, analyzes the results and produces the best version of the projected object.

3   Results

The developed programs are carried out for a pigsty - a fattener for 500 animals at the age of 6 months. the following calculations: calculation of air exchange and heat balance of the premises for the winter and summer periods in the year at outdoor temperatures from –30 °C to +35 °C; calculation of technological modes of air purification from harmful gases (ammonia, hydrogen sulphide, carbon dioxide and dust) by various sorbents; calculation of the temperature of the internal air depending on the temperature of the outside air for different types of operation of ventilation and heating (cooling) of the farm; in total, 13 of the most common variants were considered (see Table 2) – We have 7 options for the winter period of the year and 6 options for the summer (option 4 – air conditioning system (SCR) without an air dryer, option 5 - hard to fully complete); All the below listed calculations are performed for 13 options;

Table 2. The Variants of the microclimate in the farms in winter and summer

№ variant	Technological process. Winter season	№ variant	Technological process. Winter season
B0	Planned	B0	Planned
B1	Heating (Ot) – biological (feed); ventilation (PV) – infiltration		
B2	Ot – biological (feed); PV – infiltration; exhaust ventilation (VV) – natural	B2L	Colding (Ohl) – no; PV – natural; VV – natural
B3	Ot – biological (feed), water heating (boiler); PV – infiltration; VV – natural		
B4	Ot – biological (feed), electric heater; PV – infiltration, mechanical – cleaning of air from gases; air ducts; VV – natural	**B4L**	Ohl – irrigation chamber (water-chemical solution); PV – infiltration, mechanical - cleaning of air from gases; air ducts; VV – natural
B5	Ot – biological, electric heater; PV– infiltration, mechanical - air is drying, purification air from gases, air ducts; VV – natural	**B5L**	Ohl – irrigation chamber (water-chemical solution); PV – infiltration, mechanical - air drying, purification air from gases, air ducts; VV – natural
B6	Ot – biological, electric heater; PV– infiltration, mechanical - type CFO, air ducts; VV – natural	B6L	Ohl – no; PV – infiltration, mechanical - type CFO, air ducts; VV– natural
B7	Ot – biological, electric heater; PV– infiltr., Mechanical. - type PVU; VV – mechanical - type PVU	B7L	Ohl – no; PV – infiltr., Mechanical.-type PVU; VV – mechanical - type PVU

Table 3. Total average specific energy consumption for pigs up to 6 months. at various variants of microclimate systems

Index	Variants of microclimate systems					
	B.2 + 2L	B.3 + 2L	B.4 + 4L	B.5 + 5L	B.6 + 6L	B.7 + 7L
Total average specific energy consumption of pigs up to 6 months, J/(kg weight gain)	153,41	117,86	**51,85**	**45,34**	109,50	108,68

- The calculation of animal's weight gain and feed (energy) consumption during the standing time of the internal air temperature, taking into account the relative humidity of the indoor air, the concentrations of carbon dioxide and ammonia in the air inside the farm during the winter and summer periods of the year;
- the calculation of average specific indicators for the winter and summer periods of the year is invariant: weight gain (M, kg/s/goal); feed consumption (K, kg/s/goal); energy consumption of feed and additional energy for equipment (Q, J/s/goal); reduced costs (CS, rub./kg) and comparison of options.

Based on the results of the calculations in the Table 3, we can see how the feed consumption varies by a variant.

4 Discussion

Analyzing the data of Table 3, we can see that the minimum specific energy consumption has the following microclimate systems: when pigs are raised to 6 months - options 5 (45.34 J/(kg weight gain)) and 4 (51.85 J/(kg weight gain)).

5 Conclusions

The mathematical models developed of these dependencies it made possible to obtain a mathematical model of the microclimate formation system for livestock premises. An algorithm and a computer program for calculation that it allows selecting and then optimizing, by the method of sequential analysis, the technological and energy indicators of selected microclimate technologies in livestock buildings.

References

1. Gosudarstvennaya programma razvitiya sel'skogo hozyajstva i regulirovaniya rynkov sel'skohozyajstvennoj produkcii, syr'ya i prodovol'stviya na 2013–2020 gody: utverzhdena postanovleniem Pravitel'stva Rossijskoj Federacii ot 14 iyulya 2012 goda N 717 [The state program of development of agriculture and regulation of the markets of agricultural products, raw materials and food for 2013–2020: is approved by the order of the Government of the Russian Federation of July 14, 2012 N 717]. (in Russian)
2. Strategiya ustojchivogo razvitiya sel'skih territorij Rossijskoj Federacii na period do 2030 goda: utverzhdena rasporyazheniem Pravitel'stva Rossijskoj Federacii ot 2 fevralya 2015 goda N 151-r [Strategy of sustainable development of rural areas of the Russian Federation for the period till 2030: approved by the order of the Government of the Russian Federation of February 2, 2015 N 151-p]. (in Russian)
3. Baroti, I.: EHnergosberegayushchie tekhnologii i agregaty na zhivotnovodcheskih fermah [energy-Saving technologies and units on livestock farms]/ I. Baroti, p. Rafan. M.: Agropromizdat, 227 p. (1988). (in Russian)
4. Beloglazova, T.N.: Mnogovariantnoe proektirovanie kompleksa inzhenernyh sistem obespecheniya mikroklimata: Na primere cekhov holodnoj obrabotki metallov [Multivariate design of complex engineering systems for the microclimate: the case of the shops of the cold treatment of metals]: Diss. … kand. Techn. science/T.N. Beloglazova. Perm, 230 p. (2000). (in Russian)
5. Borodin, I.F.: EHnergosberegayushchaya tekhnologiya formirovaniya optimal'nogo mikroklimata v zhivotnovodcheskih pomeshcheniyah [energy-Saving technology of optimal microclimate in livestock buildings]/Borodin, I.F., Rudobashta, S.P., Samarin, V.A., Samarin, G.N.: Technological and technical support for the production of livestock products: scientific papers of VIM. Part 2, vol. 142. M: VIM, p. 113 (2002). (in Russian)

6. Draganov, B.H.: Teplotekhnika i primenenie teploty v sel'skom hozyajstve [heat Engineering and application of heat in agriculture]/B.H. Draganov, A.V. Kuznetsov, S. p. Rudobashta. M.: Agropromizdat, 463 p. (1990). (in Russian)
7. Swan, A.A.:. Metodologiya proektirovaniya optimal'nyh sistem formirovaniya sredy obitaniya v pomeshcheniyah intensivnogo zhivotnovodstva i pticevodstva [a methodology for the design of optimal systems the formation of habitat in areas of intensive livestock and poultry production]: author. of Diss...Dr. of tech. Sciences/A.A. Swan. Minsk: TSNIIMESH of the USSR, 36 p. (1991). (in Russian)
8. Metodicheskie rekomendacii po opredeleniyu ehkonomicheskoj ehffektivnosti i ispol'zovaniya v sel'skom hozyajstve kapital'nyh vlozhenij i novoj tekhniki. [Methodical recommendations on determination of economic efficiency and use in agriculture of capital investments and new equipment]. HP: NIFTINESS, 58 p. (1986). (in Russian)
9. Mishchenko, S.V.: Matematicheskie modeli mikroklimata zhivotnovodcheskih pomeshchenij [Mathematical models of microclimate of livestock buildings]/Mishchenko, S.V., Ivanova, V.M.: Mechanization and electrification of agriculture, no. 12, p. 18–21 (1987). (in Russian)
10. Onegov, A.P.: Gigiena sel'skohozyajstvennyh zhivotnyh [Hygiene of farm animals]/A. P. Onegov, I.F. Khrabustovskyi, V.I. Chernykh. M.: Kolos, 400 p. (1984). (in Russian)
11. Popyrin, L.S.: Matematicheskoe modelirovanie i optimizaciya teploehnergeticheskih ustanovok [Mathematical modeling and optimization of thermal power plants]/L.S. Popyrin. M.: Energy, 416 p. (1978). (in Russian)
12. Samarin, G.N.: Upravlenie sredoj obitaniya sel'skohozyajstvennyh zhivotnyh i pticy: monografiya [Habitat management of farm animals and birds: monograph]/G.N. Samarin. Velikie Luki: FGOU VPO velikolukskaya state agricultural Academy, 286 p. (2008). (in Russian)
13. WATT Executive Guide to world Poultry Trends. The statistical reference for poultry executive, 45 p. (2009/10)
14. Kessel, H.W.: Warmetauscher im Stall/H.W. Kessel. - Agrartechnik International 58 (2006)
15. Stauffer, L.A.: Ventilation heat recovery with a heat pipe heat exehanger/L.A. Stauffer/ Agricultural Energy, ASAE pyblication, vol. 1 (2001)
16. Evaporative cooling system: Poult. Int. 41(3), 49 (2002)
17. Cutowski, W.: Ochrona Powietrza/W. Cutowski (2007)
18. Özmen, A., Weber, G.-W., Batmaz, İ.: The new robust CMARS (RCMARS) method. In: International Conference 24th Mini EURO Conference "Continuous Optimization and Information-Based Technologies in the Financial Sector" (MEC EurOPT 2010), 23–26 June 2010, Izmir, Turkey (2010)
19. Abraham, A., Steinberg, D., Philip, N.S.: Rainfall forecasting using soft computing models and multivariate adaptive regression splines. IEEE SMC Trans. 1, 1–6 (2001)
20. Ben-Tal, A., Nemirovski, A.: Lectures on Modern Convex Optimization: Analysis, Algorithms, and Engineering Applications. MPR-SIAM Series on optimization. SIAM, Philadelphia (2001)
21. Ben-Tal, A., Nemirovski, A.: Robust optimization—methodology and applications. Math. Program. 92(3), 453–480 (2002)
22. Fabozzi, F.J., Kolm, P.N., Pachamanova, D.A., Focardi, S.M.: Robust Portfolio Optimization and Management. Wiley, Hoboken (2007)
23. Taylan, P., Weber, G.-W., Yerlikaya, F.: Continuous optimization applied in MARS for modern applications in finance, science and technology. In: ISI Proceedings of 20th Mini-EURO Conference Continuous Optimization and Knowledge-Based Technologies, Neringa, Lithuania, pp. 317–322 (2008)

Blockchain Technology in Smart City: A New Opportunity for Smart Environment and Smart Mobility

F. Orecchini[1,2], A. Santiangeli[1,2], F. Zuccari[1,2],
Alessandra Pieroni[3(✉)], and Tiziano Suppa[2]

[1] CARe - Centeer for Automotive Research and Evolution, Guglielmo Marconi
University, via Plinio 44, 00193 Rome, Italy
{f.orecchini,a.santiangeli,f.zuccari}@unimarconi.it
[2] DIS – Department of Sustainability Engineering, Guglielmo Marconi
University, via Plinio 44, 00193 Rome, Italy
[3] Department of Innovation and Information Engineering, Guglielmo Marconi
University, via Plinio 44, 00193 Rome, Italy
a.pieroni@unimarconi.it

Abstract. The main and challenging evolutionary process that leads towards the so-called Smarter City is still represented by the optimization of the Intelligent Systems for the Quality of Life (QoL) and the Quality of Services (QoS), and by the integration of heterogeneous IT services and different networks. The scientific interest in this field consists of promoting innovative solutions by use of Information and Communication Technology (ICT) able to collect, analyse and obtain value added from a large amount of data generated by several sources: IoT devices, sensor networks, wearable devices as well as sensor grids, widespread within the urban environment. Furthermore, creating synergies and integrating different enabling technologies and informative platforms, remains the most challenge to overcome in order to obtain the optimization of the services and the quality of life. The Smart City involves the implementation of digital strategies that are necessary people-centred and lead into high technology-based innovations to build more capacities and opportunities. In this context, this paper intends to investigate the possibility to integrate the innovative and multi-purpose Blockchain Technology in the smart city evolutionary process, and in particular in the Smart Environment and Smart Mobility by allowing renewable energy sources traceability and by providing information about the kind of energy used to refuel, for example, the selected vehicle. The expected result consists of more user awareness regarding the environmental and energy sustainability.

Keywords: Information technology · Smart city · Blockchain
Smart energy grid · Smart mobility · Traceability
Connected and autonomous vehicles · Electric vehicles
Vehicle electricity charging

© Springer Nature Switzerland AG 2019
P. Vasant et al. (Eds.): ICO 2018, AISC 866, pp. 346–354, 2019.
https://doi.org/10.1007/978-3-030-00979-3_36

1   Introduction and Related Works

The definition of Smart City is still an ambiguous concept. Indeed, a Smart City is described as an idealistic city, where the quality of life and the quality of services for citizens are improved by combining ICT, innovative technologies and new urban infrastructures [1], as seen in Fig. 1. The main goal of the Smart City evolutionary process consists of considering a user-centric vision, and accounting urban issues from the perspective of the citizen's needs, that involves the engaging of the citizens in the city management. In other words, the Smart City may be defined as a complex system where human and social capital heavily interact, supported by innovative technology-based solutions. In this scenario, the Internet of Things (IoT) represents the main enabling-technology for the smart city evolutionary process [2–5].

Fig. 1. Smart city vision

One of the main goal that a smart city evolutionary process should address consists in obtaining a better use of renewable resources, reducing wastes by safeguarding the environment and improving the citizens quality of life. Thus, an eco-sustainable prospective has to be at the base of any city transformation approach. This vision consists of promoting a respectful urban and industrial development, able to address current needs without compromising the capacity of future generations.

A smart city, on the other hand, may be considered being built on six fundamental pillars: Smart Economy (competitiveness), Smart People (Social and Human Capital), Smart Governance (Participation), Smart Mobility (Transport and ICT), Smart Environment (Natural resources) and Smart Living (Quality of Life), as shown in Fig. 2. An eco-sustainable approach should be applied in several layers, such as mobility, environment, social services and urban requalification [6, 7].

In this scenario, the integration of different technologies and different ICT systems, seems to be the most challenge to overcome. This paper intends to propose an innovative solution that foreseen the use of Blockchain technology to allow the citizen to carry out the join to the so-called Intelligent Energy Systems, in the smart city environment. Furthermore, the proposed approach aims to improve the smart city sustainable mobility, by allowing the citizens to choose not only the nearest charging points for the electric vehicle but also the ones that is served by sustainable and renewable energy or which offer the possibility of bi-directional exchange of electricity with the grid [8–11].

In the discussed context, the Blockchain may be considered as an emerging ICT technology that offers new opportunities and provides the transparent and user-friendly applications needed for realizing the process of energy consumption [12, 13]. Indeed, a Blockchain consists of a digital contract that allows a partner to conduct, and invoice, a transaction (e.g. a sale of electricity) directly (in a peer-to-peer way) with another partner. On the other hand, the peer-to-peer concept requires that all transactions must be stored on a computer that is part of a network composed by suppliers and customers who participate in transactions.

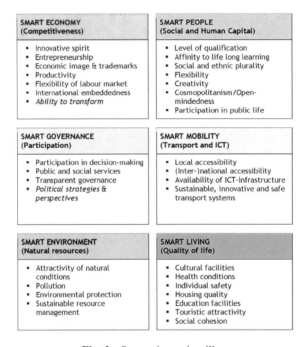

Fig. 2. Smart city main pillars

In the following sections, the proposed approach that foreseen the integration of Blockchain technology in the Smart Environment (one of the main pillars of the Smart City context) will be illustrated and in particular a description of the advantages that could be reached by using Blockchain technology in the sustainable mobility will be discussed (Sect. 2); in Sect. 3 the use of the Blockchain in the Smart Mobility context will be investigated. Finally, conclusion and final remarks will be given in Sect. 4.

2 Blockchain and Smart Environment: Energy Infrastructures

The European modern system of energy production aims, as medium-term, differentiating sources with respect to the final use and, furthermore, being able to guarantee as much as possible the "self-sustainability" of energy by means of renewable resources, while limiting and gradually reducing the consumption of non-renewable natural resources with particular reference to the fossil ones.

The above-mentioned goal has to be combined with the planetary issue of resources exhaustion, the main topics of "Global Warming", and the emissions in the atmosphere of climate-altering substances, which should be contained within a certain threshold in order to minimize the ecological impact on the planet (carbon footprint).

As said above, it is increasingly important having the possibility to choose energy sources with respect to the end-use, having the feasibility, at the same time, to select the correct "energy mix" locally available (also an integrated domestic photovoltaic system) in order to minimize consumption and provide a limited production of climate-altering substances [14]. The aim of this work consists of using Blockchain technology to accessing, visualizing and activating the energy management systems with respect to the sources available on predefined geo-spatial coordinates, in selected time intervals.

Concerning the sustainable mobility, the possibility for the user to select renewable sources by means of Blockchain technology, should provide a reduction of the ecological impact on the whole system.

The possibility to trace precisely the amount of energy that has been used for specific needs, furthermore, represents an essential instrument to understand the "trend" of consumption over the years (both from the qualitative and the quantitative point of view) and to be able to rationally act on energy production infrastructures in order to guarantee the necessary energy balancing in relation to the resources locally available.

2.1 The Advantages of Using Blockchain Technology in the Sustainable Urban Mobility

In this section, the advantages of introducing the Blockchain technology in the urban sustainable mobility will be discussed.

Million users daily use decentralized IT systems for communication and business exchanges, and this trend is in direct contradiction with the IT centralized systems that are used to secure them. Assumed that the decentralization trend is fed by the Internet distributed communications system, where no central node acts as information gatekeeper, it seems mandatory having a new approach to security based on a distributed network architecture. This new distributed approach is called Blockchain. The Blockchain is an emerging technology, originated from a small community of cryptographers but actually is predicted to have wide impacts on many sectors in society. In particular, the Blockchain has transformed the well-known third paradigm warranty and provide a way in which the transactions take place in a peer-to-peer network, by means of a distributed database (ledger) in which all the involved nodes belonging to the Blockchain network not only act as "witnesses" of transaction but also guarantee its integrity.

As said above, this paper introduces an innovative use of the Blockchain Technology applied in the sustainable urban mobility systems. The advantages of this approach consist of providing the system with the resilience and the security characteristics that the Blockchain is able to offer by design.

In other worlds, the use of this emerging technology will allow several improvements, such as:

Maximizing the use of "value chains" in order to track energy transactions for loans, realizing in real time the rates of the various types of energy resources available, notifying their origin;

Adopting the "Distributed Chain" of Blockchain Technology for the replication and sharing of data between the various nodes and blocks of the network in order to differentiate their uses and to minimize attempts at potential "tampering" and/or "attack" of the network;

Having a network that offers the widest guarantees and certifications of consumption and emissions data through the system invariability characteristic, aiming and ensuring reliability and traceability of data over time [15];

Offering the possibility to access in real time to an "un-encrypted" and "shared" network with continuous access to energy data by the entire network in order to allow the use and management in real time of the various trends in energy consumption and the various options of energy mix available locally providing the user with a range of hypotheses for choosing the energy mix that can be used in a given time interval;

Proposing the possibility to access to tracking systems, by locating and by identifying "in real time" the charging points suitable for the type of energy performance (in terms of mileage or recharge times, for example), starting from those closest to the user.

Managing the possibility of carrying out energy "trading" with a view to making available to the grid any energy produced from renewable sources in public or private places with the aim of managing and optimizing the available energy resources in relation to uses in the defined period of time (hour, day, week, month, etc.).

3   Blockchain and Smart Mobility

Concerning the use of Blockchain technology in sustainable mobility context, several applications may be identified that are characterized by the specific purpose they are designed to. First of all, the newest developments in the mobility concept (Smart Mobility) foreseen a historical shift from the concept of "ownership" to the concept of "use" of the vehicle. For this reason, various digital platforms are increasingly widespread, offering intermodal mobility and integrated logistics services "customized" to join users' needs.

3.1   Car Sharing

One of the most widespread forms of integrated sustainable mobility is the car sharing one, which always includes a shared IT platform that allows users access for managing vehicle information (e.g.: vehicle localization). The user joins the IT digital platform to collect information about, for example, the closest vehicle, according to precise requirements, such as:

- The battery state of charge for Battery Electric Vehicles (BEVs);
- The battery autonomy, according to travel needs;
- The vehicle size and performance, according to the required transport typology (number of users or products to be carried).

The use of Blockchain technology in car sharing applications may allow to collect and share the above-mentioned information across the network. In particular, the Blockchain nodes are composed by the vehicle final user (that join the network) and by each vehicle that automatically shares the information regarding the status of the vehicle (battery status, performance, equipment, etc.) and its spatial position.

3.2   IoT and Smart Mobility

The concept of Smart Mobility together with the different possible Blockchain applications allow the development of specific solutions that are characterized by data validation and disambiguation, and by their security and reliability over the time. These features are obtainable by implemented the communication and validation instruments that are granted in the Blockchain networks and by collecting and managing the great amount of data gathered by IoT devices spread across the smart urban environment and inside the vehicle.

In this context, an important Blockchain contribution in the Smart Mobility field can be provided in different applications, such as: Electric Vehicle Charging Points, Fleet Management, Autonomous Drive, V2I and V2V.

3.2.1   Electric Vehicle Charging Points

In this specific application use case, it may be assumed that the Smart Mobility user could choose the energy mix of vehicle refuelling (for example 100% from Renewables). In this case, the Blockchain would guarantee the compliance of the supply with the choice made by certifying the supplied data [15].

The opportunity for the final user to choose the type of energy with which the vehicle to be use should has been refuelled will generate an inducted market that would increase the diffusion of the use of renewable sources in transportation field, that is a fundamental sustainability factor [16].

3.2.2   Fleet Management

Many Fleet Management systems are based on real-time remote monitoring of vehicles. The use of Blockchain network in this context would guarantee data reliability and would therefore allow immediate reactions in case of alarms, especially in case of considerable first level signals, such as "Crash" or "Failure" (with these signals they are indicated a dangerous impact on the safety of the driver or deep damages of the vehicle).

3.2.3   Autonomous Drive

The data reliability and safety for an Autonomous Drive [17] has to be considered essential, otherwise conflicts could arise that could compromise passengers' safety. Blockchain technology could guarantee reliability and trustworthiness accordingly by design.

3.2.4   V2I e V2V – Interconnected Vehicles

The technological and infrastructural development in Mobility is evolving towards integrated IT systems that allow data and information exchange and remote controls, between vehicles and infrastructures [17].

In this context, it is essential to create information chains, linked to each other, supported by reliable, resistant and shared protocols being also accessible in "real time".

These information chains can be managed, regulated and implemented through certified data transaction systems, such as the Blockchain Technology that allows the validation of integrated mobility processes within ICT integrated framework able to process and manage Big Data with the increasing performance requirements that such systems involve.

Specifically, a powerful data management network could easily allow:

- The exchanging and the processing of data by integrating Blockchains between multiple interconnected networks in order to validate and register the connection between vehicles and infrastructure (V2I), and provide vehicles with reliable information such as: road signs, potential hazards, any unforeseen events (traffic, accidents, queues, weather conditions, etc.).
- The interaction between vehicles (V2V) in order to actively exchange information useful for vehicle management in terms of assisted driving and autonomous driving (ADAS systems, LIDAR, RADAR, on-board sensors, accelerometers, gyroscopes, etc.).
- Use of digital technology of "Self Learning" for vehicles and their components, such as to create an adaptive vehicle "Set Up" according to the drive "routines", i.e.: vehicle routes processed through the "hasc" (pre-established algorithms) self-learning of the vehicle while driving (distances, speed, optimization of energy use, driving styles, etc.).

4   Conclusions

The main challenge in the Smart City context remains the optimization of the Intelligent Systems and the IT Complex Systems (that are involved in the citizens Quality of Life (QoL) and the Quality of Services (QoS)). Furthermore, the scientific interest in this field consists of integrating heterogeneous IT services, different networks and IT Platforms. This scenario should involve the promotion of innovative solutions by use of Information and Communication Technology (ICT) able to collect, analyse and obtain value added from a large amount of data generated by several sources: IoT devices, sensor networks, wearable devices as well as sensor grids, widespread within the urban environment.

In this context, this paper has presented an innovative approach to investigate the possibility to integrate the innovative and multi-purpose Blockchain Technology in the smart city evolutionary process, and in particular in the Smart Environment and Smart Mobility by allowing renewable energy sources traceability and by providing information about the kind of energy used to refuel, for example, the selected vehicle.

In particular, a detailed description of the main advantages that could be reached by using Blockchain technology in the sustainable mobility have been presented, in particular in the Smart Environment and Smart Mobility by allowing renewable energy sources traceability and by providing information about the kind of energy used to refuel, for example, the selected vehicle. The main expected result of the proposed approach consists of more user awareness regarding the environmental and energy sustainability, so providing a reduction of the ecological impact on the whole system.

References

1. Pieroni, A., Iazeolla, G.: Engineering QoS and energy saving in the delivery of ICT services (2016)
2. Pieroni, A., Scarpato, N., Brilli, M.: Performance study in autonomous and connected vehicles a industry 4.0 issue. J. Theor. Appl. Inf. Technol. **96**(2) (2018)
3. Zanella, A., Bui, N., Castellani, L.Vangelista, Zorzi, M.: Internet of things for smart cities. IEEE Internet Things J. **1**(1), 22–32 (2014)
4. Perera, C., Zaslavsky, A., Christen, P., Georgakopoulos, D.: Sensing as a service model for smart cities supported by Internet of Things. Trans. Emerg. Telecommun. Technol. **25**(1), 81–93 (2014)
5. Pieroni, A., Scarpato, N., Brilli, M.: Industry 4.0 revolution in autonomous and connected vehicle a non-conventional approach to manage big data. J. Theor. Appl. Inf. Technol. **96**(1) (2018)
6. Manville, C., et al.: Mapping smart cities in the EU (2014)
7. Giffinger, R.: Smart cities Ranking of European medium-sized cities. **16**, 13–18 (2007)
8. Orecchini, F., Santiangeli, A.: Beyond smart grids - the need of intelligent energy networks for a higher global efficiency through energy vectors integration. Int. J. Hydrog. Energy (2011). ISSN 0360-3199. https://doi.org/10.1016/j.ijhydene.2011.01.160
9. Benevolo, C., Dameri, R.P., D'Auria, B.: Smart mobility in smart city. In: Torre, T., Braccini, A., Spinelli, R. (eds) Empowering Organizations. Lecture Notes in Information Systems and Organisation, vol 11, pp 13–28. Springer, Cham (2016). Print ISBN 978-3-319-23783-1, Online ISBN 978-3-319-23784-8
10. Zahedi, A: Smart grids and the role of the electric vehicle to support the electricity grid during peak demand. In: Lamont, L., Sayigh, A. (eds.) Application of Smart Grid Technologies, 1st edn. Case Studies in Saving Electricity in Different Parts of the World. Elsevier Inc. (2018). eBook ISBN 9780128031438, Paperback ISBN 9780128031285. https://doi.org/10.1016/B978-0-12-803128-5.00013-1. Imprint: Academic Press, 30th May 2018
11. García-Villalobos, J., Zamora, I., San Martín, J.I., Asensio, F.J., Aperribay, V.: Plug-in electric vehicles in electric distribution networks: a review of smart charging approaches. Renew. Sustain. Energy Rev. **38**, 717–731 (2014). ISSN 1364-0321. https://doi.org/10.1016/j.rser.2014.07.040

12. Zuccari, F., Santiangeli, A., Orecchini, F.: Simulation of operating conditions of a home energy system composed by home appliances and integrated PV powerplant with storage. Energy Procedia **101**, 456–463 (2016). ISSN 1876-6102. http://dx.doi.org/10.1016/j.egypro.2016.11.058

13. Pieroni, A., Scarpato, N., Di Nunzio, L., Fallucchi, F., Raso, M.: Smarter city: smart energy grid based on blockchain technology. Int. J. Adv. Sci. Eng. Inf. Technol. **8** (2018). ISSN 2088-5334

14. Alam, M.T., Li, H., Patidar, A.: Bitcoin for smart trading in smart grid. In: IEEE Workshop on Local and Metropolitan Area Networks, 2015–May (2015)

15. Orecchini, F., Santiangeli, A., Zuccari, F., Dell'Era, A.: The concept of energy traceability: application to EV electricity charging by Res. Energy Procedia **82**, 637–644 (2015). ISSN 1876-6102. https://doi.org/10.1016/j.egypro.2015.12.014

16. Orecchini, F., Santiangeli, A., Valitutti, V.: Sustainability science: sustainable energy for mobility and its use in policy making. Sustainability **3**, 1855–1865. ISSN 2071-1050, Indexed in Google Scholar. https://doi.org/10.3390/su3101855

17. Gerla, M., Lee, E.-K., Pau, G., Lee, U.: Internet of vehicles: From intelligent grid to autonomous cars and vehicular clouds. In: 2014 IEEE World Forum on Internet of Things (WF-IoT), March, 2014. Electronic ISBN 978-1-4799-3459-1. https://doi.org/10.1109/wf-iot.2014.6803166

Trend Detection Analyses of Rainfall of Whole India for the Time Period 1950–2006

Sanju R. Phulpagar[✉], Sudhansu S. Mohanta, and Ganesh D. Kale

Civil Engineering Department, Sardar Vallabhbhai National Institute
of Technology, Surat 395007, Gujarat, India
sanjupl213@gmail.com, sudhansusekhar94@gmail.com,
kale.gd@gmail.com

Abstract. Climate change is certain. Significant warming in 20th century's second half, emanated into extreme shift in hydrology of India. Evaluation of precipitation trends is critically crucial for a country such as India, whose economy and food security are based on timely water availability. Long time period instrumental rainfall series is important for climate studies. Therefore, in the present work, trend analyses of longest instrumental rainfall series of whole India is performed for annual, monthly and seasonal temporal scales for the period 1950–2006. Dependency of data is assessed by application of autocorrelation plot or correlogram. MK test or MK-CF$_2$ test is applied for evaluation of trend significance for independent or dependent data, respectively. SS test is applied for evaluation of trend magnitude. Also, ITA plot and SC are used to aid the results of trend analyses. The results showed existence of significant increasing trend in April month rainfall series.

Keywords: Whole India · Trend detection · Non-parametric tests

1 Introduction

Warming of climate system is without any ambiguity and from 1950s numerous of the observed alterations are unusual over decennium to millennia. The ocean and atmosphere have warmed, the quantities of ice and snow have reduced, sea level has increased and green house gases concentrations have increased [4]. In most contemporary studies, it was observed that, significant warming in the 20th century's second half, emanated into a extreme shift in the hydrology of an agrarian based country such as India [13]. Assessment of precipitation trends is critically crucial for a country such as India, whose economy and food security are relied on timely water availability [8]. Long-term instrumental rainfall series is important for climate studies [20]. Sontakke et al. [17] has carried out trend analyses of longest instrumental rainfall series of all India by application of 9 point Gaussian low pass filter along with truncated end points. Sontakke et al. [17] has not performed trend analyses by simultaneous assessment of statistical significance, magnitude and monotonic or non-monotonic pattern of data, as performed in the present work. Therefore, in the present work, trend analyses of longest instrumental rainfall series of whole India is carried out for annual, monthly and

© Springer Nature Switzerland AG 2019
P. Vasant et al. (Eds.): ICO 2018, AISC 866, pp. 355–361, 2019.
https://doi.org/10.1007/978-3-030-00979-3_37

seasonal (winter, pre-monsoon, monsoon and post-monsoon) temporal scales for the time period 1950–2006.

2   Data

The longest instrumental area-averaged rainfall of annual, monthly and seasonal temporal scales of whole India for duration 1813–2006 is downloaded from the 'Indian Institute of Tropical Methodology (IITM), Pune' website (ftp://www.tropmet.res.in/pub/data/rain-series/8-all_ind.txt "last accessed on June 18, 2018") [3, 11, 12, 14–17]. Longest possible instrumental area-averaged annual, seasonal and monthly rainfall series of whole India and seven homogeneous zones have been prepared from highly quality-controlled rainfall data obtained from well distributed network of 316 raingauge stations [20]. Four seasons considered in the trend analyses are: winter (January–February), pre-monsoon (March–May), monsoon (June–September) and post-monsoon (October–December) as given in the data and adopted in Kale and Nagesh Kumar [6].

3   Time Period Selected for Trend Analyses (1950–2006)

Since 1950s many of the observed alterations are unprecedented over the time scales of decades to millennia [4]. Therefore, in the present work, starting year of analysis period is 1950. Burn and Elnur [1] have proposed that, minimum 25 years of data is needed for ensuring statistical variability of the trend detection results. Therefore, in the present work, data of 57 years (1950–2006) is used for trend analyses, which is greater than 25 years, which assures the statistical validness of results of trend analyses.

4   Methodology

In the present work, 5% significance level is considered for evaluation of trend significance. In the present work, the assumption of data independency is evaluated by application of correlogram or autocorrelation plot. Non-parametric Mann-Kendall (MK) test [6, 7] or MK test with correction factor 2 (MK-CF$_2$ test) [18, 19] is applied for evaluation of trend significance for independent or dependent data respectively. Sen's Slope (SS) test [2, 5, 6] is applied for evaluation of trend magnitude. A smoothing curve (SC) can assist inference by accentuating the general association between the variables [9]. Therefore, in the present work, SC is used for inferring about relationship between the variables. Innovative trend analysis (ITA) plot is used for identification of monotonic trend, otherwise a combination of various trends or existence of trend free portions [10]. Therefore, in the present work also, ITA plot is used for evaluation of monotonic trend or non-monotonic trend or presence of trend free portions (nature of trend). So, in the present work, SC with ITA plot are applied to aid the results of trend analyses.

5   Results

In the present work, trend analyses of longest instrumental rainfall series of whole India is carried out for annual, monthly and seasonal temporal scales for 57 years (1950–2006). Correlogram or autocorrelation plot, ITA plot and SC are presented only for significant trend due to space restrictions.

5.1   Trend Analysis of Annual Rainfall Series of Whole India

Trend analysis results of annual rainfall series are given in Table 1. Significant trend is not found in annual rainfall series of whole India for the period 1950–2006.

Table 1. Trend analysis results of annual rainfall series of whole India (1950–2006)

Time series	SS value (mm/year)	Nature of trend	Statistical test	Trend is significant or not
Annual	−0.258	Non-monotonous	MK-CF$_2$	Not-significant

5.2   Trend Analyses of Seasonal Rainfall Data of Whole India

The results of trend analyses of seasonal rainfall data of whole India are given in Table 2. Significant trend is not detected in any of the seasonal rainfall series of whole India for the time period 1950–2006.

Table 2. Trend analyses results of seasonal rainfall series of whole India (1950–2006)

Time series	SS value (mm/year)	Nature of trend	Statistical test	Trend is significant or not
Winter	0.069	Non-monotonous	MK-CF$_2$	Not-significant
Pre-monsoon	0.243	Non-monotonous	MK	Not-significant
Monsoon	−1.083	Non-monotonous	MK	Not-significant
Post-monsoon	0.177	Non-monotonous	MK-CF$_2$	Not-significant

5.3   Trend Analyses of Monthly Rainfall Data of Whole India

The results of trend analyses of monthly rainfall data are given in Table 3. In April month rainfall series, significant increasing trend is detected by application of MK-CF$_2$ test while magnitude of trend as evaluated by SS test is 0.135 mm/year. The April month rainfall series is non-monotonous in nature. Significant trend is not detected in any other monthly rainfall series (other than April month rainfall series) of whole India for the time period 1950–2006.

Table 3. Trend analysis results of monthly rainfall data of whole India (1950–2006)

Time series	SS value (mm/year)	Nature of trend	Statistical test	Trend is significant or not
January month	−0.038	Non-monotonous	MK-CF$_2$	Not-significant
February month	0.132	Monotonous	MK-CF$_2$	Not-significant
March month	−0.001	Non-monotonous	MK-CF$_2$	Not-significant
April month	0.135	Non-monotonous	MK-CF$_2$	Significant
May month	0.045	Non-monotonous	MK	Not-significant
June month	0.362	Non-monotonous	MK	Not-significant
July month	−0.586	Non-monotonous	MK	Not-significant
August month	−0.349	Non-monotonous	MK-CF$_2$	Not-significant
September month	−0.328	Non-monotonous	MK	Not-significant
October month	0.065	Non-monotonous	MK	Not-significant
November month	0.069	Non-monotonous	MK	Not-significant
December month	0.010	Non-monotonous	MK	Not-significant

Autocorrelation plot of April month rainfall series for the time period 1950–2006 is shown in Fig. 1. Figure 1 shows that, corresponding data is dependent and so MK-CF$_2$ is applied to April month rainfall series.

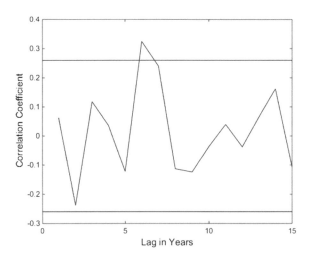

Fig. 1. Correlogram of April month rainfall time series (1950–2006)

For analyzing the nature of the trend, ITA plot of April month rainfall time series (1950–2006) is used and it is shown in Fig. 2. From Fig. 2, it is found that nature of trend is non-monotonic.

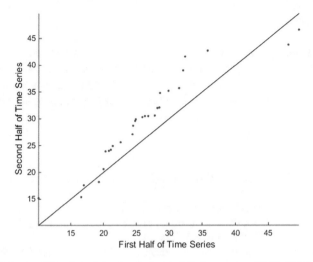

Fig. 2. ITA plot of April month rainfall time series (1950–2006)

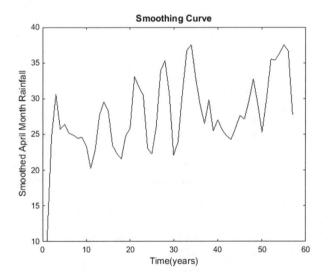

Fig. 3. SC of April month rainfall time series (1950–2006)

SC of April month rainfall time series (1950–2006) is shown in Fig. 3.

The April month rainfall time series (1950–2006) has shown significant positive trend which is aided by increasing pattern present in the data series as observed in corresponding ITA plot (as shown in Fig. 2) and increasing pattern present in data series as observed in the corresponding SC (as shown in Fig. 3) respectively.

6   Conclusion

Significant increasing trend is detected in April month rainfall time series of whole India for the period 1950–2006, if the same trend persist in the future, it may result into the flooding in April month. The attribution of significant trend detected in April month rainfall time series can be carried out in future by using appropriate attribution method.

References

1. Burn, D.H., Elnur, M.A.: Detection of hydrologic trend and variability. J. Hydrol. **255**, 107–122 (2002)
2. Deka, R.L., Mahanta, C., Pathak, H., Nath, K.K., Das, S.: Trends and fluctuations of rainfall regime in the brahmaputra and barak basins of assam, India. Theor. Appl. Climatol. (2012). https://doi.org/10.1007/s00704-012-0820-x
3. Eliot, J.: Occasional Discussions and Compilations of Meteorological Data: India and the Neighbouring Countries. India Meteorol. Mem. **14**, 709 (1902)
4. IPCC: Summary for Policy Makers. In: Stocker, T.F., Qin, D., Plattner, G.-K., Tignor, M., Allen, S.K., Boschung, J., Nauels, A., Xia, Y., Bex, V., Midgley, P.M. (eds.) Climate Change 2013: The Physical Science Basis. Contribution of Working Group I to the Fifth Assessment Report of the Intergovernmental Panel on Climate Change. Cambridge University Press, Cambridge, United Kingdom and New York, NY, USA (2013)
5. Kale, G.D., Nagesh Kumar, D.: Selection of step change and temporal trend detection tests and data processing approaches. In: Proceedings of Hydraulics, Water Resources, Coastal and Environmental Engineering (HYDRO 2014), pp. 1306–1316 (2014)
6. Kale, G.D., Nagesh Kumar, D.: Trend detection analysis of seasonal rainfall of homogeneous regions and all India, prepared by using individual month rainfall values. Water Conserv. Sci. Eng. (2018). https://doi.org/10.1007/s41101-018-0047-5
7. Khaliq, M.N., Ouarda, T.B.M.J., Gachon, P., Sushama, L., St-Hilaire, A.: Identification of hydrological trends in the presence of serial and cross correlations: a review of selected methods and their application to annual flow regimes of canadian rivers. J. Hydrol. **368**, 117–130 (2009)
8. Kumar, V., Jain, S.K., Singh, Y.: Analysis of long-term rainfall trends in India. Hydrol. Sci. J. **55**(4), 484–496 (2010)
9. Kundzewicz, Z.W., Robson, A.J.: Detecting trend and other changes in hydrological data. World Climate Programme Data and Monitoring, WMO/TD-No. 1013 (2000)
10. Sen, Z.: Innovative trend analysis methodology. J. Hydrol. Eng. ASCE **17**(9), 1042–1046 (2012)
11. Singh, N., Pant, G.B., Mulye, S.S.: A statistical package for constructing representative area-averaged rainfall series and its updating using selected stations data. In: Proceedings of National Seminar on Use of Computers in Hydrology and Water Resources, vol. 2, pp. 122–134. Indian Institute of Technology, Delhi (1991)
12. Singh, N.: Optimizing a network of stations over india to monitor summer monsoon rainfall variation. Int. J. Climatol. **14**, 61–70 (1994)
13. Sonali, P., Nagesh Kumar, D.: Review of trend detection methods and their application to detect temperature changes in India. J. Hydrol. **476**, 212–227 (2013)
14. Sontakke, N.A., Singh, N., Pant, G.B.: Optimization of the raingauges for a representative all-India and subdivisional series. Theoret. Appl. Climatol. **47**(3), 159–173 (1993)

15. Sontakke, N.A., Pant, G.B., Singh, N.: Construction of all-India summer monsoon rainfall series for the period 1844–1991. J. Clim. **6**(9), 1807–1811 (1993)
16. Sontakke, N.A., Singh, N.: Longest instrumental regional and all-India summer monsoon rainfall series using optimum observations: reconstruction and update. Holocene **6**(3), 315–331 (1996)
17. Sontakke, N.A., Singh, N., Singh, H.N.: Instrumental period rainfall series of the Indian region (1813–2005): revised reconstruction, update and analysis. Holocene **18**(7), 1055–1066 (2008)
18. Verma, R., Kale, G.D.: Trend detection analysis of gridded PET data over the tapi basin. Water Conserv. Sci. Eng. (2018). https://doi.org/10.1007/s41101-018-0044-8
19. Yue, S., Wang, C.: The Mann-Kendall test modified by effective sample size to detect trend in serially correlated hydrological series. Water Resour. Manage. **18**, 201–218 (2004)
20. Indian Institute of Tropical Methodology (IITM), Pune. ftp://www.tropmet.res.in/pub/data/rain-series/8-all_ind.txt

Multi-criteria Decision Making Problems in Hierarchical Technology of Electric Power System Expansion Planning

N. I. Voropai[✉]

Melentiev Energy Systems Institute, Irkutsk 664033, Russia
voropai@isem.irk.ru

Abstract. This paper deals with expansion planning problem of large electric power systems. The initial complicate multi-criteria problem is presented as hierarchical set of step-by-step solved sub-problems. Each sub-problem is characterized its own number of criteria which usually is only the part of initial set of criteria. Qualitative illustration of relations between criteria of initial problem and sub-problems is presented. Some generalization of results is given.

Keywords: Electric power system · Expansion planning
Multi-criteria problem

1 Introduction

Expansion planning of large electric power systems (EPS) embracing vast territories is very difficult problem due to great number of significant factors. Current technology for large EPS expansion planning includes several groups of sub-problems which specify the structure and operation of EPS stage by stage. For example, the first stage deals with determination of the necessary number and types of generation units and their locations, the second is selection of new transmission lines of the main grid, the third stage deals with the study the reliability and operating conditions of EPS variants, the last is determination of the principles and structure of EPS control [1, 2, etc.]. The paper [3] suggests the overview of current state of the problem of EPS expansion planning. Liberalization and deregulation conditions, and market specifics in the EPS expansion planning process are discussed. Modern models and methods for generation and transmission expansion planning are analysed. Holistic planning which is based on system ideology for market environment is discussed.

Taking into account above mentioned multi-stage process of EPS expansion planning, the paper [4] presents formalized interpretation of a hierarchical technology for planning of large EPS expansion. The stages of this hierarchical technology are interrelated by transformation procedures of EPS variants, system models, their parameters and chosen criteria. Suggested hierarchical technology considers general formalized approaches to hierarchical design schemes and numerical decomposition methods [5–8, etc.].

© Springer Nature Switzerland AG 2019
P. Vasant et al. (Eds.): ICO 2018, AISC 866, pp. 362–368, 2019.
https://doi.org/10.1007/978-3-030-00979-3_38

This paper deals with new aspects of interrelations between the hierarchical technology of large EPS expansion planning and structure of sets of criteria on each stage of hierarchical process of decision making. Chapter 2 presents main ideas of hierarchical technology following [4]. Chapter 3 suggests possible significant criteria for each subject of relations (stakeholder), which is involved into decision making process by EPS expansion planning. Chapter 4 includes the discussion about structures of criteria on each stage of hierarchical expansion planning technology and interrelations between sets of criteria. This discussion is based on consideration real criteria, which are used usually by expansion planning of large EPS. Chapter 5 deals with some generalization of presented results. And Chap. 6 has concluding remarks.

2   Main Ideas of Hierarchical Technology

Usually EPS expansion planning problem consists in choice of the most preferable system variant (scenario of expansion) from a set of alternatives with respect to different criteria and in determining the most preferable values of system parameters for the chosen variant.

Thus, let $X = \{X_1, X_2, \ldots\}$ be a set of alternatives (system variants); $x \in X$, $X_i = \{x_{i1}, x_{i2} \ldots\}$ – a set of system parameters; $\Phi = \{\Phi_1, \Phi_2, \ldots\}$ – a set of preference relations when making choice, which are determined by the problem content, composition and essence of criteria. Then the problem of choice can be formulated in sufficiently general form as

$$X_o = opt(X, \Phi), \quad x_o = opt(x, X, \Phi), \tag{1}$$

where opt means the preference mentioned before, rationality or optimality of choice though it can be any other kind of choice procedure.

The required scale of detailed mathematical description of large EPS often is extremely high and initial problem of choice in form (1) appears to be practically unsolvable.

Introduce $m + 1$ levels of initial problem description. To determine these levels let us first of all determine sets of preference relations at each level as

$$\Phi^m \rightarrow \Phi^{m-1} = \varphi_{m-1}(\Phi^m) \rightarrow \cdots \rightarrow \Phi^0 = \varphi_0(\Phi^1). \tag{2}$$

Arrows in (2) point refinement of preference relations from the upper level of description to the lower one, their itemization with respect to composition and content of sub-problems, criteria, key (optimized) parameters, etc. Practically the chain (2) of the sets of preference relations is formulated on the basis of researcher's knowledge and experience in the considered area.

Let us introduce an interrelated aggregate of the system models which present the system structures and states:

$$x^0 \rightarrow x^1 = opt\left(f_1\left(x^0 \Phi^1\right)\right) \rightarrow \cdots \rightarrow x^m = opt\left(f_m\left(x^{m-1}, \Phi^m\right)\right). \tag{3}$$

The arrow motion in (3) means a sequential stepwise aggregation of the system model which can be made optimally in a general case. It is understood that the system model at some description level i and a set of preference relations at the same level should correspond to each other.

Introduce a sequential disaggregation of the system model taking into account (3) and on its basis:

$$x^m \rightarrow x^{m-1} = opt\left(f'_{m-1}\left(x^m, \Phi^{m-1}\right)\right) \rightarrow \cdots \rightarrow x^0 = opt\left(f'_0\left(x^1, \Phi^0\right)\right). \tag{4}$$

The conceptual sense of (4) is similar to (3).

Now let us solve the sequence of choice sub-problems:

$$x_o^m = opt(f'_m(x^m, \Phi^m); F_m(X, \Phi^m))$$
$$\downarrow$$
$$x_o^{m-1} = opt\left(f'_{m-1}\left(x_o^m, \Phi^{m-1}\right); F_{m-1}\left(X^m, \Phi^{m-1}\right)\right),$$
$$\downarrow$$
$$\vdots$$
$$\downarrow$$
$$x_o^0 = opt\left(f'_0, \left(x_o^1, \Phi^0\right); F_0\left(X^1, \Phi^0\right)\right) \tag{5}$$

In (5) F means transformation of a set of alternatives, whose composition can change from the upper level to the lower one, both by eliminating some of them and by considering possible additional alternatives at the intermediate levels. Thus, transformation of alternatives can be written in the form:

$$X_o^m = F_m(X) \rightarrow X_o^{m-1} = F_{m-1}\left(X_o^m\right) \rightarrow \cdots \rightarrow X_o^0 = F_0\left(X_o^1\right). \tag{6}$$

As the result, we determined the best alternative X_o^0 and its the best parameters x_o^0.

3   Different Criteria of Choice

We deal with an analysis of different possible variants in organizational structure of power industry [9]. These variants include: a regulated monopoly at all levels; interaction of vertically integrated EPS at an open access to the main grid; a single buyer-seller of electricity (an electric network company) with competition of generating companies; competition of generating companies and free choice of electricity supplier by selling companies or/and consumers, when the main grid renders only transportation services; in addition to conditions of two previous cases – competition of selling companies in electricity supply to concrete consumers; intermediate and mixed variants based on considered ones.

In decision making process on EPS expansion different groups of subjects of relations have own, largely non-coincident, interests, that are expressed by the corresponding criteria [10]:

(1) Electricity producers or/and sellers (vertically integrated, generating or selling companies, an electric network company as the single buyer-seller of electricity) and also subjects of power industry rendering electric power services in the electricity market (maintenance of active and reactive power reserves, provision of system reliability, etc.) are interested in own profit maximization as a result of their business.

(2) Electricity consumers (selling companies of different levels, concrete consumers) are interested in minimization of the price for electricity bought (in the wholesale or/and retail markets), provision of electricity quality and power supply reliability.

(3) Interests of authorities (federal and regional) are related with maximization of payments into budgets of the corresponding levels, minimization of the environmental impact of electric power facilities, provision of the energy security of the country and regions, etc.

(4) External investors (banks, juridical and natural subjects) are interested in minimization of the period for return of their investments in electric power installations, maximization of dividends, etc.

Above mentioned criteria we will use in the following analysis.

4 Qualitative Case Study

The general problem of EPS expansion planning can be divided into three following groups of problem [10]:

- The state strategies and programs for development of electric power industry and EPS (the federal, interregional or regional levels);
- Strategic plans for development of power supply companies (vertically integrated, generating or network, etc.);
- Investment programs and projects of electric power installations (power plants, substations, transmission and distribution lines.

Let us take into consideration the state strategy for development of the Unified Energy System (UES) of Russia as electric power infrastructure of federal level. The state strategy determines the sets of new power plants and transmission lines, which are necessary to build for expansion of the UES. The strategy gives also the state policy for expansion of such an important infrastructural system and mechanisms of realization for this policy.

The state strategy has to take into account all significant interests and corresponding criteria for each subject of relations in the process of this complicate system expansion planning. Let us consider the following criteria for initial statement of the UES expansion planning problem in general form (1):

1. Investment costs;
2. Current costs;
3. Budget taxes;
4. Ecology penalties;
5. Electricity price;
6. Reliability of power supply;
7. Stability and survivability of the UES.

The UES expansion planning problem is examined usually as following hierarchy of sub-problems [10]:

(A) Determination of optimal volume of new power plant installations and their locations on different parts of territory;
(B) Determination of optimal development of main electrical grid;
(C) Study and ensuring of power supply reliability for consumers;
(D) Operating conditions study of future UES (system stability, transfer capability margins of ties, requirements to emergency control system development, etc.);
(E) Comprehensive analysis of recommended decisions for UES expansion.

As for main subjects of relations, taking into account current organizational structure of Russian electric power industry, we will consider:

I.   Generating companies;
II. Main network company;
III. Consumers;
IV. State government.

Generating companies have the interest in realization of sub-problem (A) by minimization of following criteria: investment costs (1.) into new generating units, current costs (2.) because operation of all generating units in expanded UES, budget taxes (3.) from new generating installations and ecology penalties (4.) due to operation of expanded set of generating units. Main network company is interested in minimization of investment costs (1.) for new transmission lines, current costs (2.) which deals with operation of expanded main network and budget taxes (3.), by optimization of main network development (sub-problem B). As for consumers, these subjects of relations are interested in maximization of power supply reliability (criterion 6.) by solving sub-problem (C) and in minimization of electricity price (criterion 5.) which is estimated during realization of sub-problem E). And last but not least, governmental bodies are interested in maximization of budget taxes (criterion 3.) by solving sub-problems (A), (B), (C), stability and survivability of the UES (criterion 7.) by realization of sub-problem D).

We can see, that the initial problem in general form (1) has 7 criteria, but each sub-problem includes reduced set of criteria: sub-problem (A) – 4, sub-problem (B) – 3, sub-problem (C) – 2, sub-problem (D) – 1, sub-problem (E) – 1. As the result, each sub-problem is more simply for decision making then initial problem. Next step of sub-problem simplification can be in transformation of some criteria into equivalent constraints. Such a transformation is possible, for example, for criteria 6 and 7. One more positive result of this initial problem decomposition consists in separation of different kinds of models on different sub-problems which relate each other by aggregation-disaggregation procedures (3) and (4) – see Chap. 2.

5 Generalization of Results

In more general case, the need can be to introduce new criteria (which are not required so far or can hardly be taking into account at the upper level), solve new additional sub-problems, separate new optimized parameters, etc., i.e. the need is to extend the initial set of preference relations, can arise at the intermediate levels of the problem description. For instance, sometimes the strategy of EPS expansion does not include above mentioned sub-problem (D), which we considered for UES expansion strategy. The inclusion of this sub-problem into consideration relates to sense of suggested discussion.

On the other hand, it appears to be unnecessary to consider all criteria and sub-systems of the initial set of preference relations Φ at the upper level m. At the same time, the relation $\Phi^{m-1} = \varphi_{m-1}(\Phi^m)$ in (2) means that at level $m - 1$ in description of the set of preference relations Φ^{m-1}, account is taken of the fact that the corresponding set Φ^m has been already determined at level m and, hence, the description of Φ^{m-1} can be simplified. Thus, each set of preference relations Φ^i turns out to be simpler than the initial set. At the same time, the composition and content of sets Φ^i at all levels of the problem description become richer than the initial set Φ.

As the result, in discussed sense the considered hierarchical scheme of choice turns out to be richer than the direct solution of initial problem (1). The final alternative X_o^0 by (6) and its parameters x_o^0 by (5), in principle, are differ from X_o and x_o in (1) (if problem (1) could be solved). Since at the intermediate levels of the problem description composition and content of preference relations were specified, a set of the considered alternatives can be extended. On the whole, integral description (model) of the system turns out to be richer than at the direct problem statement in form (1).

The presented hierarchical scheme of choice can be modified, for instance, in case of a hierarchical structure of the planned system, when it consists of some relatively independent sub-systems, and the problem of choice first is solved for the system as a whole as a coordination one and then each sub-system specifies its choice based on its own criteria as well. Passing from the upper level to the lower one the problems (5) and (6) are parallelized at some level among independent partners (sub-systems) and are solved by them independently. Then auxiliary operations (2–4) require the corresponding parallelizing.

6 Conclusions

Representation of technology for electric power system expansion planning as a hierarchical multi-level procedure allows the system variants under development with their parameters to be connected in a single logical sequence from level to level, and also transformations of EPS models to be represented as sequences of aggregation and disaggregation.

This enables one to structurize and put in order a set of problems being solved at electric power system expansion planning, and formulate quite clearly the requirements to the system of methods realizing such a hierarchical technology of EPS expansion planning.

Several objective advantages of discussed hierarchical technology are:

- In many cases this technology is constructive way to solve initial complicate problem in the comparison with practically unsolvable initial statement;
- Hierarchical technology has the possibility to consider many significant factors more detaily then by initial statement;
- Each level of hierarchical technology includes reduced sub-set of criteria based on general set of them, what is more realistic for solving considered sub-problem;
- As the result, the decision which we have after hierarchical technology application is more deeper and richer than the direct solution of initial complicate problem.

Acknowledgement. This study was performed for the project III.17.4.2 in the program of basic researches of Siberian Branch of the Russian Academy of Sciences #AAAA-A17-117030310438-1.

References

1. Sullivan, R.: Power System Planning. McGraw Hill, New York (1977)
2. Belyaev, L., Voitsekhovskaya, G., Saveliev, V.: System Approach in the Management of Power Industry Development. Nauka, Novosibirsk (1980, in Russian)
3. Voropai, N.: Power system expansion planning – State of the problem. Global J. Technol. Optim. **8**(2), 1–7 (2015)
4. Voropai, N.: Hierarchical technology for electric power system expansion planning. In: IEEE Budapest Power Tech. Budapest Proceedings, Hungary, pp. 131–135 (1999)
5. Mesarovic, M., Macko, D., Takahara, Y.: Theory of Hierarchical Multilevel Systems. Academic Press, New York, London (1970)
6. Shahidehpour, M., Fu, Y.: Benders Decomposition in Restructured Power Systems. IEEE Tutorial, New York (2005)
7. Krasnoschekov, P., Morozov, V., Popov, N.: Hierarchical design schemes and numerical decomposition methods. Izvestiya RAN, Teoriya i Sistemy Upravleniya, (5), 80–89 (2001, Russian)
8. Novikov, D., Petrakov, S., Fedchenko, K.: Decentralization of planning mechanisms in active systems. Remote Control **6**, 143–155 (2000)
9. Hant, S., Shuttleworth, G.: Competition and Choice in Electricity. March & McLennan, Chichester, England (1995)
10. Voropai, N., Podkovalnikov, S., Trufanov, V.: Electric Power System Expansion Planning: Methodology, Models, Methods, their Applications. Nauka, Novosibirsk (2015, in Russian)

Insight into Microstructure of Lead Silicate Melts from Molecular Dynamics Simulation

Thanh-Nam Tran[1(✉)], Nguyen Van Yen[4], Mai Van Dung[2,3], Tran Thanh Dung[3], Huynh Van Van[1], and Le The Vinh[1]

[1] Faculty of Electrical and Electronics Engineering, Ton Duc Thang University, No. 19 Nguyen Huu Tho Street, Tan Phong Ward, District 7, Ho Chi Minh City, Vietnam
{tranthanhnam, huynhvanvan, lethevinh}@tdtu.edu.vn
[2] Institute of Applied Materials Science, Vietnam Academy of Science and Technology, No. 01 TL29 Street, Thanh Loc Ward, District 12, Ho Chi Minh City, Vietnam
[3] Thu Dau Mot University, No. 6, Tran van on Street, Phu Hoa Ward, Thu Dau Mot City, Binh Duong Province, Vietnam
[4] Hanoi University of Science and Technology, No. 1, Dai Co Viet Street, Hanoi, Viet Nam

Abstract. In this study, the analysis on structure has been performed for lead silicate ($PbSiO_3$) melt. The structural heterogeneity of the melt was analyzed via the void-simplex, cation-simplex and oxygen-simplex. The densification of the melt is obtained by decreasing the radius and changing the number of void-simplex. We show that a number of large interstitial site for oxygen and cation present in the liquid.

Keywords: Molecular dynamics · Structure heterogeneity · Lead silicate melts

1 Introduction

Lead silicate glass $PbSiO_3$ are widely used in a variety of applications such as photonics and the electronic industry, low-melting glassy materials, and radiation shielding materials [1–5]. This type of glass, thus, has been the subject for numerous theoretical and experimental studies in the recent decades. The structure of lead silicate glass has been investigated by NMR [6–10], x-ray and infrared (IR) spectroscopy [11–21], and molecular dynamics simulation [22–26]. Although many efforts have been made to clarify the local environment of the Pb+2 ion, the role of Pb+2 in a silicate network, the structural characteristics, the polymorphism and dynamics in $PbSiO_3$ has still been in debate. Imaoka et al. [14] proposed a model consisting of chains or ribbons of PbO_3 trigonal pyramids. Wang and Zang [15] suggested that in the structure where PbO content was up to 40 mol%, SiO_2 is the main glass former, and polymeric chains of PbO_4 pyramids are connected through silicate tetrahedra in the glass. In addition, Fayon et al. [6] reported that Pb atoms form covalent PbO_4 and PbO_3 pyramids over a large compositional range. Meanwhile, Rybicki et al. [21] suggested that the PbO_4 groups are the dominant structural units for any concentration of PbO, and at low

© Springer Nature Switzerland AG 2019
P. Vasant et al. (Eds.): ICO 2018, AISC 866, pp. 369–375, 2019.
https://doi.org/10.1007/978-3-030-00979-3_39

concentrations, co-existence of the PbO_4 and PbO_3 units is possible. They also stated that the connectivity of the silicate network is broken at around 45% of PbO content, and, finally, they reported that the PbO_n pyramids form a continuous network, mainly edge-sharing, even at relatively low PbO concentrations (below 20%).

In this study, the numerical analysis of the melted lead silicate has been performed in order to characterize the structural heterogeneity of the melt. The validity of the analytical model has been judged by comparing the obtained results with previous works.

2   Computational Method

Molecular dynamic (MD) simulation has been carried out for lead silicates systems (5000 atoms) at temperature of 3200 K and pressure in the range of 0–35 GPa. In this simulation, the Born—Mayer potential has been used [23, 24]. The software used for calculation, analysis, and visualization was developed by the authors. It was written in C language and run on Linux platform. We use the Verlet algorithm to integrate the equations of motion with an MD step of 1.6 fs. This value assures the requirement for accurately integrating the Newtonian equations of motion in order to track atomic trajectories, and keeping the computational cost reasonable. The initial configuration has been obtained by randomly placing all atoms in a simulation box. This sample is equilibrated at a temperature of 6000 K for about 105 MD steps, and then it is compressed to the pressure ranging from 0 to 35 GPa and relaxed for about 106 MD steps. After that the models at different pressures are cooled down to the temperature of 3200 K with the rate of 1013 Ks^{-1}. A consequent long relaxation (about 107 MD steps) has been done in the NPT ensemble (constant temperature and pressure) to obtain the equilibrium state. To calculate dynamical properties, we use an NVE ensemble. In order to improve the statistics, the measured quantities such as the coordination number and partial radial distribution function are computed by averaging over 1000 configurations separated by 10 MD steps.

Micro-heterogeneity of silicate liquid can be clarified through overcoordinated TO_x or O_n unit. The result obtained by this way provides useful and interesting information about the network structure. For instance, many works [27–31] establish the relationship between dynamics and degree of polymerization. However the structural heterogeneity is not only described by over-coordinated unit, but also concerns the fluctuation in free volume and chemical composition. To calculate the simplex all sets of four atoms are taken and a circum-sphere (CSP) of the tetrahedron forming by those atoms has been drawn. Let R_{CSP} be the radius of CSP. Then we test the condition that the CSP does not contain any atom. Next we calculate the number of atoms N_{CSP} which are determined as a number of atoms located from the center of CSP at distance of $R_{CSP} \pm 0.1$ Å. Such, a set of the N_{CSP}-simplexes is found in the constructed model. To clarify those issues we conduct the analysis on void-simplex, oxygen-simplex and cation-simplex as shown in Fig. 1. The void-simplexes provide the spatial distribution of void for the liquid. Two other simplexes give the space regions where only oxygen or cation atoms present.

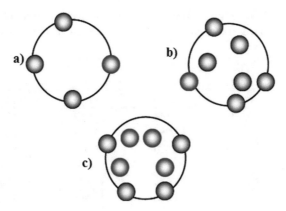

Fig. 1 The schematic illustration of simplex: The void-simplex (a); the oxygen-simplex (b); the cation-simplex (c). The black, blue sphere represents the oxygen and cation, respectively; the black circle represents the simplex.

3   Results and Discussion

To assure the reliability of MD models of liquid $PbSiO_3$, we have investigated pair radial distribution functions $g_{Pb-Pb}(r)$, $g_{Pb-O}(r)$ and $g_{O-O}(r)$. As shown in Table 1, the simulation result is in good agreement with experimental data [13, 21, 24] in terms of the position of the first peak of PRDF.

Table 1   The structure characteristics of lead silicate. r_{Pb-Pb}, r_{Pb-O}, r_{O-O} is the position of the first peak in PRDF.

Bond-length	This work	Experimental works		
		[13]	[21]	[24]
r_{Pb-Pb}, Å	3.64	3.86	-	3.82
r_{Pb-O}, Å	2.26	2.73	2.30	2.27
r_{O-O}, Å	2.64	-	2.62	-

The characteristics of void-simplex detected in two configurations are presented in Table 2. The 130 type which consists of one oxygen, three silicon and zero lead, has a largest m_{VS} corresponding to the radii of 2.48 Å. The radius of 112 and 040 void-simplex are decreased by 18.78–29.92%. From Table 1 it can be seen that the densification of melt is caused by decreasing the radius of void-simplex. This result is consistent with the distribution of simplex radius shown in Fig. 2.

As depicted in Fig. 2, the graph shifts to the left for high-pressure configuration. Moreover, the low-pressure configuration contains a number of large void-simplex with radius bigger than 2.5 Å. In contrast, the amount of these void-simplexes in high-pressure configuration is almost equal to zero. The densification also leads to the increase of m_{VS} for 022, 031, 121 and 220 void-simplex. However, the m_{VS} for 040,

Table 2. The characteristics of void-simplex; m_{VS} is the average number of void-simplex per atom; R_{VS} is the mean radius of void-simplex. a, b and c is the number of oxygen, silicon and lead, respectively.

abc	Low-pressure configuration		High-pressure configuration	
	m_{VS}	R_{VS}, Å	m_{VS}	R_{VS}, Å
022	**0.029**	**2.160**	**0.063**	**1.650**
031	**1.375**	**2.250**	**1.658**	**1.680**
040	0.776	2.640	0.332	1.850
112	-	-	0.002	2.000
121	**0.989**	**2.170**	**1.228**	**1.700**
130	2.360	2.480	1.806	1.830
211	0.020	2.450	0.016	1.990
220	**1.290**	**2.370**	**1.403**	**1.830**
310	0.061	2.540	0.049	1.960

The bold number indicates the void-simplex which has large number m_{VS}

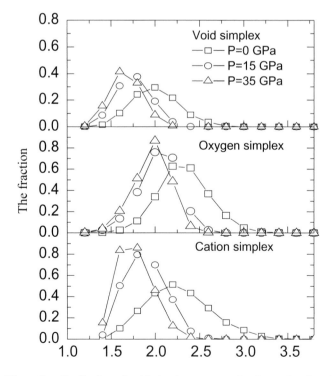

Fig. 2 The radius distribution of void-simplex, oxygen-simplex and cation-simplex.

130 and 310 type is decreased. Thus the major contribution to the densification is due to the decrease in the radius of 022, 031, 121, 220 and in the number of those void-simplex. Moreover, the number of large void-simplex significantly decreases upon compression.

The characteristics of atom-simplex are presented in Table 3. The number of atom varies from 2 to 9 for oxygen-simplex, and from 1 to 4 for cation-simplex. The radius of atom-simplex also reduces upon compression. In particular, the radius of 002, 003, 011, 300 and 400 types is decreased by 15.87–26.43%. As shown in Fig. 2, the

Table 3. The characteristics of atom-simplex; m_{VS} is the average number of atom-simplex per atom; R_{VS} is the mean radius of void-simplex. a_1, b_1 and c_1 is the number of oxygen, silicon and lead, respectively.

$a_1b_1c_1$	0 GPa		35 GPa		$a_1b_1c_1$	0 GPa		35 GPa	
	m_{VS}	R_{VS}, Å	m_{VS}	R_{VS}, Å		m_{VS}	R_{VS}, Å	m_{VS}	R_{VS}, Å
200	0.125	2.470	0.240	2.000	011	0.533	2.280	0.561	1.790
300	0.711	2.560	0.995	2.000	010	0.002	2.310	0.002	2.020
400	0.759	2.740	0.731	2.090	012	0.001	2.800	0.002	2.060
500	0.480	2.940	0.180	2.220	021	0.032	2.570	0.019	2.060
600	0.215	3.120	0.030	2.310	020	0.607	2.330	0.684	1.870
700	0.061	3.250	0.003	2.400	030	0.080	2.670	0.057	2.080
800	0.011	3.410	-	-	040	0.003	2.710	0.002	2.280
900	0.003	3.720	-	-	002	0.011	2.330	0.014	1.870
-	-	-	-	-	001	0.122	2.230	0.132	1.720

densification of the melt leads to disappearing the atom-simplex with the radius bigger than 2.5 Å.

Atom-simplex is the interstitial site for oxygen and cation. We regard the large site to atom-simplex which consists of more 6 oxygen or 3 cation atoms. For the low-pressure configuration the average number of large site per atom for oxygen and cation is 0.003 and 0.215, respectively. Because the cation has positive charge, hence if cation simplexes locate nearby, then a positive-charged cluster is formed.

The negative-charged cluster is a set of large site for oxygen where two adjacent sites have at least a common oxygen atom. By analogy the positive-charged cluster is formed by cation-simplexes.

4 Conclusions

An analysis on structure and dynamics of melted LS2 is carried out. The structure of the melt is analyzed through SC-particle and SC-cluster which consists of oxygen-part and cation-part. Adjacent SC-clusters do not have common cation, but may have common oxygen. It is shown that the densification of the melt is accompanied with decreasing the radius of core of SC-particle and number of large SC-particle.

Further, we show that the liquid comprises two types of SC-cluster. Most Si atoms belong to first type, while second type contains the majority of Pb atoms. Large SC-clusters tend to locate nearby forming space regions which represent the diffusion pathway for Pb.

Acknowledgment. This research is funded by Vietnam National Foundation for Science and Technology Development (NAFOSTED) under grant number 103.05-2017.345.

References

1. Ross, M., Stana, M., Leitner, M., Sepiol, B.: N. J. Phys. **16**, 093042 (2014)
2. Sundara Rao, M., Sanyal, B., Bhargavi, K., Vijay, R., Kityk, I.V.: J. Mol. Str. **1073**, 174–180 (2014)
3. Kopyto, M., Przybylo, W., Onderka, B., Fitzner, K.: Arch. Metall. Mater. **54**, 811–822 (2009)
4. Kharita, M.H., Jabra, R., Yousef, S., Samaan, T.: Radiat. Phys. Chem. **81**, 1568–1571 (2012)
5. Dalby Kim, N., et al.: Geochim. Cosmochim. Acta **71**, 4297–4313 (2007)
6. Fayon, F., Bessada, C., Massiot, D., Farnan, I., Coutures, J.P.: J. Non-Cryst. Solids **232–234**, 403 (1998)
7. Shrikhande, V.K., Sudarsan, V., Kothiyal, G.P., Kulshreshtha, S.K.: J. Non-Cryst. Solids **283**, 18 (2001)
8. Sudarsan, V., Shrikhande, V.K., Kothiyal, G.P., Kulshreshtha, S.K.: J. Phys. Condens. Matter **14**, 6553 (2002)
9. Fayon, F., Landron, C., Sakurai, K., Bessada, C., Massiot, D.: J. Non-Cryst. Solids **243**, 39 (1999)
10. Feller, S., et al.: J. Non-Cryst. Solids **356**, 304–313 (2010)
11. De Sousa, Meneses D., Malki, M., Echegut, P.: J. Non-Cryst. Solids **352**, 769–776 (2006)
12. Bhargavi, K., et al.: Opt. Mater. **36**, 1189–1196 (2014)
13. Takaishi, T., Takahashi, M., Jin, J., Uchino, T., Yoko, T.: J. Am. Ceram. Soc. **88**, 1591 (2005)
14. Imaoka, M., Hasegawa, H., Yasui, I.: J. Non-Cryst. Solids **85**, 393 (1986)
15. Wang, P.W., Zhang, L.: J. Non-Cryst. Solids **194**, 129 (1996)
16. Manceau, A., et al.: Environ. Sci. Technol. **30**, 1540 (1996)
17. Choi, Y.G., Kim, K.H., Chernov, V.A., Heo, J.: J. Non-Cryst. Solids **246**, 128 (1999)
18. Witkowska, A., Rybicki, J., Trzebiatowski, K., Di Cicco, A., Minicucci, M.: J. Non-Cryst. Solids **276**, 19 (2000)
19. Hoppe, U., Kranold, R., Ghosh, A., Landron, C., Neuefeind, J., Jovari, P.: J. Non-Cryst. Solids **328**, 146 (2003)
20. Mastelaro, V.R., Zanotto, E.D., Lequeux, N., Cortes, R.: J. Non-Cryst. Solids **262**, 191 (2000)
21. Rybicki, J., et al.: J. Phys. Condens. Matter **13**, 9781 (2001)
22. Rybicki, J., Ala, W., Rybicka, A., Feliziani, S.: Comput. Phys. Commun. **97**, 191 (1996)
23. Rybicka, A., Rybicki, J., Witkowska, A., Feliziani, S., Mancini, G.: Comput. Methods Sci. Technol. **5**, 67 (1999)
24. Witkowska, A., et al.: J. Non-Cryst. Solids **351**, 380–393 (2005)
25. Hemesath, E., Corrales, L.R.: J. Non-Cryst. Solids **351**, 1522–31 (2005)
26. Chomenko, K., et al.: Comput. Method Sci. Technol. **10**, 21–38 (2004)

27. Narayanan, B., Reimanis, I.E., Ciobanu, C.V.: Atomic-scale mechanism for pressure-induced amorphization of b-eucryptite. J. Appl. Phys. **114**, 083520 (2013)
28. Okuno, M., Zotov, N., Schmucker, M., Schneider, H.: Structure of SiO_2–Al_2O_3 glasses: combined X-ray diffraction, IR and Raman studies. J. Non-Cryst. Solids **351**, 1032–1038 (2005)
29. Poe, B.T., Mcmillan, P.F., Angell, C.A., Sato, R.K.: Al and Si coordination in SiO_2–Al_2O_3 glasses and liquids: a study by NMR and IR spectroscopy and MD simulations. Chem. Geol. **96**, 333–349 (1992)
30. Zhang, Z., Zheng, K., Yang, F., Sridhar, S.: Molecular dynamics study of the structural properties of calcium aluminosilicate slags with varying Al_2O_3/SiO_2 ratios. ISIJ Int. **52**, 342–349 (2012)
31. Takei, T., Kameshima, Y., Yasumori, A., Okada, K.: Calculation of metastable immiscibility region in the SiO_2–Al_2O_3 system using molecular dynamics simulation. J. Mater. Res. **15**, 186–193 (2000)

A Strategy for Minimum Time Equilibrium Targetting in Epidemic Diseases

Manuel De la Sen[1]([✉]), Asier Ibeas[2], Santiago Alonso-Quesada[1], and Raul Nistal[1]

[1] Faculty of Science and Technology, Department of Electricity and Electronics, University of the Basque Country, Campus of Leioa, Bizkaia, Spain
{manuel.delasen, santiago.alonso}@ehu.eus,
raul.nistal@gmail.com
[2] Department of Telecommunications and Systems Engineering, Universitàt Autònoma de Barcelona, UAB, Barcelona, Spain
Asier.Ibeas@uab.cat

Abstract. This paper relies on a minimum-time vaccination control strategy for a class of epidemic models. A targeted state final value is defined as a certain accuracy closed ball around some point being a reasonable approximate measure of both disease- free equilibrium points associated with the two vaccination levels used for the optimal- time control.

Keywords: Vaccination control · Bang-bang time-optimal control
Hamiltonian

1 Introduction

Biological models have received an important attention in the last years because of their importance in modeling real life problems of interest. For instance, the so-called Beverton-Holt Equation, describing the growth rates of some species which reproduce by eggs has been studied from a control theory point of view in [1, 2, 4]. Other biological models of interest are the epidemic models which describe the evolution of infectious diseases in humans, animals and plants. See, for instance, [5, 6] and references therein. Due to the nature of the biological systems, the property of positivity of the solutions under non-negative initial conditions is a requirement to be satisfied so that the model be efficient to describe a real disease. On the other hand, it is well-known that optimization techniques are important tools in the sense that the associated optimal controls minimize a suitable loss function compared to alternative choices of the controls or suboptimal controls. See, for instance, [6–9]. Typical optimal controls are the optimal regulation control, which optimizes a predesigned trade-off between the weighted states or outputs and the weighted controls, the optimal fuel consumption and the optimal time control which minimizes the necessary time to reach a targeted final state. In the case of analytical or computational difficulties for solving the optimization problems because of computational difficulties or a lack of information of the real problem to derive a sufficiently accurate model, suboptimization techniques can be pursued to search for near-optimal control solutions. Time-optimal controls are

© Springer Nature Switzerland AG 2019
P. Vasant et al. (Eds.): ICO 2018, AISC 866, pp. 376–382, 2019.
https://doi.org/10.1007/978-3-030-00979-3_40

associated with bang- bang control laws implying control switches at appropriate time instants, [7]. In fact a minimum-time control is always of a bang-bang nature. One of the typical optimization objectives is the design of optimal-time controls to reach a targeted final state, usually, an equilibrium point. The main purpose of this research is focused to the time optimal controls of an SEIR (susceptible/exposed/infectious and recovered sub-populations) epidemic model to reach certain region surrounding eventual equilibrium points. This strategy might guarantee prescribed small levels of infection in a reasonably small finite time.

2   Epidemic Model

Consider the SEIR epidemic model:

$$\dot{S}(t) = \mu - \mu S(t) - \beta I(t)S(t) - V(t); \quad S(0) = S_0 \tag{1}$$

$$\dot{E}(t) = -(\mu + \sigma)E(t) + \beta I(t)S(t); \quad E(0) = E_0 \tag{2}$$

$$\dot{I}(t) = -(\mu + \gamma)I(t) + \sigma E(t); \quad I(0) = I_0 \tag{3}$$

$$\dot{R}(t) = \gamma I(t) - \mu R(t) + V(t); \quad R(0) = R_0 \tag{4}$$

where S, E, I and R, denote respectively, the susceptible, exposed, infectious and recovered subpopulations, V is the vaccination effort and the admissible initial conditions satisfy $min(S_0, E_0, I_0, R_0) \geq 0$. The parameters are:

- μ is the natural mortality rate. Its inverse is the natural host lifespan,
- β is the disease transmission rate,
- σ is the disease latency rate. Its inverse is the average latent period,
- γ is the removal or recovery rate. Its inverse is the average infectious period.

By summing-up both sides of the above equation, one gets that the total population $N(t) = S(t) + E(t) + I(t) + R(t); \forall t \in \mathbf{R}_{0+}$ satisfies the differential equation:

$$\dot{N}(t) = \mu - \mu N(t); N(0) = N_0 = S_0 + E_0 + I_0 + R_0. \tag{5}$$

As a result of the subsequent discussion, it turns out that the above model describes the evolution of normalized subpopulations provided that $N_0 = 1$. The disease- free equilibrium point and the endemic equilibrium one are now specified and discussed:

Proposition 1. Assume that

(1) $V : \mathbf{R}_{0+} \rightarrow [0, \mu]$,
(2) $V(t) \rightarrow V^*$ as $t \rightarrow \infty$.

Then, there the solution trajectory has non-negative subpopulation solutions for all time under any non-negative initial conditions and there exists a unique disease- free equilibrium defined by

$$x^* = (S^*, E^*, I^*, R^*)^T = \left(\frac{\mu - V^*}{\mu}, 0, 0, \frac{V^*}{\mu}\right)^T \tag{6}$$

and the associated disease-free total population is $N^* = 1$.

If the limit vaccination is zeroed then $N^* = S^* = 1$ and, if $N_0 = 1$, then $N(t) = 1$ for $t \geq 0$.

If $V^* = K_V^* S^* + V_0$ then the susceptible and the recovered subpopulations are, respectively, $\frac{\mu - V_0}{\mu + K_V^*}$ and $K_V^* + V_0$.

Note from Proposition 1 that the vaccination action decreases the equilibrium susceptible subpopulation. The proof of Proposition 1 is direct by zeroing the time-derivatives in (1)–(4) with zero equilibrium values of the exposed and infectious subpopulations. Proceeding in a similar reasoning way for nonzero exposed and infectious subpopulations, the following result is obtained concerned with the endemic equilibrium point, [10].

Theorem 2. The following properties hold:

(i) Assume that $V* = K_V^* S_e^* - K_V^* \in [0\,\mu]$. Then, the endemic equilibrium point is given by the steady-state subpopulations:

$$S_e^* = \frac{(\mu + \gamma)(\mu + \sigma)}{\beta(\sigma - K_V^*)} = \frac{1}{R_0}, \quad E_e^* = \frac{(\mu + K_V^*)(\mu + \gamma)}{\beta(\sigma - K_V^*)}(R_0 - 1) \tag{7}$$

$$I_e^* = \frac{(\mu + K_V^*)}{\beta}(R_0 - 1), \quad R_e^* = \frac{(\gamma - K_V^*)}{\beta}(R_0 - 1) \tag{8}$$

and it exists and it is unique if and only if the reproduction number $R_0 > 1$ or, equivalently, that the disease transmission rate is larger than a critical value accordingly to the constraint:

$$\beta > \beta_c = \frac{(\mu + \gamma)(\mu + \sigma)}{\sigma}. \tag{9}$$

(ii) Assume that $V_0 = -K_V^*$ so that $V^* = (K_V^* - 1)S_e^* \in [0, \mu]$. If $\beta \leq \beta_c$ then the disease-free equilibrium point is the unique equilibrium point which is, furthermore, globally asymptotically stable.

Note that Theorem 2(i) is a necessary and sufficient condition for the existence of the endemic equilibrium point since if $R_0 < 1$ there are negative components of the endemic equilibrium, so that it does not exist, and if $R_0 = 1$ then it coincides with the disease- free equilibrium point.

3   Targetting the Equilibrium with Accurate Margin in Optimal Time

Now, we rewrite (1) in a compact way as a fourth order dynamic system $\dot{x}(t) = f(x(t))$ subject to $x(0) = x_0$. The minimum time loss function $J = \int_0^T dt$ is extremal under the optimal vaccination control and has an associated Hamiltonian $H = 1 + p^T(t)f(x(t))$ where $p(t)$ is the costate which is a dynamic Lagrange multiplier function which takes into account through time keeping the dynamic constraint $\dot{x}(t) - f(x(t)) = 0$, $t \geq 0$ for the optimal trajectory solution. The optimal time trajectory solution satisfies the equations:

$$\dot{x}(t) = \frac{\partial H}{\partial p}, \dot{p}(t) = -\frac{\partial H}{\partial p}, H(T) = 0 \tag{10}$$

subject to $x(0) = x_0$ and $p^T(T)x(T) = -1$. In view of (1)–(4), we can rewrite the vaccination effort as a fourth dimensional real vector

$$V_u(t) = b(V_c + u(t)) \tag{11}$$

where $b = (-1, 0, 0, 1)^T = e_4 - e_1$, where e_i is the $i - th$ unity vector of the canonic basis of R^4, $u(t)$ is an incremental scalar control and $V_c \in [0, \mu]$ is a constant vaccination scalar. Note that $V_c + u(t)$ is guaranteed to be in $[\lambda\mu, \rho\mu] \subseteq [0, \mu]$ for all time if $u(t) \in [\lambda\mu - V_c, \rho\mu - V_c]$ for $t \geq 0$ and some given design parameters $\lambda, \rho(> \lambda) \in [0, 1]$. Thus, the optimal incremental and vaccination controls and optimal vector vaccination effort are calculated as bang-bang controls based on the solution of the optimal costate as follows:

$$u_{opt}(t) = u_{opt}(t, \lambda, \rho, V_0) = \begin{cases} \lambda\mu - V_c & \text{if } p_{opt}^T(t)b > 0 \\ \rho\mu - V_c & \text{if } p_{opt}^T(t)b < 0 \end{cases} \tag{12}$$

$$V_{opt}(t) = V_{opt}(t, \lambda, \rho, V_0) = \begin{cases} \lambda\mu & \text{if } p_{opt}^T(t)b > 0 \\ \rho\mu & \text{if } p_{opt}^T(t)b < 0 \end{cases} \tag{13}$$

$$V_{uopt}(t) = V_{uopt}(t, \lambda, \rho, V_0) = \begin{cases} \lambda\mu b & \text{if } p_{opt}^T(t)b > 0 \\ \rho\mu b & \text{if } p_{opt}^T(t)b < 0 \end{cases} \tag{14}$$

The final targeted state $x(T) = x_f$ is fixed within a prescribed closed ball $\bar{B}(x^*, r)$ of radius $r > 0$ defining the prescribed accuracy degree and center x^* surrounding the disease-free equilibrium points being obtained for both constant controls saturating the bang-bang optimal one, namely, $x_i^* = \theta_i x_{i1}^* + (1 - \theta_i)x_{i2}^*$ with some $\theta_i \in [0, 1]$ for all $i \in \{1, 2, 3, 4\}$ and $x_{1,2}^*$ of components x_{ij}^* for $i \in \{1, 2, 3, 4\}$. $j \in \{1, 2\}$ being the disease- free equilibrium points for respective constant vaccination controls $V_1^* = V_{\min} = \lambda\mu$ and $V_2^* = V_{\max} = \rho\mu$. The minimal time T is that necessary for the trajectory solution to enter the ball $\bar{B}(x^*, r)$ from the given initial conditions.

4 Example

Consider the uncertainty-free SEIR model described by the following parameters $\mu^{-1} = 20\,days^{-1}$, $\sigma^{-1} = 21\,days^{-1}$, $\sigma = \gamma$ and $\beta = 0.1826\,days$. Define a ball involving all the subpopulations as follows:

$$[S_a, S_b] \times [E_a, E_b] \times [I_a, I_b] \times [R_a, R_b] = [0.22, 0.35] \times [0, 0.03] \times [0, 0.03] \\ \times [0.67, 0.75]$$

so that the minimum time will be attained when all the subpopulations lie within their corresponding intervals. In this case, the time needed to make all the subpopulations reach the defined ball is given by the immune, R, as Fig. 1 depicts, which is of 42.58 days.

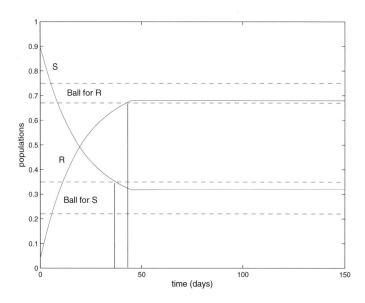

Fig. 1. Time needed to reach the defined ball for the susceptible and immune subpopulations.

Also, we consider the case when the model suffers from parametric uncertainty. In this way, the actual parameters of the model are given as above but we perform the calculation of the control law as if they were 10% higher. The following Fig. 2 displays the optimal control law obtained in this case while Fig. 3 shows the evolution of the system's trajectories. The time needed to steer the susceptible to the ball given by $[S_a, S_b] = [0.22, 0.35]$ is again of 37.42 days. It can be observed that the proposed optimal control law behaves robustly with respect to (relatively) small parametric uncertainties.

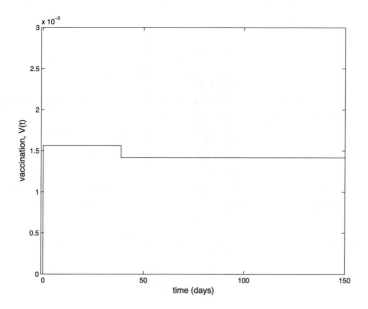

Fig. 2. Vaccination control law in the presence of uncertainties.

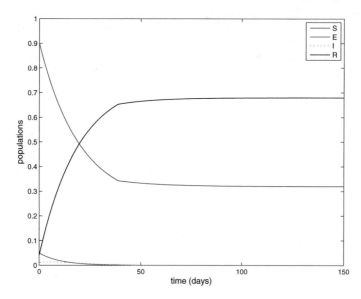

Fig. 3. Evolution of the system's trajectories in the presence of uncertainties.

Acknowledgments. This research has been supported by the Spanish Government and by the European Fund of Regional Development FEDER through Grant DPI2015-64766-R (MINECO/FEDER, UE) and by UPV/EHU by Grant PGC 17/33.

References

1. De la Sen, M., Alonso-Quesada, S.: Control issues for the beverton-holt equation in ecology by locally monitoring the environment carrying capacity: non-adaptive and adaptive cases. Appl. Math. Comput. **215**, 2616–2633 (2009)
2. De la Sen, M., Alonso-Quesada, S.: Model-matching bases control of the bevertoon-holt equation in ecology. Discrete Dyn. Nat. Soc. (2008). Article number 793512
3. De la Sen, M.: Fundamental properties of linear control systems with after-effect-I. The continuous case. Math. Comput. Model. **10**, 473–489 (1988)
4. De la Sen, M.: The generalized beverton-holt equation and the control of populations. Appl. Math. Model. **32**, 2312–2328 (2008)
5. De la Sen, M., Ibeas, A., Alonso-Quesada, S., Nistal, R.: On a new epidemic model with asymptomatic and dead-infective subpopulations with feedback controls useful for Ebola disease. Discrete Dyn. Nat. Soc. (2017). Article number 4232971
6. Karnik, A., Dayama, P.: Optimal control of information epidemics. In: 2012 Fourth International Conference on Communications Systems and Networks (COMSNETS) (2012). 1
7. Workman, M.L., Kosut, R.L., Franklin, G.F.: Adaptive proximate time- optimal servomechanisms. Discrete-time case. In: Proceedings of the American Control Conference, pp. 1548–1553 (2002). 1
8. Kim, S., Choi, D.S.: Time-optimal control of state-constrained second-order systems and an application to robotic manipulators. In: Proceedings of the American Control Conference, pp. 1478–1483 (2002). 1
9. You, K.I., Lee, E.B.: Robust near time-optimal control of nonlinear second-order systems with model uncertainty. In: Proceedings of the IEEE International Conference on Control Applications, pp. 232–236 (2000). 1
10. Keeling, M.J., Rohani, P.: Modeling Infectious Diseases in Humans and Animals. Princeton University Press, Princeton and Oxford (2008)

A Hybrid Approach for the Prevention of Railway Accidents Based on Artificial Intelligence

Habib Hadj-Mabrouk[✉]

French Institute of Science and Technology for Transport, Development and
Networks, 14-20 Boulevard Newton, 77447 Marne la Vallée, France
habib.hadj-mabrouk@ifsttar.fr

Abstract. The modes of reasoning which are used in the context of safety
analysis and the very nature of knowledge about safety mean that a conventional
computing solution is unsuitable and the utilization of artificial intelligence
techniques would seem to be more appropriate. Our research has involved three
specific aspects of artificial intelligence: knowledge acquisition, machine
learning and knowledge based systems (KBS). Development of the knowledge
base in a KBS requires the use of knowledge acquisition techniques in order to
collect, structure and formalizes knowledge. It has not been possible with
knowledge acquisition to extract effectively some types of expert knowledge.
Therefore, the use of knowledge acquisition in combination with machine
learning appears to be a very promising solution. This paper presents the result
of these two research activities which are involved in the methodology of safety
analysis of guided rail transport systems.

Keywords: Rail transport · Safety · Accident scenarios
Knowledge acquisition · Machine learning · Expert system

1 Introduction

One of the research activities which is currently in progress at the French institute
IFSTTAR relates to the certification of automated public transport systems and the
safety of digital control systems. Independently of the manufacturer, the experts of
IFSTTAR carry out complementary analyses of safety. They are brought to imagine
new scenarios of potential accidents to perfect the exhaustiveness of the safety studies.
In this process, one of the difficulties then consists in finding the abnormal scenarios
being able to lead to a particular potential accident. It is the fundamental point which
justified this work. Our study took place within this context and aimed to design and
create a software tool to aid safety analysis for automated people movers in order to
appraise the suitability of proposed protection equipment. The purpose of this tool is to
evaluate the completeness and consistency of the accident scenarios which have been
put forward by the manufacturers and to play a role in generating new scenarios which
could be of assistance to experts who have to reach a conclusion.

© Springer Nature Switzerland AG 2019
P. Vasant et al. (Eds.): ICO 2018, AISC 866, pp. 383–394, 2019.
https://doi.org/10.1007/978-3-030-00979-3_41

This article describes a contribution to improving the usual safety analysis methods used in the certification of railway transport systems. The methodology is based on the complementary and simultaneous use of knowledge acquisition and machine learning. We used the ACASYA software environment to support the safety analysis aid methodology. ACASYA aims to provide experts with suggestions of potential failures which have not been considered by the manufacturer and which are capable of jeopardizing the safety of a new rail transport system. In more formal terms, the methodology of safety analysis aid is based on two models: a generic accident scenario representation model, which is based on a static and a dynamic description of a scenario and a model of the implicit reasoning of the expert which involves three major activities, namely the classification, evaluation and generation of scenarios:

- The first level (CLASCA) relates to finding the class to which a new scenario which has been suggested by the manufacturer belongs. The purpose behind this is to provide the expert with historical scenarios which are partially or completely similar to the new scenario.
- The second level (EVALSCA) of processing considers the class which CLASCA has deduced that the scenario belongs in order to evaluate the consistency of the manufacturer's scenario. The purpose of the EVALSCA module is to compare the list of summarized failures (SFs) which are suggested in a manufacturer scenario to the list of stored historical SF in order to stimulate the formulation of hazardous situations which have not been anticipated by the manufacturer. This evaluation task draws the attention of the expert to any failures which have not been considered by the manufacturer and which might jeopardize the safety of the transport system. It may thus promote the generation of new accident scenarios.
- The two levels are supplemented by a third level (GENESCA) which makes use of the static description and the dynamic description of the scenario (the Petri model) and three reasoning mechanisms, namely, induction, deduction and abduction. Generation of a new scenario is based on injecting an SF which the previous level has defined as being plausible into a specific sequencing of the change in marking of the Petri net.

In view of the scale of the problem the design and construction of the demonstration model of the ACASYA system concentrated on the first two levels of processing (classification and evaluation of scenarios).

2   Contribution of Artificial Intelligence to the Acquisition of Railway Safety Knowledge

As part of its missions of expertise and technical assistance, IFSTTAR evaluates the files of safety of guided transportation systems. These files include several hierarchical analysis of safety such as the preliminary hazard analysis (PHA), the functional safety analysis (FSA), the analysis of failure modes, their effects and of their criticality (AFMEC) or analysis of the impact of the software errors. These analyses are carried out by the manufacturers. It is advisable to examine these analyses with the greatest care, so much the quality of those conditions, in fine, the safety of the users of the transport systems. Independently of the manufacturer, the experts of IFSTTAR carry out complementary analyses of safety. They are brought to imagine new scenarios of potential accidents to perfect the exhaustiveness of the safety studies. In this process, one of the difficulties then consists in finding the abnormal scenarios being able to lead to a particular potential accident. It is the fundamental point which justified this work. Our study took place within this context and aimed to design and create a software tool to aid safety analysis for automated people movers in order to appraise the suitability of proposed protection equipment. The purpose of this tool is to evaluate the completeness and consistency of the accident scenarios which have been put forward by the manufacturers and to play a role in generating new scenarios which could be of assistance to experts who have to reach a conclusion. Experts may find it very difficult to describe in clear terms the stages of reasoning which they go through in order to make decisions. Such a description requires experts to undertake a long process of thought which will enable them to explain the unconscious aspect of their activities. The success of a knowledge based systems (KBS) project depends on this difficult and sometimes painful task. In view of the complexity of the knowledge of experts and the difficulty which they have in explaining their mental processes there is a danger that the extracted knowledge will be either incorrect, incomplete or even inconsistent. A variety of research in Artificial Intelligence (AI) is in progress in an attempt to understand this problem of the transfer of expertise. Research is currently taking place in two major independent areas:

- Knowledge acquisition, which aims to define methods for achieving a better grasp of the transfer of expertise. These methods chiefly involve software engineering and cognitive psychology [1, 2],
- Machine learning, which involves the use of inductive, deductive, abductive or analogical techniques in order to provide the KBS with learning capacities [3, 4, 5, 6].

3 Expert System Based on Machine Learning to Assist in the Analysis and Evaluation of Railway Safety

In order to develop a KBS which aids in safety analysis we combined these two approaches and used them in a complementary way. The modes of reasoning which are used in the context of safety analysis (inductive, deductive, analogical, etc.) and the very nature of knowledge about safety (incomplete, evolving, empirical, qualitative, etc.) mean that a conventional computing solution is unsuitable and the utilization of artificial intelligence techniques would seem to be more appropriate. The aim of artificial intelligence is to study and simulate human intellectual activities. It attempts to create machines which are capable of performing intellectual tasks and has the ambition of giving computers some of the functions of the human mind - learning, recognition, reasoning or linguistic expression. Our research has involved three specific aspects of artificial intelligence: knowledge acquisition, machine learning and knowledge based systems. Development of the knowledge base in a KBS requires the use of knowledge acquisition techniques in order to collect, structure and formalizes knowledge. It has not been possible with knowledge acquisition to extract effectively some types of expert knowledge. Therefore, the use of knowledge acquisition in combination with machine learning appears to be a very promising solution. The approach which was adopted in order to design and implement an assistance tool for safety analysis involved the following two main activities:

- Extracting, formalizing and storing hazardous situations to produce a library of standard cases which covers the entire problem. This is called a historical scenario knowledge base (HSKB). This process entailed the use of knowledge acquisition techniques,
- Exploiting the stored historical knowledge in order to develop safety analysis know-how which can assist experts to judge the thoroughness of the manufacturer's suggested safety analysis. This second activity involves the use of machine learning techniques.

4   Acquisition and Formalization of Accident Scenarios

Examination of the concept of scenario revealed two fundamental aspects. The first is static and characterizes the context (Figs. 1 and 2). The second is dynamic and shows the possibilities of change within this context, while stressing the process which leads to an unsafe situation. Very schematically, guide way transit systems are considered as being an assembly of basic bricks and a new system possesses certain bricks which are shared by systems which are already known. In the context of this study the basic bricks which have currently been identified have been grouped together and the ACASYA tool finds and then exploits shared bricks in order to deduce the class to which a new scenario belongs or evaluate its completeness.

5   "ACASYA": Tool to Aid in the Safety Assessment

This article describes a contribution to improving the usual safety analysis methods used in the certification of railway transport systems. The methodology is based on the complementary and simultaneous use of knowledge acquisition and machine learning. We used the ACASYA software environment to support the safety analysis aid methodology. ACASYA aims to provide experts with suggestions of potential failures which have not been considered by the manufacturer and which are capable of jeopardizing the safety of a new rail transport system. The organization of ACASYA [7] (Fig. 3) is such that it reproduces as much as possible the strategy which is adopted by experts. Summarized briefly, safety analysis involves an initial recognition phase during which the scenario in question is assimilated to a family of scenarios which is known to the expert. This phase requires classes of scenarios to be defined. In a second phase the expert evaluates the scenario in an attempt to evolve unsafe situations which have not been considered by the manufacturer. These situations provide a stimulus to the expert in formulating new accident scenarios.

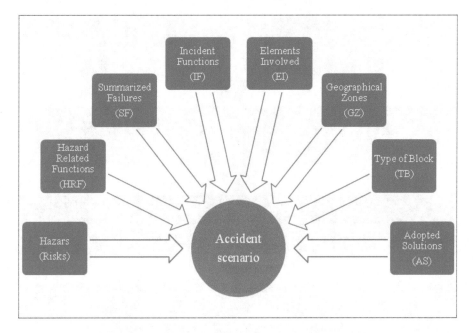

Fig. 1. Parameters which describe an accident scenario

As is shown in Fig. 3 this organization consists of four main modules. The formalization module deals with the acquisition and representation of a scenario and is part of the knowledge acquisition phase. The three other modules, CLASCA, EVALSCA and GENESCA, in accordance with the general principle which has been laid down above, deal with the problems associated scenario classification, evaluation and generation. In more formal terms, the methodology of safety analysis aid is based on two models: a generic accident scenario representation model, which is based on a static and a dynamic description of a scenario and a model of the implicit reasoning of the expert which involves three major activities, namely the classification, evaluation and generation of scenarios. The main purpose of the study is to combine these two models and make use of learning techniques in order to make the expert model as explicit as possible so that the expert process can be reproduced.

5.1 General Description of the Learning System by Classification: "Clasca"

The first level CLASCA [8] (Fig. 4) relates to finding the class to which a new scenario which has been suggested by the manufacturer belongs. The purpose behind this is to provide the expert with historical scenarios which are partially or completely similar to the new scenario. This mode of reasoning is analogous to that which experts use when they attempt to find similarities between the situations which have been described by the manufacturer's scenarios and certain experienced or envisaged situations involving equipment which has already been certified and approved.

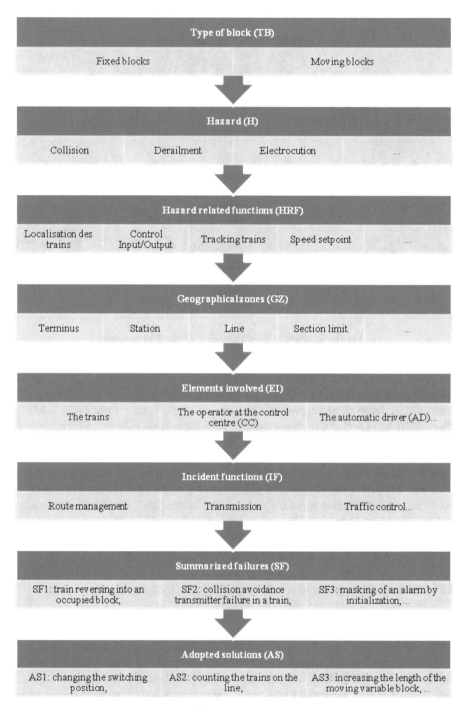

Fig. 2. List of the parameters which relate to an example of a scenario.

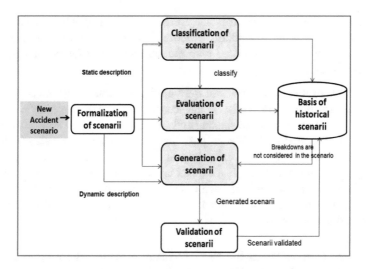

Fig. 3. Functional organization of the "ACASYA" system.

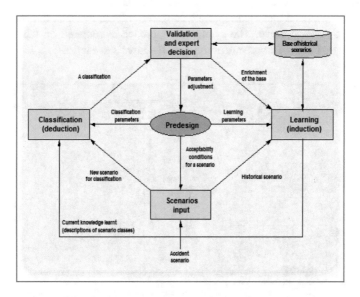

Fig. 4. General architecture of "Clasca" module

Classification of a new scenario involves the two following stages:

- A characterization (or generalization) stage for constructing a description for each class of scenarios. This stage operates by detecting similarities within a set of historical scenarios in the HSKB which have been pre-classified by the expert in the domain,

- A deduction (or classification) stage to find the class to which a new scenario belongs by evaluating a similarity criterion. The descriptors of the new scenario (static description) are compared with the descriptions of the classes which were generated previously.

This initial level of processing not only provides assistance to the expert by suggesting scenarios which are similar to the scenario which is to be dealt with but also reduces the space required for evaluating and generating new scenarios by focusing on a single class of scenarios Ck. The purpose this is to provide the expert with historical scenarios which are partially or completely similar to the new scenario. This mode of reasoning is analogous to that which experts use when they attempt to find similarities between the situations which have been described by the manufacturer's scenarios and certain experienced or envisaged situations involving equipment which has already been certified and approved.

5.2   General Description of the Expert System to Aid in the Evaluation of Safety Based on the Learning of the Rules: "Evalsca"

The second level EVALSCA (Fig. 5) of processing considers the class which CLASCA has deduced that the scenario belongs in order to evaluate the consistency of the manufacturer's scenario. The evaluation approach is centered on the summarized failures (SFs) which are involved in the manufacturer's scenario.

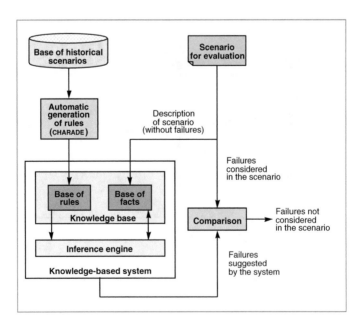

Fig. 5. General architecture of "Evalsca" module

The evaluation of a scenario of this type involves the two modules below:

- A mechanism for learning rules CHARADE which makes it possible to deduce SF recognition functions and thus generate a base of evaluation rules,
- An inference engine which exploits the above base of rules in order to deduce which SFs are to be considered in the manufacturer's scenario.

This phase of learning attempts, using the base of sixty examples which was formed previously, to generate a system of rules. The purpose of this stage is to generate a recognition function for each SF associated with a given class. The SF recognition function is a production rule which establishes a link between a set of facts (parameters which describe a scenario or descriptors) and the SF fact (Fig. 6). What is involved here is logical dependence, which can be expressed in the following form:

IF	Type of block (TB)
	And Hazard (H)
	And Hazard related functions (HRF)
	And Geographical zones (GZ)
	And Elements involved (EI)
	And Incident functions (IF)
THEN	Summarized Failures (SF)

Fig. 6. Form of SF recognition rules

A base of evaluation rules can be generated for each class of scenarios. The conclusion of each rule which is generated should contain the SF descriptor or fact. It has proved to be inevitable to use a learning method which allows production rules to be generated from a set of historical examples (or scenarios). The specification of the properties required by the learning system and a review of the literature has led us to choose the CHARADE mechanism [4]. CHARADE ability to generate automatically a system of rules, rather than isolated rules, and its ability to produce rules in order to develop SF recognition functions make it of undeniable interest. CHARADE [4] is a learning system whose purpose is to construct knowledge based systems on the basis of examples. It makes it possible to generate a system of rules with specific properties. Rule generation within charade is based on looking for and discovering empirical regularities which are present in the entire learning sample. Regularity is a correlation which is observed between descriptors in the base of learning examples. If all the examples in the learning base which possess the descriptor d1 also possess the descriptor d2 it can be inferred that d1 → d2 in the entire learning set. In order to illustrate this rule generation principle let us assume that there is a learning set which consists of three examples E1, E2, and E3.

- E1 = d1 & d2 & d3 & d4
- E2 = d1 & d2 & d4 & d5
- E3 = d1 & d2 & d3 & d4 & d6

CHARADE can in this case detect an empirical regularity between the combination of descriptors (d1 & d2) and the descriptor d4. All those examples which are described by d1 & d2 are also described by d4. The rule d1 & d2 ➔ d4 is obtained (Fig. 7).

If elements_involved = mobile_operator,
incident_functions = instructions
elements-involved = operator_in_CC.
Then   sumarized failures = SF11: invisible element on the zone of completely automatic driving,
elements_involved = AD_with_redundancy,
hazard_related_functions =train localization,
geographical_zones = terminus.

Fig. 7. A sample of some rules generated by CHARADE

The purpose of the EVALSCA module is to compare the list of SFs which are suggested in a manufacturer scenario to the list of stored historical SF (Fig. 8) in order to stimulate the formulation of hazardous situations which have not been anticipated by the manufacturer. This evaluation task draws the attention of the expert to any failures which have not been considered by the manufacturer and which might jeopardize the safety of the transport system. It may thus promote the generation of new accident scenarios.

@@   18/06/2018
    moving_block
    collision
    management_of_automatic_driving
    train_monitoring
    initialization
    terminus
    operator_at_CC
    ad_without_redundancy
    instructions
DEDUCTION:
    Summarized failure = SF19: Silent train

Fig. 8. Example result of deduction by the expert system

5.3   General Description of the Learning System by Classification: "Genesca"

The two levels of processing which have been described above make use of the static description of the scenario (descriptive parameters). They are supplemented by a third level GENESCA (Fig. 9) which makes use of the static description and the dynamic description of the scenario (the Petri model) and three reasoning mechanisms, namely, induction, deduction and abduction. Generation of a new scenario is based on injecting an SF which the previous level has defined as being plausible into a specific sequencing of the change in marking of the Petri net.

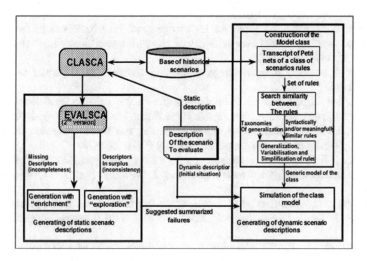

Fig. 9. General architecture of "Genesca" module

In view of the scale of the problem the design and construction of the demonstration model of the ACASYA system concentrated on the first two levels of processing (classification and evaluation of scenarios).

6 Conclusion

This paper has presented our contribution to the improvement of the methods which are normally used to analyze and assess the safety of automatic devices in guided transport systems. This contribution is based on the use of artificial intelligence techniques and has involved the development of several approaches and tools which assist in the modeling, storage and assessment of knowledge about safety. The software tools have two main purposes, firstly to record and store experience concerning safety analyses, and secondly to assist those involved in the development and assessment of the systems in the demanding task of evaluating safety studies. Currently, these tools are at the mock-up stage. Initial validation has demonstrated the interest of the suggested approaches, but improvements and extensions are required before they could be used in an industrial environment or adapted to other areas where the problem of investigating safety arises.

References

1. Gaines, B.R.: Knowledge acquisition: past, present, and future. Int. J. Hum.–Comput. Stud. (2012). http://dx.doi.org/10.1016/j.ijhcs.2012
2. Aussenac, G., Gandon, F.: From the knowledge acquisition bottleneck to the knowledge acquisition overflow: a brief French history of knowledge acquisition. Int. J. Hum.-Comput. Stud. **71**(2), 157–165 (2013)
3. Kodratoff, Y.: Leçons d'apprentissage symbolique automatique. Cepadues éd., Toulouse, France (1986)
4. Ganascia, J-G.: Agape et Charade: deux mécanismes d'apprentissage symbolique appliqués à la construction de bases de connaissances. Thèse d'État, Université Paris- sud, France (1987)
5. Ganascia, J.-G.: Logical induction, machine learning and human creativity. In: Switching Codes. University of Chicago Press (2011). ISBN 978022603830
6. Michalski, R-S., Wojtusiak, J.: Reasoning with missing, not-applicable and irrelevant meta-values in concept learning and pattern discovery. J. Intell. Inf. Syst. **39**(1), 141–166 (2012). Springer
7. Hadj-Mabrouk, H.: Contribution of learning Charade system of rules for the prevention of rail accidents. J. Intell. Decis. Technol. **11**(4), 477–485 (2017). https://doi.org/10.3233/idt-170304
8. Hadj-Mabrouk, H.: CLASCA: learning system for classification and capitalization of accident Scenarios of Railway. Int. J. Eng. Res. Appl. **6**(8), 91–98 (2016). ISSN: 2248-9622

Visual Analytics Solution for Scheduling Processing Phases

J. Joshua Thomas[1]([✉]), Bahari Belaton[2]([✉]),
Ahamad Tajudin Khader[2]([✉]), and Justtina[1]([✉])

[1] Department of Computing, School of Engineering, Computing, and Built Environment, KDU Penang University College, 32 Anson Road, 10400 George Town, Penang, Malaysia
joshopever@yahoo.com, justtinna_john@yahoo.com
[2] School of Computer Sciences, University Sains Malaysia, 11800, Penang Gelugor, Malaysia
{bahari, tajudin}@usm.my

Abstract. University Examination Timetabling Problem (UETP) is a computationally complex scheduling problem. Visual Analytics (VA) is a modern visualization supported with automated processing method. The major impulse of the method lies in its ability to integrate the key component of scientific visualization and search based heuristics in the same optimization model. This paper presents a visual analytics process (VAP) adapted for UETP. The adaption involves the human context of visual analytics on timetabling data, which are typically processed computationally with local search algorithm and then visualized and interpreted by the user in order to perform problem solving with direct interactions between the primary data, processing and visualization. The three processing phases are invoked with *user-driven* and *algorithmic-driven steering* that analyses the combined effect with automatic tuning of algorithmic parameters based on constraints and the criticality of the application for the simulations is proposed. The optimal solution for the small datasets and best overall results for the medium and large datasets are experimented.

Keywords: Examination timetabling · Interactive visualization
User-driven · Algorithm driven steering

1 Introduction

In general, examination-timetabling problem has a set of exams that must be scheduled to a set of time slots such that every exam is located in exactly one timetable, subject to certain constraints. Due to NP-hard nature, specifically, that NP-completeness arises whenever students have a wide subject choice, or examination vary in duration, or simple conditions are imposed on the choice of times for examinations or even spread through the week. It is the process of assigning exams to rooms and timeslots. The constraints are related to the preferences of lecturers and students. Although suitable time slot may be assigned to all the examinations involving one student group simultaneously, the corresponding problem for two student groups is NP-complete. There are

© Springer Nature Switzerland AG 2019
P. Vasant et al. (Eds.): ICO 2018, AISC 866, pp. 395–408, 2019.
https://doi.org/10.1007/978-3-030-00979-3_42

essential semi-automated techniques that work in the solution of exam timetables with an initial solution that can be either feasible or infeasible. Graphical representation of a departmental examination has designed [1] in there a single department showing the exam nodes the same department highlighting exams that clash with a number of other exams. The same department showing the clashes between exams. This work presents a visualization framework embedded with evolutionary algorithm to tackle complex examination timetabling problem. Initial proposed articles [2–4] are provided significantly basic contribution related to this work. The paper has organized in related work is in Sect. 2, Methodology, problem definition, are in Sect. 3. Combining visual interaction framework with evolutionary algorithm are in Sect. 4 evaluation procedure, experimental results are in Sect. 5 followed by conclusion in Sect. 6.

2 Related Work

The conventional approach on solving the examination-timetabling problem has tackled by implementing local search algorithms then evaluate its performance on using textual or statistical representation. The unconventional approach on solving examination-timetabling problem in this work has handled by implementing visual design with local search algorithms to evaluate its performance through using visual diagnosis of the results. If the results are good, enough then activate the stop condition. If the results need to be improved then use the visual cues procedures systematically to perform the local search to obtain good solution. In a recent complete survey for examination timetabling, [5] conclude. In [6] have remarkably included visualization to highlight the violations and penalties in solution with visualization methods are better in identifying proficient violations in the search space and understand the complex data generated solutions. In A-Plan [7], assignments of service technicians to customers are displayed visually and may be modified by direct manipulation. Smooth cooperative work is possible and an optimization algorithm has been integrated that facilitates semi-automatic planning. Computational steering is an investigative paradigm whereby the parameters of a running program may be altered according to what is seen in the currently visualized results of the simulation. Further exploring the solution by means of interactive visual cues and then dynamically steering the computational process helps to improve the solutions. Visualization techniques [8] and proposed an interactive optimization system to solve the course timetabling problem. Areas such as resource management, planning, scheduling, data mining, information retrieval, graph drawing and many others can be described as optimization problems. All these problems have in common the quality criterion of the optimization process, which can hardly be formalized since the problem involves a large number of user- and task-dependent constraints. Reference visualization [9] in the semi-automated analytical process, where humans and machines cooperate using their distinct complementary capabilities to obtain the most effective results. Visualize the metaheuristic algorithms [10] to solve bin packing problem has identified in literature graphically generate hybrid solution to solve 17 benchmarking instances. Another work has used modeling [11] genetic algorithm simulation, modeling of aerodynamic stall control using jet actuator to visualize the inner operations.

3   Methodology

The primary pace of applying visualization to solve an optimization problem is to model the problem in terms of scientific pipeline, which includes modelling, simulation, and visualization. The secondary pace is to specify the interactive visualization techniques of the optimization problem. In this way, an optimization with user-intervention which could change optimization goals if possible.

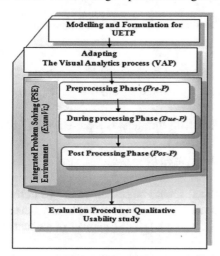

The first step includes modelling and simulating the **UETP** while the second engages in adapting **VAP** to **UETP** from visual analytics (**VA**). Three processing phase mechanisms are conducted in the integrated PSE which are designed to exemplify (i) the Preprocessing phase (**Pre-P**); (ii) during the processing phase (**Due-P**); and (iii) Post Processing phase (**Pos-P**). Every processing computationally steers each other to enhance the convergence of the solutions. The visual cues designed and developed (i.e. **MCLH, LCLH, NAVTS, AVTS**, and **PYVAL**) in each processing phases trigger the human timetabler to interact with the solutions. The results of the proposed integrated visual analytics processes (i.e. **Pre-P, Due-P** and **Pos-P**) have been compared with user-driven steering and algorithmic –driven steering. They have also been compared with [14].

3.1   Modelling and Formalization UETP Description

An uncapaciated UETP version considered in this work consists of given events assigned to given course to rooms and to timeslots according to the hard and soft constraints. The problem defines four hard constraints (HDCT1, HDCT2, HDCT3 and HDCT4) which must be satisfied for a feasible timetable; and a combination of three soft constraints (STCT1, STCT2, STCT3). The basic intention is to minimize the soft constraint violations in a feasible timetable.

3.2   Problem Formulation

The examination timetabling problem involves scheduling of a set of exams, each taken by a set of students, to a set of timeslots (periods) subject to hard and soft constraints. The main objective is to obtain a timetable that satisfies the hard constraints with the minimum penalty of the soft constraint violation. A detailed description of the problem is summarized by [5]. A timetabling solution is represented by vector x = $(x_1, x_2, ..., x_M)$ of exams, where the value of x_i is the timeslot for exam i.

Definitions for Hard and Soft Constraints. The hard constraints and soft constraint are as follows:

Definition

HDCT1: *Exam-clashing:* Every exam must be assigned to exactly one timeslot of the timetable; formulated using matrix Q, where $x(e_i, t_j) = 0$ if $\sum_{j=1}^{N_2} q_{ij} = 1$, and 1 otherwise.

Conflict matrix for exams clashes with timeslots is represented as follows:

$$\sum_{i=1}^{N_1} x(e_i, t_j) = 0. \tag{1}$$

HDCT2: *Student-clashing:* No student should be scheduled in two different places at once, i.e. any two which have students in common must not be scheduled in the same timeslot; formulated using C and Q, where $C_{i,j}$ is the number of students taking both exams e_i and e_j.

No exams with common resources are assigned simultaneously. This is represented as follows:

$$\sum_{k=1}^{N_2} \sum_{i=1}^{N_1-1} \sum_{j=i+1}^{N_1} c_{ij} \cdot q_{ik} \cdot q_{jk} = 0 \tag{2}$$

HDCT3: *Timeslot-clashing:* Certain exams must be grouped and scheduled together in the same timeslot. For instance, if exams e_i and e_j have same group, then both must be assigned in the same timeslot. This could be formulated using the matrix Q where $x(e_i, e_j) = 0$ if $g_{i,k} = g_{jk}$ and 1 otherwise.

Each examination must be assigned to timeslot only once. This is represented as follows:

$$\sum_{k=1}^{N_2} x(e_i, e_j) = 0. \tag{3}$$

HDCT4: *Timeslot-capacity:* The total number of students in all exams in the same timeslot must be less than the total capacity for that timeslot; formulated using the matrices E and Q, where $x_c(e_i, t_j) = 0$ if $\sum_{j=1}^{n_1} n_{s(e_i)} \cdot q_{ij} \leq n_c(t_j)$, and 1 otherwise.

Total number of students in the same room must be less than the capacity of the room. This is represented as follows:

$$\sum_{i=1}^{N_2} x(e_i, t_j) = 0. \tag{4}$$

STCT1: *Exam-proximity1 & 2:* No student should have two exams in adjacent timeslots on the same day, formulated using the matrices C and Q.

$$F(\sum_{i=1}^{N_1-1} \sum_{j=j+1}^{n_1} c_{i,j}.prox(t(e_i), t(e_j)))) \tag{5}$$

where $prox(t(e_i), t(e_j)) = \omega$ if $|(t(e_i), t(e_j))| = 1$, 0 otherwise; $t(e_i)$ specifies the assigned timeslot for exam e_j, ω is a weight that reflects the penalty of violating this constraint for each student, and the function $f(x)$ is a penalty function based on the total weights of students having two exams in adjacent timeslots.

- Objective Function

The objective function to evaluate a timetabling solution x for UETP can be formally defined. The objective function $f(x)$ is in Eq. (6). The value of $f(x)$ is the total *number* of soft constraints which are not met in a feasible timetable:

$$f(x) = \sum_{s \in S} (f_1(x, s) + f_2(x, s) + f_3(x, s)) \tag{6}$$

The value of $f(x)$ is referred to as the Penalty Value (PV) of a feasible timetable.

4   Visual Interaction with Evolutionary Algorithm

This section summarizes the visual analytics processes proposed for UETP. The visualization optimization methods are proposed based on the Visual analytics process: (i) a visual analytics framework (VAF) adapted from visual analytic process (VAP) for

Table 1. The UETP and optimization terms in the visualization perspective

Visualization process	Optimization	UETP
Data transformation	Generation	Scheduling
Preprocessing insight	Solution vector	Timetable solution
During produced visualization	Decision variable	Event
Preprocessing data Hypotheses	Value	Feasible room to timeslot pair
Hypotheses	Value range	Available room to timeslot pairs
User phenomena during processing Hypotheses	Objective function	Objective function formalized in Eq. 6
Interactive visual computing	Iteration	Iteration
Scientific discovery process	Optimal solution	Feasible timetable

UETP, to explore visual simulation on the data (ii) an integrated problem solving environment (PSE) *Visualisation* is developed to understand and interact with the timetabling construction processes (Pre-P, Due-P and Pos-P) to improve the solutions.

In order to bridge between the UETP and a visualization context, the relationships and correspondence are shown in Table 1. In principle, each data transformation process corresponds to an initial process where the raw data are fed into the visualization pipeline and an initial solution is generated. Likewise, in the scheduling process, a set of events is assigned with values (exams-timeslots), iteration by iteration, searching for a feasible timetable with the least number of constraint violations as determined by the objective function formalized in Eq. (6).

4.1   Visual Analytics Process Supported with Evolutionary Algorithm

Most scheduling problem including the UETP are highly constrained combinatorial optimization problem which makes it hard to tackle using a classical method [14]. The adaption of visual analytics process with the integrated problem-solving environment is proposed. The three processing phases tailored with evolutionary algorithms proposed in this work are introduced visually; each phase computationally steers and overcomes the weakness by turning the parameters and thus obtaining a good-quality solution to UETP. Adapting VA to UETP for the first time was an attempt that brought forth some success in solving small and medium timetabling instances but not for large instances. The method, called Pre-Processing Phase (**Pre-P**) had four visual cues indicators: (i) MCLH most crucial clash, (ii) LCLH least crucial clash, (iii) AVTS available time slots, (iv) NAVTS not available time slots, and (v) PRYVAL priority values obtained by visually adjusting procedures. Accordingly, the During the Processing Phase (**Due-P**) has visualized the execution initial population of the evolutionary algorithm with the genetic operators tuning the parameters of the crossover and mutation operation together with the visual cues to obtain better solution. In the process, multiple visual interaction procedures are used and both user and algorithm have to steer the parameters to improve the results substantially leading to the best solutions for the larger timetabling instances. The new eight visually guided procedures are carefully designed based on the neighbourhood structures. Post Processing Phase (**Pos-P**), nonetheless needs to be connected to the UETP search space more closely and overwrites the results and fine-tunes the visual solution to find a local optimal solution.

5   Evaluation Procedure, Experiment and Results

The experiments were designed for *User-driven steering* is a computational process by which the user can interactively explore a simulation during execution based on the visualization of the current results. A well-studied approach allows the user to give feedback to the simulation based on the visualization. *Algorithmic-driven* steering uses the evolutionary algorithm to decide application parameters to improve system and application performance. Steering enables the human timetabler to refine the ongoing simulation and see the effects of refinement immediately. The development of an integrated user-driven and automated (algorithmic) steering problem solving environment *Visualisation* for simulations, visualization, and analysis for conflicts can happen upon generation of solutions. *Visualisation* provides the user control over various visual cues, application parameters including the constraints of data for visualization. *Visualisation* reconciles between the parameters decided by the algorithm and the input parameters of the human timetabler. In the scenario, the visual analytic processing framework can give options to the user or override some of the user inputs in order to get the better solution. Most crucially, the evaluation of the PSE is analyzed in order to value the visualization for the university examination-timetabling problem (UETP). To emphasize on the PSE's effectiveness in terms of UETP, experimentation with Carter's dataset is conducted during *Pre-P* and *Due-P* phases. There are five visual parameters namely, *MCLH, LCLH, AVTS, NAVTS, PRYVAL* (see Algorithm 4.1) to evaluate the effectiveness between *user-driven steering* with *algorithmic-driven steering* incorporating the computational steering in each of the processing phases based on reducing the hard constraint violations or reconciliation of clashes between course to room to timeslot. For all sizes of problem instances (small, medium, large), adjusting the rooms and timeslots is made for seven individual cases, and the conflict matrix density is calculated based on the course allocated to the rooms multiplied by the available timeslots. The experimental results are in Sect. 5.2. The subsequent experiments have drawn attention to how the visual analytics process fits for the scheduling problem, which in this case is UETP. The implication of the experiment is to exemplify that the proposed visual analytics process (VAP) would guarantee that the human timetabler is able to improve the solution during the construction of examination timetable phases (*Pre-P, Due-P, and Pos-P*).

5.1   Data Processing Phases of VAP for UETP

Pre-P featured with interactive computational processes whereby researchers and scientists communicate with the data constraints by manipulating its visual representation. This sophisticated process of navigation allows researchers to computationally steer, or dynamically modify the exam period, timeslot computations until the hard constraint on clashes resolved or reduced. For every interaction applied on the conflicting data, the computational parameter are analyzed and tuned as a new input to the next processing stages. Figure 1(a) illustrates the MCLH from a group of students to course, room and timeslot. Figure 1(b) continuously indicates to the human timetabler on the next most crucial clash that has happened between course to rooms and timeslot. These visual cues will continue until the most crucial clash hard constraint violations have been reduced and after which the process will proceed to the next process which will be explained in the next section. To look at the line a little closer or to increase the thickness of the clashed lines for more effective visual interaction, the interface has a provided a slider and the most crucial clashed (MCLH) lines are colour coded. The following section will explain the least crucial clash lines (LCLH) from the populated initial solution.

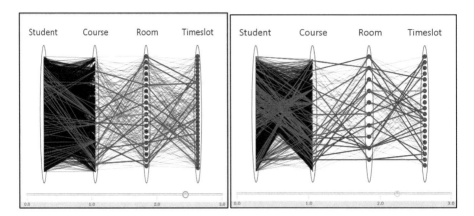

Fig. 1. (a) Most Crucial Clashes in a solution (b) Next Most crucial Clash in a different solution in the initial population (Pre-P) & (Due-P)

The "Red" line indicates the least crucial clashes. The lines exemplify the number of students affected, course id, room id, and the slot. In Fig. 2, each "Red" line signifies the least crucial clash violation of the hard constraint in the initial solution. Each clashed lines has been symbolized with the tool-tip information. The human timetabler needs to drag the mouse cursor over the "Red" least crucial clashed lines (LCLH) to update the tool-tip with the violation of constraints with the total number of students affected, course, room and the timeslot. The process will be extended based on the exams and timeslot that the human timetabler has configured while creating the initial solution.

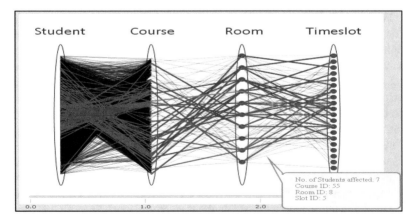

Fig. 2. Least Crucial Clash lines from the initial solution with tool-tip visual cues

Let us suppose the problem shown in Fig. 3 is the target problem from the dataset. This is classified as one of the least crucial clash. Two clashes are considered to be the most similar to the target problem. The structures of the two cases as shown in Fig. 1(a and b) are more isometric and modified, although they may not be the "good" solution for the target problem. When we look at the visualization, it can be seen that there are 2 violations of hard constraints in the solution. (1) Students cannot be in two exams at once, (2) Two exams cannot be in same room and same timeslot. In this scenario, each room has at least one exams and the exam will be assigned to a timeslot. Suppose in Room no. 10 has two least clashed lines, it violates the second hard constraint. Using the graph heuristic method, alteration is made to initial solution to minimize the clashes and this could reduce the hard constraint violation. The simple instance has confirmed that only a few alterations are needed to get solutions for the target problem on the basis of the solutions in LCLH similar instances. Instances can explore deeper knowledge in examination timetabling problem by the visual representation. Interaction that targets the alterations of every pair of events between the target problem and the altered instance(s) finds the least crucial clashes for the target problem, thus a matching relation between the events and alteration requirements is developed. The alteration requirement in the LCLH of the similarity between every event pair gives a more elaborate description with visual cues for the similarity of clashes. Thus the knowledge and experiences the human timetablers has on the clashes solutions can be exploited for

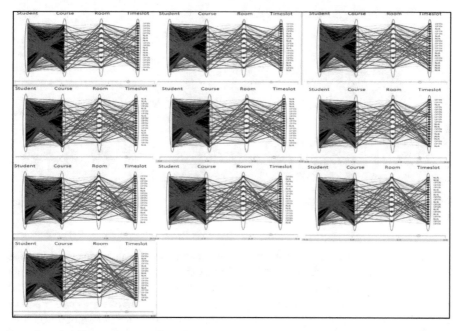

Fig. 3. *hec-s-92* dataset 81 exams, 2823 students, 10632 courses mapped to 18 timeslots with the density of 0.42 constraint violations

re-use for all the [14] problems. It is noted that the LCLH can use the NAVTS and AVTS features previously solved problems within a manner similar to that of experts in timetabling.

In *Due-P*, two modifications have been proposed: (i) the evolutionary optimizer is incorporated with *Due-P* to extend the search space to find the local optimal solution from *Pre-P*, and (ii) the visual priority values in consideration is used as a selection mechanism which selects the values and events from the best solution. This scenario (Due-P) is organized as follows firstly, the priority values is of integer values (Fig. 4) whereby the user moves the conflicted line to the highest -ve (negative) values that weighted values of constraints are reduced and if they move conflicted line to the lowest +ve (positive) the conflicts has increased. In other words, as to why highest -ve (negative) values are presumed as the most favourable is simply because the result of the interaction will reduce the overall clashes significantly! These steering mechanisms are controlled by the framework and it assists the user. Secondly, all the visual cues are invoked while the evolutionary algorithm progresses. I.e. *User-driven* and *Algorithmic-driven* steering with all visual cues able to performed.

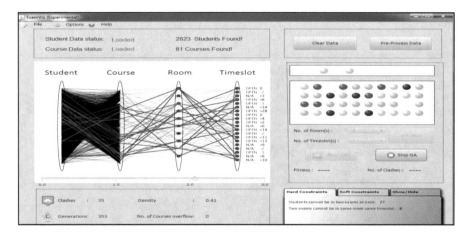

Fig. 4. Algorithmic steering on the Most Crucial Clashes (MCLH) with Priority Values (PRYVAL)

Figure 5 shows the output of the post processing phase. It can be observed that the Gaussian clustering pictorially exemplified minimization of the clashes. The textual generation from the *Visualisation* has shown the clashes from 493 have reduced to 280 with the generation of 3386 before the stopping criterion is invoked. Figure 5 shows the result of a configuration used for the visualization of the result of an optimizations run. For example, the dataset *hec-S-92* involves techniques for visualizing the state of the population or the algorithm has shown a sequence of respective output which can be highlighted in the diagram as an excel output. The exports are allowed during the

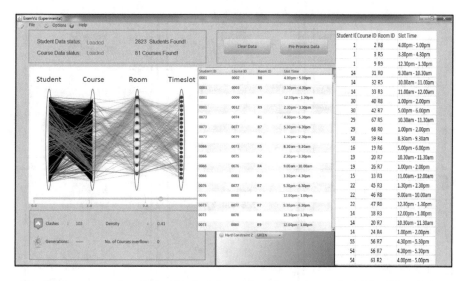

Fig. 5. Post processing phase: Tabular view, exported excel an assigned feasible solution

optimization run and can be used to access the performance of the algorithm in various strategies of the visual cues used. i.e. the output has been updated whenever an export happened in the post processing phase.

5.2 Experimental Results

The proposed method is programmed in Java technology. For better UI experience, the system has been implemented with JavaFX 2.0 which is combined with conventional Java1.6.0_18 under Windows- XP. The experiments presented here render differently depending on the size of the dataset. The performance of adapting visual analytics with steering visual representations is experimentally studied using benchmarking dataset [14] available from the link: "http://www.asap.cs.nott.ac.uk/resources/data.shtml. Each data instance is executed 7 times as maximum iteration 1000–5000 depend on the dataset. In the parameter setting, the pre-visualization constant value used in the Pre-P while β, $B_{min,}$ $B_{max,}$ were used in the Genetic operators assignment in Due-P, δ are used in *user-driven steering* interactions in Pre-P and Due-P.

The Boxplots in Fig. 6 show the distribution of results between the iterations. There are exceptions in larger datasets where almost all the solutions indicate the absence of distribution. The Boxplot distributions show that average values of the median values rather than mean values. Notably, for the small dataset and medium datasets the range of results are much better than median values. Note that the number of iterations used is 600–5000.

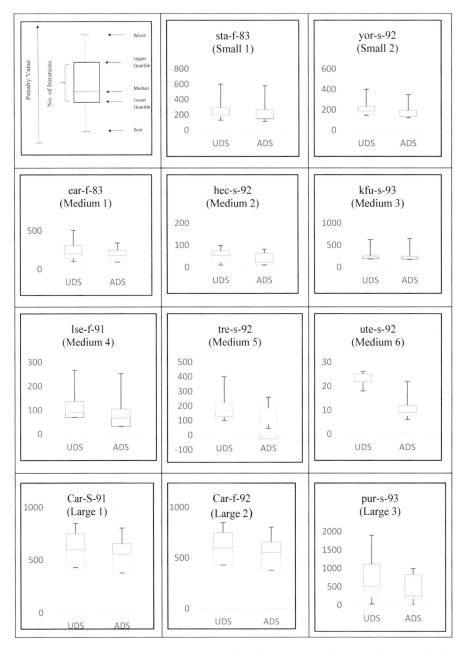

Fig. 6. A comparison between the results obtained between user-driven steering, algorithmic-driven steering with visual cue procedures

6   Conclusion

In this study, visual analytics process tangled with genetic algorithm to provide UTEP solution here, each scheduling processing phases are visually represented with intelligent visual-cues to guide the human timetabler to interact with clashes. The clashes are activated through violated hard constraints and soft constraints, which we presented in this work. The User-driven steering and Algorithmic-driven steering used to speed up the convergence and has improved GA operation in finding the optimal solution. The framework has been tested with benchmark [14] dataset and the performance are discussed. This work lead as combining Visualisation with scheduling problems, which is available for further research.

References

1. Ranson, D. Cheng, P.-H.: Graphical tools for heuristic visualization. In: Kendall, G., Lei, L., Pinedo, M. (eds.) Proceedings of the 2nd Multidisciplinary International Conference on Scheduling: Theory and Applications (MISTA), 18–21 July 2005, vol. 2, New York, USA, pp. 658–667 (2005)
2. Thomas, J.J., Khader, A.T., Belaton, B., Ken, C.C.: Integrated problem solving steering framework on clash reconciliation strategies for university examination timetabling problem. In: Neural Information Processing , pp. 297–304. Springer, Heidelberg (2012)
3. Thomas, J.J., Khader, A.T., Belaton, B.: A parallel coordinates visualization for the uncapaciated examination timetabling problem. In: Visual Informatics: Sustaining Research and Innovations, pp. 87–98. Springer, Heidelberg (2011)
4. Thomas, J.J., Khader, A.T., Belaton, B.: The perception of interaction on the university examination timetabling problem. In: PATAT 2010, p. 392 (2010)
5. Qu, R., Burke, E.K., McCollum, B., Merlot, L.T., Lee, S.Y.: A survey of search methodologies and automated system development for examination timetabling. J. Sched. **12**(1), 55–89 (2009)
6. Bonutti, A., De Cesco, F., Di Gaspero, L., Schaerf, A.: Benchmarking curriculum-based course timetabling: formulations, data formats, instances, validation, visualization, and results. Ann. Oper. Res. **194**(1), 59–70 (2012)
7. Schneider, T., Aigner, W.: A-Plan: integrating interactive visualization with automated planning for cooperative resource scheduling. In: Proceedings of the 11th International Conference on Knowledge Management and Knowledge Technologies, p. 44. ACM (2011, September)
8. Hinneburg, A., Keim, D.A.: A general approach to clustering in large databases with noise. Knowl. Inf. Syst. **5**(4), 387–415 (2003)
9. Davey, J., Mansmann, F., Kohlhammer, J., Keim, D.: Visual analytics: towards intelligent interactive internet and security solutions. In: The Future Internet, pp. 93–104. Springer, Heidelberg (2012)
10. Kroenung, L., Tauritz, D.: Visualization for Hyper-Heuristics. Front-End Graphical User Interface (No. SAND2015-2324R). Sandia National Laboratories (SNL-NM), Albuquerque, NM (United States) (2015)
11. Razaghi, R., Amanifard, N., Narimanzadeh, N.: Modeling and multi-objective optimization of stall control on NACA0015 airfoil with a synthetic jet using GMDH type neural networks and genetic algorithms. Int. J. Eng. Trans. A **22**(1), 69–88 (2009)

12. Nahavandi, N., Zegordi, S.H., Abbasian, M.: Solving the dynamic job shop scheduling problem using bottleneck and intelligent agents based on genetic algorithm. Int. J. Eng. Trans. C Asp. **29**(3), 347 (2016)
13. Lewis, R.: A survey of metaheuristic-based techniques for university timetabling problems. OR Spectr. **30**(1), 167–190 (2008)
14. Carter, M.W., Laporte, G., Lee, S.Y. Examination timetabling: algorithmic strategies and applications. J. Oper. Res. Soc., 373–383 (1996)

Investigation of Emotions on Purchased Item Reviews Using Machine Learning Techniques

P. K. Kumar[1(✉)], S. Nandagopalan[2], and L. N. Swamy[3]

[1] VTU Research Center, VTU, Belagavi, India
pandralli@gmail.com
[2] Department of IS&E, Vemana Institute of Technology, Bangalore, India
snandagopalan@gmail.com
[3] Department of MCA, VTU PG Studies, Belagavi, India
swamyln@gmail.com

Abstract. Product reviews from customers are plays vital role for the customers who want to buy the product through online. To get the knowledge of customer emotions on particular item and its features given by the owner of the product required efficient sentiments investigation. Investigating emotions of customers from huge and complex unstructured reviews is big challenge. There are lot of text mining approaches have been proposed by many researchers for understanding the different characteristics of the customer connectedness on items based on reviews. Still need a better approach for investigation of emotions which can help to improve the business. Major risks in text mining are, find out the spelling problems, links, special symbols and irrelevant phases. In this paper main objective is to framing the relationship between different emotions. For this purpose machine learning techniques are applied and evaluated

Keywords: Item reviews · Data pre-processing · Machine learning
Emotion analysis

1 Introduction

There is a huge demand for the E-commerce in the current scenario. To buy any item people will prefer online shopping and they do online search for items. Item reviews are the major factors to influence the people for buying the item. These item reviews given on the web makes major influence on customers to make decision to purchase the item. All the major E-commerce websites like Amazon, flip kart etc. given facilities to provide rating, feedbacks on purchased items. Feedback given by customers will give a major influence on other people who wants to buy the item. Customers may give positive or negative feedback depends on their perception of the item quality and service given by the providers. To identify positive, negative emotion of the customer from the given textual data in item reviews is a tough task.

P. Vasant et al. (Eds.): ICO 2018, AISC 866, pp. 409–417, 2019.
https://doi.org/10.1007/978-3-030-00979-3_43

Every day item reviews are given by the customers, so need to classify the reviews based on the emotion of the customer into positive or negative. Emotions of users are in most of the web applications such as blogs [5], health related applications [4], summary of reviews [3], features of item evaluation [2], investigation

Of items from reviews like e-commerce applications [1].

This manuscript is defined as follows: first section gives the work done on tweets emotion analysis, second section is about methodology, third section is about detailed description of classification techniques which adapted here, fourth section is about result discussion and last conclusion and further enhancement.

2   Earlier Work

Information of the products and large set of item reviews are available in e-commerce application. All are unstructured form. Useful information can extract by applying filtering technique. In prior research, in the field of phrases and expression handling with positive and negative feedback can be found in [10]. Here rule mining for getting relevant polarity have been developed. Required polarity for the sentences was developed with better techniques [9].

To set up right polarity of expressions by subjective recognition [8]. Analysis of emotion is also explored to the different languages other than English with various techniques of feature selection techniques. Different classifiers for classification of text on polarity of text are KNN, SVM, Naïve bayes etc. [7]. Compare to Naïve bayes, SVM and N-gram approaches are giving better performance [6].

Univariate approaches have been implemented for feature extraction based on information gain [15]. Researcher gives new term, maximum discrimination [MD] [14], which is selection of feature implementation using a statistical method. Multivariate methods are used for feature selection in a decision tree model [13]. Binary classification has been developed in [12]. Method for elimination of feature recursively has been implemented in [11]. Genetic algorithm for selection of features in [20]. The limitations of multivariate strategies are computationally costly. Using PMI for include extraction. Analysis of emotions is not limited to twitter information, at the same time use it for stock exchange [16], news articles [15], and political discussions [25].

3   Proposed Work

Here, new approach is by joining the NLP and machine learning model. The proposed technique contains acquiring the date, selection of features, cleaning the data etc. Need to build ML model and predict the emotions of product reviews (Fig. 1).

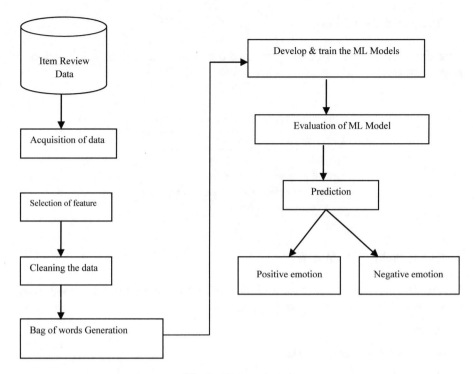

Fig. 1. Proposed work

4 Acquisition of Data and Selection of Feature

Original data used here is collected from Amazon web resource. Which is available openly [22]. It consists of different features as shown in Table 1. Identified few features such as review_rating, review_text, review_title, review_Uname.

Table 1. Features of Amazon item review data

ID	Keys
Asins	Manufacturer
Brand	Name
Categories	Prices
Color	Review_date
Date added	Review_rating
Date updated	Review_text
Dimensions	Rcview_source URL
Review_title	City
Review_Uname	Review_userProvince

5   Preparation of Data

Here original data id transformed to cleaned data. Following actions will be taken to prepare the cleaned data from the original data (Table 2).

Table 2. Action against irrelevant content

Irrelevant content	Action taken
Numbers	Removed
Special symbols	Removed
Upper case letters	Converted to lower case
Stop words	Removed

6   Generating the Words

Cleaned Item reviews are used for extraction of features using N-gram method. Used three grams method to bag of words extraction.

Example

"The Cheetah jumps above the moon"

If x = 3, the n-grams will be:

- the Cheetah jumps
- Cheetah jumps above
- jumps above the
- above the moon

7   Develop the ML Model

After extracting the bag of words next need to develop a ML Model for positive negative emotion prediction. In proposed work, Bernoulli Naïve bays, Random forest ML approaches have been used.

To get the optimal solution, divided the proposed work in to two parts such as coaching phase and anticipating phase.

Algorithm – Coaching phase

Input: Item reviews
Output: Reviews with Sentiments and training ML Model
Method:
1.   Read the reviews.csv file for coaching phase
2.   Eliminate the **irrelevant** content (Numerals, special symbols, converting to lower case, Elimination of stop words) clean data
3.   Save the data obtained from previous step into python data store
4.   Finding the sentiment for reviews which don't have ratings
for each rating of item reviews **do**
   if rating >= 4 **then**
      assign it to positive emotion
  else
      assign it to negative emotion
done
after classifying the sentiments, need to train and develop the machine learning(ML) models such as bernoulli navie bayes, decision tree and mutated random forest
for each review which is missed ratings of item reviews **do**
      Use the developed ML model to predict the sentiments based on
      item reviews.
done
5.   From Stored data repository extract the bag of words using TF and its IDF
6.   Divide the preprocessed data for training (Ratio 20 : 80)
7.   Develop a ML models for investigating item reviews namely positive and negative emotions
8.   Apply Different classifiers to train the machine learning model.
9.   Evaluate the ML models to get the performance in terms of precision recal, accuracy and F1-score.
Algorithm: Coaching Phase Ends

7.1   Implementation and Results

ML models are developed in python. Original data taken from web [21, 22]. In this item reviews data contains one thousand five hindered ninety five reviews on products of Amazon. Out of all those reviews four hundred twenty reviews are no ratings. This is tough task to find out emotions of customer where no ratings. Figure 2. Shows the example review obtained from web resource [21, 22]. After data cleaning, top twenty words from those reviews are shown in Fig. 3. And Fig. 4.

> I initially had trouble deciding between the paper white and the voyage because reviews more or less said the same thing: the paper white is great, but if you have spending money, go for the voyage. Fortunately, I had friends who owned each, so I ended up buying the paper white on this basis: both models now have 300 ppi, so the 80 dollar jump turns out pricey the voyage's page press isn't always sensitive, and if you are fine with a specific setting, you don't need auto light adjustment).It's been a week and I am loving my paper white, no regrets! The touch screen is receptive and easy to use, and I keep the light at a specific setting regardless of the time of day. (In any case, it's not hard to change the setting either, as you'll only be changing the light level at a certain time of day, not every now and then while reading).Also glad that I went for the international shipping option with Amazon. Extra expense, but delivery was on time, with tracking, and I didn't need to worry about customs, which I may have if I used a third party shipping service.

Fig. 2. Sample item review

[('not', 1115), ('fire', 759), ('kindle', 729), ('amazon', 715), ('great', 586), ('use', 564), ('sound', 525), ('one', 496), ('will', 460), ('don't', 450), ('I'm', 447), ('headphones', 442), ('tablet', 437), ('device', 399), ('echo', 397), ('tap', 393), ('screen', 359), ('well', 342), ('good', 336), ('new', 319)]

Fig. 3. Top most frequently words

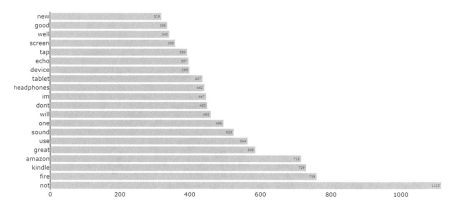

Fig. 4. Top most frequently words in item review data

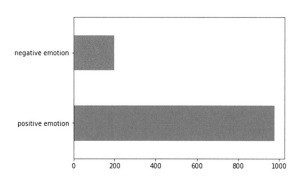

Fig. 5. Total count of positive and negative sentiments

Results if various ML algorithms are shown in Table 3. Figure 6. Displays the graphical representation of accuracy which is obtained from proposed ML algorithm. Table 4 gives the missed rating on the item. Numbers of positive negative sentiments are shown in Fig. 5.

Table 3. Accuracy of different ML techniques

	Bernoulli NB	Decision tree	Mutated random forest
Precision	0.84	0.88	0.88
Recall	0.85	0.86	0.89
Fl-score	0.81	0.82	0.87

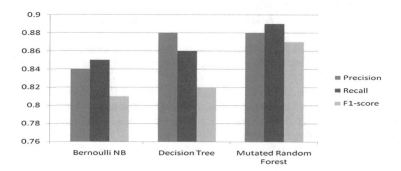

Fig. 6. Comparison of results

Table 4. Missed ratings in item review data

	review_rating	review_text
1	NaN	My previous kindle was a DX: this is my second…
2	NaN	Allow me to preface tills with a Me history…
3	NaN	Just got mine right now. Looks the s.ame as the…
4	NaN	I initially had trouble deciding between the P…
5	NaN	I am enjoying it so far. Great for reading. Ha…

8 Conclusion

Analysis of the consumer emotions based on reviews on the particular item makes more influence on business. It is important to concentrate on the quality of the service given for the item such as useness, helpfulness, utilization of the each feature of the item. In the previous work many complex approaches have been defined for sentiment analysis with different aspects of the items. Online item reviews are kind of a platform for customer to purchase any item. In this aspect there is a need of implementing new methods to manage huge volume of item reviews. These reviews are not structured and need to be pre-process. Irrelevant contents like numerals, special symbols, and stop words are eliminated from original item review data. In this research work the end users sentiment analysis of item is conducted by combining NLP and ML models. Different classifiers such as Bernoulli naive bayes and random forest have been used and mutated random forest found to be best in terms of accuracy. Further, this can be improved for find the sentiments of consumers on other items on specific feature of the item based on item reviews. More ML techniques and lexicon based approaches can be combined.

References

1. www.cnet.com, www.epinion.com. Last accessed 12 Oct 2014
2. Ojokoh, B.A., Kayode, O.: A feature-opinion extraction approach to opinion mining. J.Web Eng. **11**(1), 51–63 (2012)
3. Abulaish, M., Doja. M.N., Ahmad, T.: Feature and Opinion mining for customer review summarization. In: Chaudry, S., et al. (eds.) Pattern Recognition and Machine Intelligence. Lecture Notes in Computer Science, vol. 5909, pp. 219–224 (2009)
4. Goeuriot, L., Na, J.-C., Min Kyaing, W.Y., Khoo, C., Chang, Y.K., Theng, Y.-L., Kim, J.-J.: Sentiment lexicons for health-related opinion mining. In: Proceedings IHI'12 Proceedings of the 2nd ACM SIGHIT International Health Informatics Symposium, pp. 9–226
5. Conrad, J.K., Schilder, F.: Opinion mining in legal blogs. In: Proceedings of ICAIL '07 Proceedings of the 11th International Conference on Artificial Intelligence and Law, pp. 231 – 236
6. Ye, Q., Zhang, Z., Law, R.: Sentiment classification of online reviews to travel destinations by supervised machine learning approaches. Expert Syst. Appl. **36**(3), 6527–6535 (2009)
7. Tan, S., Zhang, J.: An empirical study of sentiment analysis for chinese documents. Expert Syst. Appl. **34**(4), 2622–2629 (2008)
8. Pang, B., Lee, L.: A sentimental education: sentiment analysis using subjectivity summarization based on minimum cuts. In: Proceedings of the 42nd annual meeting on Association for Computational Linguistics. Association for Computational Linguistics, p. 271 (2004)
9. Wilson, T., Wiebe, J., Hoffmann, P.: Recognizing contextual polarity in phrase-level sentiment analysis. In: Proceedings of the Conference on Human Language Technology and Empirical Methods in Natural Language Processing. Association for Computational Linguistics, pp. 347–354 (2005)
10. Hatzivassiloglou, V., McKeown, K.R.: Predicting the semantic orientation of adjectives. In: Proceedings of the Eighth Conference on European Chapter of the Association for Computational Linguistics. Association for Computational Linguistics, pp. 174–181 (1997)
11. Li, F., Yang, Y.: Analysis of recursive feature elimination method. In: Proceedings of the 28th Annual International ACM SIGIR Conference on Research and Development in Information Retrieval. ACM publication, pp. 633–634 (2005)
12. Sylvia Selva Rani, A., Rajalaxmi, R.R.: Unsupervised feature selection by binary bat algorithm. In: 2nd International Conference on Electronics and Communication Systems (ICECS). Published in IEEE Explore (2015)
13. Hwang, Y.-S., Rim, H.-C.: Decision tree decomposition based complex feature selection for text chunking. In: Proceedings of the 9th International Conference on Neural Information Processing, vol. 5, pp. 2217–2222 (2002)
14. Sylvia Selva Rani, A., Rajalaxmi, R.R.: Unsupervised feature selection by binary bat algorithm. In: 2nd International Conference on Electronics and Communication Systems (ICECS). Published in IEEE Explore (2015)

15. Yigit, F., Baykan, O.K.: A new feature selection method for text categorization based on information gain and particle swarm optimization. In: 3rd International Conference on Cloud Computing and Intelligence Systems (CCIS). Published in IEEE Explore 2014. Tavel, P. 2007 Modeling and Simulation Design. AK Peters Ltd

16. Xu, T., Peng, Q., Cheng, Y.: Identifying the semantic orientation of terms using s-hal for sentiment analysis. Know. Syst. **35**, 279–289 (2012)

17. Hagenau, M., Liebmann, M., Neumann, D.: Automated news reading: Stock price prediction based on financial news using context-capturing features. Decis. Support Syst. **55**(3), 685–697 (2013)

18. Zhou, L., Li, B., Gao, W., Wei, Z., Wong, K.: Unsupervised discovery of discourse relations for eliminating intrasentence polarity ambiguities. In: Presented at the 2001 conference on Empirical Methods in Natural Language Processing (EMNLP'11) (2011)

19. Cruz, F.L., Troyano, J.A., Enríquez, F., Ortega F.J., Vallejo, C.G.: Long autonomy or long delay? The importance of domain in opinion mining. Expert Syst. Appl. **40**(8), 3174–3184 (2013)

20. Abualigah, L.M., Khader, A.T, Al-Betar, M.A.: Unsupervised feature selection technique based on genetic algorithm for improving the text clustering. In: 7th International Conference on Computer Science and Information Technology (CSIT) pp. 1–6 (2016)

21. Datafiniti: Instant Access to Every Data Point on the Web, https://datafiniti.co/

22. Consumer Reviews of Amazon Products, https://www.kaggle.com/datafiniti/consumer-reviews-of-amazon-products

23. Fan, T.-K., Chang, C.-H.: Blogger-centric contextual advertising. Expert Syst. Appl. **38**(3), 1777–1788 (2011)

24. Qiu, G., He, X., Zhang, F., Shi, Y., Bu, J., Chen, C.: Dasa: dissatisfaction-oriented advertising based on sentiment analysis. Expert Syst. Appl. **37**(9), 6182–6191 (2010)

25. Maks, I., Vossen, P.: A lexicon model for deep sentiment analysis and opinion mining applications. Decis. Support Syst. **53**(4), 680–688 (2012)

Algorithms for a Bit-Vector Encoding of Trees

Kaoutar Ghazi[1]([✉]), Laurent Beaudou[2], and Olivier Raynaud[2]

[1] GREYC, Caen Normandie University, Caen, France
ghazikawtar@gmail.com

[2] LIMOS, Clermont Auvergne University, Clermont-Ferrand, France

Abstract. A bit-vector encoding is a well known method for representing hierarchies (i.e. partially ordered sets). This encoding corresponds to an embedding of a given hierarchy into a Boolean lattice whose dimension is the encoding's size. Computing an optimal bit-vector encoding, which size is called the 2-*dimension*, is an $\mathcal{NP}$-hard problem. Hence, many algorithms were designed to provide good bit-vector encoding. In this paper, we study tree hierarchies. We analyse previous algorithms for their bit-vector encoding then we point out their common strategy that led us to design a new algorithm improving all the previous ones.

Keywords: Partially ordered set · Tree · Bit-vector encoding
2-dimension · Algorithms

1 Introduction

A hierarchy can be viewed as a partially ordered set $P = (X, \leq)$, or a poset for short, where X is a set of elements and $\leq$ is an order relation (transitive, reflexive and anti-symmetric relation). A poset is often represented by a Hasse Diagram (cf Fig. 1), with an implied upward orientation, where two elements x and y from X are connected, with x drawn below y in the Hasse Diagram, if x is covered by y ($x \prec y$), which means $x \leq y$ and there is no element in between. In this case, y is called an immediate successor of x.

The main issue when handling hierarchies is storing and comparing their elements efficiently, especially when dealing with big data. For these purposes, we need to provide a compact representation of posets. In this paper, we are concerned with the *bit-vector* encoding technique, also known as a hierarchical encoding of posets [1,2,4,7,8].

Let $P = (X, \leq)$ be a poset and $\mathcal{B}_n = (2^{[n]}, \subseteq)$ be the Boolean lattice of size n.

The poset P has a bit-vector encoding of size n if it can be embedded into $\mathcal{B}_n$. In other words, if there exists a mapping ϕ between the elements of P and the elements of $\mathcal{B}_n$, such that for any two elements x and y from X, $x \leq y$ if and only if $\phi(x) \subseteq \phi(y)$.

© Springer Nature Switzerland AG 2019
P. Vasant et al. (Eds.): ICO 2018, AISC 866, pp. 418–427, 2019.
https://doi.org/10.1007/978-3-030-00979-3_44

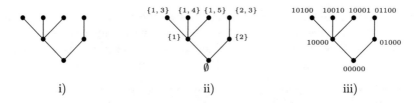

Fig. 1. (i) A Hasse Diagram of a poset P, (ii) An embedding of P into $\mathcal{B}_5$, (iii) A bit-vector encoding of P of size 5.

This encoding will assign, to each element x of P, a bit-vector V_x of size n, where its ith bit is set to one if i belongs to $\phi(x)$, and zero otherwise. So, $x \leq y$ if and only if $V_x \leq_{lexico} V_y$. For each element x of P, V_x (also $\phi(x)$) designates the code of x.

Computing an optimal bit-vector encoding of a post P, i.e. the smallest positive integer n such that P can be embedded as a subposet of $\mathcal{B}_n$, which is called the 2-*dimension* of P [11] and denoted $dim_2(P)$, is an $\mathcal{NP}$-hard problem [6]. Thereby, there is a motivation to provide algorithms that give a good bit-vector encoding for general and particular cases.

The bit-vector encoding of tree hierarchies have been the subject of several studies [1–3,5,8,9]. In this paper, we first present some results about the 2-dimension of trees (Sect. 2), then we expose the chronological improvement of the three last algorithms for trees bit-vector encoding (*Dichotomic, Polychotomic* and *Generalized Polychotomic* encodings) and point out their common processing strategy (Sect. 3). Finally, we design a new algorithm for trees bit-vector encoding and measure its performance (Sect. 4).

Note that all variables used throughout this paper are non-negative integers.

2   Trees Hierarchies

Hierarchical data having a parent-child relationship are called directed trees or simply *trees*. Formally, a tree $T = (X, \leq)$ is a poset where each element y, except the root, covers only one other element x ($x < y$). When each element in a tree is covered by only one other element, then the tree is called a *chain*.

The problem of computing the 2-dimension of trees is still an open problem. However, it is approximable for any tree (Proposition 1) and computed in polynomial time for some special tree cases (Proposition 2).

Proposition 1 *There is a polynomial-time algorithm which gives a 4-approximation of the 2-dimension when restricted to trees [6].*

Proposition 2 *Let T be a tree with n elements.*

- *If T is a chain, then $dim_2(T) = n - 1$ (Folklore).*
- *If T is an antichain (or a tree of height 1), then $dim_2(T) = sp(n)$ where $sp(n)$ is equal to $\min\{k | \binom{k}{\lfloor \frac{k}{2} \rfloor} \geq n\}$ [10].*

Until now, there is no polynomial-time algorithm that computes the 2-dimension of any tree. Despite this, there exist several algorithms for a good bit-vector encoding of trees. These algorithms provide upper bounds for the 2-dimension of trees (see Proposition 3 for trivial known bounds).

Proposition 3 (Folklore) *Let T be a tree of height h, with n elements and l leaves. Then, $\max(h, sp(l)) \leq dim_2(T) \leq n$.*

3   Analysis of Previous Algorithms for a Bit-Vector Encoding of Trees

Bit-vector encoding of trees has been first studied by Caseau in 1993, who developed the *Cmax* encoding then improved it with Habib *et al.* in 1999 by the *CHNR* encoding. In 2001, Raynaud and Thierry proposed the *Dichotomic* encoding improving the earlier ones. Algorithms of these encodings are given in [6].

For a *Dichotomic* encoding of trees, the authors designed the *Dicho* algorithm based on embedding a given tree into a binary one through the process described in [6]. This process consists in associating a weight to each element of the tree following a top-down topological order of the tree's elements. The computation of the weight of an element requires computing the weights of all its children first. If the element has no children, then it is a leaf and it weights zero. An element with only one child gets the weight of its child plus one. When it has exactly two children, it gets the maximum weight of its children plus two. Otherwise, the element has at least 3 children, let a and b be its children with smallest weights. In this case, we add a new element to the tree as a child of the current element and a parent of a and b, omit the transitive relationships, then iterate on the new tree. Finally, the size of a bit-vector encoding of the tree, computed by the *Dicho* algorithm, corresponds to the weight assigned to its root and it is the smallest weight computed by a *Dichotomic* encoding (Theorem 4). We mention that the weight assigned to each element of the tree is the size of a bit-vector encoding of the subtree rooted at this element, computed by the *Dicho* algorithm.

Theorem 4. *The Dicho algorithm generates a Dichotomic encoding of minimum size [6].*

In 2002, Filman formulated the behaviour of the *Dicho* algorithm by the function $\mathcal{D}$, defined in Formula 1, that computes the weight of an element from an increasing sequence of integers referring to the weights of its children. Let $S = [s_1, \ldots, s_n]$ be this sequence, with $s_1 \leq s_2 \leq \cdots \leq s_n$. Then,

$$\mathcal{D}(S) = \begin{cases} 0 & |S| = 0 \\ s_1 + 1 & |S| = 1 \\ s_2 + 2 & |S| = 2 \\ \mathcal{D}([s_3, \ldots, s_2 + 2, \ldots, s_n]) & \text{otherwise} \end{cases} \tag{1}$$

The process of the *Dicho* algorithm is then briefly described by the computation of the weight of each element of a given tree using the function $\mathcal{D}$

and following a top-down topological order of the elements of the tree. Thanks to this formulation, we give a formal description of a *Dichotomic* encoding in Definition 1.

Definition 1. *Let $\mathcal{A}$ be a function from increasing integer sequences to integers, and let S be an increasing sequence of integers $[s_1, \ldots, s_n]$. The function $\mathcal{A}$ is a Dichotomic encoding of S if*

$$\mathcal{A}(S) = \begin{cases} \mathcal{D}(S) & |S| \leq 2 \\ \mathcal{A}(S \backslash \{s_i, s_j\} \cup [s_j + 2]) & \forall i, j \in [\![1, n]\!] \text{ with } s_i \leq s_j \quad \text{otherwise} \end{cases} \quad (2)$$

According to Formula 1 and the following Proposition 5, we can say that the *Dicho* algorithm assigns optimum weights for leaves and elements with one child. Thus, going forward, we deal with the weight of elements with at least two children.

Proposition 5 *Let T be a tree rooted at r.*

- *If T is reduced to r, then $dim_2(T) = 0$.*
- *If r has a unique immediate successor, then $dim_2(T) = dim_2(T \backslash \{r\}) + 1$.*

In [5], Filman introduces the notion of a flat sequence as a sequence of integers $[s_1, \ldots, s_n]$ in which $s_n - s_2 < 2$. We extend this definition, then we show that the *Dicho* algorithm always generates a flat-sequence when computing the weight of an element.

Definition 2 (flat-sequence). *A sequence $[s_1, \ldots, s_n]$ is called a k-**flat-sequence** –or simply a flat-sequence– if one of these three cases occur:*

- *$s_n - s_2 < 2$ and $s_n = k$.*
- *$s_n = s_2 = k - 1$.*
- *$n = 1$ and $s_1 \leq k$.*

Definition 3 (k-Flat partitioning). *A partition of a sequence of integers S into subsequences $X_1, \ldots, X_m$ is called a k-**flat partitioning** of S by some function p if the sequence $[p(X_i)|X_i \neq [k] \text{ and } 1 \leq i \leq m]$ is a k-flat-sequence.*

Let S be an increasing sequence of integers $[s_1, \ldots, s_n]$. We observe that for any integer i with $1 \leq i \leq n$, $\mathcal{D}(S) = \mathcal{D}(X_i \cup [s_{i+1}, \ldots, s_n])$ such that X_i is a s_i-flat-sequence. We justify this observation as follows:

We already have $S = [s_1] \cup [s_2, \ldots, s_n] = [s_1, s_2] \cup [s_3, \ldots, s_n]$, with $[s_1]$ a s_1-flat-sequence and $[s_1, s_2]$ a s_2-flat-sequence by Definition 2. We assume that $\mathcal{D}(S) = \mathcal{D}(X_i \cup [s_{i+1}, \ldots, s_n])$, for some integer i, and X_i is a s_i-flat-sequence. If $X_i \cup [s_{i+1}]$ is a s_{i+1}-flat-sequence, then $\mathcal{D}(S) = \mathcal{D}(X_{i+1} \cup [s_{i+2}, \ldots, s_n])$. Otherwise, to compute $\mathcal{D}(X_i \cup [s_{i+1}, \ldots, s_n])$, the *Dicho* algorithm replaces the smallest values $a \leq b$ of the sequence $X_i \cup [s_{i+1}, \ldots, s_n]$ by $b + 2$, then iterates on the resulting sequence and repeats the same process until getting some sequence of integers $[c_1, \ldots, c_k, s_{i+1}, \ldots, s_n]$ in which c_1 and $s_{i+1} - 1$ or s_{i+1} are the smallest elements, which means $s_{i+1} - c_2 < 2$. In this case, $X_{i+1} = [c_1, \ldots, c_k, s_{i+1}]$

is a s_{i+1}-flat-sequence, and $\mathcal{D}(S) = \mathcal{D}(X_{i+1} \cup [s_{i+2}, \ldots, s_n])$. We conclude the correctness of our observation. By taking $i = n$, we confirm the existence of a s_n-flat-sequence, X_n, such that $\mathcal{D}(S) = \mathcal{D}(X_n)$. We claim that the *Dicho* algorithm generates a $\max(S)$-flat-sequence from any sequence of integers S (for $S = [s_1, \ldots, s_n]$, we have $\max(S)$ is s_n). We let F_d^S denote this flat-sequence and let $f_d(S)$ denote its size.

The process of generating a $\max(S)$-flat-sequence from a sequence S by the *Dicho* algorithm, can be also viewed as a $\max(S)$-flat partitioning of S by $\mathcal{D}$ of size $f_d(S)$, as illustrated in Fig. 2.

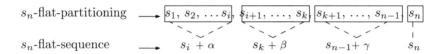

Fig. 2. A flat-partitioning generated by the *Dicho* algorithm

In general, for any integer $k \geq \max(S)$, the *Dicho* algorithm produces a k-flat-sequence from S, when computing $\mathcal{D}(S)$. Let $F_d^{S_k}$ denote this sequence and let $f_d^k(S)$ denote its size.

According to the previous observation, we can reformulate the *Dicho* algorithm's behaviour (Function 1) as follows:

$$\mathcal{D}(S) = \mathcal{D}(F_d^{[s_1, \ldots, s_i]} \cup [s_{i+1}, \ldots, s_n]) = \mathcal{D}(F_d^S) = \mathcal{D}(F_d^{S_k}) - \left\lfloor \frac{1}{|F_d^{S_k}|} \right\rfloor. \quad (3)$$

In [5, Lemma 3], Filman defines an upper bound and a lower bound of $\mathcal{D}(S)$, for S a flat-sequence, by

$$\max(S) + 2\lceil \log_2(|S|) \rceil - 1 \leq \mathcal{D}(S) \leq \max(S) + 2\lceil \log_2(|S|) \rceil. \quad (4)$$

The upper bound is reached when S is a flat-sequence with all elements, except at most one element, equal to $\max(S)$ (Proposition 6).

Proposition 6 *Let S be an increasing sequence of integers $[s_1, \ldots, s_n]$ such that $s_2 = s_n$ and $n > 1$. Then $\mathcal{D}(S) = s_n + 2\lceil \log_2(n) \rceil$.*

We deduce from Eq. 3 and Inequality (4) that, for any sequence of integers S, $\mathcal{D}(S)$ is either equal to $\max(S) + 2\lceil \log_2(f_d(S)) \rceil - 1$ or $\max(S) + 2\lceil \log_2(f_d(S)) \rceil$. As well, minimizing the value of $\mathcal{D}(S)$ can be ensured through an efficient encoding of flat-sequences. Actually, that was the idea of Filman when he proposed the *Polychotomic* encoding [5] formulated as follows.

$$\mathcal{P}(S) = \begin{cases} s_n + sp(n) & S \text{ is a flat-sequence} \\ \mathcal{P}([s_3, \ldots, s_2 + 2, \ldots, s_n]) & \text{otherwise} \end{cases} \quad (5)$$

Remark that the *Polychotomic* encoding works similarly to the *Dicho* encoding until getting a $\max(S)$-flat-sequence, then it returns

$$\mathcal{P}(S) = \max(S) + sp(f_d(S)).\qquad(6)$$

Filman has proved that $\mathcal{P}(S) \leq \mathcal{D}(S)$ for any sequence of integers S [5, Theorem 1]. This result can be also proved by checking that 1) for any integer $n > 4$, $sp(n) \leq 2\lceil \log_2(n) \rceil - 1$ (see Inequality (4) and Eq. 6), with

$$\lfloor \log_2(n) + \frac{\log_2(\log_2(n))}{2} + 1 \rfloor \leq sp(n) \leq \lfloor \log_2(n) + \frac{\log_2(\log_2(n))}{2} + 2 \rfloor,\quad(7)$$

(see [6] for a proof of (7)) and 2) $\mathcal{P}(S) \leq \mathcal{D}(S)$ for $n \leq 4$.

In accordance to Eq. 6 and knowing that sp is an increasing function, one can ask if there exists a *Dichotomic* encoding that provides, from a sequence of integers S, a $\max(S)$-flat-sequence with size less than $f_d(S)$, in order to improve $\mathcal{P}(S)$. In Theorem 7, we prove that $f_d(S)$ is the minimal value computed by a *Dichotomic* encoding (Definition 1).

Theorem 7. *Let S be an increasing sequence of integers and k be an integer greater or equal to $\max(S)$. Then, $f_d^k(S)$ is the smallest size of a k-flat-sequence generated from S by a Dichotomic encoding.*

The *Polychotomic* encoding was improved by Colomb *et al.* in 2008 [3], who proposed the *Generalized Polychotomic* encoding formulated as follows:

$$\mathcal{G}(S) = \begin{cases} s_n + sp(n) & S \text{ is a flat-sequence} \\ \mathcal{G}([s_k + sp(k), s_{k+1}, \ldots, s_n]) & \exists k \leq n - 1 \text{ with} \\ & s_k - s_2 < 2 \text{ and } s_k + sp(k) \leq s_{k+1} \\ \mathcal{G}([s_3, \ldots, s_2 + 2, \ldots, s_n]) & \text{in other cases} \end{cases}$$

$$(8)$$

The *Generalized Polychotomic* encoding is based on the same idea as the *Polychotomic* encoding, which is applying the steps of the *Dicho* algorithm until a new heaviest element would be created, i.e. a $\max(S)$-flat-sequence would be produced. Let F_g^S denote this flat-sequence and let $f_g(S)$ denote its size. Then, it returns

$$\mathcal{G}(S) = \max(S) + sp(f_g(S)).\qquad(9)$$

However, it differs from the *Polychotomic* encoding in the fact that it checks, before merging the two elements with minimum weights (i.e. run a one-step of the *Dicho* algorithm), if it is possible to merge the k smallest elements (the k first ones), when $\mathcal{G}([s_1, \ldots, s_k]) \leq s_{k+1}$.

By taking the greatest integer k such that $\mathcal{G}([s_1, \ldots, s_k]) \leq s_{k+1}$, we can reformulate the *Generalized Polychotomic* encoding as follows:

$$\mathcal{G}(S) = \begin{cases} \mathcal{P}([\mathcal{G}([s_1, \ldots, s_k]), s_{k+1}, \ldots, s_n]) & \text{with } k > 1 \text{ is the greatest integer s.t.} \\ & \mathcal{G}([s_1, \ldots, s_k]) \leq s_{k+1} \\ \mathcal{P}(S) & \text{otherwise} \end{cases}$$

$$(10)$$

The authors in [3] proved that $\mathcal{G}(S) \leq \mathcal{P}(S)$. So, the main reason why the *Generalized Polychotomic* is more efficient, is the fact that it produces a flat-sequence with smaller size (see Eqs. 6 and 9). We evaluate the difference between $f_g(S)$ and $f_d(S)$ in Theorem 8.

Theorem 8. *Let S be an increasing sequence of integers. Then, $f_d(S) - 1 \leq f_g(S) \leq f_d(S)$.*

We conclude that the common strategy used by algorithms discussed in this section consists in:

- Producing a max(S)-flat-sequence (or a max(S)-flat partitioning) from S.
- Encoding this flat-sequence.

In this paper, we focus on algorithms that provide a max(S)-flat-sequence, or a max(S)-flat partitioning of S, with smaller size. The next section presents *the Generalized Contiguous Partitioning* encoding designed for this purpose.

4   A New Algorithm for a Bit-Vector Encoding of Trees

4.1   Generalized Contiguous Partitioning Encoding

Let S be an increasing sequence of integers $[s_1, \ldots, s_n]$ and let $P = \{X_1, \ldots, X_k\}$ be a partitioning of S. The partitioning P is contiguous if for every i and j such that $1 \leq i \leq j \leq k$, we have $\max(X_i) \leq \min(X_j)$. Recall that $n \geq 2$.

The algorithm described by the function $\mathcal{C}$ (Function 11) produces a s_n-flat-sequence from a partitioning of S which is contiguous and minimal since each subsequence is maximal with respect to be encoded by at most s_n bits. Let $f_c(S)$ be the size of this partitioning.

$$\mathcal{C}(S) = \begin{cases} s_n + sp(n) & S \text{ is a flat-sequence} \\ \mathcal{C}([s_{k+1}, .., \max\{\mathcal{C}([s_1, .., s_k]), s_n\}, .., s_n]) & \exists k \text{ such that } k < n-1 \\ & \text{and } \mathcal{C}([s_1, \ldots, s_k]) \leq s_n \\ & \text{and } \mathcal{C}([s_1, \ldots, s_k, s_{k+1}]) > s_n \\ s_n + 2 & \text{in other cases} \end{cases}$$

$$(11)$$

Proposition 9 *The function $\mathcal{C}$ is not monotonous, also, it does not always improve the result returned by the function $\mathcal{G}$.*

In order to improve the *Generalized Polychotomic* encoding, we propose the *Generalized Contiguous Partitioning*, as an efficient bit-vector encoding of trees, based on the weight function $\mathcal{GC}$, by:

$$\mathcal{GC}(S) = \begin{cases} \mathcal{C}([\mathcal{G}([s_1, \ldots, s_k]), s_{k+1}, \ldots, s_n]) & \text{with } k > 1 \text{ the greatest integer s.t.} \\ & \mathcal{G}([s_1, \ldots, s_k]) \leq s_{k+1}. \\ \mathcal{C}([s_1, \ldots, s_n]) & \textit{otherwise} \end{cases}$$

$$(12)$$

The function $\mathcal{GC}$ produces a $\max(S)$-flat-sequence from S, let $f_{gc}(S)$ be its size. Then,

$$\mathcal{GC}(S) = \max(S) + sp(f_{gc}(S)). \tag{13}$$

The *Generalized Contiguous Partitioning* encoding principle is to associate a weight for each element of a tree, computed by the function $\mathcal{GC}$ with S the sequence referring to the weights of the element's children. When computing the weight of an element, this encoding embeds the given tree into another one, exactly as the *Dicho* algorithm does (merging a set of integers $s_1, \ldots, s_k$ into one integer t, means the adding of a new element with weight t as a parent of elements of weights $s_1, \ldots s_k$ respectively and as a child of their former parent in the tree). Then, it encodes the resulting tree by the *CHNR* [2] algorithm in order to provide a bit-vector encoding of the initial tree. The following two lemmas are required to prove that the *Generalized Contiguous Partitioning* algorithm for trees bit-vector encoding improves all previous ones.

Lemma 10 *Let S be an increasing sequence of integers $[s_1, \ldots, s_n]$. Then, for a given positive integer k, we have*

$$f_c(S) \leq f_c([s_1, \ldots, s_k, s_n]) - 1 + f_c([s_{k+1}, \ldots, s_n]).$$

Lemma 11 *Let S be an increasing sequence of integers $[s_1, \ldots, s_n]$.*
If there exists an integer k such that $\mathcal{D}([s_1, \ldots, s_k, s_{k+1}]) > s_n$, then

$$1 + f_d([s_{k+1}, \ldots, s_n]) \leq f_d(S).$$

4.2   Theoretical Result

In this subsection, we prove that the size of a *Generalized Contiguous Partitioning* encoding is less than the size of a *Generalized Polychotomic* encoding (Theorem 12). Recall that this size corresponds to the weight computed by the function $\mathcal{GC}$ (resp. $\mathcal{G}$) for the root of some given tree.

Theorem 12. *Let S be an increasing sequence of integers. Then,*

$$f_{gc}(S) \leq f_g(S) \quad implies \quad \mathcal{GC}(S) \leq \mathcal{G}(S).$$

4.3   Experimental Results

The *Generalized Contiguous Partitioning* encoding improves all previous algorithms for trees encoding (Theorem 12). However, to evaluate the rate of this improvement, we need to examine results of our encoding on some hierarchies, preferably natural ones. Benchmarks used for our test are hierarchies of programming languages published by Krall[1].

---

[1] http://www.complang.tuwien.ac.at/andi/typecheck/

The Table 1 lists some characteristics of these hierarchies. The size of a hierarchy (a poset) corresponds to the number of its elements while MaxChildren refers to the maximum number of the children of an element of this hierarchy. In the second table, we present, from the left to the right, the size of a bit-vector encoding of each data from the benchmarks using the *Dichotomic*, *Polychotomic*, *Generalized Polychotomic* and the *Generalized Contiguous Partitioning* encodings (Table 2).

Table 1. Hierarchies characteristics

Data	Size	Height	Max children
VisualWorks2	1957	15	181
Digitalk3	1357	14	141
NeXTStep	311	8	142
ET++	371	9	87

Table 2. Experimental results

Data	$\mathcal{D}$	$\mathcal{P}$	$\mathcal{G}$	$\mathcal{GC}$
VisualWorks2	32	30	29	**27**
Digitalk3	29	28	28	**27**
NeXTStep	20	19	17	**17**
ET++	20	20	19	**18**

Results show a slight improvement on the encoding size for hierarchies. However, this improvement provides a significant gain in term of the memory space needed to store these hierarchies. For instance, while we decrease the size of a bit-vector encoding of VisualWorks2 hierarchy by only 2 bits, we reduce the memory space needed to store this hierarchy by around 4000 bits.

5   Discussion

The computation of an optimal bit-vector encoding of posets, i.e. the 2-dimension, is an $\mathcal{NP}$-hard problem. However, for the class of trees, the complexity is unknown; it is conjectured to be polynomial [6, Conjecture 37]. Actually, the problem is solved for only a few very simple cases of trees (chains, antichains, comb trees, etc.). Thereby, previous works have been focused on approximating the 2-dimension of trees or improving their encoding.

Habib *et al.* [6] proved that the *Dicho* algorithm is a 4-approximation of the 2-dimension of trees. In the same paper, they also conjectured that this algorithm is a 2-approximation of the 2-dimension of trees and settled the case for 4-ary trees. Several algorithms were proposed in the literature to improve the *Dicho* algorithm. In this paper, we present a general framework that allows a better understanding of these algorithms and we propose a new algorithm improving all the previous ones.

References

1. Caseau, Y.: Efficient handling of multiple inheritance hierarchies. In: Proceedings of OOPSLA (1993)
2. Caseau, Y., Habib, M., Nourine, L., Raynaud, O.: Encoding of Multiple Inheritance Hierarchies. Computational Intelligence (1999)
3. Colomb, P., Raynaud, O., Thierry, E.: Generalized polychotomic encoding: a very short bit-vector encoding of tree hierarchies. In: MCO (2008)
4. Fall, A.: The Foundations of Taxonomic Encodings. Computational Intelligence (1998)
5. Filman, R.E.: Polychotomic encoding: a better quasi-optimal bit-vector encoding of tree hierarchies. In: Proceedings of ECOOP (2002)
6. Habib, M., Nourine, L., Raynaud, O., Thierry, E.: Computationel aspects of the 2-dimension of partially ordered sets. Theor. Comput. Sci. (2004)
7. Habib, M., Nourine, L.: Bit-vector encoding for partially ordered sets. In: Proceedings of the International Workshop on Orders, Algorithms, and Applications (1994)
8. Krall, A., Vitek, J., Horspool, R.N.: Near optimal hierarchical encoding of types. In: Proceedings of Ecoop (1997)
9. Raynaud, O., Thierry, E.: A quasi optimal bit-vector encoding of tree hierarchies. In: Proceedings of ECOOP 2001 Application to Efficient Type Inclusion Tests (2001)
10. Sperner, E.: Ein satz uber untermengen einer endlichen menge. Math. Z (1928)
11. Trotter, W.T.: Embedding finite posets in cubes. Discret. Math. (1975)

Dynamic Programming Solution to ATM Cash Replenishment Optimization Problem

Fazilet Ozer[1], Ismail Hakki Toroslu[1], Pinar Karagoz[1(✉)], and Ferhat Yucel[2]

[1] METU Computer Engineering Department, 06800 Cankaya Ankara, Turkey
{fazilet.ozer,toroslu,karagoz}@ceng.metu.edu.tr
[2] Intertech Bilgi Islem Ve Pazarlama Ticaret A.S., Istanbul, Turkey
Ferhat.Yucel@intertech.com.tr

Abstract. Automated Telling Machine (ATM) replenishment is a well-known problem in banking industry. Banks aim to improve customer satisfaction by reducing the number of out-of-cash ATMs and duration of out-of-cash status. On the other hand, they want to reduce the cost of cash replenishment, also. The problem conventionally has two components: forecasting ATM cash withdrawals, and then cash replenishment optimization on the basis of the forecast. In this work, for the first component, it is assumed that reliable forecasts are already obtained for the amount of cash needed in ATMs. We focus on the ATM cash replenishment component, and propose a dynamic programming based solution. Experiments conducted on real data reveal that the solutions of the baseline approaches have high cost, and the proposed algorithm can find optimized solutions under the given forecasts.

1 Introduction

According to World Bank reports[1], the number of Automated Teller Machines (ATMs) all over the world increased by about 2.5 times within the last ten years. The increase in the use of ATMs facilitates banking services for both customer and banks, especially for simple and standard services such as cash withdrawal. On the other hand, additional ATM management costs arise for banks. One of the well-known ATM management problems is *cash replenishment optimization*, which mainly focuses on how often and how much cash to be loaded to an ATM in each cash replenishment period. The problem contains two optimization criteria. First of all, banks aim to reduce the amount of *idle* cash (i.e. cash that was loaded and was not withdrawn from ATM for a period of time), since this amount of cash can not be utilized in a profitable way, thus it is considered as a loss. Therefore, it is aimed to avoid loading more amount of cash than needed. This cost is calculated as an interest lost in terms of the number of days cash stays in ATM idle. We call this cost as *interest cost*. On the other hand, loading

---

[1] https://data.worldbank.org/indicator/FB.ATM.TOTL.P5?view=chart.

© Springer Nature Switzerland AG 2019
P. Vasant et al. (Eds.): ICO 2018, AISC 866, pp. 428–437, 2019.
https://doi.org/10.1007/978-3-030-00979-3_45

small amount of cash causes *out-of-cash* ATMs, and this is an important problem that affects *customer satisfaction* considerably. Additionally, cash replenishment incurs a cost involving cash transportation and loading process to an ATM. We call this cost as *loading cost*. Hence, it is important to reduce the frequency of replenishment where possible. We call the total cost generated by interest and loading costs as *replenishment cost*. ATM replenishment optimization is based on keeping these factors balanced.

This problem can be divided into two steps: forecasting how much cash to be withdrawn each day, and finding an optimization algorithm for cash replenishment schedule. For the first step, we assume that a reliable forecast for the amount of cash to be withdrawn each day for a period of time (typically for a week) is available. There are several works focused on this first phase of the problem [1,6,7,9,10].

The focus of this work is on the second step. Given the reliable forecast, we propose a *dynamic programming* based solution for ATM cash replenishment, such that ATM is *never out-of-cash*, and the cost of replenishment and cash utilization is optimized. Assuming that maximum replenishment frequency is daily, loading only the required amount of daily cash does not create any interest cost while maximizing the loading cost. On the other hand, for the lowest cash loading frequency, such as weekly, the loading cost is minimized, but, the interest cost is maximized.

In the literature, there is a limited set of studies that are related to the ATM replenishment problem that we have introduced in this paper. In [2], linear programming approach is used for solving optimum cash replenishment routing problem of an ATM network. In [5], mixed integer programming based approach is developed to solve the cash replenishment problem for a set of ATMs where cash is supplied from another set of cash centers. In [3], ATM withdrawal forecasts are used and a simulation based optimization solution is developed for the cash replenishment decision. [4] also focuses on solving both the routing and optimum replenishment of a set of ATMs, None of these studies are the same as our problem, and, to the best of our knowledge this is the first introduction of the ATM cash replenishment optimization problem which tries to determine the optimum loading times for a given period for the given interest cost (obtained from the interest rate) and the fixed cash loading cost for each replenishment operation.

We modeled the ATM cash replenishment problem similar to the *matrix chain multiplication* problem, such that n consecutive days ATM replenishment is modeled similar to the multiplication of a sequence of n matrices. Thus, we develop a dynamic programming solution to this problem. We also present dynamic programming based optimized replenishment on a set of cases in comparison to baseline approaches.

The paper is organized as follows. In Sect. 2, we present the proposed optimization method. In Sect. 3, the details of the method is illustrated on an example. In Sect. 4, experiments conducted on a real world data set in comparison to baseline approaches are presented. The paper is concluded with an overview and future work in Sect. 5.

2   Solving ATM Cash Replenishment Optimization Problem with Dynamic Programming

Banks need to find out a way to optimize how much cash and how frequent to load cash into each ATM machine. Loading cash to an ATM has a cost independent from the amount loaded. We can reduce this cost by trying to reduce the number of replenishments. However, that means loading larger amounts each time an ATM is loaded, which generates an interestcost for each day that cash stays in the ATM. Therefore, an optimized solution tries to reduce the number of replenishment to decrease the loading cost and reduce the amount loaded into an ATM to reduce the interest cost. These two objectives are contradictory and therefore the optimum solution should do these decisions to minimize the overall cost.

Many optimization problems are solved efficiently by using dynamic programming approach. In dynamic programming, the problem is divided into subproblems and the results of these subproblems are combined to generate the results of the larger problem instance. There are several well-known problems solved with different dynamic programming algorithms. *Matrix chain product* is one of the most well-known problems solved with dynamic programming method. Its solution approach has been applied to variety of other optimization problems, such as query optimization [8].

Matrix chain multiplication problem - as the name implies - basically aims to find out the most efficient way of multiplying a sequence of matrices. In order to find the most efficient way of doing this operation, the order of the multiplications should be determined. Since, the matrix multiplication operation is associative the aim of the matrix chain product problem is to determine how to put parenthesis around the matrixpairs (input matrices or the ones obtained from previous multiplications) to execute the whole sequence of multiplication operation. Due to the associativity of the matrix multiplication operation, this parenthesization operations does not effect the result, but it effects the multiplication cost (i.e., the number of individual multiplications). Hence, we have to find out how to place parenthesis in order to keep the multiplication cost at the minimum.

Consider an instance of matrix chain product problem where each matrix i has dimensions p_i and p_{i+1}. Assume that, we want to determine the minimum cost paranthesization of a sequence of matrices i, $i+1$, ..., j to be multiplied whose cost is represented as $m[i,j]$. The recurrence relation of this matrix chain product problem is defined as follows:

$$m[i,j] = min \begin{cases} min_{i \leq k < j}(m[i,k] + m[k+1,j] + p_{i-1}p_kp_j) \ if \ i < j \\ 0 \ if \ i = j \end{cases} \quad (1)$$

Matrix chain multiplication problem can be adopted to ATM replenishment optimization problem easily. Matrix chain multiplication aims to find how to locate parenthesis to have the lowest matrix multiplication cost. On the other hand, this problem aims to find out when and how much cash to load to ATM in order to have the lowest cost. There are more parameters in our problem such as interest rate, and also the cost calculation is a bit different, but matrix chain multiplication problem can be thought as a base for this problem and thus, the same approach can be used for our problem with minor changes.

ATM replenishment optimization problem and matrix chain product optimization problem are very similar to each other. Both optimization problems between i to j (matrices from i to j or days from i to j) can be solved by exploring all pairs of smaller instances already solved optimally between i and j (that is by considering all instance pairs as i to k and $k+1$ to j for all k values between i and j). However, they also have some differences:

- In the matrix chain product problem an instance with a single matrix has no (multiplication) cost. However, in the ATM cash replenishment problem an instance of a single day has a loading cost.
- In the matrix chain product problem, there is a cost to combine two solved instances, which corresponds to the cost of multiplying two matrices. On the other hand, in the ATM replenishment problem the cost of combining two smaller solutions to generate the solution of larger problem has no associated cost. That is, if we want to determine the cost of the solution from day i to day j, and, if we already have the solutions for day i to k and day $k+1$ to j' we can just add these two solutions.
- In the matrix chain product problem only the solutions generated previously for smaller instances are needed to choose the minimum cost multiplication of matrices. However, in the ATM cash replenishment problem, for the solution of larger problem instance, in addition to the already solved smaller problem instances, we also have to consider the single loading solution of the large problem instance as well.

Let I be the accumulated interest cost matrix for keeping the cash in ATM for days from i to j and α be the loading cost, then, the recurrence relation of the ATM cash replenishment problem where $c[i,j]$ is the minimized replenishment cost for an ATM from day i to j (including j) can be defined as follows:

$$c[i,j] = min \begin{cases} min_{i \leq k < j}(c[i,k] + c[k+1,j]) \\ \sum_{r=i}^{j} I[i,j] + \alpha \end{cases} \qquad (2)$$

3   Solution Details

The recurrence relation given in the previous section defines the way the ATM replenishment optimization problem is solved. Moreover, there are some other details that need to be defined on this solution. In this section, we will describe these details and the whole solution process with a simple example. Consider a simple instance of ATM Replenishment Problem for 5 days with the following inputs:

Before calculating the values of the recurrence relation c we have to calculate the values of Interest cost matrix firstly. In order to calculate the interest costs (I values) the following process is applied:

Step 1: Calculate the total amount of money for days from 1 to n under the given interest rate r. In Table 1 the rows correspond to the days from 2 to 5, and the columns corresponds to the days from 0 to 5 where interest can be applied. An entry at row i and column j corresponds to the amount the money of day i will become with the given interest rate r in j days. The entries that are not calculated left empty. For example, for the day 2, the cash can be in the ATM at most for 1 day, if it is loaded at day 1. Thus, its interest applies for only 1 day. The calculations are performed by using Eq. 3.

$$amount\_with\_interest = amount * (1 + r)^{(number\_of\_days)} \qquad (3)$$

Table 1. Accumulated interests

Amount with interest	Day 0	Day 1	Day 2	Day 3	Day 4
Day 2 (200)	200	202			
Day 3 (100)	100	101	102.01		
Day 4 (300)	300	303	306.03	309.0903	
Day 5 (100)	100	101	102.01	103.0301	104.0604

Number of days: $n = 5$
Amount per days: [100, 200, 100, 300, 100]
Interest rate: r=0.01 *(i.e., 1% per day)*
Loading cost: $\alpha = 5$

Step 2: Calculate the interest cost. If we extract the amount, then we will find the actual interest cost. Calculations are done based on Eq. 4. Table 2 contains the interest cost for the amount of each day for the required days.

$$interest\_cost = amount\_with\_interest - amount \qquad (4)$$

Table 2. Interest costs

Interest cost	Day 0	Day 1	Day 2	Day 3	Day 4
Day 2 (200)	0	2			
Day 3 (100)	0	1	2.01		
Day 4 (300)	0	3	6.03	9.09	
Day 5 (100)	0	1	2.01	3.03	4.06

Step 3: Calculate the accumulated interest cost. When we load the cash at day 1 and then the next loading is at day 4, that means we need to load 3 days required cash at day 1 (i.e., for days 1, 2 and 3). Thus, for day 2 we will pay an interest cost for 1 day and for day 3 we will pay interest cost for 2 days, That is why we need to calculate the accumulated interest costs for loading the cash at day i until to day j. That means the next loading is on day $j + 1$.

The calculation is done by Eq. 5. In Table 3 the rows correspond to loading days, and the columns correspond to the day until which the loading is done.

$$I[i,j] = \sum_{k=1}^{j-1} \{IF \ (i+k-1 \leq n) \ THEN \ interest\_cost[i+k-1,k] \ ELSE \ 0\} \quad (5)$$

Table 3. Accumulated interest costs

I	Day 1	Day 2	Day 3	Day 4	Day 5
Day 1		2	4.1	13.19	17.25
Day 2			1	7.03	10.06
Day 3				3	5.01
Day 4					1

After the matrix I is generated we can use it to solve our main optimization problem. Table 4 shows the calculation of the optimized values for the recurrence relation c defined in the previous section for the same example.

Table 4. Optimized costs

c	Day 1	Day 2	Day 3	Day 4	Day 5
Day 1	5	7	9.1	14.1	?
Day 2		5	6	11	12
Day 3			5	8	10.01
Day 4				5	6
Day 5					5

Similar to the matrix chain product problem, the cell at row 1 and column 5 (i.e. marked with "?") is the one that we are trying to find out. Notice that the entry $c[i,j]$ has a value corresponding to the optimum solution of the problem from day i to day j. The main difference in our optimization problem is the accumulated interest cost and the loading cost that needs to be considered in the calculation. The rest is the same as the matrix chain product problem.

In the cost matrix, $c[1, 2]$ shows the minimum cost of cash replenishment from the beginning of first day until the end of the second day. While finding the minimum cost, there are two options to be compared: loading cash day by day or at once. To calculate day by day replenishment cost, sum of $c[1, 1]$ and $c[2, 2]$ can be used as shown in Eq. 6.

$$day \ by \ day[1, 2] = cost[1, 1] + cost[2, 2] \quad (6)$$

To calculate the replenishment cost for loading cash at once, sum of corresponding cell of accumulated interest cost matrix and loading cost is needed. So, day by day calculation gives total cost as 10 (=5+5) as given in Eq. 7:

$$at\,once[1,2] = \alpha + I[1,2] \tag{7}$$

At once calculation gives 7 (=5+2) as a total cost, which is smaller. Therefore, the value of c[1, 2] is 7. The calculations of the values of other entries of c matrix is done in the same way as the matrix chain product. That is, the order of calculations are done in terms of diagonals.

The last entry to be calculated is c[1, 5], which is the result of this problem instance. Calculations for c[1, 5] requires choosing the minimum of the results of Eqs. 8–12.

$$(12345) = \alpha + I[1,5] = 5 + 17.25 = 22.25 \tag{8}$$

$$(1)(2345) = c[1,1] + c[2,5] = 5 + 12 = 17 \tag{9}$$

$$(12)(345) = c[1,2] + c[3,5] = 7 + 10.01 = 17.01 \tag{10}$$

$$(123)(45) = c[1,3] + c[4,5] = 9.1 + 6 = 15.1 \tag{11}$$

$$(1234)(5) = c[1,4] + c[5,5] = 14.1 + 5 = 19.1 \tag{12}$$

(123)(45) is the optimum solution for c[1, 5] and also the value of "?" is found as 15.1 according to calculations.

4   Experiments

In this section, we present the optimized cash replenishment costs by the proposed method on four real ATMs. As the cash withdrawal prediction, we use the real withdrawal amounts. The data set contains the withdrawal amounts for about one year. We construct weekly cash replenishment plans and report the average weekly costs together with minimum and maximum weekly costs obtained. Weekly average cash replenishment costs for the baselines of daily replenishment and weekly replenishments are reported as well. All the results are presented in Tables 5, 6, 7, and 8 where costs are calculated under the interest rate of 0.01 and loading cost of 5.

Table 5. Cash replenishment costs for ATM 1

Method	Avg cost	Min	Max
Proposed method	32.89	26.2	35.0
Daily replenishment	35.0	35.0	35.0
Weekly replenishment	652.89	154.3	2086.35

Table 6. Cash replenishment costs for ATM 2

Method	Avg cost	Min	Max
Proposed method	30.52	5.0	35.0
Daily replenishment	35.0	35.0	35.0
Weekly replenishment	1463.48	5.0	2889.60

Table 7. Cash replenishment costs for ATM 3

Method	Avg cost	Min	Max
Proposed method	34.35	15.0	35.0
Daily replenishment	35.0	35.0	35.0
Weekly replenishment	5355.05	187.8	13704.0

Table 8. Cash replenishment costs for ATM 4

Method	Avg cost	Min	Max
Proposed method	32.73	15.0	35.0
Daily replenishment	35.0	35.0	35.0
Weekly replenishment	1472.53	316.5	4403.09

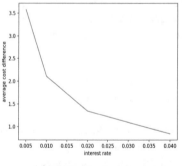

(a) Daily Replenishment

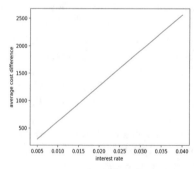

(b) Weekly Replenishment

Fig. 1. Cash replenishment cost difference versus interest rate

As seen in the tables, the proposed method generates schedules with much lower costs especially with respect to weekly replenishment. Daily replenishment schedule generates the same cost due to transportation on each day. On the other hand, weekly replenishment cost varies depending on the amount of money required for the whole week (Fig. 1).

Another analysis that we conducted is on the amount of the difference between the optimized cost by the proposed method and the costs by the baseline approaches under varying interest rate. As expected, the difference increases

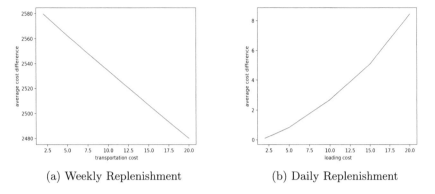

(a) Weekly Replenishment   (b) Daily Replenishment

Fig. 2. Cash replenishment cost difference versus loading cost

for weekly replenishment as the interest rate increases, whereas it approaches to 0 for daily replenishment. We can expect a contrary behavior for varying loading cost (Fig. 2).

5   Conclusions and Future Work

In this work, we present a dynamic programming based approach for optimizing the ATM cash replenishment problem. We assume that the withdrawal amount predictions are available, and focus on scheduling the cash replenishment in order to optimize the cost. The proposed approach is based on well-known matrix chain multiplication solution through defining a mapping between matrices to be multiplied and daily ATM cash requirements. The proposed solution guarantees that the ATM is never out-of-cash, and the cost due to transportation and idle cash is minimized. Note that there is a trade-off between transportation cost and idle cash cost and the solution has to find a balance between these two cost factors.

The proposed solution is applied on a set of real world ATM data in comparison to baselines of daily and weekly replenishments. The results reveal that the straightforward strategies can not find optimized cost, furthermore the cost of the replenishment plan is much higher than the optimized cost, especially for weekly replenishment. This work can be further extended by including daily withdrawal predictions. Another line of research direction can be considering ATM groups as in [6].

References

1. Andrawis, R.R., Atiya, A.F., El-Shishiny, H.: Forecast combinations of computational intelligence and linear models for the nn5 time series forecasting competition. Int. J. Forecast. **27**(3), 672–688 (2011)
2. Anholt, V.R.G., Coelho, L.C., Laporte, G., Vis, I.F.A.: An inventory-routing problem with pickups and deliveries arising in the replenishment of automated teller machines. J. Trans. Sci. **50**, 1077–1091 (2016)

3. Baker, T., Jayaraman, V., Ashley, N.: A data-driven inventory control policy for cash logistics operations: an exploratory case study application at a financial institution. Decis. Sci. **44**(1), 205226 (2013)
4. Bati, S., Gozupek, D.: Joint optimization of cash management and routing for new-generation automated teller machine networks. IEEE Trans. Syst. Man Cybern. Syst. 1–15 (2017)
5. Chotayakul, S., Charnsetthikul, P., Pichitlamken, J., Kobza, J.: An optimization-based heuristic for a capacitated lot-sizing model in an automated teller machines network. J. Math. Stat. **9**(4), 283288 (2013)
6. Ekinci Y., Lu, J.-C., Duman, E.: Optimization of atm cash replenishment with group-demand forecasts. Expert Syst. Appl. (2014)
7. Kalchschmidt, M., Verganti, R., Zotteri, G.: Forecasting demand from heterogeneous customers. Int. J. Oper. Prod. Manag. **26**(6), 619–638 (2006)
8. Kossmann, D., Stocker, K.: Iterative dynamic programming: a new class of query optimization algorithms. ACM Trans. Database Syst. (TODS) **25**(1), 43–82 (2000)
9. Teddy, S., Ng, S.: Forecasting atm cash demands using a local learning model of cerebellar associative memory network. Int. J. Forecast. **27**(3), 760–776 (2011)
10. Venkatesh, K., Ravi, V., Prinzie, A., den Poel, D.V.: Cash demand forecasting in atms by clustering and neural networks. Eur. J. Oper. Res. **232**(2), 383–392 (2014)

Prediction of Crop Yields Based on Fuzzy Rule-Based System (FRBS) Using the Takagi Sugeno-Kang Approach

Kalpesh Borse and Prasit G. Agnihotri[✉]

Civil Engineering Department, Sardar Vallbhbhai National Institute of Technology, Surat, India
kalpeshborse22@gmail.com, pga@ced.svnit.ac.in

Abstract. Predicting the effects of climate change on crop yields requires a model and its parameters, how crops respond to weather. Predictions from different models often disagree with the climatic variables and its impact. A common approach is to use statistical models trained on historical yields and some simplified measurements of weather, such as growing season average temperature and precipitation. Climate change is really concern to the entire world. Its direct impact on the crop growth and yield is very important to understand. In the present study Fuzzy logic crop yield model was developed by considering different climate change variables. Temperature, Rainfall, evaporation, humidity parameters are considered for the crop yield model. Model is being developed by considering the 15-year crop yield data and same period for the climatic variables. Triangular membership function is being adopted in the fuzzy model. In this study a fuzzy rule-based system (FRBS) using the Takagi Sugeno-Kang approach has been used for the developing the crop yield model. Model is validated by coefficient of correlation, and found that, there is more than 0.9 coefficient of correlation between observed and evaluated yield.

Keywords: Takagi Sugeno-Kang approach · Crop yield · Climate change
Fuzzy logic

1 Introduction

Agriculture is one of the main sectors to be impacted by different sources like climatic changes, soil attributes, seasonal changes etc., Crop yield prediction is based on various kinds of data collected and extracted by using data mining techniques different sources which are useful for growth of the crop. It is an art of forecasting crop and the quantity of yield in advance i.e., before the harvest actually takes place. Predicting the crop yield can be extremely useful for farmers. They can contract their crop prior to harvest, if they have an idea of the amount of yield they can expect which gives often securing a more competitive price than if they were to wait until after harvest. The involvement of experts in prediction of crop yield leads to issues like lack of knowledge about natural events, negation of personal perception and fatigue etc. such issues can be to overcome by using the models and decision tools for crop yield prediction. Likewise, industry can do better planning the logistics of their business as the benefit from yield predictions.

© Springer Nature Switzerland AG 2019
P. Vasant et al. (Eds.): ICO 2018, AISC 866, pp. 438–447, 2019.
https://doi.org/10.1007/978-3-030-00979-3_46

In India it is possible to cultivate large number of crops due to diverse climatic conditions. Among these crops Rice (*Oryza sativa* L.) is important food crops of the country. The total area under cultivation of Rice 43.6 (22.6% of gross cropped area) and the total production of Rice is 94.1 million tonnes during the year 2014–15 constituting about 42.9% of total food production (Economic Survey 2014–2015). Crop failure on account of drought or flood will have a severe repercussion not only on the country's economy but also on food security.

The forecasting of crop yield may be done by using three major objective methods (i) biometrical characteristics (ii) weather variables and (iii) agricultural inputs [1]. These approaches can be used individually or in combination to give a composite model. Forecasting is a significant aid in effective and efficient planning and is more important aspect for a developing economy such as ours so that adequate planning exercise is undertaken for sustainable growth, overall development and poverty alleviation. In management and administrative situations, the need for planning is great because the lead time for decision making ranges from several years (for capital investments) to a few days or hours (for transportation or production schedules) to a few seconds (for telecommunication routing or electrical utility loading). Forecast of crop yield is of immense utility to the government and planners in formulation and implementation of various policies relating to food procurement, storage, distribution, price, import-export etc.

Several studies have been carried out to forecast crop yield using weather parameters [2, 3]. However, such forecast studies based on statistical models need to be done on continuing basis and for different agro-climatic zones, due to visible effects of changing environmental conditions and weather shifts at different locations and areas. Therefore, there is a need to developed area specific forecasting models based on time series data to help the policy makers for taking effective decisions to counter adverse situations in food production.

In recent years soft computing techniques like, artificial neural network (ANN), fuzzy logic, genetic algorithm and chaos theory have been widely applied in the sphere of prediction and time series modelling. Adaptive neuro-fuzzy inference system (ANFIS) which is integration of neural networks and fuzzy logic has the potential to capture the benefits of both these fields in a single framework. ANFIS utilizes linguistic information from the fuzzy logic as well learning capability of an ANN.

This study was carried out to develop ANFIS based crop yield model based on the climatic parameters. ANFIS model was developed by considering different combination of climatic parameters. Detail explanation was given in the methodology section below.

2 Materials and Methods

2.1 Study Area

The present study was carried out to develop forecasting models for predicting the yield of Kharif rice at Nashik taluka of Nashik district of Maharashtra state in India. It is bounded on the north by the Mahadeo Hills, the Satmala Hills, on the north-west by the

Ajanta Range, on the west by the North Sahyadri range of the Western Ghats, on the east and south-east by the Eastern Ghats and on the south by the Balaghat Range. It is located at 20.33°N latitude and 73.25°E longitudes. It has a dry season from early October to mid-June and a wet season from June to early October. The annual average rainfall of this region is about 713.50 mm, which is subjected to large variation. Yearly yield data of rice (kg/ha) for 27 years (w.e.f. 1987–88 to 2014–15) was collected from the Department of Agriculture, Maharashtra State. The Fig. 1 shows the location map of the study area.

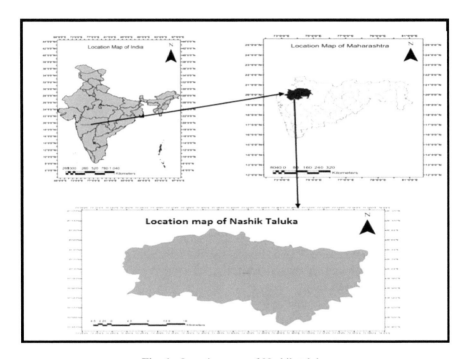

Fig. 1. Location map of Nashik taluka

The time series daily weather data of 27 years (from 1987–88 to 2014–15) were collected from WRDHP, Nashik (M. S). Five weather parameters were included in this study; namely average daily temperature (T °C), average daily relative humidity (Rh %), average daily total rainfall (P), average weekly number of rainfall days (n) and average daily pan evaporation (E). However, daily weather data related to kharif (the autumn harvest also known as the summer or monsoon crop in India) crop seasons starting from a fortnight before sowing up to one month before harvest were utilized for the development of models in the present study therefore, the weather data for rice crop (Kharif season), from May 21 (about a fortnight before sowing) to October 22 (one month before harvest) in each year were employed.

3   Adaptive Neuro-Fuzzy Inference Systems (ANFIS)

ANFIS is a fuzzy algorithm that is based on Sugeno-Kang (TSK) fuzzy inference system [4, 5]. ANFIS is the powerful soft computing technique works based on the principle of two powerful computing techniques that is Artificial Neural Network. ANFIS utilizes linguistic information from the fuzzy logic as well learning capability of an ANN for automatic fuzzy if-then rule generation and parameter optimization. An ANFIS input interference panel in MATLAB is shown in Fig. 2.

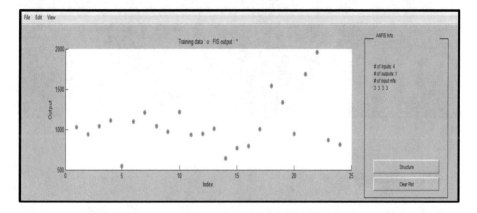

Fig. 2.   ANFIS input panel

ANFIS has the lot of advantages over the individual computing tool such as FUZZY system and artificial neural network. In the FUZZY system making rules is very important, complete predictions are based on the how one makes the rules. If the parameters are involved more it will become more completed to make all such rule. ANFIS solves this issue by taking the help of ANN, ANN optimizes the parameters and makes the rules. These rules are fetch into the FUZZY inference system. Due to overhead effect ANFIS costs high computing cost. ANFIS presents some linearity with respect to some of its parameters, hence it increases the overhead of computation process without increasing the efficiency. Optimization of fuzzy rules are also not so efficient as compared to manual rules making hence predictions are subjected to more uncertainty. Structure of ANFIS is as shown in the Fig. 3.

From the figure it can be seen that, there are four Inputs. All the four inputs are divided into three membership function each. And there is only single output. In the present study out is Yield and inputs are different combination of parameters.

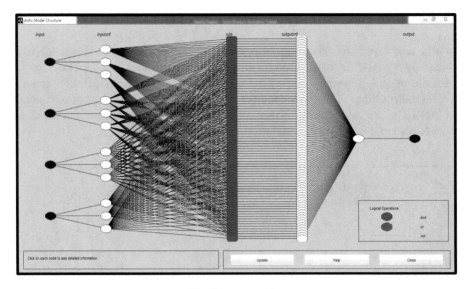

Fig. 3. Fuzzy rules

4 Methodology

Climatological data was first statistically checked for the outliers and missing data analysis. Outliers are checked at the 95% significance level and outliers are removed. After removing outliers, missing data analysis was carried out for the all the climatological data using inverse distance weightage method.

4.1 Missing Data Analysis

Data for the period of missing rainfall data could be filled using estimation technique. The length of period up to which the data could be filled is dependent on individual judgment. Generally, rainfall for the missing period is estimated either by using the normal ratio method or the distance power method.

Distance power method:

The rainfall at a station is estimated as a weighted average of the observed rainfall at the neighbouring stations. The weights are equal to the reciprocal of the distance or some power of the reciprocal of the distance of the estimator stations from the estimated stations. Let D_i be the distance of the estimator station from the estimated station. If the weights are an inverse square of distance, the estimated rainfall at station A is:

$$P_A = \frac{\sum_{i-s}^{n} P_t/D_i^2}{\sum_{t-1}^{n} 1/D_i^2} \tag{1}$$

Note that the weights go on reducing with distance and approach zero at large distances. A major shortcoming of this method is that the orographic features and spatial distribution of the variables are not considered. The extra information, if stations are close to each other, is not properly used. The procedure for estimating the rainfall data by this technique is illustrated through an example. If A, B, C, D are the location of stations discussed in the example of the normal ratio method, the distance of each estimator station (B, C, and D) from station (A) whose data is to be estimated is computed with the help of the coordinates using the formula:

$$\text{Di}^2 = \left[(x - xi)^2 + (y - yi)^2 \right] \tag{2}$$

Where x and y are the coordinates of the station whose data is estimated and xi and yi are the coordinates of stations whose data are used in estimation. In the present study Distance power method is adopted as it is most recommended by many researchers.

4.2   Model Input

Models parameters are selected based on the parameters which affects the crop growth. For the present study parameters are purely climatic. Many researchers have considered the parameters other than the climatic but the problem with the other parameters not easily available or very long experimentations are required. This study demonstrates the developing the crop yield model based purely based on the climatic parameters such as: Average rainfall (mm), Maximum Rainfall (mm), Total Rainfall (mm), Minimum Avg. Temperature (°C), Maximum Avg. Temperature (°C), Average Temperature (°C), Average Evaporation (mm), Relative humidity (%) are studied and combination was tried to formulate the best model.

Amongst all the listed parameters, different combinations are tried to predict the crop yield and most effective parameters are selected for the prediction of crop yield. Variation of crop yield over the period of time are as shown in the Fig. 4.

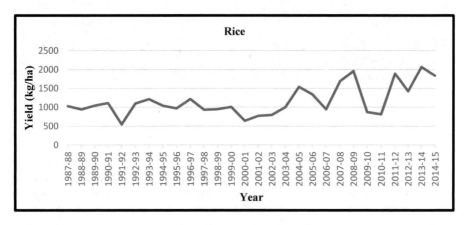

Fig. 4. Rice yield over the period of time.

5 Result and Discussions

Amongst the 8 available Fuzzy membership function, Product of two sigmoidal function (psigmf) was used and hybrid optimization was applied while training the variables. Two different cases are considered based on the types of the variables are considered for the training and best fit parameter selection. All two cases are explained in the subsequent section.

5.1 Model Training

Case 1:

First Model was formulated by considering the Average rainfall, Maximum avg. Temperature, average evaporation and relative humidity found that, after 1000 epoch the error in the model was 0.012173 as shown in the Fig. 5 and model was optimized at the epoch 2.

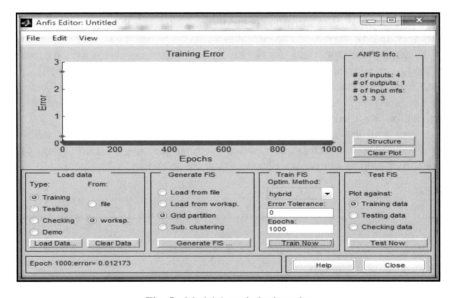

Fig. 5. Model 1. optimised results

Case 2:

The second model was developed based on the Maximum Rainfall (mm), Minimum Avg. Temperature (°C), Maximum Avg. Temperature (°C), Average Evaporation (mm), Relative humidity (%). the designated error was zero while developing the model and reached to zero at epoch 2. as shown in the Fig. 6. From the model. Model characteristics were as shown in the fig. model is very efficient as compared to case 1. Because it has the lesser error 0.00995 from the same number of epochs.

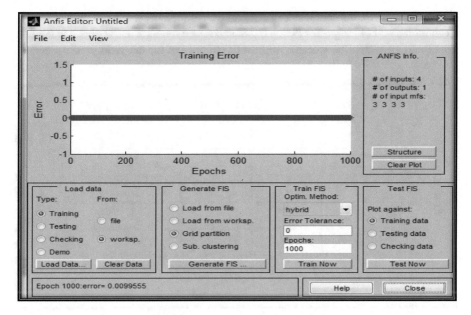

Fig. 6. Model 2. optimized results

From all above discussion it has been concluded that model two of the case 2 is the best performing model for predicting the yield in the present study area.

5.2   Model Validation

By selecting the independent set of data of the parameters mentioned in the case 2, predicted the crop yield and results are as shown in the fig below (Fig. 7).

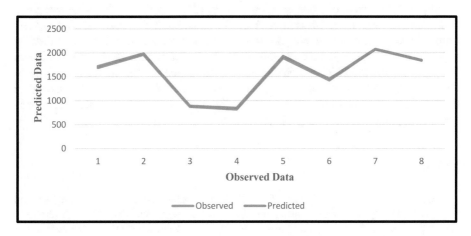

Fig. 7. Model performance

Figure 8 shows the coefficient of correlation between observed and predicted yield it can be seen that r^2 is greater than the 0.8. and hence model 2 of the case 2 is the final model and respective parameters are best fit parameters for the yield prediction.

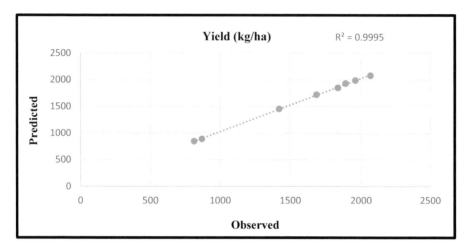

Fig. 8. Coefficient of correlation between observed and predicted yield

6 Conclusion

In the present study two ANFIS models were developed for the crop yield prediction based on climatological parameters such as Average rainfall (mm), Maximum Rainfall (mm), Total Rainfall (mm), Minimum Avg. Temperature (°C), Maximum Avg. Temperature (°C), Average Temperature (°C), Average Evaporation (mm), Relative humidity (%) are studied and combination was tried to formulate the best model. Amongst all the parameter combination such as Maximum Rainfall (mm), Minimum Avg. Temperature (°C), Maximum Avg. Temperature (°C), Average Evaporation (mm), Relative humidity (%) are performing to the greater accuracy. The second Model was performing at the accuracy of 0.0099 kg/hectare accuracy. Hence model number two was recommended for the prediction of yield in the present study. Coefficient of correlation between observed and predicted yield it can be seen that r^2 is greater than the 0.8. And hence model 2 of the case 2 is the final model and respective parameters are best fit parameters for the yield prediction.

References

1. Agrawal, R., Jain, R.C., Mehta, S.C.: Yield forecast based on weather variables and agricultural inputs on agro-climatic zone basis. Ind. J. Agri. Sci. **71**(7), 487–490 (2001)
2. Huda, A.K.S., Ghildyal, B.P., Tomar, V.S., Jain, R.C.: Contribution of climate variables in predicting rice yield. Agric. Met. **15**, 71–86 (1975)

3. Chowdhury, A., Sarkar, M.B.: Estimation of rice yield through weather factors in a dry sub-humid region. Mausam **32**(4), 393–396 (1981)
4. Jang, J.S.R., Sun, C.T., Mizutani, E.: Neuro-fuzzy and soft computing, a computational approach to learning and machine intelligence. Prentice Hall, NJ, USA (1997). ISBN 0-13-261066-3
5. Loukas, Y.L.: Adaptive neuro-fuzzy inference system: an instant and architecture-free predictor for improved QSAR studies. J. Med. Chem. **44**(17), 2772–2783 (2001)
6. Jang, J.-S.R.: ANFIS: adaptive network-based fuzzy inference systems. IEEE Trans. Syst. Man Cybern. **23**, 665–685 (1993)
7. Kim, B., Park, J.H., Kim, B.S.: Fuzzy logic model of Langmuir probe discharge data. Comput. Chem. **26**(6), 573–581 (2002)
8. Nayak, P.C., Sudheer, K.P., Rangan, D.M., Ramasastri, K.S.: A neuro-fuzzy computing technique for modelling hydrological time series. J. Hydrology **291**, 52–66 (2004)
9. Takagi, T., Sugeno, M.: Fuzzy identification of systems and its application to modeling and control. IEEE Trans. Syst. Man Cybern. **15**, 116–132 (1985)
10. Zadeh, L.A.: Fuzzy Sets. Inf. Control **8**, 338–353 (1965)
11. Kokate, K.D., Sananse, S.L., Kadam, J.R.: Pre-harvest forecasting of rice yield in Konkan region of Maharashtra. J. Maha. Agric. Univ. **25**(3), 289–293 (2000)
12. Jones, A.J.: New tools in non-linear modeling and prediction. CMS **1**(2), 109–149 (2004)

Land Use Land Cover Analysis of Khapri Watershed in Dang District Using Remote Sensing (RS) and Geographical Information System (GIS)

Ashish Guruji$^{(\boxtimes)}$ and Prasit Agnihotri

Sardar Vallabhbhai National Institute of Technology, Surat, Gujarat, India
ashishguruji@gmail.com

Abstract. India has two important resources, cultivable land and fresh water. Land use and land cover constitutes key ecological data for many scientific, exploratory resource management and policy purposes, as well as for a range of human activities. The land use type has considerable impact on the nature of runoff and related hydrological characteristics. In this paper, attempt was made to know land cover and land use (LULC) by analyzing Satellite images. Agriculture, Built up, Dense Forest, Sparse Forest and Water Body are different categories of Land Use and Land Cover found out from satellite image of 2018, 2009 and 1997. From satellite image analysis of 2018 and 1997, it is found that Agriculture area increase by 21%. Dense Forest decrease continuously but Sparse forest increase steadily. Ground Truth has been verified for Agriculture and Dense Forest using GPS instrument.

Keywords: Land use land cover (LULC) · Landset-8 · Landset-5
QGIS · GPS

1 Introduction

Hydrology is essentially a science based on imperfect observations in a complex and sometime discontinuous domain. The range of hydrologic information is tremendous. The upper range includes the vast scale of the meteorological phenomena, where information is based on sampling at relatively few points in a great range of space and time. One of major requirement of human being is water and prosperity of any region will be measured using availability of water. For development of any area, water availability is considered as one of the prime need.

River basin considered as the basic hydrological unit for planning and development of water resources. There are 12 major river basins with catchment area of 20000 km^2 and more comprising total 25.3 lakh km^2 in India. There are 46 medium river basin with catchment area between 2000 and 20000 km^2 comprising total of 2.5 lakh km^2 [8].

The change in land use type has considerable impact on the nature of runoff and related hydrological characteristics[1].Satellite images are used to quantify the various parameters of land use by applying various image-processing techniques [2]. Decadal land-cover changes using satellite data may be measured [4].

© Springer Nature Switzerland AG 2019
P. Vasant et al. (Eds.): ICO 2018, AISC 866, pp. 448–454, 2019.
https://doi.org/10.1007/978-3-030-00979-3_47

Land use and land cover is a key factor in understanding the relations of human activities with the environment [5]. Land cover maps represent earth cover reality due to natural or political changes. Satellite images are utilized to analyze land cover and land use (LULC) maps to understand changes in hydrologic response of an area. LULC may be extracted through the process of classification which is categorization of all pixels in a digital image into one of various land cover classes [6].

The Dang district in Gujarat state gets rainfall of around 2400 mm per year, but still it suffer acute drinking water problem in few months of the year. The Dang district receives most of the rainfall from the South West monsoon from June to September.

In this paper, Land Use and Land Cover of Khapri watershed has been analyzed using satellite image of three time scale namely 2018, 2009 and 1997. There are five classes identified in the watershed mainly, Agriculture, Built up, Dense Forest, Sparse Forest and Water Body.

2 Study Area

The important tributaries of the Ambica river are Kapri, Kaveri and Kharera rivers. The Kapri river rises at an altitude of 1030 m in Sahyadri hill range in Ahwa taluka of Dang district in the state of Gujarat and joins the river Ambica near village Milan at an elevation of 100 m. The length of river Kapri is about 80 km. The Kapri catchment up to its confluence with Ambica river is spread over an area of 537 km^2 which is about 19% of the total catchment of the Ambica basin. The river Khapri river lies between 20°37′34″ and 20°49′1″ North latitude and 73°28′1″ and 73°49′44″ East longitude (Fig. 1).

The river Ambica basin lies between 20°31′ and 20°57′ North latitude and 72°48′ and 73°52′ East longitude with a drainage area of 2830 km^2. The Valsad, Dangs and Surat Districts of Gujarat and a small portion of the Nasik district of Maharashtra falls in the basin. The basin extends over an area of 2830 km^2 out of which 102 km^2 lies in Maharashtra while 2728 km^2 is in Gujarat. The effective drainage area of the river is 2685 km^2 since 145 km^2 area near the mouth is low-lying, marshy and cannot be beneficially utilized.

The climate of the basin is characterized by a hot summer and in general dry except during the Southwest monsoon season. The maximum, minimum temperatures observed vary from 32 to 40 °C and 25 to 8 °C respectively. The forest area in the basin is 87006 ha. Soil of Ambica basin can be broadly classified into three groups i.e. Laterite soil, deep black soil and alluvial soil. The basin can be divided into two prominent physiographic zones. The eastern part comes under rugged mountain chains of the Saputara Hills and descends on the western side to the edge of the uplands of Surat district. This region is placed at a general elevation of 1050 to 100 m. The western part, barring the coastal plain, is essentially in the sub sahyadrin zone of hills and valleys generally below 100 m elevation. Deccan traps and intermediate amphitheatres have developed out of the alluvial debris washed from the hills. The lower reaches of the basin up to the coastal margins are mainly alluvial plains [10].

Total geographical area of the district is about 1.7 lakhs hectares. It is noteworthy to find that only 33% of the geographical area is under cultivation in the district [7].

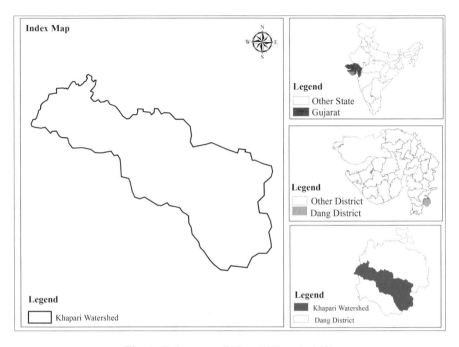

Fig. 1. Index map of Khapri Watershed [3]

3   Methodology

In this study, https://earthexplorer.usgs.gov, United State Geological Survey, Earth Explorer web site is used for satellite data from 1997 to 2018. [9] The collected data were compiled and analyzed systematically by keeping in view of the objectives of the study. GIS and Remote Sensing techniques are used for the visual analysis and interpretation of the images. Landsat-8 and Landset-5 data was used and downloaded from website for analysis. For Landset-8, Band combination 5-4-3 used and for Landset-5, 4-3-2 was used respectively in Fig. 2.

Shape file of Khapri watershed has been uploaded which was prepared from Bhuvan online portal. On the basis of my interest area satellite data short listed as per availability of path and without cloud cover. Three decade interval data selected for analysis of land use and Land cover. Year selected are 2018–2009 and 1997.

Three satellite images on 31-3-2018, 22-3-2009 and 17-2-1997 are selected for Land Use Land Cover analysis. Work has been done in QGIS environment.

Land Cover: Surface covered naturally. Water, Grass land, Deciduous Forest, impervious surfaces, bare soil comes under Land Cover.

Land Use: How the land is utilized? It is a process of development and conservation practices in the area. Agriculture land, Urban and recreation area comes under Land Use (Fig. 3).

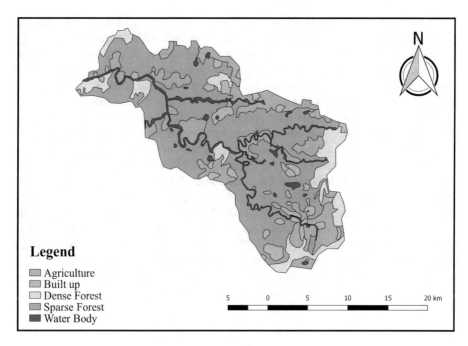

Fig. 2. Land use & land cover for satellite image 31-3-2018

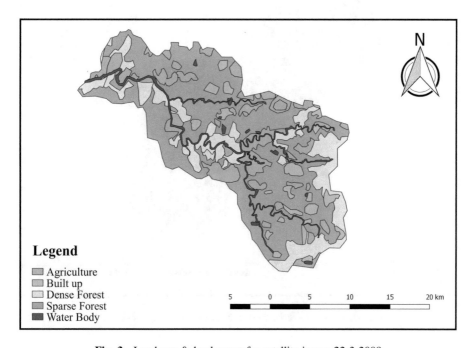

Fig. 3. Land use & land cover for satellite image 22-3-2009

Ground control point in terms of GPS (Global Positioning System) points collected during visit of Dang on 18-10-2016. In this hand held GPS instrument used. This instrument is taken on site and by entering proper details; it will give latitude and longitude of that location. This instrument can be connected to computer using USB cable. These points are superimposed on satellite image or Khapri watershed map prepared in QGIS. Dense forest and river identified with GPS locations 3.

LULC analysis for Satellite image on 22-3-2009 has been carried out. Now as the image for 2018 and 2009 are from different satellite namely Landsat-8 and Landsat-5 respectively. The FCC (False Colour Combination) has been modified and image verified with ground control points collected on 29th and 30th October, 2017. This GPS points include Soil sample points and Check dams. Soil Sample points indirectly show agricultural land in the area. So, for LULC analysis five categories used. After completion of analysis of LULC, Map showing LULC with Legend has been prepared by using proper colour combination. Using this LULC, curve number should be derived for estimating runoff. This curve number is based on weighted area of respective Land use and Land cover. For doing LULC, visual interpretation on screen digitization technique along with field data in form of. GPS points are super-impose on satellite image for accurate identification of Land Use/Land Cover class [3] (Fig. 4).

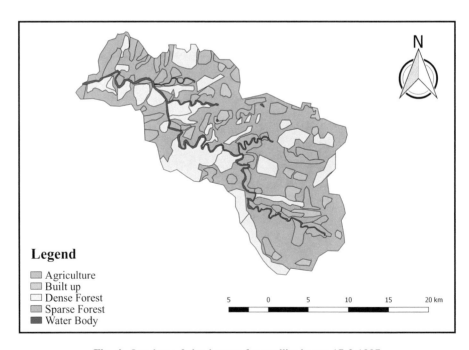

Fig. 4. Land use & land cover for satellite image 17-2-1997

4   Results and Discussion

Following table depict the result for Land Use and Land Cover in Khapri watershed for year 2018, 2009 and 1997 (Table 1).

Table 1. Land use land cover area (ha) in Khapri Watershed

Category	Area under diff land use classes (ha)		
	Mar-2018	Mar-2009	Feb-1997
Agriculture	9110.73	7027.05	7529.58
Built up	2881.8	2467.68	1966.9
Dense forest	4967.26	9098.56	10566.37
Sparse forest	26468.03	25253.79	24557.67
Water body	2622.38	2209.04	1435.67
Total	46050.2	46056.12	46056.19

From satellite image analysis of 2018 and 1997, it is found that Agriculture area increase by 21%. Built up Area increase gradually from 1997, 2009 and 2018. With respect to 1997, Built up area increase by 25.4% and 46.5% respectively. Dense Forest decrease continuously but Sparse forest increase steadily. Agriculture and Dense Forest has been checked by using GPS instrument for their location.

Land Use and Land Cover of watershed will help in generating Curve Number in SCS (Soil Conservation Service) method.

References

1. Amruth Chand, B., Varija K.: Quantifying the effect of land use land cover change in Pavanje river Basin and its effect in Pavanje River. In: 20th International Conference on Hydraulics, Water Resources and River Engineering, IIT Roorkee, Roorkee (2015)
2. Duulatov, E., Xi, C., Kurban, A., Ndayisaba, F., Monoldorova, A.: Detecting land use/land cover change using landsat imagery: Jumbal district, Kyrgyzstan. Int. J. Geoinf, 1–7 (2016). ISSN 1689-6576
3. Guruji, A.L.: Land use and land cover analysis of Khapri watershed in Dangs District using RS ad GIS. Bhaskaracharya Institute for Space Applications and Geo-informatics, Department of Science & Technology, Government of Gujarat, Gandhinagar, India (2018)
4. Kotha, M., Kunte, P.D.: Land-cover change in Goa-An integrated RS-GIS approach. Int. J. Geoinf., 37–43 (2013). ISSN 1689-6576
5. Matkar, P.S., Zende, A.M.: Land use/land cover changes patterns using geospatial techniques-Satara district, Maharastra, India: a case study. In: Hydro-2017 International Conference of Hydraulics, Water Resources and Costal Engineering, L.D. College of Engineering, Ahmedabad, Gujarat, India, pp. 1837–1846 (2018)
6. Shilpi, S.R.M.: Change detection and hydrologic responses simulation for LULC change using remote sensing and GIS. In: 20th International Conference on Hydraulics, Water Resources and River Engineering, IIT Roorkee, Roorkee (2015)

7. Final District Agriculture Plan (Dangs), Water and Power Consultancy Services (India) Limited (2013)
8. Water Resources Information System of India, http://www.india-wris.nrsc.gov.in/wrpinfo/index.php?title=Basins
9. United State Geological Survey EarthExplorer, https://earthexplorer.usgs.gov/
10. Physical Features, http://nwda.gov.in/upload/uploadfiles/files/9904183747.pdf

Recovery Method of Supply Chain Under Ripple Effect: Supply Chain Event Management (SCEM) Application

Fanny Palma[1(✉)], Jania A. Saucedo[1], and José A. Marmolejo[2]

[1] Universidad Autónoma de Nuevo León, Monterrey, Mexico
{fanny.palmavz, jania.saucedomrt}@uanl.edu.mx
[2] Universidad Panamericana, Cd de México, Mexico
jmarmolejo@up.edu.mx

Abstract. Facing the new challenges that arise in supply chains (SC) such as globalization itself, implies not only managing a geographical expansion across the planet, but also depending on third parties, whether these suppliers, partners, members, distributors or customers themselves, whether they are just around the corner or on the other side of the planet, managing a network dependent on third parties becomes increasingly complex and vulnerable to a ripple effect (RE) in the event of disruption, ensure good performance will depend on everyone working collaboratively. Therefore, it is important to look for methods or tools that reduce the vulnerability of the RE into a SC and, in turn, promote collaboration and visibility. Methodologies such as Supply Chain Event Management (SCEM) include processes and systems that alert to unplanned changes, also called disruptive events (DE), in supply lines or other processes in order to respond with alternatives, through its five functionalities: monitoring, notification, simulation, control and measurement of SC activities. This helps to ensure the good performance of a company, of its entire SC and to diminish the RE of global networks.

Keywords: Supply chain · Collaboration · Ripple effect · Simulation

1 Introduction

Successful and correct supply chain management (SCM) is not only about planning, process management and business linkages, it also means continuously seeking improvements to meet new challenges. Among these challenges are: globalization, competition in the market fiercer and the forefront of Industry 4.0, where the flow of information and data produced by companies must be broken down into useful and valuable information, with the contribution that simulation provides when evaluating the effectiveness of the decisions that are proposed to execute without affecting the real system. Likewise, SCM, like any other process, requires plans to achieve its objectives. However, during the implementation of the plan, unforeseen deviations usually occur. These deviations, also called disruptive events (DE), can affect compliance times, jeopardize customer satisfaction and generate high damage repair costs. The creation of information and communication technologies (ICTs) applied to the SC is a resource

© Springer Nature Switzerland AG 2019
P. Vasant et al. (Eds.): ICO 2018, AISC 866, pp. 455–465, 2019.
https://doi.org/10.1007/978-3-030-00979-3_48

that we must take full advantage of, because through them we can have accurate and timely information about what is happening at the SC, in other words, have visibility, react in time with immediate corrective actions and ensure not only the good performance of a company, but also that of all members of the network collaboratively.

In recent years we have seen intense changes in the way we manage a network with a globalized SC, where geographical boundaries have succumbed, but which, in turn, requires intermediaries within it; this dependence on third parties makes management increasingly complex and vulnerable, justified by the need to reduce costs in the face of the great competition in the market and to question the validity of non-core process administrations. All the above makes a ripple effect (RE) disruption more latent with an impact on the economic performance of a global SC more latent.

> Finally, to bridge the gap that still exists between Supply Chain Integration (SCI) and SCM that is still far from being proactive, analyze, collaborate and resolve deviations that can be transcendental and of great effect, which requires tools or methodologies that serve as drivers to the application of these philosophies.

1.1 Current Issues in Logistics and SC

Concepts such as SCM and SCI have become increasingly important, as classic approaches to business management focus especially on intra-organizational optimization measures that only improve the performance of a company, while SCM and SCI break the perspective of divided improvement and try to reduce inefficiencies with inter-organizational measures throughout the value chain, in other words, one concept is focused on the coordination of logistics processes and the other is aimed at integrating these processes among network members in order to overcome the challenges that arise.

Globalization does not necessarily mean that a company spreads geographically across the planet, it also means dependence on third parties (call these third parties): suppliers, partners, logistic operators and even customers) that may be close or far from each other, the above, due to more fierce competition in the market, such dependence on third parties makes management more complex, less controllable and more vulnerable, however, this is justified by the need to reduce costs and increase profits, since these do not occur exclusively in production savings or increased sales respectively, the SC is formed by, among others, costs of supply, distribution, transport, storage and recovery; any break in any of these links or mismanagement generates costs that have a negative impact and must be considered.

Another aspect that leads to dependence on third parties is the questioning of the validity of non-core business process administrations, which refers to the management of processes that are not the main business line or activity of a company on which its vital income resides, but which are essential to give added value to the SC and which necessarily fall to third parties. All the above makes a Ripple Effect (RE) disruption with potential impact on the economic performance of a global SC more latent, since such a disruption would affect other levels of the chain itself. It is here where we must take maximum advantage of the tools at the forefront of Industry 4.0, where a flow of information and data produced by companies makes it difficult to process and analyze, then all this information should be broken down into useful and valuable information, where tools such as Data mining that with the contribution that simulation provides an effective evaluation of the decisions that are executed without affecting the real system.

1.2   Ripple Effect on the SC

Just as there are many types of disruptions in the SC, there are also many publications on its management, mostly with the generalization of the disruptive impact on operational and strategic economic performance, as well as on recovery, in the face of the latter there is a spread throughout the SC, whose literature proposes such a spread as: Ripple Effect on SC [1].

The RE manifests itself when a DE occurs but with two explicit particularities: (1) DE do not remain localized in a single process or content in one part of the SC, they spread to other levels or nodes of the SC; (2) disruption impacts the performance of the overall SC; (3) the occurrence of this type of DE is very low, hence its low predictability and expected contingency actions (Fig. 1 shows these characteristics); therefore, their risk and stabilization should be assessed at an early stage after DE is proactively presented and further, if possible, at the design and planning stages. Such an impact could include late deliveries of orders and even failure to deliver them; loss of market share and reputation; loss of some revenue, profits and even decline in stock performance, the latter a devastating cost to many companies [1], which can often lead to bankruptcy.

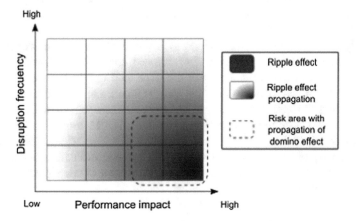

Fig. 1. Ripple effect.

In this RE theme, the pioneers in examining the term in depth were Ivanov, Sokolov and Dolgui, who defined it as: the result of the disruptive propagation of an initial interruption to other stages of the global SC, in the production and distribution networks [2].

If not controlled, the RE tends to spread to other levels of the chain, thus, the consequences worsen and the cost of repairing damage increases (Fig. 2).

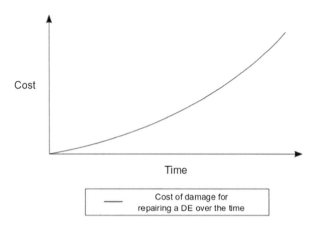

Fig. 2. DE's cost.

2   Disruption Management

As mentioned above, at the time of becoming a more complex SC, the likelihood of not achieving the planned performance objectives increases. Therefore, it is essential that organizations look for methods or tools that reduce the vulnerability of SC, if possible, to take these steps from design and planning or, in their redesign. Organizations are also required to understand the dependency among SC members, identify risks, consequences and their impact.

Supply chain risk management (SCRM) is the identification, assessment and prioritization of risks followed by the synchronized and cost-effective application of resources to reduce, control and control the likelihood and/or impact of unfortunate events [3]. A typical SCRM includes four components: identification, response, monitoring and evaluation [4]. Supply chain event management (SCEM) has been implemented as a disruption management tool. The SCEM consists of timely identification of SC disruptions; analysis of these and alerts on what interruptions have occurred or may occur; notification of decision makers; evaluation of possible contingency and correction actions; and, finally, development of control actions to restore the operability of the SC.

2.1   Supply Chain Event Management

The involvement of relationships with other companies around the world in the SC, as well as the dependence on third parties, suggests integrating all members of the SC collaboratively for proper management.

SCEM is defined as: processes and systems that alert companies to any unplanned changes in supply lines or other events so that they can respond with alternatives [5].

Additionally, a DE is defined as: those non-deterministic events that trigger problems and state transitions of some type of object. Disturbances, interruptions, malfunctions and other concepts to describe a negative impact are referred to this concept [6]. These, in addition, can be propagated through many levels of a system.

The SCEM methodology is composed of 5 functions: monitoring, reporting, simulation, control and measurement.

Monitoring: Is a continuous activity, which involves detecting relevant DEs; in this function, unplanned changes are observed in the execution of a plan: these changes may be: resource availability; order progress; order specification changes, i.e. original modifications to the original order specification such as start time, quantity or completion time, not because of adjusting the order in response to another DE [7]. There are two types of monitoring: reactive and preventive.

Monitoring in the SC starts in a variety of circumstances. Four categories of relevant situations are described [6], which are:

- Alert trigger: when an alert is received for the occurrence of a DS from a CS member who identifies it.
- Status request: an explicit request is made for information on the status of an order, an ED is identified.
- Random trigger: A random selection of orders is made. These are then monitored, and a disturbance is identified.
- Probabilistic trigger: when knowledge of a certain order is available, there is sufficient information to predict a high probability that this order will be affected by a disruption.

Notification: This function refers to timely informing the relevant decision-makers when an ED [8] is detected. This allows process managers to intervene immediately and directly to contain risk and to contain major disruptions in SC performance, which means reducing the RE; otherwise, the passage of time without acting represents the increased cost of repairing the damage to a DE, also intended to provide relevant information through a level of care that is both time-sensitive (priority level) and cost-sensitive (severity level). Then with the support of Fuzzy Logic methodology we obtain the index alert. Fuzzy Logic is a tool eficaz to address the problem of knowledge representation in an environment of uncertainty, vagueness and imprecision, this is based on the semantics of a membership value. In this semantics, a proposition is interpreted as a system of elastic restrictions, and the reasoning is seen as a propagation of restrictions [9]. (Figures 1 and 2 the index alert and the rules for obtaining it with Fuzzy Logic).

Simulation: Refers to evaluating the consequences of DE by modeling specific management actions and analyzing trends [6]. The simulation allows to estimate improvement tactics without disturbing the functioning of the real system; to create hypotheses about certain actions and validate them; to analyze the impact and understand the holistic system. Simulation is an alternative to optimization for SC analysis. The simulation models in this topic are not restricted by rigid mathematical structures. Almost any SC problem can be coded as a simulation object with a set of parameterized behaviors.

Control: Once the hypotheses have been evaluated as alternative solutions supported by the simulation, corrective actions are implemented to reduce or eliminate the RE. In this phase it is necessary to validate that the implemented actions obtained according to the simulation and the planned result is the desired result [10].

Measurement: The most convenient way to comply with the last of the five functionalities is to apply a continuous and updated storage of event data based on specific data from the DE presented and the necessity of adequate IT support for performance measurement in inter-organizational cooperation [11], for which the performance indicators are required, adjusted and updated or, if new indicators are required, the SCEM methodology cycle continues as shown in Fig. 3 (Figs. 4, 5 and 6).

Alert	Action must be taken immediately
Critical	Critical fulfillment condition with high impact
Error	Significant fulfillment problems with medium impact
Warning	Slight fulfillment problems with low impact
Notice	Mainly normal fulfillment without negative impact

Fig. 3. Category for alert index.

The perspective of this methodology focuses on the visibility and collaboration of logistics processes between the different companies that make up the CS, which should allow early recognition of disruptions or ED in the flow of material, goods and information, so that control measures can be taken proactively before risking compliance with the plan and therefore SC performance.

3 Recovery Method of Supply Chain Under Ripple Effect: Case Study

For the realization of a recovery method of SC under RE, we will rely on SCEM methodology, described by Knickle & Kemmeter [8], and Zimmerman [6].

One sector that has stood out for remaining at the forefront of technology and operational methods is the automotive sector. This is a sector that contributes great profits to the GNP in those countries where the different assembly plants are located, but it is also the one with the greatest losses due to a disruption since they work under just-in-time (JIT) and just-in-sequence (JIS) manufacturing schemes, where a disruption can have great consequences and transcendence in the different levels of the SC, a characteristic of the RE.

The case study focuses on a manufacturing company that assembles vehicles, with an investment of nearly 3.5 billion dollars and a production capacity of approximately 250,000 units annually. The distribution of finished units is destined for the American continent and the rest of the world, as well as national destinations, being the multimodal form the means of transport for international destinations and land for domestic destinations. The distribution begins in a logistics yard, where the shipment of finished

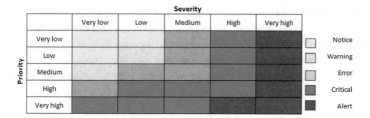

Fig. 4. Set of rules for fuzzy logic alert index.

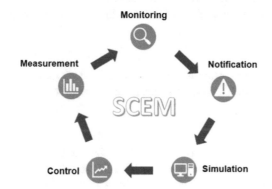

Fig. 5. Functions of Supply Chain Event Management.

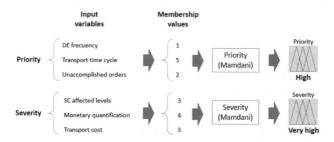

Fig. 6. Fuzzy logic process for index alert.

units is planned and executed under 3 scenarios, via rail for the United States (west) and Canada destinations; via land for domestic destinations; and via land to a port that in turn distributes to the rest of the world and the United States (east). Only for national distribution is direct delivery to the dealers, for the rest the shipments arrive at another logistics yards.

Two approaches to action are established: a reactive and a proactive approach. The reactive approach aims to adjust the SC processes in the presence of unexpected events for which there is no similar or reference historical event. The proactive approach considers possible DE that have already occurred for which there are action or contingency plans, the basis of the proposed methodology is that in the SCEM control function, it feeds new reference parameters to perform predictive monitoring, so that in the future the DEs can be avoided. In the simulation function this is done with the help of Anylogic® software simulation which allows not only to visualize the SC processes, but also to observe the impact of disruptions and recovery times in different scenarios.

The problem arises from the damage and total loss of some units finished in an assembly plant, as a result of intense rain that caused a flood, it is worth mentioning that this is not a common case in the region. Another antecedent is that the assembly process had a good production rate, but the over-inventory was stored in places not suitable for it, since the next process, which is the final inspection of the finished units did not materialize, the correct thing would have been to inspect all the units and move them to a logistic yard where they would remain safe until their shipment.

3.1   Monitoring

The type of monitoring carried out for this case is reactive, of the situation category "alert trigger", since the DE are identified by a SC member.

3.2   Notification

The type of monitoring carried out for this case is reactive, of the situation category "alert trigger", since the DE are identified by a SC member.

Based on the fuzzy logic, we determine the priority and severity of the DE, taking as input variables the circumstances of a DE with a RE and the indicators of the affected logistics process(es). By default, the indicators according to the characteristics of the latter, frequency of ED, affected levels of SC and the effect on SC performance are also considered.

Then, through a set of rules like:

```
IF < <alert ≫
THEN < <distribution list alert ≫
THEN < <send notification ≫
END.
```

The recipients will receive the notification with the level of < ≪alert ≫.

The purpose of notification is to inform decision makers and corrective action makers, to give a sense of urgency to the disruption, and to ensure that it reaches those to whom it is addressed with the appropriate gift of command to take timely action.

3.3   Simulation

A simulation model is made under system dynamics paradigm, from the OEM inspection yard to the carrier and dealers, with the respective flow parameters through the SC links (in this case the distribution process). There are several controls such as: timetables, loading rate for each destination, where the optimum should be selected to maximize a general performance indicator K_P (Performance) taking into account the priority coefficients λ of the key performance indicators $K_1,..., K_j$ (e.g., service level, logistics costs, penalties for delays, empty unused equipment, among others).

Next, the disruption in the inventory of finished units available to be transferred to the logistics yard for later shipment is added to the simulation model $z(t)$, so that the impact of the DE in question can be observed and the recovery times to take immediate and accurate corrective actions can be observed.

The simulation is carried out with the help of AnyLogic software, from which three action scenarios are derived, this without considering the recovery of production that is not within the scope of the case study.

- Scenario 1: recovery actions taken involve operating with costly additional emergency transport without priority to any destination.
- Scenario 2: recovery actions taken involve operating with costly additional emergency transport giving priority to the most lost-lose destination in the event of failure.
- Scenario 3: recovery actions taken involve operating with normal transport costs giving priority to the destination with the greatest loss in case of failure.

As can be seen in the analysis, the scenario 2 is the most promising according to the simulation as shown in Fig. 7, where additional expensive emergency transport is used instead of the normal costs and with the best stabilization time of the processes. In this case, the total gain in all three is obviously reduced with respect to undisturbed operation of a DE, but it is also the one with the best total gain (Figs. 8 and 9).

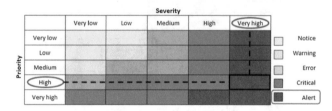

Fig. 7. Index alert for priority = high and severity = very high.

3.4   Control

In this function, corrective actions are implemented that result from the best alternative proposed by the simulation, it is important that through a method of monitoring the implementation of these actions, validate that what was desired to obtain according to the simulation and what was planned is the result obtained in an effective and efficient manner, in which case it is also important to standardize the actions taken. Otherwise, the process parameters are recalibrated at this stage in order not to affect the performance of the SC.

3.5   Measurement

From this last stage, the necessary changes or adjustments are obtained for the KPI's, these will feed an information's repository of the indicators for a new monitoring with

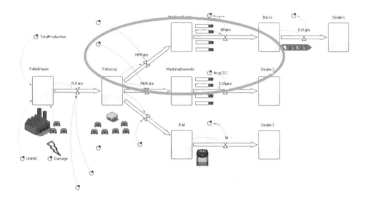

Fig. 8. Simulation model.

Scenario 1

Customer	Total profit
Ocean	51.09%
North America	18.75%
Domestic	12.12%
	81.96%

Scenario whitout disruption

Customer	Total profit
Ocean	62.21%
North America	23.65%
Domestic	14.14%
	100.00%

Scenario 2

Customer	Total profit
Ocean	56.13%
North America	19.50%
Domestic	12.91%
	88.54%

Scenario 3

Customer	Total profit
Ocean	49.88%
North America	14.50%
Domestic	9.96%
	74.34%

Fig. 9. Scenario assessment.

sufficient information that allows to predict or recognize a possible disturbance early on, thus continuing with the SCEM cycle.

4   Conclusions

The simulation aims to provide SC executors and planners with new tools to support their decision making without putting the real system at risk, estimating the impact of DE on the total economic performance of the SC and suggesting corrective measures for stabilization and recovery, seeking to make them efficient and effective. The results of this work can be used in the future to establish preventive measures. Also, proactive and reactive approaches to attack the RE can be compared with a perspective of flexibility by considering in the future to what extent the disruption of the DE may occur without affecting the implementation of the initial plan.

References

1. Hendricks, K.B., Singhal, V.R.: Association between supply chain glitches and operating performance. Manag. Sci. 695–711 (2005)
2. Dolgui, A., Ivanov, D., Sokolov, Y.B.: Ripple effect in the supply chain: an analysis and recent literature. Int. J. Prod. Res. **56**, 1–2, 414–430 (2018)
3. Council of Supply Chain Management Professionals. Supply Chain Management Terms and Glossary (2013). www.cscmp.org
4. Vanany, I., Zailani, S., Pujawan, N.: Supply chain risk management: literature review and future research. Int. J. Inf. Syst. Supply Chain Manag. **2**(1), 16–33 (2009)
5. Ijioui, R., Emmerich, H., Ceyp, M., Diercks, Y.W.: Supply Chain event management als strategisches Unternehmensführungskonzept, pp. 3–13. Physica Verlag (2007)
6. Zimmermann, R.: Agent-based Supply Network Event Management, 1st edn. Switzerland, Birkhäuser Basel (2006). Smith, T.F., Waterman, M.S.: Identification of common molecular subsequences. J. Mol. Biol. **147**, 195–197 (1981)
7. Heusler, K., Stölzle, W., Bachmann, Y.H.: Supply chain event management. Grundlagen, Funktionen und potenzielle Akteure. WiSt Heft, pp. 19–24 (2006)
8. Knickle, K., Kemmeter, Y.J.: Supply Chain Event Management in the Field: Success with Visibility. Garner (2002)
9. Yager, R.R., Zadeh, L.A.: An Introduction to Fuzzy Logic Applications in Intelligent Systems. USA, Springer Science and Business Media (2012)
10. Nissen, V.: Supply Chain Event Management. Wirtschaftsinformatik **44**(5), 477–480 (2002)
11. Küppers, S., Ewers, C.: Supply Chain Event Management in der Pharmaindustrie: Status und Möglichkeiten, pp. 85–102. Physica-Verlag (2007)

Investigating the Reproducibility and Generality of the Pilot Environmental Performance Index (EPI 2006)

Tatiana Tambouratzis[1(✉)], Angela Mathioudaki[1,2], and Kyriaki Bardi[1]

[1] Department of Industrial Management & Technology,
University of Piraeus, 18534 Piraeus, Greece
`tatianatambouratzis@gmail.com`,
`tzela_m@hotmail.com`, `kiki_mpardi@hotmail.com`
[2] Department of Electrical and Electronic Engineering,
National Technical University of Athens, 9,
Heroon Polytechneiou Street, 15780 Athens, Greece

Abstract. The reproducibility and generality of the Pilot Environmental Performance Index (EPI) 2006 is investigated. Initially, the most accurate means of deriving the Pilot EPI 2006 scores and rankings of the countries which participated in the creation of the index from the various constructs of the Pilot EPI 2006 dataset is identified using a variety of traditional and computational intelligence tools. Use-all and leave-one-out cross-validation are subsequently employed for recreating the index values of the participating countries from the entire, as well as from parts only of the, dataset; parametric as well as non-parametric methodologies are used to this end, consequently establishing the accuracy with which the Pilot EPI 2006 scores and rankings of participating countries can be predicted from these relationships. The optimal means (combination of traditional and computational intelligence methodologies) of estimating the Pilot ESI 2006 scores and rankings of the participating countries is tested and verified; the results raise some questions concerning the reproducibility and generality of the Pilot EPI 2006.

Keywords: Environmental sustainability · Participating country
Country of interest · Environmental performance index · Construction
Consistency · Reproducibility · Generality · Polynomial approximation
Computational intelligence · General regression artificial neural networks

List of Acronyms

(ANN)	Artificial neural network
(BO)	Broad objective
(COI)	Country of interest
(CV)	Cross-validation
(EPI)	Environmental performance index
(ES)	Environmental sustainability

© Springer Nature Switzerland AG 2019
P. Vasant et al. (Eds.): ICO 2018, AISC 866, pp. 466–475, 2019.
https://doi.org/10.1007/978-3-030-00979-3_49

(GRNN)	General regression artificial neural network
(PC)	Policy category
(PTD)	Proximity-to-target tata
(RD)	Raw data

1 Introduction

The interest in environmental sustainability (ES) has risen considerably since the late 1990's, with a number of international organisations assembling and processing sets of pertinent ES-related parameters which collectively express "how the environment is fairing" [1] at the country level. These parameters are collected in as consistent a manner as possible over all the interested ("participating") countries and, subsequently, combined into a uniformly created cumulative (singleton) index value which expresses the level of ES attained by each participating country. The gradual – and ultimately maximal - hierarchical compression of the collected data allows direct ES-based cross-country comparisons to be made between the participating countries, while it further facilitates the improvement of the ES level of any country of interest (COI) via the adoption of ES-based policies of other participating countries which demonstrate (a) similar sets of values with the COI for a number of pertinent parameters, yet (b) higher ES levels.

In this piece of research, the consistency of the Pilot Environmental Performance Index (EPI) 2006 [2] is investigated in terms of the accuracy with which the ES levels (expressed via the reported Pilot EPI 2006 scores) and rankings can be reproduced as well as predicted. Cross-validation (CV) [3] is implemented to this end, concurrently establishing (a) whether the Pilot EPI 2006 scores and rankings of the 133 countries which participated in the creation of the Pilot EPI 2006 can be faithfully reproduced from the original data, and (b) how accurately the Pilot EPI 2006 scores of non-participating countries of interest can be predicted from the same relationship. Mathematical and computational intelligence-based methodologies are compared and combined for selecting the most accurate unified approximation methodology that is – at the same time - capable of successfully generalising to novel (yet with compatible data) countries.

This contribution is organised as follows: Sect. 2 describes the structure of the Pilot EPI 2006 in terms of participating countries, constructs, transitions between constructs [2] and scores/rankings; Sect. 3 establishes the optimal combination of methodologies and constructs for reproducing the Pilot EPI 2006 scores and rankings of the participating countries as well as for predicting those of novel countries with data that is compatible to that of the participating countries according to the Pilot EPI 2006 hierarchy; the set of optimal methodologies (one for each approximation) is used in a concerted manner in Sect. 4 for verifying the Pilot EPI 2006 and for, subsequently, pinpointing some weaknesses in index construction which lead to inconsistencies in the duplication of the scores and rankings of the participating countries; finally, Sect. 5 summarises the findings and points towards some additional questions that it will be of interest to answer in ensuing research.

2 The Pilot EPI 2006

Launched in 2006, the Pilot EPI 2006 constitutes the first outcome of the collaboration between (a) the Yale Center for Environmental Law and Policy, (b) the Columbia Center for International Earth Science Information Network, and (c) the World Economic Forum, towards the construction of a universal index that communicates ES at the country level in a maximally comprehensive and transparent manner. The pertinent characteristic of the Pilot EPI 2006 (attested by its continued publication every other year to this day and the influence it still exerts globally in ES-related matters) stems from the fact that - unlike most ES indices which are based on inter-country comparisons - the particular index independently measures the distance of each participating country from a common target that represents absolute ES, thereby constituting a transparent and objective supportive tool for setting the optimal environmental policies of not only each participating country, but also at a more global level.

The Pilot EPI 2006 is based on the data of the 133 countries appearing in [2, p. 27] and is built according to the five-construct/three-level hierarchy detailed below.

2.1 Constructs

The five Pilot EPI 2006 constructs involve:

– The Raw Data (RD), namely 16 parameters which have been derived according to the "Millenium Development Goals" [4], and whose ranges and target values have been set in a uniform manner over all the participating countries.
– The Proximity-to-Target Data (PTD), with each PTD derived by processing the corresponding RD via (i) independent scaling in the [0 100] range; (ii) value reversal for rendering all the PTDs unidirectional and with lower/higher values denoting lower/higher ES, respectively; (iii) 5% cut-off of the extreme, trimming the lowest and highest values of each PTD (including the values that are below the minimum and those that exceed the target maximum) to the minimum and maximum set values, respectively; (iv) Box-Cox transformations [4] for independently pseudo-normalising each PTD.
– Six Policy Categories (PCs), derived from (I) principal component analysis (PCA [5]) for producing the Environmental Health, Biodiversity & Habitat, and Sustainable Energy PCs; (II) expert judgment for producing the Air Quality, Water Resources, and Productive Natural Resources PCs.
– Two Broad Objectives (BOs), each compiled from the set of co-relevant PCs, with all PCs which are used in the creation of a BO being assigned the same importance.[1]
– The Pilot EPI 2006 value, which constitutes the sum of the two BOs or - equivalently - the sum of the six PCs[2].

---

[1] As the BOs do not constitute a functional part of the Pilot EPI 2006 hierarchy [2], but are rather conceptual, they are not considered further in this investigation.

[2] The use of weights between PCs and BOs as well as between BOs and the Pilot EPI 2006 (as shown in Table 2), denotes that the relationship between PCs and the Pilot EPI 2006 is linear.

2.2   Transitions

The three transitions of the Pilot EPI 2006 hierarchy involve [2]:

(a)   A one-to-one relationship between each RD and the corresponding PTD, with some of the underlying relationships being non-linear.
(b)   A linear combination of the 16 PTDs for producing the six PCs, where some PTDs contribute to more than one PC.
(c)   A linear relationship between the PCs for producing the Pilot EPI 2006.

3   Verifying the EPI Hierarchy of the Pilot EPI 2006

The Pilot EPI 2006 hierarchy of Fig. 1 is systematically investigated for establishing whether the Pilot EPI 2006 values of the 133 participating countries (listed in [2]) can be reproduced, thus assessing the consistency of the index. Subsequently, 10-fold CV is performed for testing the accuracy of evaluating the Pilot EPI 2006 value of a novel (non-participating) country through (a) its RD values (provided that these are compatible with those of the participating countries) and (b) the optimal relationships between constructs. For the purposes of generality and homogeneity in handling the problem space, the countries are sorted in ascending order of their EPI values and each, namely the ith $(i = 1, 2, \ldots, 10)$ fold is created using the ith, $i + 10$th, ... sorted countries.

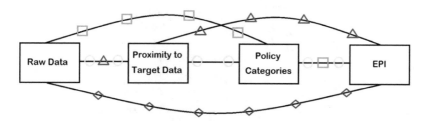

Fig. 1.   The four complete paths between RD and the Pilot EPI 2006, marked by **a** yellow circles (three elementary transitions replicating [2]), **b** red triangles (two transitions, one elementary, one compound), **c** green squares (two transitions, one of them compound and the other one elementary), and **d** blue diamonds (a single compound transition, composed of the three elementary transitions).

3.1   Test Paths

All possible means of deriving the Pilot EPI 2006 are explored, with the four complete paths of Fig. 1 being of primary interest. For completeness of the investigation, and in order to independently evaluate the reproduction accuracy of every transition, each partial path is also exhaustively investigated.

3.2   Methodologies

Whenever available (either in the form of "weights", or via some other explicit qualitative expression/description), the methodology of [2] used for deriving the relationship between each pair of neighbouring constructs has been replicated (marked as "[2]" in the corresponding fields of Table 1). Complementary to replication, the relationships between all possible pairs of constructs have also been reproduced via (a) parametric approaches in the form of polynomial approximation [6] of 1st through to 5th degrees[3] and (b) non-parametric approaches in the form of general regression artificial neural networks (GRNNs) [7]). The most accurate methodology (marked as "best" in the corresponding fields of Table 1) is subsequently employed for reproducing the relationships between constructs.

Table 1. The 13/eight combinations of transitions and methodologies implementing the four complete/seven partial paths between constructs of the Pilot EPI 2006, as these are shown in Fig. 1. The complete/partial paths are marked as dark/light entries, respectively. Both the duplication of the evaluation procedure described in [2] and the optimal (parametric or non-parametric) approximation of each relationship appear in the Table as "[2]" and "best", respectively.

RD_0	Best	PTD_1				Partial
RD_0	best		PC_1			Partial
RD_0	best					EPI_2 (complete)
		PTD_0	[2]	PC_2		Partial
		PTD_0	best	PC_3		Partial
		PTD_0	best			EPI_3
		PTD_1	[2]	PC_4		Partial
		PTD_1	best	PC_5		Partial
RD_0	best	PTD_1	best			EPI_4 (complete)
				PC_0	[2]	EPI_5
				PC_0	best	EPI_6
RD_0	best			PC_1	[2]	EPI_7 (complete)
RD_0	best			PC_1	best	EPI_8 (complete)
				PC_2	[2]	Partial
				PC_2	best	Partial
				PC_3	[2]	Partial
				PC_3	best	Partial
RD_0	best	PTD_1	[2]	PC_4	[2]	EPI_{12} (complete)
RD_0	best	PTD_1	[2]	PC_4	best	EPI_{13} (complete)
RD_0	best	PTD_1	best	PC_5	[2]	EPI_{14} (complete)
RD_0	best	PTD_1	best	PC_5	best	EPI_{15} (complete)

---

[3] The extensive range of polynomial degrees employed amply accommodates for any rounding errors and/or non-linearities of each relationship.

The GRNN model is of particular interest for reproducing these relationships as - complementary to polynomial approximation - it provides:

- Simple and computationally efficient training, which is completed after a single presentation of the training data. The non-parametrically produced regression surface is capable of approximating the Bayes optimal for an appropriately chosen value (or range thereof) of the single GRNN tunable "spread" parameter ($\sigma \in [0\ 1]$), which regulates the influence that the output of each training pattern has on the test inputs based on the distance between the test input and each training pattern.
- Universal function approximation potential; given a sufficiently representative (of the problem space) set of training patterns - and a (range of appropriate) σ value(s) -, the created GRNN output surface is capable of practically perfect approximation/ generalisation, i.e. of accurately reproducing the input-output hypersurface that represents the problem space. This is expected to hold under the 10-fold cross-validation scheme implemented next since: (a) the available data constitute the universe of the problem (pattern space created by the 133 countries participating in the Pilot EPI 2006), and (b) the EPI EPI 2006 construct values and scores have been derived in a uniform manner over the entire set of countries [2]. For more details on GRNN construction and operation, the interested reader is referred to [7].

4   Results

The findings derived from replicating/approximating the relationships between pairs of constructs are detailed next; each construct is used - in turn - as the output of the respective transitions, with each of the preceding constructs being independently used as input.

4.1   PTD from RD

In agreement with the generally non-linear nature of the transformations described in [2, 8] for deriving the PDTs from the corresponding RDs, the GRNN proves to be a strong approximation tool, constituting the best and second-best methodology in 8 and 3 (out of the 16) predictions, respectively; the 5th degree polynomial is found best four times, and the 1st and 4th degree polynomials twice each.

In terms of prediction accuracy (under 10-fold CV), the INDOOR and OVRFSH RD to PTD relationships are perfectly recreated via 1st and 5th degree polynomials, respectively; the same applies to the OZONE, PWI, and PACOV relationships via 5th, 1st and 5th polynomials, respectively, and the NLOAD relationship via a 5th degree polynomial. The remaining relationships involve significant approximation errors, i.e. the PTDs cannot be reproduced in a satisfactory manner from the corresponding RDs by either parametric or non-parametric means. In most of these cases, the GRNN significantly outperforms the alternative methodologies.

Table 2. The performance of the best and second-best methodologies for approximating the pairwise relationships between EPI construct RD.

relationship	best methodology	second-best methodology
RD - > PTD	[8] (case III)	
RD - > PC		
ENV_HEALTH	3rd degree polynomial	2nd degree polynomial
AIR	5th degree polynomial	3rd degree polynomial
WATER	3rd degree polynomial	2nd degree polynomial
BIODIVERSITY	GRNN	2nd degree polynomial
RESOURCE_MGT	GRNN	2nd degree polynomial
ENERGY	3rd degree polynomial	2nd degree polynomial
RD - > TO EPI	**1st degree polynomial**	**GRNN**
PTD - > PC	[8] (case II)	
PTD - > TO EPI	**best methodology**	**second best methodology**
	Yale Method	GRNN
PC - > TO EPI	**best methodology**	**second best methodology**
	Yale method	1st degree polynomial

4.2   PC from RD

The GRNN constitutes the most appropriate methodology for recreating the non-linear relationships[4] between RD and PCs; it is found best in approximating three out of the six PCs, with the third-degree polynomial being found best in the remaining cases. This result is in agreement with the non-linear nature of the transformations described in [2] for implementing the first leg of the relationship, as also mentioned above. However, none of the approximations is found sufficiently accurate under 10-fold CV.

4.3   Pilot EPI 2006 from RD

This relationship is confirmed to be linear[5]; the fact that the GRNN is second-best serves to confirm the universal approximation capabilities of this ANN paradigm. However, the clearly non-linear nature of the PTD from RD relationship is incompatible with the linear nature of the RD/EPI relationship.

4.4   PC from PTD

The relationship that holds between PTDs and related PCs is found mostly linear, which is in general agreement with the linear nature of the transformations between the

---

[4] According to [2], the relationship results from a non-linear and a linear relationship between the RD and PTD and between PTD and PC, respectively.

[5] This is expected as the relationship constitutes the combination of two linear relationships.

two constructs described in [2]; the weights appearing in [8, 9] optimally as well as accurately approximate the values of four (AIR, WATER, BIODIVERSITY and ENERGY) out of the six PCs. For the remaining two PCs: the 1st degree polynomial outperforms the Yale weights for ENV_HEALTH, with both approximations - however - being highly accurate; the GRNN is also found more accurate than the Yale method for RESOURCE_MGT, even though neither methodology is capable of approximating the underlying relationship to a satisfactory degree.

4.5   Pilot EPI 2006 from PTD

Linearity also holds for this relationship.

4.6   Pilot EPI 2006 from PC

This relationship is linear, with the linear approximation verifying the weights reported in [2].

4.7   General Comments

It can be concluded that - as also supported in [2] and shown in Fig. 1 - the relationships between PTD and PC, as well as between PC and Pilot EPI 2006 scores, are both linear; as expected, the relationship between PTD and Pilot EPI scores is also confirmed to be linear. Conversely, the relationship between RD and PTD is non-linear for 14 out of the 16 approximations, as is - with no exception - the relationship between RD and PC. It is, thus, not clear how the observed non-linearities are reduced to such an extent as to render the relationship between RD and Pilot EPI 2006 scores a clearly linear one.

In terms of comparisons between the 15 EPI approximations shown collectively in Table 3, EPI2 proves to be the most consistent in duplicating the Pilot EPI 2006 scores, being in fact an order of magnitude more accurate than the next-best approximations

Table 3. The accuracy of the various Pilot EPI 2006 full reconstructions in terms of errors in scores and rankings.

Errors EPI	Mean	Min	Max	Std	Max rank
2	0.032	0.000	0.082	0.020	8
4	0.857	0.005	6.590	1.121	26
7	0.694	0.002	2.620	0.565	8
8	0.711	0.001	4.580	1.081	13
12	0.788	0.005	2.718	0.607	8
13	1.099	0.009	5.705	1.180	22
14	1.132	0.021	6.170	1.181	2
15	1.998	0.005	12.462	1.988	26

(EPI7 and EPI12). In terms of ranking accuracy, EPI14 is clearly superior, managing to retain the ranking order despite its relatively reduced accuracy in EPI 2006 score prediction, with EPI2, EPI7 and EPI12 being jointly second best.

5   Conclusions

The reproducibility and generality of the Pilot Environmental Performance Index (EPI) 2006 has been investigated. The most accurate means of duplicating the Pilot EPI 2006 scores and rankings of the participating countries from the various constructs of the Pilot EPI 2006 dataset have been identified from an assortment of combinations of traditional (parametric) and computational intelligence-based (GRNN) methodologies. The index has been duplicated via an exhaustive set of partial and complete paths between the various Pilot EPI 2006 constructs and the corresponding scores. The optimal means (combinations of the optimal methodologies per transition between constructs) of estimating the Pilot EPI 2006 scores and rankings have been tested and verified, with the results raising some questions concerning the reproducibility and generality of the Pilot EPI 2006. It is clear that further investigation is needed for implementing a unified methodology that can duplicate the step-by-step transition from one the level of the Pilot EPI 2006 hierarchy to the next. However, some light has been shed to the strengths and weaknesses of index as well as of the alternative prediction methodologies.

References

1. Hammond, A., Ariannse, A., Rodenburg, E., Bryant, D., Woodward, R.: Environmental indicators: a systematic approach to measuring and reporting on environmental policy performance in the context of sustainable development, World Resource Institute Report (1995)
2. Yale Center for Environmental Law & Policy: Center for International Earth Science Information Network (CIESIN), Pilot 2006 Environmental Performance Index (2006)
3. Devijver, P.A., Kittler, J.: Pattern Recognition: A Statistical Approach. Prentice-Hall, London, GB (1982)
4. http://www.un.org/millenniumgoals/bkgd.shtml
5. Box, G.E.P., Cox, D.R.: An analysis of transformations (with Discussion). J. Roy. Stat. Soc. B **26**, 211–252 (1964)
6. Jolliffe, I. T.: Principal component analysis. In: Springer Series in Statistics, 2nd edn. Springer, New York (1986) (ISBN 0-387-95442-2)
7. Atkinson, K.A.: An Introduction to Numerical Analysis, 2nd edn. John Wiley and Sons (1988)
8. Specht, D.F.: A General Regression Neural Network. IEEE Trans. Neural Netw. **2**, 568–576 (1991)

9. Tambouratzis, T., Bardi, K.S., Mathioudaki, A.G.: How Reproducible - and Thus Verifiable - Is the Environmental Performance Index?, Recent Advances in Environmental Sciences and Financial Development. In: Proceedings of the 2nd International Conference on Environment, Ecosystems and Development (EEEAD 2014) Athens, Greece, 28th–30th Nov 2014, pp. 27–33 (2014)
10. Tambouratzis, T., Bardi K.S., Mathioudaki A.G.: Comprehending the pilot environmental performance index. In: Proceedings of the 15th UK Workshop on Computational Intelligence (UKCI 2015), Exeter, UK, 7th–9th Sept 2015

New Smart Power Management Hybrid System Photovoltaic-Fuel Cell

Mohammed Tarik Benmessaoud[1(✉)], A. Boudghene Stambouli[1,2],
Pandian Vasant[4], S. Flazi[1], H. Koinuma[2,3], and M. Tioursi[1]

[1] Electrical and Electronics Engineering Faculty,
University of Sciences and Technology of Oran,
USTO-MB Algeria, BP 1505 EL M'Naouer, Oran 31000, Algeria
tarik.benmessaoud@univ-usto.dz
[2] Sahara Solar Breeder Foundation International, Tokyo, Japan
http://www.ssb-foundation.com
[3] Graduate School of Frontier Sciences, Tokyo University, Tokyo, Japan
[4] Universiti Teknologi Petronas, 32610 Seri Iskandar, Perak, Malaysia

Abstract. Currently, energy consumption in the planet is high, public awareness of energy consumption, environmental protection and steady progress in
the deregulation of conventional energy, distributed generation systems (based
on hydrogen) have attracted increased interest. Fuel cell (FC) base and high-
temperature systems also have great potential in future single-source or multi-
source (hybrid- HSE) applications due to their rapid technological development
and numerous benefits such as well, as the appreciable efficiency, low emission
(greenhouse gases) and flexible modular technology. Dynamic models for the
main components of the system, namely the photovoltaic (PV) energy conversion system, fuel cells, electrolysers, electric power interconnection circuits,
protective battery, storage tank of hydrogen, the gas compressors are developed.
Two renewable energy modes, a photovoltaic field and a solid oxide fuel cell
(SOFC) or Dynamics of proton exchange membrane fuel cells (PEMFC) with
hydrogen storage system for generating part of system electrical energy is
presented. Feasibility of using fuel cell (FC) for this system is evaluated by
means of simulations. The electrical dynamic model, temperature change and
dual layer capacity effect are considered in all simulations. Photovoltaic system
(PV) output current is connected to the bus. Using a MPPT (maximum power
point tracker) which is an electronic DC to DC converter that optimizes the
match between the solar array and utility grid. The proposed system utilizes an
electrolyser (EL) to generate hydrogen and a tank for storage. Therefore, there is
no need for batteries. Moreover, the generated oxygen could be used in FC's
system and other applications. Moreover, such as the photovoltaic system, it is
possible to connect fuel cell (FC) output voltage to DC bus alternatively.
A controller model is presented to control flow of energy of system, hydrogen
and oxygen to FC and improve transient and steady state responses of the output
voltage to load disturbances. Simulations are carried out via
MATLAB/SIMULINK and results show that the load tracking and output
voltage are acceptable.

© Springer Nature Switzerland AG 2019
P. Vasant et al. (Eds.): ICO 2018, AISC 866, pp. 476–486, 2019.
https://doi.org/10.1007/978-3-030-00979-3_50

Keywords: Modeling · Optimization · Hybrid system energy (HSE) Photovoltaic (PV) · Electrolyzer (EL) · Fuel cell (FC) · Hydrogen

1 Introduction

Energy is a vital issue for social and economic development of any country. According to Renewables 2015 Global Status Report, 81% of worldwide energy demand is supplied by conventional energy sources like coal, natural gas, crude oil, etc. The conventional fossil fuel energy sources such as petroleum, natural gas, and coal which meet most of the world's energy demand today are being depleted rapidly. Also, their combustion products are causing global problems such as the greenhouse effect and pollution which are posing great danger for our environment and eventually for the entire life on our planet [1]. Many alternative energy sources including photovoltaic, thermal solar, wind, diesel system, fuel cell, micro turbine and gas turbine can be used to construct a hybrid energy system. There are many applications in the world, powered by autonomous systems of electricity generation. These generators use local renewable sources. There are photovoltaic panels, wind turbines and microturbines. Electricity from renewable sources is intermittent, dependent on climatic conditions. These renewable generators are coupled to a storage system ensuring continuous energy availability. The photovoltaic generator is the main source of our application. Generally, energy storage is provided by batteries. The batteries have very good yields, of the order of 80–85%, and a very competitive price, if one considers the lead technology. But their disadvantages are multiple (Self-discharge, Lifetime, maintenance and safety). The operating constraints described above require that the size of the batteries be compared with the power of the photovoltaic generator as a function of the autonomy of the storage system. To improve the photovoltaic systems, while maintaining their quality of respect for the environment, we decided to integrate a hydrogen system to store energy in the long term. Indeed, the gas can be produced by an electrolyser, stored without significant loss whatever the duration of the storage, and then converted into electricity in a fuel cell. In this paper, a stand-alone hybrid renewable energy system comprising of photovoltaic (PV), fuel cell (FC) and electrolyzer (EL) is projected. PV is the prime energy source of the system to take complete benefit of renewable energy, and the FC electrolyzer combination is used as a long-term storage system and a backup unit. The particulars of the system configuration and the features of the major system components are also deliberated in the paper. A complete power management strategy is planned for the system to synchronize the power flows among the different power sources. Simulation studies have been carried out to validate the system performance under different situations using practical load profile and real weather data. The paper is systematized as follows. The system configuration is conferred and the inclusive power management strategy for the hybrid system in Sect. 2. Section 3 gives the system component characteristics via modelization. Section 4 gives the simulation results. Conclusion of the paper is given in Sect. 5.

2 Hybrid Renewable Energy Systems HRES: PV-EL-FUEL CELL

Hybrid renewable energy systems (HRES) are becoming popular as stand-alone power systems for providing electricity in remote areas due to advances in renewable energy technologies and subsequent rise in prices of petroleum products. A hybrid energy system, or hybrid power, usually consists of two or more renewable energy sources used together to provide increased system efficiency as well as greater balance in energy supply [2, 3].

Addit Fig. 1 shows the system configuration for the proposed stand-alone hybrid energy system, photovoltaic - fuel cell. The renewable PV power is taken as the primary source while the FC via electrolyzer combination is used as a backup and a long-term storage system.

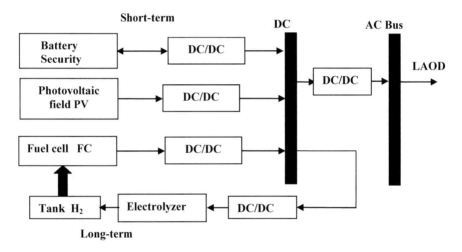

Fig. 1. Concept of an autonomous energy system based on hydrogen technology PV-FC.

All the energy systems are connected in parallel to a common DC bus line through appropriate power electronic interfacing circuits. When there is excess solar generation available, the electrolyzer is activated to initiate hydrogen production, which is delivered to hydrogen storage tanks at low pressures. When there is deficit in power generation, the FC will start to produce energy using hydrogen from the reservoir tanks. In order to compensate for the electrical response time and to attenuate the power fluctuations in the DC bus, a safety battery bank has been integrated. Figure 1 shows a Hybrid power systems in our case are designed to generate electrical power by using two power generating devices such as photovoltaic and fuel cell (high temperature and temperature base). Such systems may range from the small system capable of supplying power to a large system that can power a system. Hybrid feed systems provide energy to remote communities, particularly in developing countries, where the national grid is not economically and technically viable. The first hybrid energy systems, consisting of photovoltaic generators and fuel cells, were installed in the 1990s in Germany and the United States [4–6]. The combination of PV and FC benefits from reduced battery capacity among other benefits. However, for better performance of the PV-FC hybrid system, good solar irradiation potential must be on the site. Influencing factors, photovoltaic generator capacity, fuel cell capacity, capacity of storage device (number of security batteries), hydrogen storage device capacity and safety, storage site Generation (distance between the plant and the consumer), etc. An important role in the operation, maintenance and cost of the PV/FC hybrid system.

3 Power Management Strategy

The main requirements of the proposed power management strategy for the stand-alone hybrid power system are to satisfy the load demand under varying weather conditions and to manage the power flow while ensuring efficient operation of the various energy systems time. The variability inherent in the solar generation produces variability in the operation of the fuel cell, the electrolyser and the battery. Since the stable operation of these energy systems is vital for efficiency, service life and cost, the management strategy should mitigate the effects of power fluctuations on their mode of operation. In this implementation, a simple hysteresis control is used for this purpose. The key decision parameters for the proposed hysteresis control are shown in Fig. 2. The management strategy should primarily use the power generated by the photovoltaic system as a priority to satisfy the load demand. Any excess power should be used to produce hydrogen by electrolysis of the water and the fuel cell and/or the battery must satisfy any shortage of energy. In the case of a power generation deficit ($P_{System} < 0$), the deficit is covered by the fuel cell [7].

If $V_{Bus} \leq V_{FC\_start}$ and $V_{H2} > V_{H2\_min}$, the fuel cell is selected to provide the energy required to meet the load demand. If $V_{FC\_start} < V_{Bus} < V_{FC\_end}$ and in the previous step, the fuel cell was still running and there is a power shortage ($P_{System} < 0$), the FC will not stop and continue to operate until the V_{Bus} reaches Limit V_{FC_end}.

In the range $V_{FC\_start} < V_{Bus} < V_{FC\_end}$ and provided neither the electrolyzer nor the fuel cell are operating, the battery is charged ($P_{System} > 0$) or discharged ($P_{System} < 0$) according to the net power level. Figure 3 shows the detailed logic diagram of the proposed energy management strategy.

4 System Modeling

The HRES examined in this study includes fuel cells and solar panels on the supply side and a hydrogen storage system with a bank of security storage bays. The integrated main controller decides to store excess power by electrolysis or to provide deficient power with the fuel cell plant according to the supply-demand scenario. The standard models available in the literature for fuel cell subsystems, photovoltaic, electrolyser and gas production model (hydrogen and oxygen) and hydrogen storage are used. These simple models have been chosen because they are sufficiently accurate to provide an estimate of average power on an hourly basis without having to deal with the

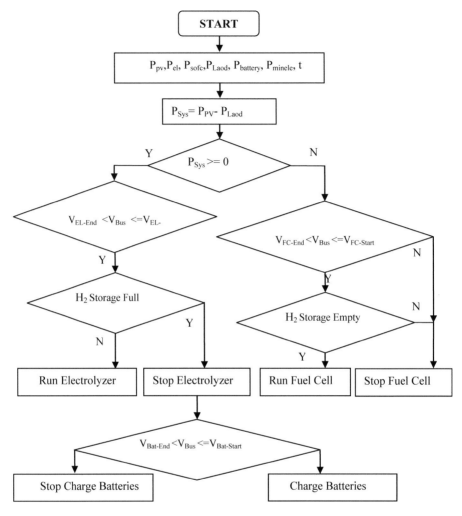

Fig. 2. Logical block diagram for the power management strategy.

complexities associated with other more complex physical models. System components must be dimensioned. The following parameters must be defined, the peak power of the photovoltaic field, rated power components, electrolyser and fuel cell, rated capacity storage batteries and storage volume of gas. Assumptions and criteria will be detailed in the following paragraphs. The nominal power of the fuel cell system is set so that it can in any case ensure the supply of power to the load:

$$\text{Photovoltaic system} : P_{PV} = P_{LAOD} \quad P_{PV} = P_{LAOD} = 5kW$$
$$\text{The electrolyzer system} : P_{nomEL} = P_{cretePV}$$
$$\text{The Fuel Cell System} : P_{nomFC} = P_{MaxLAOD}$$

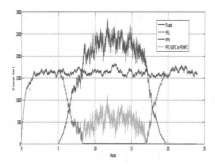

Fig. 3. Evolution of the powers exchanged on the DC bus for a particular day of operation of the HRES-PV-PAC system.

To optimize system performance, simple rules of operation can be issued; PV provides the maximum power available. PV power is favored over that of the fuel cell that serves as extra the electrolyzer cell and must never operate simultaneously. Many electrical configurations are available: the DC bus, AC bus and all intermediate configurations. The AC bus provides modularity important but the overall performance of the system may be low. The only way to control the power supplied by each of the two generators, the PV and the fuel cell, is to integrate the two power converters feeding a continuous floor. In addition, the floor must be continuously high voltage to reduce ohmic losses. The architecture is one selected electrical high voltage DC bus phases to which are connected components via voltage converters [8]. To obtain the maximum power of the PV array, it simply sets the voltage of the solar field avoiding implementing a complex control strategy (MPPT) [9]. The input voltage is the nominal voltage of the solar field. The following table explains the instruments used:

Designation	Power (W)	Instrument
Photovoltaic field	6 KWp	20 × 300 Wp
Fuel cell SOFC	5 KW	1 Stack – 5 kW
Fuel cell PEMFC	5 KW	10 Stack × 500 W
Electrolyser	5 KW	1 Stack – 5 KW

5   Analysis of HRES: PV-H$_2$-PAC Dynamic Behavior

In this section, we will analyze the dynamic behavior of HRES under climatic variations (illumination) in order to study the effectiveness of control strategies enabling renewable energy sources to optimize energy efficiency. The inputs of the model are the climatic conditions and the load demanded as a function of time. All instrument models, photovoltaic, fuel cell, electrolyser and voltage converter, are detailed in the paper [10]. The complete system allows to calculate the operating points of each component during a full year. This model consists of many algebraic loops. The battery and the electrolyser are powered. Since the electrical model is the voltage as a function of the current, the operating point must be searched for the required power. The other algebraic loop is that of the PMU. Indeed, at time t, the power of the PV, its voltage and the power demanded by the load are known. Simulink is the ideal tool to solve this type of problem.

5.1   Simulation Outputs

Once the parameters in the component parameter tables are entered and configured and the input profiles are defined, the simulation of one day or one year of operation can be started. The output variables of the simulator are numerous. For each component, the input and output powers are recorded over the entire simulation period. It is therefore possible to monitor the evolution of the energy efficiency of each component and evaluate the various energy losses in the system. Figure 3 shows the evolution of the powers involved in the PV–H_2–PAC, DC bus over a simulation day.

Around 18 h, the production of the PV field is no longer surplus (the sunlight is no longer sufficient), the supply of the electrolyser system and consequently the production of hydrogen stop. But it still allows the user to feed until it no longer even suffices. The fuel cell system (PEMFC or SOFC) then starts (around 19 h) to supply the complement and finally the total demand when the sunshine is zero. Figure 4 shows the evolution of the quantities of hydrogen and oxygen consumed in the fuel cell (Fig. 5).

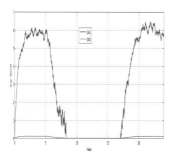

Fig. 4. Evolution of the quantities of hydrogen and oxygen consumed in the fuel cell.

From 0 to 8 h, the fuel cell system (PEMFC or SOFC) starts (around 19 h). The hydrogen consumed by the fuel cell system is appreciable. Figure 4 shows the comparison of the evolution of the voltages exchanged at the DC bus of the PEMFC and SOFC system from 0 to about 8 h. The PEMFC voltage is higher than the SOFC voltage at the start 0–1 h), then the SOFC voltage is higher or nearer to the PEMFC voltage (from 1 to 6 h), then the PEMFC voltage is increased and exceeds the SOFC voltage.

5.2   Analysis of the PV-EL-PAC System

We will evaluate the influence of the various parameters of the system in order to determine the performance of the current system. The influence of the load profile is

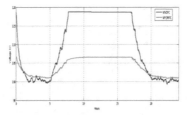

Fig. 5. Evolution of the voltages exchanged on the DC bus operating the PEMFC system and SOFC.

also studied. The selected location: ORAN-Algeria. Average sunlight is 5.2 kWh/m^2 per day in a 45° inclined plane (inclination of photovoltaic panels). Figure 6 describes the evolution of the sunshine during the year [11]. The operating efficiency of the system is the ratio of the annual energy consumed by the load to the energy produced by the solar panels. Since the solar surface (30 m^2) is not modified in any system studied, the photovoltaic energy produced during a year of operation in Oran is constant, equal to 10248.8846 kWh.

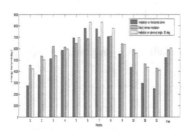

Fig. 6. Average daytime sun exposure in Oran (inclination 45°) as a function of the month.

The difference between the photovoltaic energy and the energy consumed by the user corresponds to the losses in the system. Intrinsic consumption could not be taken into account for lack of empirical models. Indeed, the central energy conversion and management apparatus, the electrochemical components and the PMU, should have a significant intrinsic consumption. Several loads were tested (P_{LAOD} = 1200 W and P_{LAOD} = 1400 W) to evaluate the behavior of the system (Figs. 7 and 10). The key decision parameters for the proposed hysteresis control are shown in Fig. 11. The second part passes through the hydrogen storage system and is reduced by losses in electrochemical cells related to energy and faradic (P_{ELi}-Losses). The sum of these two energies is further diminished by the loss in the converters ($P_{LAODc\text{-}Losses}$) and the loss due to the intrinsic consumption (electrochemical components). The overall efficiency of the system is thus 0.68. Half of the energy produced by the PV field (PmpptPV) is supplied to the end user, the rest is lost (Figs. 8 and 9).

Over a year of operation, only 51.04% of the electricity produced by the PV field is directly supplied to the user. This quantity of electricity has little loss and must

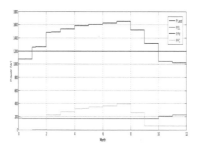

Fig. 7. Evolution of the power exchanged at the DC bus for one year of operation (P_{LAOD} = 1200 W.

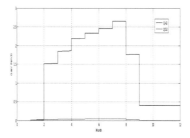

Fig. 8. Evolution of the quantities of hydrogen and oxygen produced by an electrolyser.

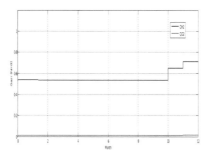

Fig. 9. Evolution of the quantities of hydrogen and oxygen consumed by the PAC for one year

therefore be maximized in order to increase the overall efficiency of the system. The energy passing through the storage components suffers significant losses due to the conversion efficiency of the batteries, the electrolyser and the fuel cell. The intrinsic consumption of electrochemical components is very important in a hybrid energy system (P_{FCi}-Losses, P_{ELi}-Losses, P_{PVi}-Losses). In our study, it will be conceded that the intrinsic consumption is constant, because we do not have a model that expresses the intrinsic consumption of the electrochemical components in value; we consider that the power consumed by the auxiliary components varies between 5 and 15% of the Overall system power. In our system, the generation of thermal energy is periodic; it depends on

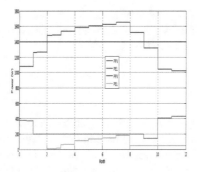

Fig. 10. Evolution of the power exchanged at the DC bus for one year of operation (P_{LAOD} = 1400 W).

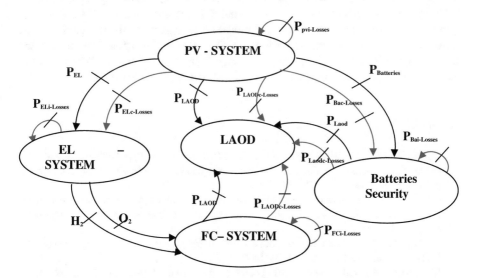

Fig. 11. Representation of the operation of the stand-alone hybrid power system.

the profile of the load and the meteorological conditions. In addition the thermal generation is linked to the use of the SOFC fuel cell and the electrolyser during the hours of daily operation, generally unfavorable, this case is obtained for example during the month of December and January. The SOFC fuel cell used has a low thermal capacity (740 J Kg^{-1} K^{-1}). The maximum temperature reached 1220 °C (the operating temperature is about 1173 °C) corresponds to the maximum energy demanded by the load, i.e. 5 kW. Simulation point of view, compared with other fuel cell systems, e.g. PEMFC. In our case, the thermal capacity of a tubular SOFC is important. For the electrolyser, the maximum temperature reached 60 °C, the average annual heat output of the system is 404.5 W. The annual heat output of the system is 13,440 kWh. The efficiency of the system is increased by 14.2% thanks to the thermal production. The generation of hydrogen by our system depends on the profile of the load and the meteorological conditions. The average daily value of hydrogen consumed by the fuel cell during the year is of the order of 0.0776 Nm^3. On the other hand, the averaged quantity generated by the electrolyser is of the order of 0.315 Nm^3.

6   Conclusion

The simulations carried out using the developed tool allow an analysis of the PV-PAC system operation and the influence of its parameters. The trends in this paper can be summarized as follows; The realization of a hybrid system of energy generation based on solar and electrochemical energy without storage of the battery is possible in Oran, the theoretical yield reached is appreciable because the solar seasonal is important. The improvement of the storage system, notably through the use of electrochemical components, the case of the SOFC high-temperature fuel cell, makes it possible to obtain performances comparable to the conventional systems based on PEMFCs used. The thermal energy in our case allows an additional average gain of 10–20% on the overall performance of the system, depending on the load and weather conditions. Heat generation, advantages of the hydrogen storage system. However, the lifespan of fuel cells is currently the technical and economic constraint of the PV-PAC system.

References

1. http://www.ren21.net/wp-content/uploads/2015/07/REN12GSR2015_Onlinebook_low1.pdf
2. Deshmukh, M.K., Deshmukh, S.S.: Modeling of hybrid renewable energy systems. Renew. Sustain. Energy Rev. **12**, 235–249 (2008)
3. Issaadi, Wassila, Khireddine, Abdelkrim, Issaadi, Salim: Management of a base station of a mobile network using a photovoltaic system. Int. J. Renew. Sustain. Energy Rev. (Elsevier). **59C**, 1570–1590 (2016)
4. Lehman, A., et al.: Operating experiences with photovoltaic-hydrogen energy system. Int. J. Hydrog. Energy. **22**(5), 465–470 (1997)
5. Goetzberger, A., et al.: The PV/Hydrogen/Oxygen – system of the self-sufficient solar house Freiburg, pp. 1152–1158. IEEE (1993)
6. Meurer, et al: PHOEBUS – An autonomous supply system with renewable energy: six years of operational experience and advanced concepts. Sol. Energy **67**(1–3), 131–138 (1999)
7. Pinto, P.J.R., Rangel, C.M.: A Power Management Strategy for a Stand-Alone Photovoltaic/Fuel Cell Energy System for a 1 kW Application, Hydrogen Energy and Sustainability - Advances in Fuel Cells and Hydrogen Workshop; 3º Seminário Internacional Torres Vedras, Portugal, 29–30 April 2010
8. Bernal-Agustın, Jose L., Dufo-Lopez, Rodolfo: Simulation and optimization of stand-alone hybrid renewable energy systems. Renew. Sustain. Energy Rev. **13**, 2111–2118 (2009)
9. Salas, V., Olıas, E., Barrado, A., Lazaro, A.: Review of the maximum power point tracking algorithms for stand-alone photovoltaic systems. Sol. Energy Mater. Sol. Cells **90**, 1555–1578 (2006)
10. Benmessaoud, M.T.: Modelling and simulation of a stand-alone hydrogen photovoltaic fuel cell hybrid system, renewable and alternative energy: concepts, methodologies, tools, and applications. https://doi.org/10.4018/978-1-5225-1671-2.ch016 (2017)
11. http://re.jrc.ec.europa.eu/pvgis/

Deep Convolutional Network Based Saliency Prediction for Retrieval of Natural Images

Shanmugam Nandagopalan[1](✉) and Pandralli K. Kumar[2]

[1] Department of Information Science and Engineering, Vemana Institute of Technology, Bangalore, India
snandagopalan@vemanait.edu.in
[2] Department of MCA, Research Resource Centre, VTU, Belagavi, India
pandralli@gmail.com

Abstract. Content Based Image Retrieval (CBIR) is an important area in the field of image processing and analysis. A novel method is proposed in this paper in order to retrieve visually similar images. The method uses the visual attention model to extract the saliency map of a given image with the help of *SalNet* algorithm. It is based on deep learning methods which have shown that many difficult computer vision problems can be solved by machine learning algorithms and more specifically by Deep Convolution Neural Networks (DCNNs). Using this model, first the saliency region or segment is detected from the images and then the traditional visual features such as color histogram, texture, Histograms of Oriented Gradients (HOG), etc. are computed and are stored in the feature database. Using saliency detection will make our retrieval process easier and accurate as the salient regions in the image is automatically detected. Hence, we can retrieve the most visually similar images with respect to a given query image, because saliency regions exactly map what humans visually perceive. The experimental dataset contains 1000 images including horses, elephants, food, African people, etc. from WANG database. Our results show that the proposed method is efficient and accurate compared to other previously existed models.

Keywords: CBIR · DCNN · SalNet · Deep learning · Saliency

1 Introduction

Image retrieval from large image databases has been the prime research area since many years. However, most of the research in this area was towards traditional feature extraction (color, texture, etc.) especially image global features [13]. The large-scale DCNNs can effectively learn end-to-end from a large amount of labelled images in a supervised learning mode [6, 8]. The increasing amount of digitally produced images, for example Facebook, WhatsApp, Instagram, etc., requires new methods to archive and access this data. Conventional databases allow for textual searches on meta data only. Content Based Image Retrieval (CBIR) [14] is a technique which uses visual contents, normally called as features, to search images from large scale image databases according to users' requests in the form of a query image.

© Springer Nature Switzerland AG 2019
P. Vasant et al. (Eds.): ICO 2018, AISC 866, pp. 487–496, 2019.
https://doi.org/10.1007/978-3-030-00979-3_51

The main challenge in CBIR systems is the ambiguity in the high-level (semantic) concepts extracted from the low-level (pixels) features of the image [2]. Approaches based on one specific algorithm (e.g., color, texture or shape) can work effectively only on specific types of images. When different types of images are input to these systems their performance is degraded. For example, approaches based on color histogram take into account only the visual contents relating to colors and ignore shape and texture [10]. Hence, in our approach we integrate all dominant features that are common across a variety of datasets.

In this paper, we propose a novel method of introducing saliency map as one of the features to reduce the semantic gap. Given an input query image, visually similar images are retrieved from a large image database by extracting new feature(s) and computing the Euclidean distance. Our approach introduces new architectures and improvements in salient object detection with deep convolution network algorithm, *SalNet*, executed through "*Algorithmia*" which provides machine intelligence to build smarter applications.

The paper is structures as follows. Section 2 presents the previous and recent works using convolutional networks for saliency prediction and detection in the area of image retrieval. Section 3 introduces the proposed architecture for image retrieval. This section further explains the visual attention method of saliency detection from a given natural image, extraction of other features – color, texture, etc., - image ranking based on the distance calculation, etc. Experimental results are provided in Sect. 4 where we prove that our method outperforms other techniques that exists. Sections 5 and 6 are dedicated for conclusion and references respectively.

2   Related Work

Extensive research work has already been done on CBIR and still new methods are being explored. However, little work is done on image retrieval using visual attention techniques. Alex Papushoy and Adrian G. Bors have developed a new method by finding saliency for each image region of an image, as it would be perceived by a human observer, and this in turn is used for image retrieval [2]. The authors have applied Graph Based Visual Saliency (GBVS) method to estimate the saliency in image regions.

Junting Pan et al., proposed a novel method for finding the saliency by addressing the problem with a completely data-driven approach by training a convolutional neural network (*convnet*) [9]. Deep learning is used in videos as well for saliency determination. Souad Chaabouni et al., developed algorithms for the detection of salient areas in natural video by using the new deep learning techniques [4]. Hollywood dataset was used for experimental purposes. A combination of saliency and SIFT algorithms were used by D. R. Dhotre et al., in their work on CBIR. The proposed approach in this paper was to combine the feature extraction algorithm; SIFT with the Saliency Detection technique in order to provide relevant image output [5].

In [6, 7], authors have applied successfully visual attention model to solve the CBIR problem. Again Coral and GHIM datasets were used for ranking the image similarity.

3   Feature Extraction Methods

In this section we describe the various feature descriptors that are used in building the image retrieval system. With these feature extraction methods, the CBIR architecture can be built based on the pipes-and-filter design pattern. The filters are various image manipulation task and the pipes are the collaborators of these tasks. Here, data source is the original input image database and the data sink is the feature database. In order to implement our proposed system, Python and OpenCV tools were used under Visual Studio 2017 framework. For saliency detection, deep learning algorithms of *Algorithmia* have been employed.

Here, a brief note on various image descriptors and saliency are presented in the following sub-sections. Basically our work focusses on saliency map, color, texture, HOG, and daisy feature descriptors for image similarity and ranking.

3.1   Deep Learning Algorithm for Saliency Detection

Deep learning has emerged as a new field of research in machine learning, providing learning at multiple levels of abstraction for mining the data such as images, sound and text. Although, it is hierarchically created usually on the basis of neural networks, deep learning presents a philosophy to model the complex relationships between data. Since recently, deep learning has become the most exciting field which attracts many researchers [4]. The predictive power of Deep Convolutional Neural Networks (DCNN) is interesting for the use in the problem of prediction of visual attention in visual content, i.e. saliency of the latter.

 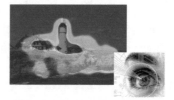

Fig. 1.   a Sample image **b** Saliency map of (**a**)

The salience (also called saliency) of an item – be it an object, a person, a pixel, etc. – is the state or quality by which it stands out relative to its neighbors. Saliency detection is considered to be a key attentional mechanism that facilitates learning and survival by enabling organisms to focus their limited perceptual and cognitive resources on the most pertinent subset of the available sensory data. This deep learning algorithm automatically detects salients for any given image. Take a look at the picture shown in Fig. 1.

For any human eye, the following objects may be seen in Fig. 1a: lighthouse, house, and rocks. So the eye-fixation map or saliency map of this image may look like the one shown in Fig. 1b. There are two commonly used convolution networks namely shallow and deep [9].

In our work, the *SalNet* API of *Algorithmia* is called by storing the input image in the cloud directory data://.my after creating an account. The saliency image of the input image is saved in the *SalNetTest* folder. The API is called through the Python function and is given below:

```python
def saliency():
    client = Algorithmia.client("simMkQEb8NNVdePGWn8sRr5/N1/1")
    # Create testing directory if it doesn't exist
    if not client.dir("data://.my/SalNetTest").exists():
        client.dir("data://.my/SalNetTest").create()

    input = {"image": "data://.my/sng/107100.jpg",
             "location": "data://.my/SalNetTest/107100.png",
             "saliencyLocation": "data://.my/SalNetTest/107100_saliencyMatrix.json"
    }
    result = client.algo("deeplearning/SalNet/0.2.0").pipe(input).result
    print (result)
```

Note that "image": is the input image file location, "location": is the output file location, and "saliencyLocation": is the jason output file location.

3.2   Color Feature

In image retrieval systems color histogram is the most commonly used feature. The main reason is that it is independent of image size and orientation. Also it is one of the most straight-forward features utilized by humans for visual recognition and discrimination. Statistically, it denotes the joint probability of the intensities of the three color channels [14]. Since, these features are not sensitive to rotation, translation and scale changes, they could be applicable for CBIR systems [1].

After finding the saliency region/segmented object from the input image, we invoke OpenCV function *cv2.calcHist*() to extract color histogram of all channels. That is, for each color component R, G, and B, the frequency of each color index from 0 to 255 is calculated and raised in the form of histogram value for each color component, and is written as a vector.

```
hist  =  cv2.calcHist([img], [0, 1, 2], None,
         [8, 8, 8], [0, 256, 0, 256, 0, 256])
```

The histogram above shows the number of pixels for every pixel value, from 0 to 255. In fact, we used 8 values (bins) to show the above histogram. It could be 8, 16, 32, 256, etc. and OpenCV uses *histSize* to refer to bins.

3.3   GLCM Texture Feature

Texture analysis is the mostly used method in image processing. It is possible to get knowledge about segmentation and classification of spatial parameters in images by texture analysis. Texture analysis is frequently utilized in medical image processing, remote sensing, and control systems. Features of texture can be extracted with variety of methods such as statistics, geometry, model-based, and signal processing etc.

In this section, we illustrate texture features being calculated using grey level co-occurrence matrices (GLCMs) with the help of Scikit-image which is an image processing toolbox for SciPy. A GLCM is a histogram of co-occurring greyscale values at a given offset over an image. First the input color image is transformed into a grayscale image, and its gray-level co-occurrence matrix is computed. A grey level co-occurrence matrix is a histogram of co-occurring greyscale values at a given offset over an image. The output of this *glcm* is fed to another function greycoprops(glcm, "contrast") to calculate the contrast feature. Similarly, all the other texture features can be calculated as shown below:

```
def texture(gray):
    # Texture features
    glcm = feature.greycomatrix(gray, [1, 2],
            [0,np.pi/2],normed=True, symmetric=True)
    contrast = feature.greycoprops(glcm, 'contrast')
    homo = feature.greycoprops(glcm, 'homogeneity')
    diss = feature.greycoprops(glcm, 'dissimilarity')
    eng  = feature.greycoprops(glcm, 'energy')
    corr = feature.greycoprops(glcm, 'correlation')
    ASM  = feature.greycoprops(glcm, 'ASM')
    ent  = entropy(gray)
```

All these features are concatenated to form a single feature vector for the input image *gray*.

3.4   Histogram of Gradients (HOG) Feature

To compute the HOG, we adopt the following steps: (a) global image normalization (b) computing the gradient image in *x* and *y* (c) computing gradient histograms (d) normalizing across blocks (e) flattening into a feature vector.

Now finding HOG features from *skimage* package of Python is shown below:

```
hog_data, hog_image = feature.hog(gray, orientations=8,
pixels_per_cell=(32, 32), cells_per_block=(1, 1),
visualise=True)
```

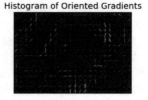

Input image   Histogram of Oriented Gradients

Fig. 2. Input color image, grayscale, and HOG image.

In Fig. 2, the input color image, its grayscale, and the oriented gradients are shown. The hog function also gives *hog_data* as given above code which is stored as part of the feature vector.

3.5   Daisy Feature

In this section we shall show how to extract DAISY feature descriptors densely for the given image. DAISY is a feature descriptor similar to SIFT (Scale-Invariant Feature Transform) formulated in a way that allows for fast dense extraction. Typically, this is practical for bag-of-features image representations.

Using this algorithm, it is possible to get much reduced error rates compared to the SIFT algorithm and also to reduce computational cost and descriptor storage requirements. The DAISY algorithm consists of the five building blocks: Feature Detector, Summation, Robust normalize, PCA, quantize and compress. Only the first three steps are required for all applications.

Typically, the input is a square monochrome image patch and the output is a vector of bytes. This feature is extracted from the *skimage* package of Python as given below:

```
daisy_data, daisy_img = feature.daisy(img, step=180,
radius=58, rings=2, histograms=6,orientations=8,
visualize=True)
```

After execution of this function, we get two daisy descriptors as shown in Fig. 3. The *daisy_data* is a vector of real values, 0.00628139, 0.00757163, 0.0074997, 0.0061575, 0.0058985.

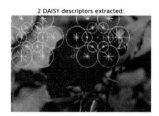

Fig. 3. DAISY descriptors for the input image of Fig. 2.

4   Proposed Architecture

To accomplish the solution for the problem of retrieving similar images from a database of images efficiently and accurately, a novel architecture is proposed and is shown in Fig. 4.

We have two separate set of tasks called "offline" and "online".

During the offline phase (shown in blue color shaded boxes), the images from the database are taken for preprocessing and feature extraction which yields to a vector. Then all these vectors are stored in the feature database. Similarly, when a query image (online phase – shown as red shaded boxes) is given as input where images similar to this are to be retrieved, same steps are carried out and the distance is calculated using Euclidean distance formula. If the distance, d is less than a fixed threshold value, all such images are ranked accordingly and outputted. Various features of the images are extracted and stored in the database as explained in the previous section.

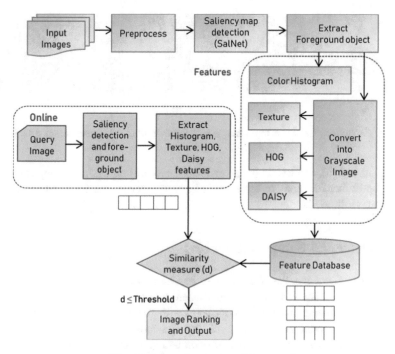

Fig. 4. Proposed system architecture.

When the above steps are applied on WANG database and Holiday dataset, we get the outputs as indicated in Figs. 5 and 6 respectively. Here, few categories of the database images are considered for object detection and extraction for further processing.

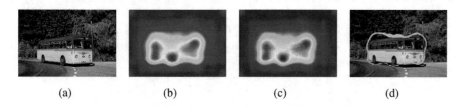

(a)   (b)   (c)   (d)

Fig. 5. WANG database **a** Input image **b** Saliency output **c** Merged output **d** Object

In Fig. 5, (a) shows the input image from WANG database (bus category), (b) shows the saliency image (the foreground object is the point of interest), (c) is after

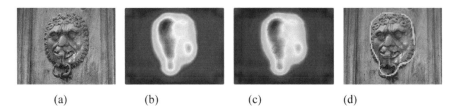

(a)	(b)	(c)	(d)

Fig. 6. Holiday dataset **a** Input image **b** Saliency output **c** Merged output **d** Object

merging (a) and (b), and finally (d) gives the detected object. From this object the color, texture, and other features are extracted. The same processes are shown for an image taken from Holiday dataset.

5   Experimental Setup and Results

For our experiment, the WANG database has been used as the main dataset and in addition, the Holiday dataset has also been considered for accuracy evaluation. The WANG database consists of 1000 images with 10 categories African people, roses, horses, elephants, monuments, dinosaurs, etc. The entire application has been coded in Python, version 3.6.3, and a number of open source tools were used: OpenCV APIs, *skimage*, *Algorithmia*, Visual Studio 2017 (Community Edition), etc. The entire code runs on a computing environment having Intel Core i5-7200U CPU, 8 GB RAM, and Windows 10 Operating System.

5.1   Results

We have conducted experiments by considering different query images and obtained the output ranking of the retrieved images. These output images are shown in Fig. 7.

Fig. 7.   Retrieved images for WANG database – query is top left

For some query images, say, the system retrieves dissimilar images. When we run the program with holiday dataset, our system's response is good and is shown in Fig. 8.

The first image is the query image and one false hit which is the last image – i.e. 117100.jpg.

0_query.jpg   108100.jpg   108101.jpg   108200.jpg   108201.jpg   108202.jpg   117100.jpg

Fig. 8. Output from Holiday dataset, query image is top left.

5.2 Retrieval Performance – Precision and Recall

The precision and recall are used to measure performance of retrieval. Recall is used to measure the system's ability to retrieve all the images that are relevant, while precision is used to measure the system's ability to retrieve only the images that are relevant [13, 14]. The equation of the recall and precision are illustrated in the following:

$$Precision = \frac{No.\ of\ relevant\ images\ retrieved}{Total\ No.\ of\ images\ retrieved}$$

$$Recall = \frac{No.\ of\ relevant\ images\ retrieved}{Total\ No.\ of\ relevant\ images\ in\ the\ Database}$$

We have conducted experiments to measure precision and recall on both WANG and Holiday datasets for our proposed approach. Its results are shown in Table 1.

Table 1. Results of precision and recall values obtained through our proposed model

Dataset	Query image	Precision	Recall
WANG	202.jpg	0.58	0.60
	438.jpg	1.0	0.97
	938.jpg	0.67	0.80
Holiday	108100.jpg	0.80	0.90

In can been that the performance of the proposed approach yields a better accuracy compared to other results reported already.

6 Conclusion

Saliency based object detection and feature extraction is the major focus of this research work. There are two primary reasons for selecting the saliency region detection using DCNN, first it is efficient, second visually similar images would match with the saliency regions of an image. With the precision and recall values that we have achieved it is proved that the goal is achieved. The shortcoming of the present work is that when the query image is changed or dataset is modified, there could be slight degradation in the performance. Future work can concentrate on local features with more low level features so that the accuracy of retrieval may be enhanced.

References

1. Gautama, A., Bhatiaa, R.: A novel method for CBIR using ACO-SVM with DTCWT and color features. In: International Conference on Processing of Materials, Minerals and Energy, July 29, 30, 2016, Ongole, Andhra Pradesh, India (2017)
2. Papushoy, A., Bors, A.G.: Visual attention for content based image retrieval. In: Azzopardi, G., Petkov, N. (eds.) Computer Analysis of Images and Patterns. CAIP 2015, Valetta, Malta. Lecture Notes in Computer Science, vol. 9256. Springer, Cham (2015)
3. Bastan, et al.: Active Canny: edge detection and recovery with open active contour models. IET Res. J., 1–11 (2017)
4. Chaabouni, S., Benois-Pineau, J., Hadar, O., Amar, C.B.: Deep Learning for Saliency Prediction in Natural Video (2016)
5. Dhotre, D.R., Bamnote, G.R., Gadhiya, A.R., Pathak, G.R.: CBIR using saliency mapping and sift algorithm. Int. J. Sci. Eng. Res. **7**(2) (2016)
6. Awad, D., Mancas, M., Riche, N., Courboulay, V., Revel, A.: CBIR-based evaluation framework for visual attention models. In: 23rd European Signal Processing Conference (EUSIPCO), IEEE (2015)
7. Liu, G.-H.: Content-based image retrieval based on visual attention and the conditional probability. In: International Conference on Chemical, Material and Food Engineering (CMFE-2015) (2015)
8. Pan, H., Jiang, H.: A Deep Learning Based Fast Image Saliency Detection Algorithm (2016)
9. Pan, J., Sayrol, E., Giro-i-Nieto, X., McGuinness, K., O'Connor, N.E.: Shallow and deep convolutional networks for saliency prediction. In: IEEE Conference on Computer Vision and Pattern Recognition (2016)
10. Iqbal, Kashif, Odetayo, Michael O., James, Anne: Content-based image retrieval approach for biometric security using colour, texture and shape features controlled by fuzzy heuristics. J. Comput. Syst. Sci. **78**, 1258–1277 (2012)
11. LAIB, L., Ait-Aoudia, S: Efficient Approach for Content Based Image Retrieval Using Multiple Svm in Yacbir, ACSIT, SIPM, CMIT, pp. 19–29, (2016)
12. Setia, L., Ick, J., Burkhardt, H.: SVM-based relevance feedback in image retrieval using invariant feature histograms. In: Proceedings of the IAPR Workshop on Machine Vision Applications, 2005, pp. 542–545 (2005)
13. Rejito, Juli, Abdullah, Atje Setiawan, Akmal, Setiana, Deni, Ruchjana, Budi Nurani: Image Indexing using Color Histogram and k-means Clustering for Optimization CBIR in Image Database. J. Phys. Conf. Series **893**, 012055 (2017)
14. Nandagopalan, S., Adiga, B.S., Deepak, N.: A universal model for content-based image retrieval. Int. J. Comput. Inf. Eng. **2**(10) (2008)
15. Sun, X., Pan, W., Wang, X., Yuan, W.: Image retrieval based on saliency detection in the application of the guide system. In: International Conference on Information Sciences, Machinery, Materials and Energy (ICISMME 2015) (2015)
16. Zdziarski, Z., Dahyot, R.: Feature Selection Using Visual Saliency for Content-Based Image Retrieval, ISSC 2012, NUI Maynooth, June 28–29 (2012)

Optimization of Parameters and Operation Modes of the Heat Pump in the Environment of the Low-Temperature Energy Source

Evgenia Tutunina[1]([📧]) [iD], Alexey Vaselyev[1] [iD], Sergey Korovkin[2], and Sergey Senkevich[1]

[1] Federal State Budget Scientific Institution "Federal Scientific Agroengineering Center VIM", 109428 Moscow, Russia
tutuninaev@gmail.com, vasilev-viesh@inbox.ru,
sergej_senkevich@mail.ru
[2] JSC Atomenergoproekt, 107996 Moscow, Russia
svkorovkin@mail.ru

Abstract. One of the options to reduce the energy consumption of agricultural products is the use of heat and cold supply schemes based on heat pump units. Ecological requirements are tightening. The role of environmental friendly coolants is increasing. The most environmentally friendly coolant is ice, which can be produced with the use of refrigeration machines (heat pumps). However, the production of ice in heat pump systems requires additional energy costs for its removing from the heat exchange surface. The freezing of the heat exchanger leads to the worsening of productivity of the entire heat pump. The article deals with the optimization of the process of creating artificial ice in the membrane heat exchanger of the heat pump unit. The basis of the study was the experiment conducted in accordance with the Box-Benken plan for three factors. The experimental data was analyzed and the regression equation was made. A mathematical model of the ice generation rate in the membrane heat exchanger of the heat pump from the volume of cooled water and the time of filling the space under the membrane with the refrigerant was obtained during the study. The response surface was plotted according with the obtained equation.

Keywords: Optimization · Heat pump · Ice
Membrane (flexible) heat exchanger

1 Introduction

It is necessary to take into account new increased requirements for environmental safety along with the need to improve the energy efficiency of new technological installations. The most problematic substances of refrigeration equipment from the point of view of environmental safety are coolants. The most environmentally friendly coolant is ice, which can be produced with the use of refrigeration machines (heat pumps) [1–3].

© Springer Nature Switzerland AG 2019
P. Vasant et al. (Eds.): ICO 2018, AISC 866, pp. 497–504, 2019.
https://doi.org/10.1007/978-3-030-00979-3_52

The heat pump uses electricity to transfer heat from less heated volume to more heated one.

The use of heat pumps is most effective in such technological processes, where both high temperature of the coolant and low temperature of the refrigerant are required. In this case, the use of low-temperature sources of thermal energy can improve the efficiency of cooling the product with obtaining additional thermal energy. The use of heat pumps in the technological processes of agricultural production reduces their energy intensity. Low-temperature energy sources limits their use. The fact is that the heat pumps work well only at a cold source temperature above the −5 °C. The efficiency of the heat pump drops sharply in severe frosts, and the heat exchange surfaces are covering with ice crust [4].

Due to the long winter frosts in Russia, the use of water from nearby ponds is problematic, and the soil freezes to a depth of one and a half meters and more. In addition, more than half of the territory of Russia is in the permafrost zone, where the use of the ground heat exchangers is impossible in principle.

Of course, these factors do not eliminate the use of heat pumps for heating, but significantly increase the cost of their use. The use of heat exchangers with variable geometry of the heat exchange surface allows changing the situation. In such devices, it became possible to change the shape of the membrane with a given frequency and amplitude, which provides constant self-cleaning of the heat exchange surface from the ice layer and allows using water (or any water-containing liquid medium) with a temperature of +4 °C and below as an energy source.

2 Main Part

2.1 Research Purpose

The research purpose is the optimization of the parameters and operating modes of the heat pump in the environment of the low temperature energy source. To achieve this goal, the following tasks were solved:

1. an experimental design of a heat pump unit with a membrane heat exchanger was made;
2. a preliminary experiment was performed to identify the factors affecting the ice formation;
3. an experiment plan was chosen and the results of the experiment were obtained;
4. a model of the ice generation rate in the membrane heat exchanger of the heat pump was obtaining;
5. the optimal values of significant factors were found.

2.2 Materials and Methods

The novelty of the method lies in the use of a new type of heat exchanger – membrane (flexible) heat exchanger in heat pump unit.

Let us explain the terms used in the proposed material. Membrane is a thin flexible plate fixed on the perimeter, designed for separating of two cavities with different pressures or separating of the closed cavity from the common volume, as well as for conversing of pressure changes into linear move and vice versa [5].

Membrane heat exchanger is a heat exchanger with the membrane heat transfer surface.

The membrane geometry varies depending on the pressure difference of the media that the membrane separates.

It will forms an ice layer on the membrane if one of the media will be water, and another one will be intermediate coolant cooled to a negative temperature (glycol solution, alcohol). This process is known as freezing of the heat exchange surface. The main difference of a heat pump unit with a membrane heat exchanger is the process of ice generation on a vibrating heat exchange surface [6–8].

If it will be organized the process of periodic changes in the geometry of the membrane in the heat exchanger, the formed ice will exfoliates from the membrane and it will break in some modes of generation. In this case, we obtain the possibility to change the thickness of the freezing ice layer.

The process of changing the geometry of the membrane can be implemented by feeding and removing the coolant from the cavity under the membrane. Experimental studies are necessary to determine the modes of ice generation that will provide the required performance of the installation.

An experimental installation was designed and made to determine the optimal operation mode of the membrane heat exchanger in the mode of ice generation, which includes:

- freon compressor with 300 W electric power;
- freon condenser;
- freon evaporator;
- ethylene glycol tank;
- circulation pump;
- time relay;
- membrane heat exchanger with a diameter of 0.36 m.

The freon evaporator is immersed in a tank with ethylene glycol for cooling ethylene glycol to a temperature −10 °C. Ethylene glycol from the tank is supplied to the membrane heat exchanger by a circulation pump, cools the water there and flows back into the tank by gravity. Membrane vibrations are carried out by turning on and off the circulation pump. The duration of the switching periods of the circulation pump is set by the time relay included in the electrical circuit of the installation.

The main parameters that determine the operation of the installation are:

1. Electric power of the compressor;
2. Temperature of ethylene glycol;
3. Rate of ice generation in the membrane heat exchanger;
4. Volume of water in the membrane heat exchanger;
5. The initial temperature of the water in a membrane heat exchanger;
6. The frequency of membrane oscillations.

The ice generation rate w_ice is selected as a response. Independent factors were determined for carrying out experimental studies. It was selected the following factors:

the duration of the coolant supply under the membrane τ_post; the volume of water V for freezing ice; the initial water temperature Tv_ish for freezing ice.

The basis of the research was a second-order non-composite plan for three factors (The Box–Benkin plan) [9]. This plan is a defined sample of strings from a complete 3 K type factor experiment. In this regard, each variable varies only at three levels: +1, 0, −1. The use of non-position plans with only three levels of variation of factors simplifies and reduces the cost of the experiment. A second-order non-position plan for 3 factors is implemented to obtain a mathematical description of the studied process in the form of a second-degree polynomial (Table 1).

Table 1. Levels and intervals of factors variation

Factor name	Factor designation	Coded designation	Variation interval	Natural values corresponding to the levels of coded factors		
				Upper +1	Main 0	Lower −1
The duration of refrigerant supply under the membrane	τ_post, min	X_1	2	6	4	2
Water volume	V, l	X_2	1	4	3	2
Initial water temperature	Tv_ish, °C	X_3	12	28	16	4

The experiment was performed as follows. The time relay is set to the duration of the supply of ethylene glycol under the membrane. The compressor and pump are turned on. After the unit is put into the operation (the temperature of ethylene glycol reaches −5 °C÷−12 °C), the required volume of water of the required temperature is filled into the volume (body) of the membrane heat exchanger.

The change in the levels of factors variation in the experimental setup was made as follows:

- the duration of coolant supply under the membrane τ_post – is set using a time relay;
- the volume of water V – the required amount of water is measured by dimensional glass;
- initial water temperature Tv_ish – is controlled by mercury thermometer;

2.3   Experiment Results

It was carried out the regression analysis based on the results of experimental data. Calculations were made in the program MathCAD, the charts was plotted in Excel.

After checking the significance of the regression coefficients and excluding the insignificant ones, the mathematical model in natural parameters takes the following form:

$$y(x_1, x_2, x_3) = 5.888 - 0.5515x_1 - 1.282x_2 + 0.04925x_1^2 + 0.184x_2^2 \qquad (1)$$

It was checked the adequacy of the model using the Fisher's criterion. The table value of the criterion is 19.3 and the calculated value is 2.819. Since 2.819 < 19.3, the obtained equation adequately describes the experimental data.

It follows from Eq. (1) that the initial temperature of the water in the ice freezing tank was an insignificant factor. It has no significant effect on the speed of ice freezing. This provision is very important for technology, because the temperature of the water that needs to be cooled can vary widely in the processing agricultural products.

The response surface plotted using the Eq. (1) is shown in Fig. 1.

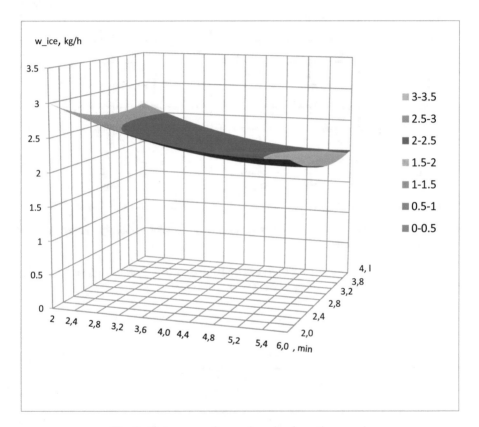

Fig. 1. Response surface w_ice - ice formation speed.

The presented surface allows us to say that the highest rate of ice formation occurs with a small amount of cooling water. This is in good agreement with studies of other authors, which showed the effectiveness of freezing water when spraying [10–12]. This position allows determining the dimensions of the chamber for freezing ice depending on the needs of the process.

Recommendations on the speed of filling the sub-membrane space with coolant also follow from the type of response surface. So filling the sub-membrane space in 2 min gives the greatest performance of freezing ice.

It should be noted that at a high rate of freezing of ice its thickness would be the smallest. The volume of ice forms layers in the process of freezing. This contributes to its rapid dissolution when cooling water. This allows speeding up the process of preparation of the coolant for use in the process.

Figure 2 shows the dependence of the ice layer thickness on the duration of supply /removing of the coolant under the membrane.

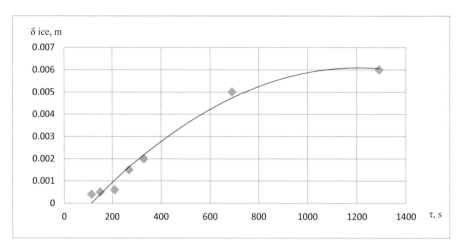

Fig. 2. The dependence of the ice layer thickness on the duration of supply /removing of the coolant under the membrane.

The different total area of ice layers is generated in the membrane heat exchanger per hour depending on the mode (Fig. 3).

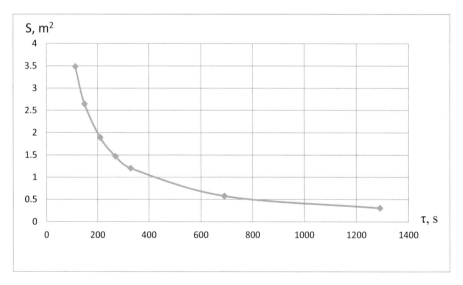

Fig. 3. The dependence of the total area of ice layers on the duration of supply /removing of the coolant under the membrane.

3 Conclusion

The use of low-temperature sources for obtaining energy from them with the help of heat pumps is limited due to freezing of the system and reducing the efficiency of its operation.

The solution of this problem is possible by using intermediate heat exchangers with a flexible membrane.

The use of membrane heat exchangers allows regulating the speed of ice freezing for technological processes and influencing on the structure of the ice.

The maximum speed of ice generation in the membrane heat exchangers of the heat pump corresponds to the duration of the refrigerant supply mode of 2 min and the volume of water in the heat exchanger of 2 L.

Thus, the design of the membrane (flexible) heat exchanger allows to simultaneously solving the following tasks:

- remove the thermal energy of the water-ice phase transition;
- freeze the required amount of ice for cooling agricultural products;
- it is possible to obtain ice of different thickness with different total surface area depending on the mode (duration of supply/removing of the coolant under the membrane).

References

1. Marinyuk, B.T., Ugolnikova, M.A.: Dynamics of a generation of water ice on tubular elements of ice generators. Refrigerating Equip. **12**, 44–47 (2016)
2. Zimin, A.V.: Systems of accumulation of cold with use of binary ice. Refrigerating Equip. Technol. T. **51**(4), 17–20 (2015)
3. Shepenin, A.I.: Analysis results of modern methods of accumulation of cold and modeling of a namorozka of ice in the block of cold accumulators of coiled type. In The Collection: Youth and the 21st Century - 2015 Materials V of the International Youth Scientific Conference: in 3 volumes. Editor-in-chief: A.A peas, pp. 198–201 (2015)
4. Vasilyev, G.P., et al.: Technical solution for protection of heat pump evaporators against freezing the moisture condensed. MATEC Web Conf. **40**, 05002 (2016)
5. Interstate standard 28466-90. Horns and signal whistles. General specifications, Moscow (1991), 1 p. (In Russian)
6. Korovkin, S.V., Tutunina E.V.: Use of the generator of "liquid" ice with the membrane heat exchanger for milk cooling. Innov. Agric. **4**(19), 115–119 (2016)
7. Patent 2490567 Russian Federation, F25C1/12 F25C1. Method of generating ice/ Korovkin S.V., Vinokurov N.P., Tutunina E.V. - № 2012138395/13; appl. 10.09.2012; publ. 20.08.2013 Bul. № 23.–3p.: fig
8. Sushentseva, A.V., About Korovkin, S.V.: Invention "Heat transfer with variable geometry of the heat transfer surface". Youth scientific and technical bulletin # 06, June, 2012, http://sntbul.bmstu.ru/doc/475952.html, Last accessed 22 June 2018
9. Spiridonov A.A. Planirovanie ehksperimenta pri issledovanii tekhnologicheskih processov. Moscow: Mashinostiroenie, 1981. – 184 p. (Planning of experiment at research of technology processes. Moscow: Mashinostiroenie, 1981. – 184 p.)

10. Mouneer, T.A., El-Morsi, M.S., Nosier, M.A., Mahmoud, N.A.: Heat transfer performance of a newly developed ice slurry generator: A comparative study. Ain Shams Engineering Journal **2010**(1), 147–157 (2010)
11. Huang, Ch-N, Ye, Y.-H.: Development of a water-mist cooling system: a 12,500 Kcal/h air-cooled chiller. Energy Reports **1**, 123–128 (2015)
12. Wa, G.: Partial freezing by spraying as a treatment alternative of selected industrial wastes. National Library of Canada (1998)

A Data Confidentiality Approach to SMS on Android

Tun Myat Aung$^{(\boxtimes)}$, Kaung Htet Myint, and Ni Ni Hla

University of Computer Studies, Yangon, Myanmar
tma.mephi@gmail.com, kolynn.2013@gmail.com,
ni2hla@ucsy.edu.mm

Abstract. Short Message Service (SMS) is a text messaging service component of mobile communication systems. It uses standardized communications protocols to exchange short text between mobile devices. SMS does not have any built-in procedure to offer security for the text transmitted as data. Most of the applications for mobile devices are designed and developed without taking security into consideration. In practical use, SMS messages are not encrypted by default during transmission. Therefore, a data confidentiality approach to SMS on Android will be developed in the paper. It includes design, implementation and confidentiality measurement of RC4 stream cipher for SMS data confidentiality on mobile networks.

Keywords: SMS security · Data confidentiality · Mobile application
Cryptogarphy · RC4 stream cipher

1 Introduction

Data confidentiality is a protection of data from unauthorized disclosure. It is the most common aspect of information security. It not only applies to the storage of information, also applies to the transmission of information. We need to protect our sensitive information from malicious actions during transmission of Short Message Service (SMS). For data confidentiality security service, an encipherment security mechanism can be used.

Short Message Service (SMS) is a mechanism of delivery of short messages over the mobile networks. It is a store and forward way of transmitting messages to and from mobiles. The message (text only) from the sending mobile is stored in a central short message center (SMS) which then forwards it to the destination mobile. Global System for Mobiles (GSM), Code Division Multiple Access (CDMA) and Time-Division Multiple Access (TDMA) are supporting SMS transmission.

The primary purpose of SMS is to deliver text messages from one mobile device to another. It provides many benefits to our everyday life. But, it is now considered as a safe and secure tool when sensitive information is transmitted using the typical SMS services. Nowadays, there are many possible threats on SMS, therefore, it is important not only to prevent the SMS message from being illegally intercepted by illegal sources but also to ensure the origin of the message from the legitimate sender.

© Springer Nature Switzerland AG 2019
P. Vasant et al. (Eds.): ICO 2018, AISC 866, pp. 505–514, 2019.
https://doi.org/10.1007/978-3-030-00979-3_53

Cryptography is the science of information and communication security. Cryptographic system transforms a plaintext into a cipher text using a key generated by a cryptographic algorithm. RC4 is a stream cipher that is used to protect internet traffic as part of the Secure Socket Layer (SSL). RC4 stream cipher is used to protect data confidentiality for SMS transmitted over mobile networks.

The purpose of this paper is to provide data confidentiality during SMS message transmission period in order to prevent the SMS message from being illegally intercepted by illegal sources and to ensure the origin of the message from the legitimate sender. The structure of this paper is as follows. The Sect. 2 includes basic concepts of SMS technology, SMS mobile network communication system, introduction to cryptography and RC4 cipher. In Sect. 3, we discuss design and implementation of mobile applications that are used to protect data confidentiality of SMS message transmitted on mobile networks. The Sect. 4 describes how statistical tests suite is used to measure data confidentiality. Finally, in Sect. 5 we conclude our discussion by describing data confidentiality level of pseudorandom number sequence generated by RC4 cipher and by suggesting RC4 cipher should be used for data confidentiality of SMS message transmitted on mobile network communication system.

2   Background Theory

2.1   Basic Concepts of SMS Technology

SMS messages are created by mobile phones or other devices (e.g. personal computers). These devices can send and receive SMS messages by communicating with the GSM network. All of these devices have at least one MSISDN number. They are called Short Messaging Entities (SMEs). The SMEs are the starting points (the sender) and the end points (the receiver) for SMS messages. They always communicate with a Short Message Service Center (SMSC) and never communicate directly with each other [3]. An SME can be a mobile telephone. An SME can also be a computer equipped with a messaging software, such as Ozeki NG - SMS Gateway, which can communicate directly with the SMSC of the service provider. Depending on the role of the mobile phone in the communication, there are two kinds of SMS messages: Mobile-originated (MO) messages and Mobile-terminated (MT) messages. MO messages are sent by the mobile phone to the SMSC. MT messages are received by the mobile phone. The two messages are encoded differently during transmission. The functions of Short Message Service Center (SMSC) are shown in Fig. 1 [5].

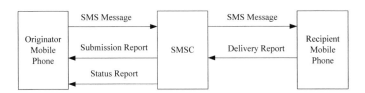

Fig. 1. The functions of Short Message Service Center (SMSC)

2.2 SMS Mobile Network Communication System

SMS messages are transmitted over the Common Channel Signaling System 7 (SS7). SS7 is a global standard that defines the procedures and protocols for exchanging information among network elements of wire line and wireless telephone carriers [2]. These network elements use the SS7 standard to exchange control information for call setup, routing, mobility management, etc. Figure 2 shows the mobile network architecture for SMS communication. Conceptually, the general SMS mobile network architecture consists of two segments that are central to the SMS model of operation: the Mobile Originating (MO) part, which includes the mobile handset of the sender, a base station that provides the radio infrastructure for wireless communications, and the originating Mobile Switching Centre (MSC) that routes and switches all traffic into and out of the cellular system on behalf of the sender. The other segment, the Mobile Terminating (MT) part, includes a base station and the terminating MSC for the receiver, as well as a centralized store-and-forward server known as SMS Centre (SMSC). The SMSC is responsible for accepting and storing messages, retrieving account status, and forwarding messages to the intended recipients [5].

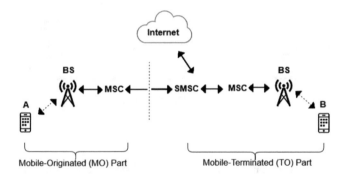

Fig. 2. Mobile network architecture for SMS communication

2.3 Cryptography

Cryptography is the science of using mathematics to encrypt and decrypt data. Cryptography enables you to store sensitive information or transmit it across insecure networks like the Internet so that it cannot be read by anyone except the intended recipient. Encryption is the process of converting ordinary information (called plain text) into unintelligible gibberish (called cipher text). Decryption is the reverse, in other words, moving from the unintelligible cipher text back to plain text.

Cryptographic algorithms can be divided into:

- Symmetric key algorithms.
- Asymmetric key algorithms.

Symmetric key algorithms have the property that same secret keys are used for encryption and decryption. It is also called as private key algorithms. Asymmetric key algorithms use two different keys: public key for encryption and private key for decryption.

There are two types of symmetric-key algorithm: *block cipher* and *stream cipher*. (1) Stream Cipher - In a stream cipher, encryption and decryption operate on the basis of one symbol (a bit or byte) at a time, (2) Block Cipher - In a block cipher, encryption and decryption operate on the basis of a block of symbols of particular size [1]. The general concept of a stream cipher is shown in Fig. 3.

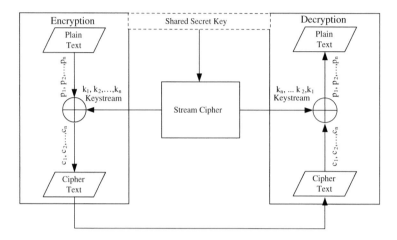

Fig. 3. Stream cipher

2.4   RC4 Cipher

RC4 cipher is one of the most used software-based stream ciphers in the world. Rivest Cipher 4 was designed by Ron Rivest in 1987 and is known as RC4 cipher. The general logic design structure of RC4 cipher is shown in Fig. (4). It is a standard of IEEE 802.11 within WEP, Wireless Encryption Protocol, and generates a keystream. This stream cipher consists of two parts.

- key-scheduling algorithm (KSA).
- Pseudo-random generation algorithm (PRGA).

The key-scheduling algorithm (KSA) is used to initialize the permutation in the array box "S". The length of key is number of bytes in the key and is in the range 1 to 256. First, the array "S" is initialized to the identity permutation. The array box "S" is then processed for 256 iterations in a similar way to the main PRGN, but also mixes in bytes of the key at the same time. The key-scheduling algorithm (KSA) [1] is listed below.

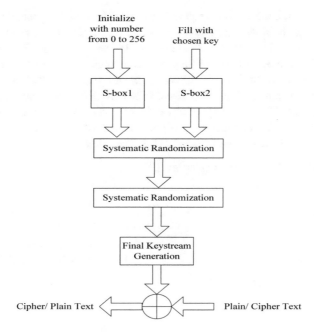

Fig. 4. General logic design structure of RC4 cipher

Key-Scheduling Algorithm (KSA)

```
begin

for i from 0 to 255

    S[i] := i

endfor

j := 0

for i from 0 to 255

    j := (j + S[i] + key[i mod keylength]) mod 256

    swap values of S[i] and S[j]

endfor

end
```

For as many iterations as are required, the pseudo-random generation algorithm (PRGA) modifies the state and generates a byte of the keystream. In every iteration, the PRGA increments i, looks up the *i*th element of the array box "S", S[i], and adds that to j, swaps the values of S[i] and S[j], and then uses the sum S[i] + S[j] (modulo 256) as an index to obtain a third element of array box "S", (the keystream value K below) which is bitwise exclusive ORed (XORed) with the next byte of the plain text to generate the next byte of cipher text. Each element of the array box "S" is exchanged with another element at least once in each of 256 iterations. The Pseudo-Random Generation Algorithm (PRGA) [1] is listed below.

Pseudo-Random Generation Algorithm (PRGA)

```
begin

i := 0

j := 0

while GeneratingOutput:

    i := (i + 1) mod 256

    j := (j + S[i]) mod 256

    swap values of S[i] and S[j]

    K := S[(S[i] + S[j]) mod 256]

    output K

endwhile

end
```

3   Design and Implementation

First, we implement RC4 stream cipher by using *Key-Scheduling Algorithm* (KSA) and *Pseudo-Random Generation Algorithm* (PRGA) in Java programming language. Then we implement two android mobile applications: *SendSMS* and *ReceieveSMS*. *SendSMS* mobile application is used for the sender to transform from SMS plain text to cipher text, and to send it confidentially to the receiver through SMS mobile network communication system while *ReceieveSMS* mobile application is used for the receiver to receive cipher text that is passed through SMS mobile network communication system and to transform from cipher text to SMS plain text.

The design for implementation of two android mobile applications, *SendSMS* and *ReceieveSMS,* is shown in Fig. 5. For *SendSMS* mobile application, at first password is used in RC4 cipher to generate keystream and it is XORed with SMS plain text to output cipher text. The *Sending* process sends the cipher text to the phone number accepted by this application. Correspondingly, in *ReceieveSMS* mobile application the *Receving* process receives the cipher text from the phone number accepted by this application. Then the cipher text is XORed with the keystream generated by RC4 cipher that uses the same password to output original SMS plain text.

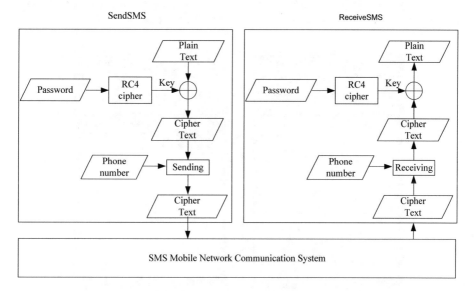

Fig. 5. Design for implementation

The general data flow diagram of these two mobile applications is shown in Fig. 6. *SendSMS* mobile application accepts SMS plain text, password and phone number of the receiver as inputs and outputs cipher text. The cipher text is passed through mobile network communication system. *ReceieveSMS* mobile application accepts cipher text that passed through mobile network communication system, password and phone number of the sender as inputs and outputs SMS plain text.

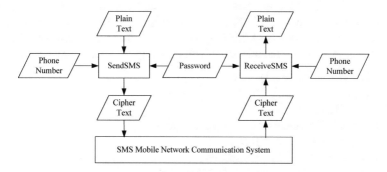

Fig. 6. General data flow diagram

The system user interfaces for *SendSMS* and *ReceieveSMS* mobile applications is shown in Fig. 7. *SendSMS* mobile application is used at the side of the sender and *ReceieveSMS* mobile application is used at the side of the receiver. The sender must input phone number of the receiver, password and SMS message to *SendSMS* mobile application and press *Send Message* button. The receiver must input phone number of the sender and the same password used by the sender to *ReceiveSMS* mobile application and press *Receive Message* button. Then SMS message of the sender is appeared in the display window screen of *ReceiveSMS* mobile application.

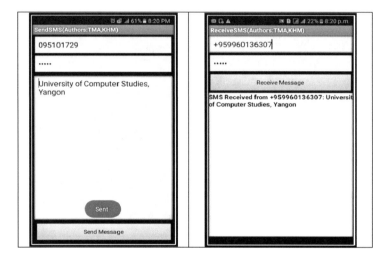

Fig. 7. System user interfaces

4   Confidentiality Measurement

RC 4 stream cipher generates keystream that is pseudorandom number sequence. The pseudorandom number sequence can be used for data confidentiality mechanism during data transmission. The quality of this data confidentiality mechanism depends on the randomness of pseudorandom number sequence generated by RC4 stream cipher. The randomness of pseudorandom number sequence can be measured by using following statistical tests recommend by NIST, National Institute of Standards and Technology. These tests focus on a variety of different types of non-randomness that could exist in a sequence. Some tests are decomposable into a variety of subtests. The 15 tests are:

1. The Frequency (Monobit) Test,
2. Frequency Test within a Block,
3. The Runs Test,
4. Tests for the Longest-Run-of-Ones in a Block,
5. The Binary Matrix Rank Test,
6. The Discrete Fourier Transform Test,
7. The Non-overlapping Template Matching Test,

8. The Overlapping Template Matching Test,
9. Maurer's "Universal Statistical" Test,
10. The Linear Complexity Test,
11. The Serial Test,
12. The Approximate Entropy Test,
13. The Cumulative Sums Test,
14. The Random Excursions Test,
15. The Random Excursions Variant Test.

A statistical test is formulated to test a specific *null hypothesis* (H0). The null hypothesis under test is that the pseudorandom number sequence being tested is *random*. Associated with this null hypothesis is the alternative hypothesis (Ha), that is, the pseudorandom number sequence is *not random*. For each applied test, a decision or conclusion is derived that accepts or rejects the null hypothesis, i.e., whether the pseudorandom number generator is (or is not) producing random values, based on the pseudorandom number sequence that was produced [4].

For each test, a relevant randomness statistic must be chosen and used to determine the acceptance or rejection of the null hypothesis. Under an assumption of randomness, such a statistic has a distribution of possible values. A theoretical reference distribution of this statistic under the null hypothesis is determined by mathematical methods. During a test, a test statistic value is computed on the pseudorandom number sequence being tested. This test statistic value is compared to the critical value. If the test statistic value exceeds the critical value, the null hypothesis for randomness is rejected. Otherwise, the null hypothesis (the randomness hypothesis) is not rejected (i.e., the hypothesis is accepted) [4].

Each test is based on a calculated test statistic value, which is a function of pseudorandom number sequence. If the test statistic value is S and the critical value is t, then the Type I error probability is $P(S > t \| Ho$ is true$) = P($reject $Ho \mid H0$ is true$)$, and the Type II error probability is $P(S \leq t \| H0$ is false$) = P($accept $H0 \mid H0$ is false$)$. The test statistic is used to calculate a *P-value* that summarizes the strength of the evidence against the null hypothesis. For these tests, each *P-value* is the probability that a perfect random number generator would have produced a pseudorandom number sequence less random than the pseudorandom number sequence that was tested, given the kind of non-randomness assessed by the test. If a *P-value* for a test is determined to be equal to 1, then the pseudorandom number sequence appears to have perfect randomness. A *P-value* of zero indicates that the pseudorandom number sequence appears to be completely non-random. A significance level (α) can be chosen for the tests. If *P-value* α, then the null hypothesis is accepted; i.e., the pseudorandom number sequence appears to be random. If *P-value* $< \alpha$, then the null hypothesis is rejected; i.e., the pseudorandom number sequence appears to be non-random. The parameter α denotes the probability of the Type I error. Typically, α is chosen in the range [0.001, 0.01] [4].

An α of 0.001 indicates that one would expect one sequence in 1000 pseudorandom number sequences to be rejected by the test if the sequence was random. For a *P-value* $\geq$ 0.001, a pseudorandom number sequence would be considered to be random with a confidence of 99.9%. For a *P-value* $<$ 0.001, a pseudorandom number sequence would be considered to be non-random with a confidence of 99.9% [4].

An α of 0.01 indicates that one would expect one sequence in 100 pseudorandom number sequences to be rejected. A *P-value* ≥ 0.01 would mean that the pseudo-random number sequence would be considered to be random with a confidence of 99%. A *P-value* < 0.01 would mean that the conclusion was that the pseudorandom number sequence is non-random with a confidence of 99% [4].

5   Conclusion

The pseudorandom number sequence generated by RC4 stream cipher is measured by the statistical test suite developed by NIST. According to P-value of each test, the pseudorandom number sequence may be considered to be random with a confidence of 99%. Moreover, RC4 stream cipher possesses better performance among stream ciphers. Therefore, we suggest that the pseudorandom number sequence generated by RC4 stream cipher should be used for data confidentiality of SMS message transmitted on mobile network communication system.

References

1. Forouzan, B.A.: Cryptography and Network Security. International Edition, McGrawHill, ISBN: 978-007-126361-0 (2008)
2. Agoyi, M., Seral, D.: SMS security: an asymmetric encryption approach. In: IEEE 6th International Conference on Wireless and Mobile Communications (2010)
3. Medani1, A.G., Zakaria, O., Zaidan, A.A., Zaidan, B.B.: Review of mobile short message service security issues and techniques towards the solution. Sci. Res. Essays Acad. J. 6(6), ISSN 1992-2248 (2011)
4. NIST: A Statistical Test Suite for Random and Pseudorandom Number Generators for Cryptographic Applications. National Institute of Standards and Technology Special Publication 800–822 (2010)
5. Katankar, V.K., Thakare, V.M.: Short message service using SMS gateway. Int. J. Comput. Sci. Eng. 02(04) (2010)

Analysis of Attribute-Based Secure Data Sharing with Hidden Policies in Smart Grid of IoT

Nishant Doshi[(✉)]

Department of Computer Science and Engineering, Pandit Deendayal Petroleum University, Raisan, Gandhinagar, Gujarat, India
doshinikki.backup@gmail.com

Abstract. Apropos to communication in conventional computer network or in electronics, in electrical field, the smart grid is the approach to transfer the electricity through distribution of network. It not only improves the security, reliability of electric system but also the communication of it too. Recently J Hur in IEEE Tran. on Smart Grid proposed an approach to provide an efficient and secure scheme in smart grid using attribute based analogy. The author claim that the proposed scheme is also provide the receiver anonymity which is one of the crucial feature of smart grid. However, in this paper, we claim that the scheme presented by Hur, fails to provide receiver anonymity as well as it is susceptible to the key escrow attack.

Keywords: Smart grid · Policy-based Data Sharing · Privacy
Security · Communication

1 Introduction

Consider the world where we have all the technologies who drive the world to the galaxy and beyond. All this technology can be made possible in timely manner with electricity. Thus, electricity plays a vital role in what the technologies we are using in today's life. Smart grid is the technology in which the interconnected network works intelligently to transmit and distribute the electricity through network. Figure 1 depicts the schematic of smart grid environment. Even with the help of Internet of Things, the smart grid becomes in huge demand now-a-days. More details of smart grid is available in [4–7].

On the other end, security is important in entire network. Symmetric key cryptography is a technique in which sender and receiver share the same secret key. Public key cryptography is a technique in which sender having two keys i.e. private key (to decrypt data) and public key (to encrypt data). However, in multicast scenario, the overhead will be more on sender side. Thus, attribute based cryptography is a technique in which selected set of receipients who's satisfies the criteria will be able to decrypt the data. The research attempt in this domain is more explored in the [8–22].

In [1], the author has presented the scheme for smart grid, based on the construction of Bethencourt et al.'s [2] Ciphertext Policy Attribute Based Encryption (CP-ABE)

© Springer Nature Switzerland AG 2019
P. Vasant et al. (Eds.): ICO 2018, AISC 866, pp. 515–521, 2019.
https://doi.org/10.1007/978-3-030-00979-3_54

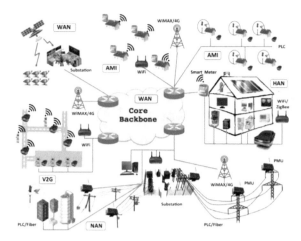

Fig. 1. Smart grid communication architecture [3]

scheme. In [1], the author claims that the proposed scheme provides hidden policy with partial decryption and secure against the key escrow attack.

The entities involved in the scheme are as follows:

- **Key Generation Center (KGC):** It generates the secret key for users as well as system's public and private parameters.
- **Storage Center (SC):** It stores the ciphertext generated by the sender. It also does the partial decryption.
- **Sender:** It encrypts the document based on obfuscated access policy.
- **Receiver:** It decrypts the ciphertext if attributes within the secret key satisfy the access policy of ciphertext.

In [1], the author claims (p. 2172, right column, Sect. 1.1) the following:

- The policy is hidden from SC and KGC.
- The KGC is not decrypt the document, thus the scheme is secure against the key escrow.

However, we claim that, in a typical environment, the above claims become false and lead to the system failure.

In the next section, we review the scheme of [1]. In Sect. 3, we have given our security analysis in a typical scenario. The conclusion and references are at the end.

Review of Hur's Scheme

For brevity, the detailed discussion of algorithms is omitted. The detailed discussion is in the [1].

KGC-Setup

It is run by KGC. Let G_0, G_1 be the bilinear group and Z_p be the set of prime order p. g is the generator for group G_0. $e : G_0 \times G_0 \to G_1$ is the bilinear function. KGC selects $\alpha, \beta \in_R Z_p$. KGC sets $h = g^\beta$. KGC selects two hash functions $H : \{0, 1\}^* \to G_0$,

$H1 : G_1 \rightarrow \{0,1\}^{logp}$. The Master Public Key (MPK) and Master Secret Key (MSK) are as follows

$$MPK = \{G_0, G_1, e, H, H1, e(g,g)^\alpha, h\} \quad MSK = \{\beta, g^\alpha\}$$

SC-Setup

This is run by SC. It selects $\gamma \in_R Z_p$. Here ID_{SC} is publicly universal ID of the SC. The private and public parameters of SC are as follows:

$$PK_{SC} = g^\gamma$$
$$SK_{SC} = H(ID_{SC})^\gamma$$

KeyGen

This is run by KGC to generate the secret key for user u with attribute list L. KGC selects $r, r_j \forall j \in L$.

$$SK_u = \{D = g^{\frac{\alpha+r}{\beta}}, \forall j \in L : D_j = g^r \cdot H(j)^{r_j}, D'_j = g^{r_j}, D''_j = H(j)^\beta$$

Encrypt

It is run by the sender with access policy W and Message M. The sender generates $a \in_R Z_p$ and computes $K_S = e((g^\gamma)^a, H(ID_{SC}))$.

$$CT = (T, C' = M \cdot K_S \cdot e(g,g)^{\alpha S}, C = h^S,$$
$$\forall y \in Y : C_y = g^{q_y(0)}, C'_y = H(\lambda)^{q_y(0)}).$$

Sender sends the $CT' = \langle CT, ID_a, g^a \rangle$

GenToken

This is run by the receiver. After receiving g^a from SC, the algorithm computes $s_j = e\left(g^a, H(j)^\beta\right)$ on SC, the algorithm computes d of alogorithms. the detailed one is in the [1]. Then, it generates $\tau \in_R Z_p$ and computes token for attribute set $\wedge$ as follows:

$$TK_{\wedge,u} = \left(\forall j \in \wedge : I_j = H_1(s_j), (D_j)^\tau, \left(D'_j\right)^\tau\right).$$

Partial-Decrypt

This is run by the SC. It computes $K_S = e(g^a, SK_{SC}) = e(g^a, H(ID_{SC})^\gamma)$ and $C'' = \frac{C'}{K_S} = M \cdot e(g,g)^{\alpha S}$ in CT. The SC sends the $CT' = <C'', C = h^S, A >$ to the receiver, where $A = DecreyptNode\left(CT, TK_{\wedge,u}, R\right) = e(g,g)^{r\tau s}$.

Decrypt

It is run by the user u. The user gets message M by computing $C'' / \left(\frac{e(C,D)}{(A)^{\frac{1}{\tau}}}\right)$.

1.1   Paper Organization

This paper is organized as follows. In Sect. 2, we have given the analysis on Hur's scheme. Finally, we concluded in Sect. 3. References are at the end.

2   Analysis of Hur's Scheme

Before presenting our security analysis on Hur's scheme, we take the following illustration to support our claim.

Let us assume the typical college management system consisting of *students*. Assume that there are *nine* departments in the college such as computer, electrical, mechanical, etc. In addition, there are 480 students (120 students within each year) within each department.

Claim 1: Fails to Provide Receiver Anonymity

The value of parameter $D'' = H(j)^\beta$ of each *mechanical* student is the same as it depends on j and β only, and they are constant for a particular attribute (say "branch = mechanical"). Same for other branches.

Now considering a model in which compromising one student from 120 is not difficult. In general, the possible complexity for launching such attack is $O(j)$ as there are j attributes in the system. The actual complexity is less than $O(j)$ as one user contains more than one attributes in a secret key. From now onwards, we assume that attacker playing the role of corrupt student from each branch.

With this assumption, attacker can get the value D'' for every attribute in the system.

The attacker runs the following algorithm to know the policy from GenToken as follows. The SC knows the g^a from the tuple (ID_a, g^a, CT) given by u_a.

Run by : Attacker
Input : $\forall j \in \wedge, I_j = H_1(s_j), a$ and U=all attributes set.
Output : Policy details.

Policy-list=$\emptyset$.
For j in $\wedge$
Do   For i in U
        Do   If $(H_1(e(g^a, D_i'')) = I_j)$
             Then   Add attribute i in policy-list.
                   Break.
        Done
Done
Output Policy-list.

With $x = |\wedge|$ and $y = |U|$, the complexity of finding the policy details is $O(xy)$.

Although attacker cannot able to decrypt the ciphertext, but attacker knows the list of attributes used in the policy, which reveal the plain attributes used in the policy. Thus, the Hur's system fails to maintain receiver anonymity.

In addition, author mentioned in Sect. 3.2 (policy privacy) that the authorized receivers cannot be able to know the policy. However, in Sect. 5.3.4, receiver gives subset of attribute set S. It is not specified that how receiver selects this subset. If receiver knows policy than it breaches the policy privacy. If not, than how to select subset of the SC or decryption algorithm requires the exact match of attribute in CT and in SK. Without knowing policy how one can match is not defined.

Note: Attacker only compromises the minimal users having secret key and combining them covers all attributes of the system. There is no need for the sender to reveal any information to attacker or KGC or SC. Therefore, the insider attack where the valid user can give its secret information for encryption and hiding policy is not required by the attacker. In fact, the attacker assumed here is considered in all ABE systems.

Claim 2: KGC can Decrypt any Message

KGC can generate secret key for any user in system. KGC can do the complete decryption like normal user. Here SC is assumed as honest-but-curious as in Hur's scheme. The task of the SC during partial decryption is to check if received token is appropriate for respective ciphertext, if yes than run Partial-Decrypt otherwise fails.

Therefore, the proposed system is key escrow as KGC can generate any secret key like valid user (receiver) and do decryption in same way.

3   Conclusion

Recently, in *IEEE Transactions on Parallel Distributed Systems*, Junbeom Hur [1] proposed an interesting Attribute-Based Secure Data Sharing with Hidden Policies in Smart Grid scheme that is claimed to hide the policy. However, in this paper, we demonstrate that unfortunately, their scheme is not able to hide the policy with the minimal set of users collude. At present, it is still a challenging open problem to construct an Attribute-Based Secure Data Sharing with Hidden Policies in the Smart Grid scheme.

References

1. Hur, J.: Attribute-based secure data sharing with hidden policies in smart grid. IEEE Trans. Parallel Distrib. Syst. **24**(11), 2171–2180 (2013)
2. Bethencourt, J., Sahai, A., Waters, B.: Ciphertext-policy attribute-based encryption. In: IEEE Symposium on Security and Privacy, SP 2007, pp. 321–334 (2007)
3. Vijayanand, R., Devaraj, D., Kannapiran, B.: A novel dual euclidean algorithm for secure data transmission in smart grid system. In: IEEE International Conference on Computational Intelligence and Computing Research (ICCIC), pp. 1–5. (2014)

4. Khurana, H., Hadley, M., Lu, N., Frincke, D.A.: Smart-Grid security issues. IEEE Secur. Priv. **8**(1), 81–85 (2010)
5. Kim, Y., Thottan, M., Kolesnikov, V., Lee, W.: A secure decentralized data-centric information infrastructure for smart grid. IEEE Comm. Mag. **48**(11), 58–65 (2010)
6. Bobba, R., Khurana, H., AlTurki, M., Ashraf, F.: PBES: A policy based encryption system with application to data sharing in the power grid. In: Proceedings of the International Symposium on Information, Computer, and Communications Security (ASIACCS 2009) (2009)
7. The Cyber Security Coordination Task Group: Smart grid cyber security strategy and requirements. NIST Technical Report Draft NISTIR 7628 (2009, Sept)
8. Cheung, L., Newport, C.: Provably secure ciphertext policy ABE. In: Proceedings of the ACM Conference on Computer and Communications Security, pp. 456–465 (2007)
9. Goyal, V., Jain, A., Pandey, O., Sahai, A.: Bounded ciphertext policy attribute-based encryption. In: Proceedings of the International Colloquium on Automata, Languages and Programming (ICALP), pp. 579–591 (2008)
10. Liang, X., Cao, Z., Lin, H., Xing, D.: Provably secure and efficient bounded ciphertext policy attribute based encryption. In: Proceedings of the International Symposium Information, Computer, and Communications Security (ASIACCS), pp. 343–352 (2009)
11. Yu, S., Ren, K., Lou, W.: Attribute-based content distribution with hidden policy. In: Proceedings of the Workshop Secure Network Protocols, pp. 39–44 (2008)
12. Nishide, T., Yoneyama, K., Ohta, K.: Attribute-based encryption with partially hidden encryptor-specified access structure. In: Proceedings of the Sixth International Conference on Applied Cryptography and Network Security (ACNS 2008), pp. 111–129 (2008)
13. Boneh, D., DiCrescenzo, G., Ostrovsky, R., Persiano, G.: Public key encryption with keyword search. In: Proceeding of the Theory and Applications of Cryptographic Techniques Annual International Conference on Advances in Cryptology (Eurocrypt 2004), pp. 506–522 (2004)
14. Boneh, D., Franklin, M.K.: Identity-based encryption from the weil pairing. In: Proceedings of the Annual International Cryptology Conference Advances in Cryptology (CRYPTO 2001), pp. 213–229 (2001)
15. Boneh, D., Waters, B.: Conjunctive, subset, and range queries on encrypted data. In: Proceedings of the Theory of Cryptography Conference (TCC 2007), pp. 535–554 (2007)
16. Boyen, X.: A tapestry of identity-based encryption: practical frameworks compared. Int. J. Appl. Cryptogr. **1**, 3–21 (2008)
17. Goyal, V., Pandey, O., Sahai, A., Waters, B.: Attribute-based encryption for fine-grained access control of encrypted data. In: Proceedings of the ACM Conference on Computer and Communications Security, pp. 89–98 (2006)
18. Kate, A., Zaverucha, G., Goldberg, I.: Pairing-based onion routing. In Proceedings of the Privacy Enhancing Technologies Symposium, pp. 95–112 (2007)
19. Katz, J., Sahai, A., Waters, B.: Predicate encryption supporting disjunctions, polynomial equations, and inner products. In: Proceedings of the Theory and Applications of Cryptographic Techniques 27th Annual International Conference on Advances in Cryptology (EUROCRYPT 2008), pp. 146–162 (2008)
20. Lewko, A., Okamoto, T., Sahai, A., Takashima, K., Waters, B.: Fully secure functional encryption: attribute-based encryption and (hierarchical) inner product encryption. In: Proceedings of the Theory and Applications of Cryptographic Techniques Annual International Conference on Advances in Cryptology (EUROCRYPT 2010), pp. 62–91 (2010)

21. Ostrovsky, R., Sahai, A., Waters, B.: Attribute-based encryption with non-monotonic access structures. In: Proceedings of the ACM Conference on Computer and Communication Security, pp. 195–203 (2007)
22. Sahai, A., Waters, B.: Fuzzy identity-based encryption. In: Proceedings of the Theory and Applications of Cryptographic Techniques Annual International Conferrence Advances in Cryptology (Eurocrypt), pp. 457–473 (2005)

A Single Allocation *P*-Hub Maximal Covering Model for Optimizing Railway Station Location

Sunarin Chanta[✉] and Ornurai Sangsawang

King Mongkut's University of Technology North Bangkok, 129M.21, Noenhom,
Muang 25230, Prachinburi, Thailand
{sunarin.c, ornurai.s}@fitm.kmutnb.ac.th

Abstract. In this paper, we propose an optimization model for determining locations of railway stations. The objective is to maximize the covered the number of expected passengers that can be covered. We focus on the case study of locating optimal stations of high speed train on the north route railway line of Thailand, which is on the government plan to be built. We considered the number of expected passengers based on the real passengers traveling during year on the line. Two types of coverage are considered, which are the condition on time to go to a station and the condition on total travelling time. To solve this problem, we developed Simulated Annealing. Computational results showed that the proposed metaheuristic algorithm found the high quality solutions in reasonable time.

Keywords: Maximal covering · Hub location · High speed train
Simulated annealing

1 Introduction

As the world population growth increases, public transportation becomes one of the most important development challenges. High Speed Rail (HSR) is an efficient alternative for rapid mass transits that has been developed in many countries around the world. HSR network is a solution of mass transportation, since it offers a type of transport that fast, convenience, reliable, save, ecological, and environment friendly, which connected cities across countries or between continents. However, the cost of building HSR is very expensive. So a design process on the routes, station locations, number of trips, which according customers' demand is necessary.

In this paper, we propose a way to determine optimal station locations on a HSR using a maximal hub covering location model. The objective is to maximize the number of passengers travel by HSR, under the limitations of the number of stations,

---

S. Chanta—Please note that the AISC Editorial assumes that all authors have used the western naming convention, with given names preceding surnames. This determines the structure of the names in the running heads and the author index.

capacity of the stations, and coverage distance. The case study of the north route railway line, in Thailand was presented. The number of real passengers travelled during year was used to estimate the expected number of HSR passengers. Since the problem is complex and belong to a type of combinatorial problems, we develop the Simulated Annealing (SA) for solving this problem. The results can be found in reasonable time.

2   Literature

Hub location problem deals with locating facilities (hub) and allocating point of demands (node) to facilities on a network, in order to provide a transshipment service, routed between origin-destination. The objective is to minimize the total transportation cost of moving flows from origin node to destination node via hub. Figure 1 showed the structure of hub location problem, where k, m represented hubs and i, j represented nodes. Hub location problems are widely studied and can be classified in many types of problems such as single allocation and multiple allocation, uncapacitated and capacitated servers. Moreover, they may classify based on their objective such as median problem, center problem, and covering problem.

O'Kelly [1] has presented the first quadratic integer programing formulation for the single allocation *p*-hub median problem, where the number of hub is fixed at p and the objective is to minimize total cost of the system that associated with traveled distance. Campbell [2] developed the first linear integer programming formulation for the single allocation *p*-hub median problem. His formulation has $n^4 + n^2 + n$ variables and $n^4 + 2n^2 + n + 1$ linear constraints. Skorin-Kapov, et al. [3] proposed a mixed integer programing formulation with $n^4 + n^2$ variables and $2n^3 + n^2 + n + 1$ linear constraints. Ernst and Krishnamoorthy [4] proposed an improved linear integer programming formulation with $n^3 + n^2$ variables and $2n^2 + n + 1$ linear constraints.

Later, the single allocation hub location problem with fixed costs was formulated by O'Kelly [5], where the number of hub is not fixed and the objective is to minimize the total cost of the system that composited of operating cost (associated with traveled distance) and fixed cost (or cost of establishing hub). Campbell [6] introduced capacitated multiple allocation *p*-hub median problem, where capacity of each hub is limited. Campbell [6] also introduced the hub center problem, where the objective is to minimized the maximum cost of the origin-destination pair and the hub covering problem, where the objective is to maximize the demand that can be covered. Later, Kara and Tansel [7] proposed the single allocation *p*-hub covering problem, and then Wagner [8] improved the *p*-hub covering location problem with bounded path lengths. Hwang and Lee [9] proposed the uncapacitated single allocation *p*-hub maximal covering problem for solving a problem using several instances from CAB (civil Aeronautics Board) data set. Two heuristics were proposed; distance based allocation and volume based allocation methods. Peker and Kara [10] presented the *p*-hub maximal covering problem with gradual decay function as partial coverage. They proposed an improved mixed integer programming model and compared to the previous versions. The model was tested on CAB and Turkish Network (TR) data sets. For more details on hub location problems, see Amur and Kara [11] and Farahani, et al. [12].

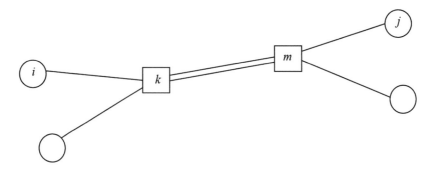

Fig. 1. Single allocation p-hub location problem

3   A Single Allocation P-Hub Maximal Covering Model

In this paper, we developed the model based on the single allocation p-hub maximal covering model by considering two conditions of coverage, which adjusted from Campbell [6] and Peker and Kara [10]. The first coverage is on the travel time to go to the station, since it is an important factor that effect on the decision of passengers to travel by HSR or not. The second coverage is on the total travel time on the trip, since passengers may not want to travel for a long time, and they whether have their deadline or acceptable total travel time. So these two conditions have to be satisfied by passengers in order to be counted as covered demand. The formulations of the single allocation p-hub maximal covering model with two conditions on coverage we proposed are detailed as follows.

Index and notations

i	= index of origin, where $i = 1, 2, ..., n$
j	= index of destination, where $j = 1, 2, ..., n$
k	= index of start station location, where $k = 1, 2, ..., n$
m	= index of end station location, where $m = 1, 2, ..., n$
n	= maximum number of nodes (demand locations)
p	= maximum number of hubs (HSR stations)
t	= coverage on time to go to a station (the maximum travelling time that people willing to go to the nearest railway station)
α	= speed factor when travelling by HSR
β	= coverage on total travelling time (the maximum travelling time that people allowing on their trip)

Parameters

W_{ij} = flow of demand between origin i and destination j

c_{ij} = travel time between private car or other public transportation origin i and destination j

c_{ij}^{km} = travel time between origin i and destination j via stations k and m, where $c_{ij}^{km} \leq c_{ij}$, and $c_{ij}^{km} = c_{ik} + \alpha c_{km} + c_{mj}$; c_{ik} = travel time by private car or other public transportation from origin i to station j, c_{km} = travel time by HSR from station k to station m, c_{mj} = travel time by private car or other public transportation from station m to destination j

a_{ij}^{km} = covered matrix with two conditions

= 1 if $c_{ij}^{km} \leq \beta$ and $c_{ik} \leq$ t, otherwise 0

λ_{ij} = $\max_{km} a_{ij}^{km}, \forall i, j$

Decision Variables

Z_{ij} = fraction of flow routed from node i to j that is covered

X_{ik} = 1 if node i is assigned to hub k, otherwise 0

X_{kk} = 1 if node k is selected to be hub, otherwise 0

Maximize

$$\sum_{i=1}^{n}\sum_{j=1}^{n} W_{ij}Z_{ij} \tag{1}$$

Subject to

$$Z_{ij} \leq \sum_{k=1}^{n} a_{ij}^{km} X_{ik} + \lambda_{ij}(1 - X_{jm}); \forall i, j, m \tag{2}$$

$$\sum_{k=1}^{n} X_{ik} = 1; \forall i \tag{3}$$

$$\sum_{k=1}^{n} X_{kk} = p \tag{4}$$

$$X_{ik} \leq X_{kk}; \forall i, k \tag{5}$$

$$X_{11} = 1 \tag{6}$$

$$X_{nn} = 1 \tag{7}$$

$$X_{ik} \in \{0.1\}, \forall i, k \tag{8}$$

$$Z_{ij} \geq 0; \forall i, j \tag{9}$$

The objective is to maximize the total demand between each origin-destination pair covered by predefined coverage distance. Note that the coverage distance is the longest distance that people will travel to their nearest HSR station. Constraints (2) determine fraction of flow between node i and j that is covered with hubs k and m ($X_{ik} = X_{jm} = 1$). Constraints (3) ensure that a node is assigned to exactly one hub. Constraint (4) limits the number of stations at p. Constraints (5) ensure that node is only assigned to hub. Constraints (6)–(7) force start station and end station to be open. Constraints (8)–(9) define sign of decision variables.

4 Simulated Annealing

Simulated annealing (SA) is a probabilistic based algorithm to solve combinatorial optimization problems which was developed by Kirkpatrick, et al. 1983 [13]. The algorithm simulates physical annealing process of metals and accepts non-improving changes with some probability to escape the trap of local optima. According to the Metropolis criterion, the new solution is accepted if a random number, r, generated from uniform distribution [0,1] is no larger than the acceptance probability.

4.1 Solution Representation

An initial solution consists of two arrays: location and allocation parts. The lengths of arrays indicate the number of nodes. The location part, a hub node is allocated to itself, represents the node is a hub. The allocation part indicates the assignments of the non-hub nodes to hub node. For instance, the array in Fig. 2 represents $n = 10$, $p = 3$, where n = number of nodes, p = number of hubs. Non-hub nodes 1, 2, and 4 are connected to hub node 3, non-hub nodes 4 is allocated to hub node 5 and non-hub nodes 9 and 10 are connected to hub node 8.

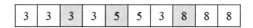

Fig. 2. Solution representation of simulated annealing algorithm

4.2 Neighborhood Operators

Neighborhood operators are based on location and allocation parts which in turn can change opening and closing stations and allocation nodes to different cluster. To generate a new solution, there are three operators are called intra - cluster move, inter - cluster move, and reallocation. In the intra - cluster move, a hub and a node in the same cluster are randomly selected and changed a hub with a node. The inter - cluster operator considers two randomly a hub and a node in different cluster and performs replacing a hub with a node to the different cluster. The reallocation operator considered a node is chosen at random to allocate at a new position hub. The details of SA procedure are showed in Fig. 3.

The temperature T_i in iteration i is reduced by a cooling schedule of $\rho = 0.98$ and the initial temperature is 2000, based on initial empirical results. When the same solutions appear in $2np$ iterations, reheating is used to avoid local optima.

```
begin
        Initial solution creation s₀;
        Check capacity restriction;
        Set initial temperature T₀ and cooling rate ρ;
While )stopping criterion is not achieved( do

        generate random number r unif [0,1)
        If (r < 0.30) then
             intra - cluster move;
        If (r > 0.60) then
             inter - cluster move;
        Else reallocation;
        If ( ΔE < 0 ) then
             s* = s;
        Else if ( random[0,1)  > exp⁽⁻ᴰᴱ/ᵅ· ᵀ⁾ then
             s* = s;
        Update temperature    Tᵢ₊₁ = ρTᵢ

End while
```

Fig. 3. Procedure of simulated annealing algorithm

5 Computational Experiments

5.1 A Case Study of HSR

Our case study is taken from the HSR in Thailand. The aim is to find the optimal station location of the north route railway line, which routed from Bangkok to Chiang Mai, with total of 751 km. The candidate HSR stations are the existing stations on the line, which total of 54 stations ($n = 54$) as showed in Fig. 4. Note that in the north route line, there are more than 100 existing stations; but in this study, we selected only 54 stations to be the candidate HSR stations. We neglected the small and low demand stations. The number of passengers traveled on the north railway line during year, 2014 was used to estimate the expected demand of the HSR line. We assumed that HSR will operate at speed 250 km./h., while travelling by other mode of transportation will take 60 km./h. So, we set the speed factor (α) at 0.24. The coverage on the total travel time (β) was set at 180 min, while the coverage on the time to go to assigned station (t) was set at 50 min. The number of stations (p) was varied from 5 to 15.

5.2 Experiments and Computational Results

To see the difference from solving the problem with or without condition on the coverage, we conducted 2 experiments with different parameter setting on the coverage, which detailed as follow; Experiment I) the total travel time on trip not exceed

180 min. without condition on time to the assigned HRS, Experiment II) total travel time on trip not exceed 180 min. and time to the assigned HRS not exceed 50 min. In each experiment, the number of stations (p) was varied from 5 to 15, which total of 14 cases. All experiments were solved on an Intel Core i5-2410 M CPU 2.3 GHz. with 6 GB of RAM. The mathematical programming formulations in Sect. 3 were implemented and solved by optimization software; Optimization Programming Language (OPL) 12.7. Then, the optimal solutions were compared to the solutions obtained by the proposed SA algorithm as detailed in Sect. 4, which coded in C language. The results of both techniques were reported in Tables 1 and 2, with open stations, coverage percentage, and run times. The results showed that OPL obtained optimal solution for all cases also with a range of run times between 360 to 1363 s, while SA also found optimal solution for all cases with a range of run times between 5 to 20 s. Based on these experiments, we see that with condition on the coverage OPL tended to find the optimal solution faster. In the opposite side with condition on the coverage SA tended to find the optimal slightly longer.

Fig. 4. Map of railway station locations on the north route line, Thailand

Table 1. Results of experiment I (*t* = free)

#Stations	Open station	Coverage (%)	Time(s)	Time(s)
(*p*)	(X_{kk})		OPL	SA
5	1,22,35,48,54	97.96	1363	5
6	1,18,32,43,50,54	99.55	610	7
7	1,9,23,36,47,51,54	99.93	1068	4
8	1,8,19,29,35,48,52,54	100.00	440	4
9	1,7,16,25,35,45,50,53,54	100.00	416	7
10	1,8,17,24,31,39,41,48,52,54	100.00	340	9
15	1,2,8,14,21,22,31,39,40,48,50, 51,52,53,54	100.00	337	19

Table 2. Results of Experiment II (*t* = 50)

#Stations	Open station	Coverage (%)	Time(s)	Time(s)
(*p*)	(X_{kk})		OPL	SA
5	1,22,33,43,54	95.64	561	6
6	1,16,29,37,48,54	96.48	656	7
7	1,16,29,34,43,50,54	99.65	390	9
8	1,16,28,34,43,49,51,54	99.95	384	9
9	1,8,16,28,34,43,49,51,54	99.99	472	9
10	1,11,18,29,34,43,50,51,53,54	100.00	406	10
15	1,8,9,16,22,25,28,29,33,40,46, 49,51,53,54	100.00	363	20

6   Conclusion and Future Research

We proposed a hub covering location problem, which considered two conditions on the coverage; travel time on node to hub and travel time on the trip routed from node to hub, hub to hub, and hub to node. The case study of HSR in Thailand was presented. The number of real passengers travelled during year was used to estimate the expected number of HSR passengers. Because of the complexity, we developed the Simulated Annealing to solve the problem. The results showed that the proposed metaheuristic algorithm yielded the quality solution with reasonable time.

For future research, we interest to extend the maximal *p*-hub covering model to have multiple allocation, which allows passengers on a node can go different stations. The consideration on linking more than one mode of transportation is also an advantage that reflects the real customers' behavior, which allows people to choose their preferred transportation mode.

Acknowledgments. This work is financially supported by King Mongkut's University of Technology North Bangkok under the grant number KMUTNB-GOV-58-49.

References

1. O'Kelly, M.E.: A quadratic integer program for the location of interacting hub facilities. Eur. J. Oper. Res. **32**, 393–404 (1987)
2. Campbell, J.F.: Location and allocation for distribution for distribution systems with transshipments and transportation economies of scale. Ann. Oper. Res. **40**, 77–99 (1992)
3. Skorin-Kapov, D., Skorin-Kapov, J., O'Kelly, M.E.: Tight linear programming relations of uncapacitated p-hub median problems. Eur. J. Oper. Res. **74**, 582–593 (1996)
4. Ernst, A.T., Krishnamoorthy, M.: Efficient algorithms for the uncapacitated single allocation p-hub median problem. Locat. Sci. **4**, 139–154 (1996)
5. O'Kelly, M.E.: Hub facility location with fixed costs. Pap. Reg. Sci. **71**, 292–306 (1992)
6. Campbell, J.F.: Integer programming formulations of discrete hub location problems. Eur. J. Oper. Res. **72**, 387–405 (1994)
7. Kara, B.Y., Tansel, B.C.: The single-assignment hub covering problem: models and linearizations. J. Oper. Res. Soc. **54**(1), 59–64 (2003)
8. Wagner, B.: Model formulations for hub covering problems. J. Oper. Res. Soc. **59**, 932–938 (2008)
9. Hwang, Y.H., Lee, Y.H.: Uncapacitated single allocation p-hub maximal covering problem. Comput. Ind. Eng. **63**, 382–389 (2012)
10. Peker, M., Kara, B.: The p-hub maximal covering problem and extensions for gradual decay functions. Omega **54**, 158–172 (2015)
11. Alumur, A., Kara, B.Y.: Network hub location problems: the state of the art. Eur. J. Oper. Res. **190**, 1–21 (2008)
12. Farahani, R.Z., Hekmatfar, M., Arabani, A.B., Nikbakhsh, E.: Hub location problem: A review of models, classification, solution techniques, and applications. Comput. Ind. Eng. **64**, 1096–1109 (2013)
13. Kirkpatrick, S., Gelatt, C.D., Vecchi, M.P.: Optimization by simulated annealing. Science **220**, 671–680 (1983)

Path-Relinking for Fire Station Location

Titima Srianan and Ornurai Sangsawang[✉]

King Mongkut's University of Technology North Bangkok,
129 Moo 21 Neonhom, Mueng, Prachinburi, Thailand
titima.srianan@gmail.com, ornurai.s@fitm.kmutnb.ac.th

Abstract. In this paper, we propose the path relinking to solve the maximal covering location problem in which the demand points correspond to number of employees in industrial factories in Rayong, a province on the east coast of the Gulf of Thailand. According to the statistics for the fire in Thailand from 2012–2016; fire events tended upwards continuously from 2012 at 58.15 percent. Studying data on the fire to factories over the country in 2014–2015, Rayong province was high ranked fire occurrences which result in massive destruction. Path relinking algorithms are compared on the real-life instances to illustrate the efficiency of the proposed method.

Keywords: Path relinking · Maximal covering location problem
Fire stations

1 Introduction

Disaster can be classified based on the formation into 2 types as the natural disaster; flood, storm, fire, occupational hazard, drought, avalanche, plague; and the man-made disaster, including traffic hazard, abnormally cold weather, civil disorder, electricity hazard, hazardous material and technological disaster.

Fire is a disaster caused by both nature and humanity in a form of negligent fire and consequently spread out making the tremendous loss of property and life as well as the overall economic losses.

In Thailand, 2012–2016. A total fire incidents 6,051 times practically occurred in many provinces of Thailand. This made damage to 49,438 people as 266 dead and 1,045 injured, furthermore, 344 factories and 41 cases of department stores and buildings resulting in the total damage value of 3,403,211,372 Baht. Considering a tendency to fire among the factories and department stores and buildings, there is a tendency for fires, particularly the factories with 58.51 percent increase on average as shown Fig. 1.

Please note that the AISC Editorial assumes that all authors have used the western naming convention, with given names preceding surnames. This determines the structure of the names in the running heads and the author index.

© Springer Nature Switzerland AG 2019
P. Vasant et al. (Eds.): ICO 2018, AISC 866, pp. 531–538, 2019.
https://doi.org/10.1007/978-3-030-00979-3_56

Number of fires

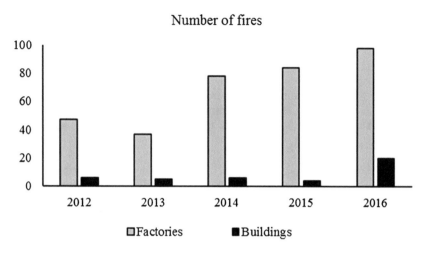

Fig. 1. Diagram of number of damage to construction year 2012–2016 (Source: Disaster relief bureau department of disaster prevention and mitigation, 2017)

Rayong province is a main conglomeration of different industries in Thailand. There are 8 developed industrial estates in the province which hold the rank 1 in the country which include the large petrochemical plants. (Source: Industrial Estate Authority of Thailand). From the statistic on factories fire in the country, 2014–2015, the province has ranked as the top 5 in frequency of fire. Furthermore, the number of industrial plants continually increased from 2013 to 2017.

The propose of this research is to locate fire stations with maximal covering factories in two industrial cities in Rayong. In order to reduce losses in large manufacturing sector with large number of workers, we consider number of employees in each factory as demands. The path relinking algorithms are proposed to solve the Maximal Covering Location Problem, MCLP. To locate fire stations, the alternative fire locations comprise fire operating services, police stations and bus terminals. The fire vehicle is specified to leave from parking place to an incident scene within the standard time, 8 min or 8-km radius.

The Maximal Covering Location Problem, MCLP was analyzed that a location can provide a service within a given or covered radius Richard and Charles [1]. Later, MCLP was broadly used to study various location problems. The hub covering location problem was firstly proposed by O'kelly [2]. The problem types were categorized as hub covering location problem with minimum cost, focusing on economics and the hub covering location problem based on the client's demand focusing on the level of service Campbell [3]. It was applied to determine an appropriate location of emergency medical service unit by considering the severity of the patient's condition for finding the optimal parking location. Regarding the case study area, it was found that the improvement of the emergency medical service location according to the newly proposed mathematical model resulted in the better service covering the patient's demand as percent increased Ployphun and Sunarin [4]. As a consequence of improving the emergency medical service system, the new location by calculations covered the call service area lowering the call from the client than the current location, but the higher service competence in flood condition Neeranuch et al. [5]. Furthermore, it was found that the man-made disasters had increased exponentially. Thus, the facility location problem became the required approach to deal with emergency humanitarian logistical problems Chawis et al. [6], Alan [7] indicated the importance of re-planning and re-evaluating the system for service expansion and planning strategies in order to solve maximal coverage location problem for fire prevention. A model was created to support fire station in California, which 9 existing fire stations unmet demands. As the non-systematic design required 18 fire stations for 80% demand coverage, by contrast, the systematic design gave only 16 stations providing 80.26% demand coverage. Philippe et al. [8] utilized the supportive application for making a decision on fire station placement with managing emergency incidents in Belgium consequence. This analysis completely conducted by the software, MapInfo GIS, including 3 sections as; first section incident identification, second section parameter specification, and third section a click to analyze dealing problems, find the optimal option. This application had been used to revitalize the national affairs, fire station initiated since 2010.

Regarding Part Relinking algorithm, it was a effective metaheuristics to find out the solutions originally proposed by Glover and Laguna [9]. This technique aimed at exploring "Path" among the given solution sets (2 sets, in general) obtaining a new elite solution. Afterward, it was applied with AP data for solving the uncapacitated multiple allocation hub location problem Pe'rez [10]. A design of Path Relinking applying in the hub airport location of Chinese aerial freight flows was proposed by Qu and Weng [11]. The devised methods for finding the solution of hub high-speed rail station location was proposed into 2 algorithm types which were the Genetic algorithm and Path Relinking Algorithm. By comparison the analysis results between both methods, the Path Relinking Algorithm technique provided the optimal solution Titima et al. [12]. Moreover, designing transport network with Tabu search, the Path Relinking method was added to Tabu search method revealing that Path Relinking method improved and increased the efficiency of exploring the solutions [13].

2   Maximal Covering Location Problem

The mathematical model is used mathematically to solve the maximal covering location problem proposed by (Richard Church, Charles ReVelle, 1974) [2]. Here the characteristics are.

$$\text{Maximize} \quad \sum_i w_i Z_i \tag{1}$$

$$\text{Subject to} \quad \sum_{j \in N_i} X_j \geq Z_i \quad ; \forall i \tag{2}$$

$$\sum_j X_j = P \tag{3}$$

$$X_j \in \{0,1\} \quad ; \forall_j \tag{4}$$

$$Z_i \in \{0,1\} \quad ; \forall i \tag{5}$$

Where

w_i represents population of zone i

Z_i represents factory i

X_j represents fire station j

N_i represents set of location in coverage of zone i

P represents number of fire stations.

The objective is to provide maximized coverage of the industrial factory population in Rayong for each fire station. Constraints (1) ensures that the populations in the area i are covered or serviced by selecting location j as the fire station. For Constraints (2), limitation on the number of opened locations is equal to P. Constraints (3) numerical limit, 1 if the fire station j is located and zero, otherwise. Constraints (4) numerical limit, 1 when industrial factory j is covered within the restricted distance, and zero otherwise.

3 Path Relinking for the Maximal Covering Location Problem

A method for finding the solution proposed by Fred Glover and Manuel Laguna (1993) [11]. It proposes investigative searching "Path" among local optima in order to intensify and diversify the search space. This research employs path relinking without random hub and Path Relinking with random hub techniques.

In Fig. 2, the procedure of a path relinking algorithm for the maximal covering location problem is shown. We firstly generate two elite solutions, the initial solution X' and the guiding solution X''. Then, indicates different values between the X' and X''. Then, for each pair of position i of elite solutions (X' and X'') a path is built to connect to the guiding solution. At the position i of X' and X'', a random station is also inserted to generate a new solution and to expand the search space. If the new solution has a better coverage than the current incumbent, then the incumbent and its objective are updated.

The procedure of the path relinking is followed by these steps: Firstly, select two elite solutions from the pool of population to be the initial solution X'' and the guiding solution X''. Build a path by replacing a different chromosome between X' and X'' in each position to be the solution C_1 and C_2 as shown in Fig. 3. For the path relinking with a random hub, an alternative station is randomly inserted at the same position of the chromosome to increase the search space. This procedure is repeated until no different value between the initial solution X'' and the guiding solution X''. Then, update the elite set of solutions until no further improvement is found.

```
Objective function f(X), X = (X₁, X₂, ... Xₙ)
Initialize population P;
Select two elite solutions X' and X''
Identify difference between X' and X''
repeat
        while ( |Δ(X', X'')| > 0 ) do
        for each ( i < P) do
                        Pick an index position i of (X', X'')
                        Build a path from X' to X''
                        Apply a random hub at position i of (X', X'')
                        If ( f(S) > f(S*) ) then
                        S* = S;
                        i = i + 1;
                end for
        end while
update the pool of elite solutions
```

Fig. 2. Pseudo-code for path relinking with a random hub

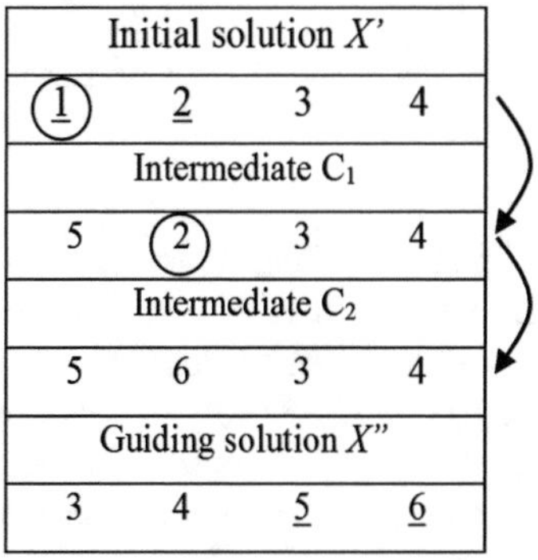

Fig. 3. Relinking path between X' and X''

4   Computational Results

This research carried out a comparative experiment to evaluate the efficiency of Path Relinking without random hub and Path Relinking with random hub. They were applied to solve the maximal covering location problem for locating fire stations in two cities, Amphoe Pluak Daeng and Nikhom Phatthana in Rayong province. We take into account the number of employees in 175 industrial factories as the demand points. In this research, we consider alternative fire stations including the current fire stations (node 1, 2, 3, 6), police stations (node 4 and 8) and bus terminals (node 5 and 7). To evaluate coverage rate, we consider different number of hubs, 3, 4 and 5 nodes.

Table 1. Population coverage of current fire stations

		Current fire stations	
P	Stations	Coverage (employees)	Covering (%)
4	1 2 3 6	45,152	92.18

Table 1 shows the population coverage of current fire stations in two cities. Fire department of two cities currently operates four fire stations (1, 2, 3 and 6) which covered 45,152 employees or 92.18%.

Table 2 reports the results on the real world instance of both algorithms. From the results, the path relinking with random hub outperforms path relinking without random hub in terms of the coverage functions.

Table 2. Percentage of coverage of the PR and the PR with random hub

	PR			PR with random hub		
P	Stations	Coverage (employees)	Covering (%)	Stations	Coverage (employees)	Covering (%)
3	1 2 3	39,396	81.59	1 2 6	42,332	86.43
4	2 3 4 8	40,841	83.38	1 2 5 6	46,432	94.80
5	1 2 3 4 7	45,559	93.02	1 2 3 5 6	46,432	94.80

In case of P = 3, the path relinking algorithm results in maximal covered employees by stations 1, 2 and 3 as total coverage of 39,396 employees or 81.59. Meanwhile, the opened fire stations obtained from the path relinking with random hub are 1, 2 and 6 with total coverage of 42,332 employees or 86.43%. In case the value of P = 4, the path relinking algorithm results in maximal covered employees by stations 2, 3, 4 and 8 as total coverage of 40,841 employees or 83.38%. The opened fire stations obtained from the path relinking with random hub are 1, 2, 5 and 6 with total coverage of 46,432 employees or 94.80%.

For P = 5, the solution obtained from the path relinking with random hub outperforms that of the path relinking without random hub in terms of percentage of coverage. From the results of the path relinking with random hub, locating 4 or 5 fire stations reaches the same percentage of coverage demands. Figure 4 shows coverage of demands for locating the proposed fire stations with P = 4 (1, 2, 5 and 6).

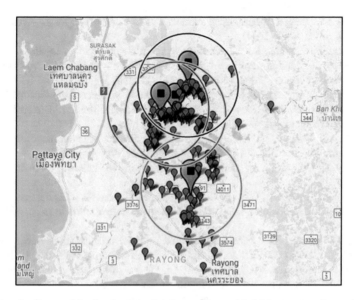

Fig. 4. Geographic distribution of 4 fire stations with 94.8 percent of covering

5 Conclusion

This research proposed locating fire stations to cover employees' in factories of two cities, Amphoe Pluak Daeng and Nikhom Phatthana in Rayong province. The performance of the path relinking algorithms is investigated to solve the maximal covering location problem. We consider the covered distance within 8 km to guarantee effective response. Candidate fire stations take into account the current local fire stations and high population density area such as bus terminals and police stations.

By comparison, the path relinking without random hub and the path relinking with random hub, the results found that the path relinking with random hub outperforms the path relinking without random hub. The results obtained from the path relinking with random hub can improve demands covering compared with the current location of fire stations. Future works could be enhanced performance of the algorithms with different relinking strategies and compared with various metaheuristics on this problem. It may be possible to consider realistic constraints such as partial coverage and road conditions in different time.

Acknowledgements. This research was funded by King Mongkut's University of Technology North Bangkok. Contract No. KMUTNB-61-GOV-01-64.

References

1. Church, R., Velle, C.R.: The maximal covering location problem. Pap. Reg. Sci. **32**(1), 101–118 (1974)
2. O'kelly, M.E.: A quadratic integer program for the location of interacting hub facilities. Eur. J. Oper. Res. **32**(3), 393–404 (1986)

3. Campbell, J.F.: Integer programming formulations of discrete hub location problems. Eur. J. Oper. Res. **72**, 1–19 (1994)

4. Srikijakarn, P., Chanta, S.: Determining appropriate emergency medical service unit locations by considering severity of patient's condition: a case study in Muang district, Prachin Buri province. J. KMUTNB **25**(2), 243–254 (2015)

5. Lomnak, N., Limsuwan, T., Samorngarm C., Chanta, S..: Improving emergency medical service: a case study Chao Praya Abhaibhubejhr Hospital. In: The 2015 Operations Research Network of Thailand, pp. 42—47. Bangkok (2015)

6. Boonmee, C., Arimura, M., Asada, T.: Facility location optimization model for emergency humanitarian logistics. Int. J. Disaster Risk Reduct. **24**, 485–498 (2017)

7. Murray, A.T.: Optimising the spatial location of urban fire stations. Fire Saf. J. **62**, 64–71 (2013)

8. Chevalier, P., Thomas, I., Geraets, D., Goetghebeur, E., Janssens, O., Peeters, D., Plastria, F.: Locating fire stations: an integrated approach for Belgium. Socio-Econ. Plan. Sci. **46**, 173–182 (2012)

9. Glover, F., Laguna, M.: Tabu search, modern heuristic techniques for combinatorial problems. In: Reeves, C. (ed.) pp. 70–150. Blackwell Scientific Publishing, Oxford (1993)

10. Pérez, M.P., Rodríguez, F.A., Vega, J.M.: On the use of the path relinking for the p-hub median problem. Lecture Notes in Computer Science, pp. 155–164 (2004)

11. Qu, B., Weng, K.: Path relinking approach for multiple allocation hub maximal covering problem. Comput. Math Appl. **57**, 1890–1894 (2009)

12. Srianan, T., Sangsawang, O., Chanta, S.: Path-relinking for high-speed rail station location. In: The 2016 operations research network of Thailand, pp. 259–265. Nakorn Ratchasima (2017)

13. Nguyen, V.P., Prins, C., Prodhon, C.: A multi-start iterated local search with tabu list and path relinking for the two-echelon location-routing problem. Eng. Appl. Artif. Intell. **25**(1), 56–71 (2012)

Modeling and Optimization of Flexible Manufacturing Systems: A Stochastic Approach

Gilberto Pérez Lechuga[1](✉) and Francisco Martínez Sánchez[2]

[1] División de Investigación, Desarrollo e Innovación, Universidad Autónoma
del Estado de Hidalgo, Pachuca, Mexico
glechuga2004@hotmail.com
[2] Escuela Superior de Apan, Universidad Autónoma del Estado de Hidalgo,
Pachuca, Mexico
marzan67@gmail.com

Abstract. This document describes the development an analysis of the main optimization methods used in flexible manufacturing systems (FMS) problems. The results are obtained from a random sample of more than one hundred documents published from 1986 to 2018 dedicated to the optimization of FMS. These are classified by their applications in the most significant fields of this area in analytical methods and heuristic methods. The discussion is based on the importance of this branch of engineering in the economic activity of a country and is motivation to continue doing research in it. The analysis also addresses some aspects of computational complexity found in ordinary optimization as well as the most common NP-hard problems of the FMS. The statistical results obtained are presented and the virtual future of the FMS is projected as well as the new challenges and opportunities to venture into this important field of human activity.

Keywords: Mathematical programming · Heuristics
Computational complexity · Analytical methods

1 Introduction

A flexible manufacturing system (FMS) is defined as the integration of manufacturing or assembly processes, material flow and computer communications and control. The objective is to let the production floor respond rapidly and economically to changes in system operations. The concept of flexibility in FMS has attained significant transcendence in meeting the challenges for a variety of products with shorter lead-times, together with higher productivity and quality [64].

The manufacturing industry is the global foundation of the economic power of the richest and most powerful nations in the world. These are the ones that control most of the world production of manufacturing technology. That is, it is not enough to have factories and produce more goods, but you have to know how to build the machinery that makes merchandise. The key is to know how to build the "means of production" [65].

© Springer Nature Switzerland AG 2019
P. Vasant et al. (Eds.): ICO 2018, AISC 866, pp. 539–546, 2019.
https://doi.org/10.1007/978-3-030-00979-3_57

This document discusses the fields of manufacturing and related areas that, due to their importance, have received great attention in terms of mathematical modeling and optimization. This proposal is not intended to be a review of the state of the art in terms of mathematical modeling of manufacturing systems, rather, it presents a general overview of the methods used to address these problems from a quantitative perspective for engineers, practitioners, entrepreneurs, mathematicians and operations researchers.

2   A Classification of Problems in Manufacturing

In practice there are hundreds of classifications and sub-classifications of manufacturing. Most of them focus on the processes of physical-chemical transformation that materials undergo to be transformed into a finished product by adding value to the process. Others focus on the administrative aspects of planning control and logistics of operations in the processes. A classification of manufacturing problems is [14, 43]:

1. *Design problems of the operation*: refers to the design, configuration, operation and control of the electrical, electronic, software and hardware systems involved in the process. The most important are: (a) Design and planning, (b) Scheduling and control, (c) Cellular manufacturing, (d) Automatic systems. The most important references for planning problems are [4, 8, 22, 30, 39, 53, 75, 78, 96]; for scheduling and control [5, 10, 24, 60, 93]; for cellular manufacturing [1, 2, 7, 15, 25, 28, 37, 50, 51]; for automatic systems [6, 14, 18, 26, 44, 48, 98].
2. *Modern manufacturing problems*: Modern manufacturing refers to cutting-edge technology that has impacted in recent times the traditional methods and procedures with which it has been working. The most important is: (a) Additive manufacturing: The most important references for this field are [13, 35, 54, 55, 68, 70, 71]
3. *Planning, control, logistics and operation problems*: This is the soft focus of the SMF management and organization problem. The most important are: (a) Enterprise resource planning (ERP), (b) Material and requirement planning (MRP), (c) Manufacturing resource planning (MRP-II), (d) *Lean* manufacturing (Just in Time manufacturing), (e) Agile manufacturing, (f) Inventory management, (g) Planning, control and integration of production, (h) Supply chain management. The most important references for ERP are [20, 27, 40, 41, 46, 80]; for MRP [36, 38, 57, 58, 62, 79, 89, 91, 92]; for MRP II [21, 26, 88]; for planning control and integration of production [12, 17, 45, 57, 63, 67, 77, 85]; for supply chain management [42, 59, 66, 74, 81]; for lean manufacturing [9, 16, 31, 61, 73, 76, 81, 82, 94, 95]; for agile manufacturing [3, 33, 34, 43, 69]; for inventory management [11, 19, 23, 30, 52, 82, 83, 88, 99];

3   Results of the Analysis

The results obtained show that only 16.94% of the authors use heuristics in the solution of their optimization instances for manufacturing problems. The most representative percentages of the analytical methods are: Mixed integer linear programming 10.20%,

Fuzzy techniques 8.16%, Nonlinear programming 7.14%, Simulation 7.14%, Stochastic programming 7.14%, Linear programming 5.10%, Finite element method 3.06%, Multicriteria programming 3.06%. The rest is shared among other methods. These suggest that most authors prefer to use traditional optimization methods for the analysis of manufacturing systems. In the same way, the heuristic methods were distributed as follows: Genetic algorithms 30%, Simulated annealing 25%, Cuckoo search 15%, Tabu search 10%, Swarm intelligence 10%, Bee based algorithms 5%, Flower pollination algorithms 5%. In relation to the heuristic, and according to the Pareto of the sample, 80% of the sampled works apply Genetic algorithms, Simulated annealing, Cuckoo search, and Tabu search as main tools in the optimization of their models. The sampling done also indicates the great dispersion of methods and techniques currently available to solve an instance from different perspectives. Finally, and although there is a great attraction for traditional methods of linear programming and its variants, fuzzy techniques, and traditional approaches to stochastic programming, undoubtedly, heuristic methods are a tool that is here to stay.

4 Conclusions

In this proposal an approximation to the frequency of use and typology of methods used in the MFS has been elaborated. These results not only summarize the current research activities in this field (e.g. main research directions, algorithms analytical and heuristic) employed, but also indicate existing deficiencies and potential research directions through proposing a research agenda. Findings of this analysis can be used as the basis for future research in industry.

The results obtained in this study suggest that even with the great development in terms of heuristics, traditional analytical algorithms are still widely used in the optimization of instances associated with flexible manufacturing problems. The development and application of new heuristic techniques to solve models of great computational complexity is a challenge that researchers of this important branch must assume. It is interesting to note that in none of the sample values obtained is artificial intelligence used. In several real problems associated with routing, plant location, sequencing of operations, logistics and supply chain, it is common to find problems of great size and computational complexity. For example, in the case of traveling salesman problem the use of heuristics as ant's colony and/or greedy algorithms optimally solve combinatorial problems of great scale, hence the importance of its use.

A great variety of problems associated with FMS belong to the group of problems called NP-hard, so we cannot expect to be able to solve instances of practical problems of arbitrary size to optimum. Depending on the size of an instance or depending on the available CPU time, we should often be satisfied with the computation of approximate solutions and it may sometimes be impossible to evaluate the actual quality of the approximate solutions. There is a large number and variety of difficult problems, which arise in practice and need to be solved efficiently, and this has promoted the development of efficient procedures in an attempt to find good solutions, even if they are not optimal. Such methods, in which the speed of the process is as important as the quality of the solution obtained are called heuristic or approximate algorithm [47].

References

1. Ah-kioon, S., Bulgak, A.A., Bektas, T.: Integrated cellular manufacturing systems design with production planning and dynamic system reconfiguration. Eur. J. od Oper. Res. **192**(2), 414–428 (2009)
2. Albadawi, Z., Bashir, H.A., Chen, M.: A mathematical approach for the formation of manufacturing cells. Comput. Ind. Eng. **48**(1) (2005)
3. Al-Tahat, M.D., Bataineh, K.M.: Statistical analyses and modeling of the implementation of agile manufacturing tactics in industrial firms (2012)
4. Sadrazadeh, A.: A genetic algorithm with the heuristic procedure to solve the multi-line layout problem. Comput. Ind. Eng. 1055–1064 (2012)
5. Andrade, A., Santoro, M., De Tomi, G.: Mathematical model and supporting algorithm to aid the sequencing and scheduling of mining with loading equipment allocation. Rem: Rev. Esc. Minas **67**(4), 379–387 (2014). Ouro Preto
6. Andreas Hees, A., Reinhart G.: Approach for production planning in reconfigurable manufacturing systems. In: 9th CIRP Conference on Intelligent Computation in Manufacturing Engineering—CIRP ICME 2014 (2015)
7. Ariafara, S., Ismail, N.: An improved algorithm for layout design in cellular manufacturing systems. J. Manuf. Syst. **28**(4), 132–139 (2009)
8. Arostegui, M., Kadipasaoglu, S., Khumawala, B.: An empirical comparison of Tabu search, simulated annealing and genetic algorithms for facilities location problems. J. Prod. Econ. **103**, 742–754 (2006)
9. Barla, S.B.: A case study of supplier selection for lean supply by using a mathematical model. Logistics Inf. Manag. **16**(6), 451–459 (2003)
10. Baruah, S.: A scheduling model inspired by control theory. In: Proceedings of the 6th International Real-Time Scheduling Open Problems Seminar (RTSOPS) (2015)
11. Braglia, M., Multi-attribute classification method for spare parts inventory management. J. Qual. Maint. Eng. **10**(1) (2004)
12. Brah, S., Hunsucker, J., Shah, J.: Mathematical modeling of scheduling problems. J. Inf. Optim. Sci. **12** (1991)
13. Brenner, S.C., Carstensen, C.: Finite Element Methods. Wiley Online Library (2017)
14. Burnhual, S., Deb, S.: Scheduling optimization of flexible manufacturing systems using cuckoo search-based approach. Int. J. Adv. Manuf. Technol. **64**(5–8) (2013)
15. Buzacott, J.A., Shanthikumar, J.G.: Stochastic models of manufacturing systems. Prentice Hall EEUU, New York (1993)
16. Cao, D., Chen, M.: A mixed integer programming model for a two line CONWIP-based production and assembly system. Int. J. Prod. Econ. **95**(3), 317–326 (2005)
17. Cárdenas, G.E., Pérez-Lechuga, G., Tuoh-Mora, J.C., Medina-Marín, J.: Obtaining the optimal short-term hydrothermal coordination scheduling: a stochastic view point. Appl. Math. Comput. Sci. **3**(4), 361–379 (2012). ISSN 0976–1586
18. Cassandras, C.G., Lafortune, S.: Introduction to Discrete Event Systems. Springer, New York (2010)
19. Chang, C.-T., Teng, J-.T., Goyal, S.K.: Inventory lot-size models under trade credits: a review. Asia Pac. J. Oper. Res. **25**(01), 89–112 (2008)
20. Chin, C., Chen, W., Chien, F., Wang, J.J.: An AHP-based approach to ERP system selection. Int. J. Prod. Econ. **96**(1, 18), 47–62 (2005)
21. Choudhary, S.K.: Study about the types of information technology service for supply chain. In: Proceedings of the World Congress on Engineering and Computer Science 2016 Vol II WCECS 2016, 19–21 October 2016, San Francisco, USA (2016)

22. De Carlo, F., Arlo, M.A. et.al.: Layout design for a low capacity manufacturing line: a case study. Int. J. Eng. Bus. (2013)
23. Ekşioğlu, S.D., Acharya, A., Leightley, L.E., Arora, S.: Analyzing the design and management of biomass-to-biorefinery supply chain. Comput. Ind. Eng. **57**(4), 1342–1352 (2009)
24. Faizrahnemoon, M.: Mathematical modelling of the scheduling of a production line at SKF, Master's Thesis, University of Gothenburg (2012)
25. Fantahun, M.D., Chen, M.: A comprehensive mathematical model for the design of cellular manufacturing systems. Int. J. Prod. Econ. **103**(2), 767–783 (2006)
26. Fanti, M.P., Zhou, M.: Deadlock control methods in automatic manufacturing systems. IEEE Trans. Syst. **34**(1) (2004)
27. Ghapanchi, A., Jafarzadeh, M.H., Khakbaz, M.-H.: Fuzzy-data envelopment analysis approach to enterprise resource planning system analysis and selection. Int. J. Inform. Syst. Change Manage. **3**(2) (2008)
28. Ghosha, T., Senguptaa, S., Chattopadhyayb, M., Dana, P.K.: Meta-heuristics in cellular manufacturing: a state-of-the-art review. Int. J. Ind. Eng. Comput. (2011)
29. Ghosh, I.: An Intelligent hybrid multi criteria decision making technique to solve a plant layout problem. Int. J. Ind. Eng. Res. Dev. (IJIERD) **5**(3), 13–23 (2014)
30. Gumus, A.T., Guneri, A.F.: A multi-echelon inventory management framework for stochastic and fuzzy supply chains. Expert Syst. Appl. **36**(3), 5565–5575 (2009). Part 1
31. Hao, Q., Shen, W.: Implementing a hybrid simulation model for a Kanban-based material handling system. Manufacturing **24**(5), 635–646 (2008)
32. Hong, M., Payandeh, S., Gruver, W.A.: Modeling and analysis of flexible fixturing systems for agile manufacturing. In: IEEE International Conference on Systems, Man and Cybernetics. Information Intelligence and Systems (1996)
33. How does additive manufacturing work? https://www.ge.com/additive/additive-manufacturing
34. Jarimo, T., Pulkkinen, U.: A multi-criteria mathematical programming model for agile virtual organization creation. https://link.springer.com/content/pdf/10.1007%2F0-387-29360-4_13.pdf
35. Jin. J.M.: The Finite Element Method in Electromagnetic, 3rd edn. Wiley (2014)
36. Karsak, E.E., Özogul, C.O.: An integrated decision-making approach for ERP system selection. Expert Syst. Appl. **36**(1), 660–667 (2009)
37. Kia, R., Baboli, A., Javadian, N., Tavakkoli-Moghaddam, R., Kazemi, M.: Solving a group layout design model of a dynamic cellular manufacturing system with alternative process routings, lot splitting and flexible reconfiguration by simulated annealing. Comput. Oper. Res. (2012)
38. Kilic, H.S., Zaim, S.M., Delen, D.: Development of a hybrid methodology for ERP system selection: the case of Turkish Airlines. Decis. Support Syst. **66**, 82–92 (2014)
39. Khusna, D., Zawiah, S., Jamasri, J., Aoyama H.: A proposed study on facility planning and design in manufacturing process. In: Proceedings of the International Multiconference of Engineers and Computers Scientists 2010, vol. 3, Hong Kong (2010)
40. Lall, V., Teyarachakul, S.: Enterprise resource planning (ERP) system selection: a data envelopment anaysis (DEA) approach. J. Comput. Inf. Syst. **47**(1) 2006 (2016)
41. Lenny Koh, S.C., Mike Simpson, M.: Could enterprise resource planning create a competitive advantage for small businesses? Benchmarking Int. J. **14**(1), 59–76 (2007)
42. Lee, Y.H., Golinska, P., Wu, J.Z.: Mathematical models for supply chain management. Mathematical Problems in Engineering, vol. 2016, p. 4. Hindawi Publishing Corporation (2016). Article ID 6167290

43. Li, P., Shi, K., Zhang, J.: Modeling Agile manufacturing cell using object-oriented timed petri net. http://keshi.ubiwna.org/paper/Modeling%20Agile%20Manufacturing%20Cell%20using%20Object-Oriented%20Timed%20Petri%20net.pdf

44. Li, Z., Wu, N., Zhou, M.: Deadlock control of automatic manufacturing systems based on Petri Nets—A literature review. IEEE Trans. Syst. **42**(4) (2012)

45. Li X., Guo, S., Liu, Y., Du, B., Wang, L.: A production planning model for make-to-order foundry flow shop with capacity constraint. Math. Probl. Eng. **2017** (2017). Article ID 6315613

46. Liao, X., Li, Y., Lu, B.: A model for selecting an ERP system based on linguistic information processing. Inf. Syst. **32**(7), 1005–1017 (2007)

47. Liao, Y., Deschamps, F., Loures, E.F., Ramos, L.F.P.: Past, present and future of Industry 4.0- a systematic literature review and research agenda proposal. Int. J. Prod. Res. **55**(12) 2017 (2017)

48. López-Ortega, O., Pérez-Lechuga, G.: An express meta model for sharing and exchanging flexible manufacturing resources. In: World Congress Manufacturing, Rochester Institute of Technology, Rochester, New York, USA (2001)

49. Madhav, P.A.: Mathematical Modelling of Flexible Manufacturing Systems. University of Windsor (1986)

50. Madavhi, I., A., Paydar, Solymanpur, M., Heidarzade, A.: Genetic algorithm approach for solving a cell formation problem in cellular manufacturing. Expert Syst. Appl. (2009)

51. Madavhi, I., Aalaei, A., Paydar, M.M., Solymanpur, M.: Designing a mathematical model for dynamic cellular manufacturing systems considering production planning and worker assignment (2010)

52. Mandal, N.K., Roy, T.K., Maitic, M.: Multi-objective fuzzy inventory model with three constraints: a geometric programming approach. Fuzzy Sets Syst. **150**(1), 87–106 (2005)

53. Maniya, K., Bhatt, M.G.: An alternative multiple attribute decision making methodology for solving optimal facility layout design selection problems. Comput. Indus. Eng. **61**, 542–549 (2011)

54. Manzhirov, A.V., Lychev, S.A.: Mathematical modeling of additive manufacturing technologies. In: Proceedings of the World Congress on Engineering 2014, WCE 2014, vol. II, London, UK (2014)

55. Manzhirov, A.V.: Mechanical design of viscoelastic parts fabricated using additive manufacturing technologies. In: 2015 Proceedings of the World Congress on Engineering, WCE 2015, vol. II, London, UK (2015)

56. Method for controlled optimization of enterprise planning models. https://translate.google.com/?hl=es#es/en/Method%20for%20controlled%20optimization%20of%20enterprise%20planning%20models

57. Mula, J., Poler, R., García, J.P.: MRP with flexible constraints: a fuzzy mathematical programming approach. Fuzzy Sets Syst. **157**(1), 74–97 (2006)

58. Mula, J., Poler, R., García, J.P.: Models for production planning under uncertainty: a review. Int. J. Prod. Econ. **103**(1), 271–285 (2006)

59. Mula, J., Peidro, D., Diaz-Madroñero, M., Vicens, E.: Mathematical programming models for supply chain production and transport planning. Eur. J. Oper. Res. **204**(3), 377–390 (2010)

60. Norouzi, G., Heydari, M., Noory, S., Bagherpour, M.: Developing a mathematical model for scheduling and determining success probability of research projects considering complex-fuzzy networks. J. Appl. Math. **2015** (2015). Article ID 809216

61. Omar, M., Sarker, R., Othman, W.A.M.: A just-in-time three-level integrated manufacturing system for linearly time-varying demand process. Appl. Math. Model. **37**(2013), 1275–1281 (2013)

62. Pérez, L.G., Gress, E.S.H., Karelyn, A., Orán, M.G.M.: Some efficiency measures in the operation of flexible manufacturing systems: a stochastic approach. In: Proceedings of the 2009 SIAM Conference on "Mathematics for Industry", SIAM, EEUU, pp. 31–36 (2010)
63. Pérez-Lechuga, G., Flores-Rivera, C., Karelyn, A., Tarasenko, A.: A multi-period stochastic model to link the supply chain production planning in a manufacturing organization, Advances in science and statistics applications. **6**(4) 255–283 (2011). ISSN 0974-6811
64. Pérez, L.G., Pérez, R.E., Venegas, M.F.: Stochastic optimization of manufacture systems by using Markov decision processes. In: Vasant, P., Weber, G., Dieu, V.N. (eds.) Handbook of Research on Modern Optimization Algorithms and Applications in Engineering and Economics, Engineering Science Reference, EEUU. IGI Global (2015)
65. Pérez, L.G., Bujari, A.A., Venegas, M.F.: Economic aspects of the production planning in lines of manufacturing under conditions of uncertainty. In: Ramírez, C.E.C., Venegas, M.F., Herrera, F. (eds.) Modeling of Economic and Financial Phenomena: A Contemporary Vision, Editorial Castdel, S.A. de C.V, Mexico (2017)
66. Pérez-Lechuga, G.: Optimal logistics strategy to distribute medicines in clinics and hospitals. J. Math. Ind. **8**, 2 (2018)
67. Pochet, Y., Wolsey, L.A.: Production Planning by Mixed Integer Programming. Springer, New York (2006)
68. Rai, J.K., Xirouchakis, P.: Finite element method based machining simulation environment for analyzing part errors induced during milling of thin-walled components. Int. J. Mach. Tools Manuf. **48**(6), 629–643 (2008)
69. Raj, S.A., Sudheer, A., Vinodh, S., Anand, G.: A mathematical model to evaluate the role of agility enablers and criteria in a manufacturing environment. Int. J. Prod. Res. **51**(19) (2013)
70. Roberts, I.A., Wang, C.J., Esterlein, R., Stanford, M., Mynors, D.J.: A three-dimensional finite element analysis of the temperature field during laser melting of metal powders in additive layer manufacturing. Int. J. Mach. Tools Manuf. (2009)
71. Rosen, D.W.: Design for additive manufacturing: a method to explore unexplored regions of the design. https://pdfs.semanticscholar.org/101a/8f03ed20725ae71d30537e90f783a398 e86f.pdf
72. Rui, L., Xiaofei, Z., Guiwu, W.: Models for selecting an ERP system with hesitant fuzzy linguistic information. J. Intel. Fuzzy Syst. **26**(5), 2155–2165 (2014)
73. Savsar, M., Jawini, A.N.: Simulation analysis of just-in-time production systems. Int. J. Prod. Econ. **42**(1), 67–78 (1995)
74. Sawik, T.: Scheduling in Supply Chains Using Mixed Integer Programming. Wiley (2011)
75. Shah, K., Ripon, N., Glette, K., et. al.: Adaptive variable neighborhood search for solving multi- objective facility layout problem with unequal area facilities. Swarm Evol. Comput. (2012)
76. Shahabudeen, P., Krishnaiah, K., Narayanan, M.T.: Design of a Two-card dynamic Kanban system using a simulated annealing algorithm. Int. J. Adv. Manuf. Technol. **21**(10–11), 754–759 (2003)
77. Shapiro, J.F.: Mathematical programming models and methods for production planning and scheduling. In: Handbooks in Operations Research and Management Science, vol. 4, pp. 371–443 (1993)
78. Solymanpur, M., Vrat, P., Shankar, R.: Ant colony optimization algorithm to the inter-cell layout problem in cellular manufacturing. Eur. J. Oper. Res. **157**(3), 592–606 (2004)
79. Sun, L., Heragu, S.S., Chen, L., Spearman, M.L., Simulation analysis of a multi-item MRP system based on factorial design. In: 2009 Proceedings of the Winter Simulation Conference (WSC), pp. 2107–2114 (2009)
80. Talluri, S., Narasimhan, R., Nairb, A.: Vendor performance with supply risk: A chance-constrained DEA approach. Int. J. Prod. Econ. **100**(2), 212–222 (2006)

81. Tayur, S., Ganeshan, R., Magazine, M.: Quantitative Models for Chain supply Management. Kluwer's International Series (1999)
82. Thapa, G.: Optimization of just-in-time sequencing problems and supply chain logistics. Mälardalen University Press Dissertations No. 184 (2015). https://www.diva-portal.org/smash/get/diva2:851513/FULLTEXT01.pdf
83. Urek, B., Kaya, O.: A mixed integer nonlinear programming model and heuristic solutions for location, inventory and pricing decisions in a closed loop supply chain. Comput. Oper. Res. **65**, 93–103 (2016)
84. Voss, S., Woodruff, D.: Connecting MRP, MRP II and ERP- supply chain production planning via optimization models, In: Greenberg, H.J. (ed.) Tutorials on Emerging Methodologies and Applications in Operations Research. Springer Science & Business Media (2006)
85. Wang, R.C.H., Liang, T.F.: Application of fuzzy multi-objective linear programming to aggregate production planning. Comput. Ind. Eng. **46**(1), 17–41 (2004)
86. What are automated manufacturing systems? https://connell-ind.com/what-are-automated-manufacturing-systems/
87. What is agile manufacturing? https://translate.google.com/?hl=es#en/es/WHAT%20IS%20AGILE%20MANUFACTURING%3F%20Why%20is%20it%20an%20effective%20strategy%3F
88. Wazed, M.A., Ahmed, S., Nukman, Y.: Mathematical models for multistage production under commonality and uncertainty. http://ebooks.asmedigitalcollection.asme.org/content.aspx?bookid=481&sectionid=38789428
89. Wei, C.C.: Wang, J, J., A comprehensive framework for selecting an ERP system. Int. J. Project Manage. **22**(2), 161–169 (2004)
90. Yadav, A., Jayswal, S.C.: Modelling of flexible manufacturing systems: a review. Int. J. Prod. Res. **56**(7) (2017)
91. Yazgan, H.R., Boran, S., Goztepe, K.: An ERP software selection process with using artificial neural network based on analytic network process approach. Expert Syst. Appl. **36**(5), 9214–9222 (2009)
92. Yazıcı, E., Büyüközkan, G., Baskak, M.: A new extended MILP MRP approach to production planning and its application in the jewelry industry. Math. Probl. Eng. **2016** (2016). Article ID 7915673
93. Yilmaz, O.F., Suleiman, T. (eds.): Handbook of Research on Applied Optimization Methodologies in Manufacturing. IGI-Global (2018)
94. Yousefi, M., Hosseinioun, S.S., Taghikhah, F.: Determining optimum number of kanbans for workstations in a kanban-based manufacturing line using discrete-event simulation: a case of study. In: Proceedings of the 2012 International Conference on Industrial Engineering and Operations Management, Istanbul, Turkey (2012)
95. Zhang, W., Chen, M.: A mathematical programming model for production planning using CONWIP. Int. J. Prod. Res. **39**(12), 2001 (2010)
96. Zhenyuan, J., Xiaohong, L., Wei, W., et al.: Design and implementation of lean facility layout system of a production line. Int. J. Ind. Eng. **18**, 260–269 (2011)
97. Zhou, J., Yang, G., Tang, G.: The mathematical modeling of the mobile recharging facility vehicles' scheduling and its genetic algorithm solution. In: IEEE International Conference on Cyber Technology in Automation, Control, and Intelligent Systems (CYBER) (2015)
98. Zhou, M., Dicesare, F.: Petri Net Synthesis for Discrete Event Control of Manufacturing Systems. Springer Science, New York (1993)
99. Ziukov, S.: A literature review on models of inventory management under uncertainty. Bus. Syst. Econ. **5**(1) (2015)

Monkey Algorithm for Packing Circles
with Binary Variables

Rafael Torres-Escobar[1], José Antonio Marmolejo-Saucedo[2(✉)],
Igor Litvinchev[3], and Pandian Vasant[4]

[1] Faculty of Engineering, Universidad Anáhuac México. North Campus,
Av.Universidad Anáhuac 46, Lomas Anáhuac, 52786 Huixquilucan, Estado de
México, Mexico
rafael.torrese@anahuac.mx
[2] Universidad Panamericana, Facultad de Ingeniería, Augusto Rodin 498, 03920
México, Ciudad de México, Mexico
jmarmolejo@up.edu.mx
[3] Faculty of Mechanical and Electrical Engineering, Nuevo Leon State
University, 66450 Monterrey, NL, Mexico
litvin@ccas.ru
[4] Department of Fundamental and Applied Sciences, Universiti Teknologi
Petronas, Seri Iskandar, Malaysia
pvasant@gmail.com

Abstract. The problem of packing non-congruent circles within a rectangular
container is considered. The objective is to place the maximum number of
circles inside the container such that no circle overlaps with another one. This
problem is known to be NP-Hard. Dealing with these problems efficiently is
difficult, so heuristic-based methods have been used. In this paper the problem
of packing non-congruent circles is solved using the binary version of monkey
algorithm. The proposed algorithm uses a grid for approximating the container
and considering the grid points as potential positions for assigning centers of the
circles. The algorithm consists of five main routines: the climb process, watch-
jump process, repairing process, cooperation process and somersault process.
Numerical results on packing non-congruent circles are presented to demonstrate the efficiency of the proposed approach.

Keywords: Packing problem · Non-congruent circles · Metaheuristic
Binary monkey algorithm · Evolutionary computing

1 Introduction

A packing problem consists of the best arrangement of several objects inside a bounded
area named as a container. The arrangement must satisfy the technological constraints
in order to meet the business objectives.

These objectives mainly concern the maximization of the area occupied. On the
other hand, the technological constraints commonly used in packing problems are
overlapping constraints and domain constraints to ensure that no object is left outside

© Springer Nature Switzerland AG 2019
P. Vasant et al. (Eds.): ICO 2018, AISC 866, pp. 547–559, 2019.
https://doi.org/10.1007/978-3-030-00979-3_58

the container. The shape of the container may vary from a circle to square, to rectangular, etc.

Packing problems can be important components in other studies such as cutting patterns, area coverage, layout planning, loading of vehicles, assignment, sequencing of equipment and supply chain management [1–4].

Circular packing problems are difficult to solve because of their combinatorial nature in the arrangement of such objects. In fact, they are NP-hard combinatorial optimization problems [5]; that is, no procedure is able to exactly solve them in deterministic polynomial time [3].

Some packing models for circular objects are typically formulated as non-convex optimization problems; where the continuous variables are the coordinates of the objects, so they are limited to not finding optimal solutions [6–8]. The non-convexity is mainly provided by nonoverlapping conditions among circles. These conditions typically state that the Euclidean distance separating the centers of the circles is greater than a sum of their radii [9].

In recent years, heuristic methods are being used extensively which combine methods of global search and methods of local exhaustive search of local minima or their approximations [3, 4].

In almost all real-world optimization problems, it is necessary to use a mathematical algorithm that iteratively seeks out the solutions because an analytical solution is rarely available. Therefore, many evolution algorithms such as genetic algorithm, ant algorithm, and particle swarm methodology have been developed and received a great deal of attention in the literature [10].

Evolutionary computation or population-based algorithms are direct search methods, which use only objective function values to drive the search and employ more than one solution at each iteration [4].

This paper addresses the packing problem using a regular grid to approximate the container and the binary version of monkey algorithm to solve the model with noncongruent circles.

The paper is organized as follows. Section 1 describes the model for the approximate packing circles in a rectangular container. Section 2 describes the definition of the Monkey Algorithm for Packing Circles in a Rectangular Container. The numerical results with the binary monkey algorithm are shown in Sect. 3. In Sect. 4 we present our conclusions.

2   The Model

The main idea of this model was first implemented in [11], later a similar approach was used in [12, 13]. This article is a continuation of the work proposed in [14, 15].

Suppose we have non-identical circles C_k of known radius $R_k, k \in K = \{1, 2, \ldots, K\}$. Here we consider the circle as a set of points that are all the same distance R_k (not necessarily Euclidean) from a given point. In what follows, we will use the same notation C_k for the figure bounded by the circle, $C_k = \{z \in R^2 : z - z_{0k} \leq R_k\}$. Denote S_k by the area of C_k.

Let at most M_k circles C_k are available for packing and at least m_k of them must be packed. Denote by $i \in I = \{1, 2, \ldots, n\}$ the node points of a regular grid covering the rectangular container. Let $F \subseteq I$ be the grid points lying on the boundary of the container. Denote by d_{ij} the distance (in the sense of norm used to define the circle) between points i and j of the grid. Define binary variables $x_i^k = 1$ if center of a circle C_k is assigned to the point i; $x_i^k = 0$ otherwise.

In order to the circle C_k assigned to the point i be non-overlapping with other circles being packed, it is necessary that $x_i^k = 0$ for $j \in I, l \in K$, such that $d_{ij} < R_k + R_l$. For fixed i, k let $N_{ij} = \{j, l \neq j, d_{ij} < R_k + R_l\}$. Then the problem of maximizing the area covered by the circles can be stated as follows:

$$max \sum_{i \in I} \sum_{k \in K} S_k^2 x_i^k \tag{1}$$

Subject to

$$\sum_{k \in K} x_i^k \leq 1, i \in I \tag{2}$$

$$R_k x_i^k \leq \min_{j \in F} d_{ij}, i \in I, k \in K, \tag{3}$$

$$x_i^k + x_j^l \leq 1, for \ i \in I; k \in K; (j, l) \in N_{ik}, \tag{4}$$

$$x_i^k \in \{0, 1\}, i \in I, k \in K \tag{5}$$

The Eq. (1) is the objective function, maximizing the area covered by objects. Constraints (2) that at most one center is assigned to any grid point; constraints (3) that the point i can not be a center of the circle C_k if the distance from i to the closest grid point of the boundary is less than R_k; pair-wise constraints (5) guarantee that there is no overlapping between the circles; constraints (6) represent the binary nature of variables.

In (4) R_k is compared with the distance from the center i to grid points of the boundary, i.e. the boundary is represented by its grid points only. Thus, in general constraints (4) do not guarantee that a circle fits into the container. However, if the point of the boundary closest to C_k is a grid point, then (4) ensure that there are no intersections between circles and the boundary.

3 Monkey Algorithm for Packing Circles with Binary Variables

The monkey algorithm (MA) is a new type of swarm intelligence-based algorithm. It was proposed by [10] and is derived from simulation of the mountain-climbing processes of monkeys. It consists of three processes: the climb process, watch–jump process and somersault process. The climb process is designed to gradually improve the objective function value. However, MA will spend considerable computing time searching for local optimal solutions in the climbing process [16]. The watch–jump process can speed up the convergence rate of the algorithm, the purpose of the somersault process is to make monkeys find new search domains to avoid falling into local search. The algorithm has the advantages of simple structure, strong robustness, and not easy falling into local optimal solutions [16].

In this paper, we will address the packing problem for non-congruent circles with the binary version of the monkey algorithm proposed by [16] which incorporates a cooperation process and a greedy strategy. The algorithm improves the calculation accuracy and increases the convergence speed of the algorithm to a certain degree. The numerical experiment results show that the proposed algorithm has good performance in solving the approximate packing problem. It can be an efficient alternative for solving the approximate packing problem.

3.1   Coding Method

In the following, the coding of the algorithm proposed by [16] for the approximate packing problem by using a regular grid of rectangular shape is described. The idea of using a grid to approximate the solutions of a packing problem to a problem of type assignment can be observed in [12–14]. The grid is generated by discretizing the solution space, and it has the shape of a rectangular container. This grid creates a set of points in which circular objects can be assigned with the intention to maximize the space occupied. Given the length and width of a rectangular container denoted by L, W respectively, a number of M points are selected on the length of the container and a number of points N on its width, with this we have a total of $M \times N$ points covering the container as we can see in the Fig. 1.

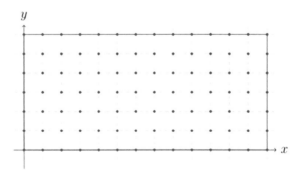

Fig. 1. Example of a grid to approximate the container.

With the $M \times N$ points covering the container, the problem of include or not a circular object in the container is similar to the knapsack problem in which the decision variable is to include or not an object in the knapsack. So in this paper an adaptation of [16] to apply the binary monkeys algorithm using variables that represent the possibility of include an object type k to a point i of the grid that covers the rectangular container is made. We refer to this problem as Approximate Packing Circles in a Rectangular Container. Let S be the population size of monkeys.

For monkey $s \in S$, its position is denoted as a vector $X_s = \left(x_{1s}^1, x_{2s}^1, \ldots, x_{is}^k\right)$ for $i \in I$, $k \in K$ and this position will be employed to express a solution of our problem, where $x_{is}^k \in \{0, 1\}$ for $i \in I$, $k \in K$, $s \in S$, the dimension of the vector is $(|I| \times |K|)$; $x_{is}^k = 1$ indicates the object is included in the container and $x_{is}^k = 0$ indicates it is not.

3.2 Generation of Initial Population

In the algorithm proposed by [16], each element of the possible solution vector (individual) is generated with a probability of 0.5 to include an element or not in the knapsack, however, there exists issues with this approach, the probability of generating infeasible solutions is very high [4]. This is one of the drawbacks of population heuristic methods in the initialization step, so care must be taken to choose a strategy that leads to obtaining feasible initial solutions. In this paper, the generation of the initial population takes place in two steps. The first one is to use the grid points that guarantee at least the inclusion of the smallest object inside the container.

The second step consists in randomly select an integer number between a and b that represent the number of objects that each individual must contain in the initial population. After determining the number of objects that each monkey will have, we select both a point and an object at random, in this way we have included an object in the container, i.e., an element of the solution. In this step, the process is repeated until the desired number of objects is reached. During the inclusion of objects to the solution list, It may be, there exist overlaps between each of the objects. If any overlap occurs, another random point is chosen to remove the overlapping object and place it in the new position, this procedure is shown in Fig. 2.

```
Require: Grid
Ensure: C     //List of objects
procedure InitialPosition(a, b)
        N ← Random Integer between (a, b)
        for i = 1 to N do
            Select a grid point at random
            Select an object Cᵢ at random
            Assign the object Cᵢ to the selected point
            while Cᵢ overlaps with another object in C do
                Select another grid point
            end while
            if object Cᵢ is outside of the container then
                Select the object Cᵢ with minimum radius
            end if
            Assign object Cᵢ to list of circles C
            Remove grid points covered by object Cᵢ
        end for
        return C: List of circles
end procedure
```

Fig. 2. Algorithm for generating the monkey population

3.3   Climbing Process

For the monkey s, its position is $X_s = \left(x_{1s}^1, x_{2s}^1, \ldots, x_{is}^k\right)$ respectively. $f(X_s)$ is the corresponding objective function value. The improved climb process is given as follows [16]:

1. Randomly generate two vectors $\Delta x_i' = \left(\Delta x_{i2}', \Delta x_{i2}', \ldots, \Delta x_{in}'\right)$, and $\Delta x_i'' = \left(\Delta x_{i2}'', \Delta x_{i2}'', \ldots, \Delta x_{in}''\right)$
2. Set $x_{ij}' = \left|x_{ij} - \Delta x_{ij}'\right|$ and $x_{ij}'' = \left|x_{ij} - \Delta x_{ij}''\right|$. Set $X_i' = \left(x_{i1}', x_{i2}', \ldots, x_{in}'\right)$, and $X_i'' = \left(x_{i1}'', x_{i2}'', \ldots, x_{in}''\right)$
3. Calculate $f(X_i')$ and $f(X_i'')$
4. If $f(X_i') > f(X_i'')$ and $f(X_i') > f(X_i)$, set $X_i = X_i'$. If $f(X_i'') > f(X_i')$ and $f(X_i'') > f(X_i)$, set $X_i = X_i''$
5. Repeat steps 1 to 4 until maximum allowable number of iterations has been reached. As we can see in Fig. 3.

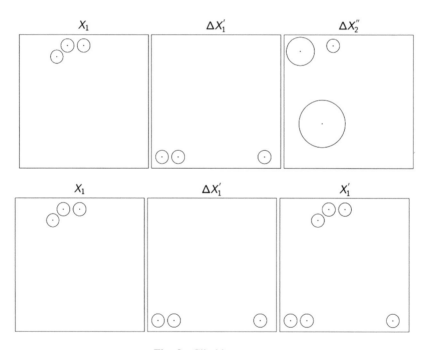

Fig. 3. Climbing process

3.4   Watch-Jump Process

When the monkeys reach the top of the mountains, each monkey will take a look and determine whether there are higher points than the current one within its sight. If yes, it will jump somewhere of the mountain from the current position and then repeat the climb process. For the monkey s, its position is $X_s = \left(x_{1s}^1, x_{2s}^1, \ldots, x_{is}^k\right)$, $s = 1, 2, \ldots, T$. The watch–jump process is given as follows:

1. Randomly generate a real y_i^k in interval $\left[x_{is}^k - b, x_{is}^k + b\right]$, $i = 1, 2, \ldots, n$, $k = 1, 2, \ldots, \ell$
2. Because of the real $y_i^k \in [-1, 2]$, if $y_i^k < p$, set $y_i^k = 0$; otherwise, set $y_i^k = 1$. Here p is an acceptance parameter for the new position.
3. Calculate $f(Y), i = 1, 2, \ldots, T$ respectively.
4. If $f(Y) > f(X_s)$, then $X_s = Y$
5. Repeat the climb process steps 1 to 4 until the maximum allowable number of iterations has been reached (Fig. 4).

3.5 Repair Process

An abnormal encode individual (who does not meet the constraint conditions) may be obtained in solving the problem using the monkey algorithm. Most evolutionary computation methods start with a random population, and then successively apply perturbations to members of the population, until the best possible solution is reached. When generating new solutions, both randomly and by perturbations, the probability of generating overlaps is high [4, 16]. The repair process consists in remove overlapping circles with the smallest radius and then try to accommodate them in free points of the grid, as we can see in Figs. 5 and 6. We need to perform the accommodate procedure (Fig. 5) until a maximum number of trials is reached, otherwise, the object is removed.

```
C: List of non-congruent circles
Cs ← ∅
procedure RemoveOverlaps(C)
    for i = 1 to |C| − 1 do
        for j = i + 1 to |C| do
            if cj overlaps ci then
                Cs ← min(cj, ci)
                Remove min(cj, ci) from C
            end if
        end for
    end for
    return Cs: List of small size circles
end procedure
```

Fig. 4. Algorithm for removing overlaps

```
C: set of circles
procedure Accomodate(C, Trials)
    C' ← RemoveOverlaps(C)
    while Trials do
        Identify free points
        Sort objects by size
        for all c ∈ C' do
            Select random point from free points
            Assign c to a selected point
            Label the selected point as "Occupied"
            C ← c
        end for
        C' ← RomoveOverlaps(C)
    end for
    end while
end procedure
```

Fig. 5. Algorithm for accommodate circles

3.6 Cooperation Process

After the climb process and the watch–jump process, each monkey will arrive at the top mountain in its neighborhood. However, they differ among all the monkeys. The purpose of the cooperation process is to make the monkeys find a better solution by cooperating with the monkey that has the best position. The monkeys will go forward along the direction of the best monkey. This process can speed up the convergence rate. Assume that the optimal position is $X^* = \left(x_i^{k*}\right)$ for $i \in I, \quad k \in K$. For the monkey s, its position is $X_s = \left(x_{1s}^1, x_{2s}^1, \ldots, x_{is}^k\right), I \in I, \quad k \in K, s \in S$. The cooperation process is given as follows:

1. Randomly generate a real number α from the interval $[0, 1]$.
2. If $\alpha < p$, then $y_i^k = x_i^k$, otherwise $y_i^k = x_i^{k*}$ for $i \in I, k \in K$ respectively. The parameter p is a factor of cooperation $p \in (0, 1)$.
3. Update the monkeys' position X_s with Y.

3.7 Somersault Process

Monkeys will reach the maximal mountaintops around their initial points after repetitions of the climb process and the watch–jump process, respectively. To find a higher mountaintop, each monkey will somersault to a new search domain. The new position is not blind to choose, but limited within a certain region, which is determined by the pivot and the somersault interval. The somersault process can effectively prevent monkeys falling into the local search. However, after many iterations, the somersault process may lose efficacy. The monkeys will fall into the local optima domain, and the population diversity will decrease. In the original MA, the monkeys will somersault along the direction pointing to the pivot, which is equal to the barycenter of all monkeys' current positions. Here, we randomly choose a monkey's position as the pivot to replace the center of all monkeys and adopt a new somersault process. For the monkey s its position is $X_s = \left(x_{1s}^1, x_{2s}^1, \ldots, x_{is}^k\right)$, $I \in I$, $k \in K, s \in S$. The improved somersault process is given as follows:

1. Randomly generate real numbers θ from the interval $[c, d]$ (called the somersault interval, which governs the maximum distance that monkeys can somersault).
2. Randomly generate an integer q, $q = 1, 2, \ldots, T$ respectively. Let the location of the monkey q be the somersault pivot.
3. Calculate $y_i^k = x_{iq}^k + \theta\left(x_{iq}^k - x_{is}^k\right)$
4. Set $y_i^k = 0$, if $y_i^k < p$; otherwise, set $y_i^k = 1$. Update the monkey's position X_s with Y. Here $p \in [0, 1]$

After repetitions of the somersault process, monkeys may reach the same domain to make the somersault process lose efficacy. In case of this problem, we set a parameter called "limit" to control monkeys running into the local optima solution. If the global optimal solution is not improved by a predetermined number of trials, the monkeys are abandoned and then reinitialized.

3.8 Stop Condition

Following the climb process, watch–jump process, greedy method optimization process, cooperation process and somersault process, all monkeys are ready for the next iteration. The condition for terminating the Binary Monkey Algorithm iteration could either be when the optimal solution has been found or when a relatively large number of iterations have been reached.

4 Experimental Analysis

In this section we present a numerical study. A rectangular grid of size Δ was determined by the M, N number of equidistant grid points on both the horizontal and vertical edge of the container, respectively. The M, N points were generated with a uniform distribution $U(50, 60)$ for both M and N. It was used a test set of 20 instances for packing the maximal number of non-congruent circular objects into a rectangular container. The values of radii were generated at random with a uniform distribution $U(2, 6)$. The size of the container was generated with a uniform distribution for the Length and Width $(L, W) \sim U(90, 100)$. The models were codified in Python 3.6 on a desktop computer with 4 GB in RAM, Intel Core i3-3470 processor at 3.2 GHz with Windows 10 operating system.

4.1   Numerical Results

We can see six columns in Tables 1 and 2. The column number one represents the number of objects inside the container. The columns number two and three represents the area of the objects inside the container and the area of the container. The column labeled as "Density" is the ratio of the area occupied by objects and the area of the container. The column number nine of the results show the objective value of the fitness function for the monkey algorithm. For the binary monkey algorithm, we use a compound fitness function that comprises three elements: density, the intensity of overlaps and intensity of objects that are outside of the container. The intensity of overlaps follows the idea of [4], and the intensity of overlaps is formulated in a similar way.

The population size for results shown in this section were 10 monkeys. For every instance, we use 30 iterations as the maximum number of iterations and a parameter "limit" of 10.

Table 1. Results with Cooperation = 0.4 and Combiner = 0.5

NumObjects	AreaObjects	AreaContainer	Density	Objective	Seconds
74	4080.1	8463	0.482	4080.1	724.8
51	3409.8	9009	0.378	3409.8	856.6
58	3591.1	8736	0.411	3591.1	140.3
77	4039.2	8910	0.453	4039.2	314.2
50	3418.5	8372	0.408	3418.5	455.4
74	3884.5	9000	0.432	3884.5	628.3
65	4128.8	9021	0.458	4128.8	769.2
83	4320.7	9021	0.479	4320.7	947.7
50	3455	8930	0.387	3455	1083.3
78	4176.8	9215	0.453	4176.8	1266.5
59	3928.7	8460	0.464	3928.7	149.4
62	3880.1	8910	0.435	3880.1	146
69	4028.5	9408	0.428	4028.5	322.4
51	3494.4	8100	0.431	3494.4	138.6
86	4032.2	9506	0.424	4032.2	205.8
57	3784.6	8930	0.424	3784.6	925.4
55	4088.5	9603	0.426	4088.5	137.8
73	4406.9	8928	0.494	4406.9	279.3
59	3679.9	8740	0.421	3679.9	423.3
69	3903.6	9212	0.424	3903.6	567.1

In Table 1 we use a value of combiner of 0.5 and for cooperation a value of 0.4, these values represent the probability of accepting an element of the best monkey the cooperating process and the probability of accepting a new position in the watch-jump process. For every instance, we can see that the maximum density that was reached by the algorithm it was 49\%. Maybe this is because of the generation of initial population is evenly distributed. We can see that the objective function has the same value of the area covered by objects, this means that at the end of the iterations we neither have overlappings nor objects outside of the container, this is important to remember because our objective function uses the value of the objects' area and the value in the overlappings as well as the value of the objects that are outside of the container.

Table 2. Results with Cooperation = 0.2 and Combiner = 0.5

NumObjects	AreaObjects	AreaContainer	Density	Objective	Seconds
78	4212.6	8463	0.498	4165.9	1294.5
51	3568.8	9009	0.396	3568.8	1457.8
54	3536	8736	0.405	3536	1614.8
76	3891.7	8910	0.437	3891.7	1840.1
65	3533.8	8372	0.422	3533.8	2008.6
98	4507.4	9000	0.501	4429	2233.4
72	4822.7	9021	0.535	4751.7	2408.2
82	4154.7	9021	0.461	4154.7	3131.7
70	4605.8	8930	0.516	4560	3297.9
73	3996.8	9215	0.434	3996.8	3511.2
77	4381.9	8460	0.518	4331.3	165.9
70	3930	8910	0.441	3930	3750.3
83	4819	9408	0.512	4761	357.9
47	2962.6	8100	0.366	2962.6	527.4
110	4175.7	9506	0.439	4175.7	262.8
52	3676.1	8930	0.412	3676.1	429
47	3658.1	9603	0.381	3658.1	594.7
53	3220	8928	0.361	3220	774.4
69	4471.5	8740	0.512	4419.2	943.2
66	3916.3	9212	0.425	3916.3	1112.8

In Table 2 we use a value of combiner of 0.5 and for cooperation a value of 0.2. For every instance in Table 2, we can see that the maximum density that was reached by the algorithm it was 49%. In the Fig. 6 we can see an example of one result after 16 iterations.

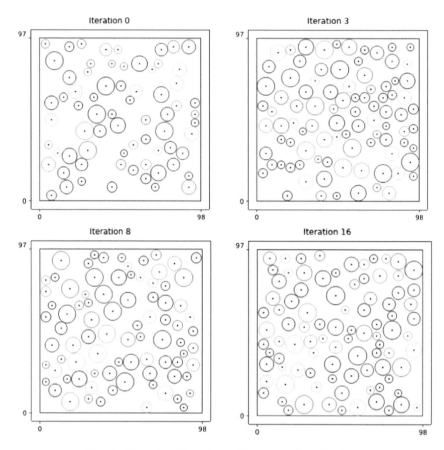

Fig. 6. Example of iterations of binary monkey algorithm

5   Conclusions

This paper proposed a binary version of the monkey algorithm proposed by [10, 16] for the approximate packing non-congruent circles in a rectangular container. The algorithm used a greedy algorithm to correct the infeasible solutions and to improve the feasibility, introduced the cooperation process to speed up the convergence rate, modified the somersault process by randomly choosing one monkey as the pivot of another to avoid falling into the local optimal solutions, and reinitialized the population if the global optimal solution was not improved after a predetermined number of generations.

The experiments show that the proposed binary monkey algorithm has strong advantages in solving the approximate packing non-congruent circles. We can use this algorithm as a generator of initial solutions for exact methods by reducing computing time in the initial steps of the optimization process.

References

1. Birgin, E.G., Bustamante, L.H., Flores Callisaya H., Martínez, J.M.: Packing circles within ellipses. Int. Trans. Oper. Res. 365–389 (2013)
2. Castillo, I., Kampas, F.J., Pintér, J.D.: Solving circle packing problems by global optimization: numerical results and industrial applications. Eur. J. Oper. Res. **191**(3), 786–802 (2008)
3. Hifi, M., M'Hallah, R.: A literature review on circle and sphere packing problems: models and methodologies. Adv. Oper. Res. **2009**, Article ID 150624, 22 (2009)
4. Flores, J.J., Martínez, J., Calderón, F.: Evolutionary computation solutions to the circle packing problem. Soft. Comput. **20**(4), 1521–1535 (2015)
5. Huang, W., Xu, R.: Two personification strategies for solving circles packing problem. Sci. China Ser. E Technol. Sci. **42**(6), 595–602 (1999)
6. Birgin, E.G.: Applications of nonlinear programming to packing problems. In: Applications + Practical Conceptualization + Mathematics = fruitful Innovation: Proceedings of the Forum of Mathematics for Industry 2014, Tokyo, pp. 31–39. Springer, Japan (2016)
7. Huang, W., Ye, T.: Quasi-physical global optimization method for solving the equal circle packing problem. Sci. China Inf. Sci. **54**(7), 1333–1339 (2011)
8. Miyazawa, F.K., Pedrosa, L.L.C., Schouery, R.C.S., Sviridenko, M., Wakabayashi, Y.: Polynomial-time approximation schemes for circle packing problems. Algorithmica **76**(2), 536–568 (2016)
9. He, Y., Wu, Y.: Packing non-identical circles within a rectangle with open length. J. Global Optim. **56**(3), 1187–1215 (2013)
10. Zhao, R., Tang, W.: Monkey algorithm for global numerical optimization. J. Uncertain Syst. **2**(3), 165–176 (2007)
11. Beasley, J.E.: An exact two-dimensional non-guillotine cutting tree search procedure. Oper. Res. **33**(1), 49–64 (1985)
12. Galiev, S.I., Lisafina, M.S.: Linear models for the approximate solution of the problem of packing equal circles into a given domain. Eur. J. Oper. Res. **230**(3), 505–514 (2013)
13. Toledo, F.M., Carravilla, M.A., Ribeiro, C., Oliveira, J.F., Gomes, A.M.: The dotted-board model: a new MIP model for nesting irregular shapes. Int. J. Prod. Econ. **145**(2), 478–487 (2013)
14. Litvinchev, I., Ozuna Espinosa, E.L.: Integer programming formulations for approximate packing circles in a rectangular container. Math. Probl. Eng. **2014**, Article ID 317697, 6 (2014)
15. Litvinchev, I., Infante, L., Ozuna Espinosa, E.L.: Packing circular-like objects in a rectangular container. J. Comput. Syst. Sci. Int. **54**(2), 259–267 (2015)
16. Zhou, Y., Chen, X., Zhou, G.: An improved monkey algorithm for a 0/1 knapsack problem. Appl. Soft Comput. **38**(2016), 817–830 (2016)

Machine Learning Applied to the Measurement of Quality in Health Services in Mexico: The Case of the Social Protection in Health System

Roman Rodriguez-Aguilar[1(✉)], Jose Antonio Marmolejo-Saucedo[2], and Pandian Vasant[3]

[1] Faculty of Engineering, Universidad Anáhuac México Norte, Mexico, D.F., Mexico
roman.rodriguez@anahuac.mx
[2] Universidad Panamericana, Facultad de Ingeniería, Augusto Rodin 498, México, Ciudad de México 03920, Mexico
[3] Faculty of Science and Information Technology, Universiti Teknologi Petronas, Seri Iskandar, Malaysia

Abstract. To propose a satisfaction indicator of users of health services affiliated to the Social Protection System in Health (SPSS). Identify the effect of the main factors that are directly related to the satisfaction level and perception of quality of health services. A machine-learning model based on Logistic Models and Principal Components was developed to estimate a satisfaction index. The survey data collected for the "SPSS 2014 User's Satisfaction Study" was used, considering a sample of 28,290 users. The proposed model shows, in general, the positive perception of quality of health services (national average 0.0756). There are factors statistically significant that influence these results, the good perception of infrastructure (OR:2.12; CI 95%:1.9–2.36); the gratuity of the service provided (OR:1.98; CI 95%: 1.42–2.76); and full medicines supply (OR:1.81; CI 95%:1.91–2.36). The proposed index can be used as an indicator for improving health care quality of the population covered by the SPSS.

Keywords: Machine learning · Logistic models · Principal components
Care quality · Quality indicators · Satisfaction · Health surveys
Mexico

1 Introduction

For The World Health Organization (WHO) a health system works well if it responds to the needs and expectations of the population meeting the following objectives [1]: to improve population's health; reduce health inequities; provide quality effective access and improve efficiency in the use of resources. On the other hand, the World Health Assembly has urged countries to promote the availability and universal access to essential goods and services for health and well-being, with special emphasis on equity.

© Springer Nature Switzerland AG 2019
P. Vasant et al. (Eds.): ICO 2018, AISC 866, pp. 560–572, 2019.
https://doi.org/10.1007/978-3-030-00979-3_59

For the latter, it is important to address effective access and quality of health services ensuring the optimal use of resources.

In the Mexican case, the health system is facing four major challenges to improve quality and effective access, reduce or contain costs and make use of economic resources more efficiently. Even though all are important, the perceived quality of health services has a major influence on patient's behaviors (satisfaction, references, choice, use, etc.) in comparison with access and cost.

In Mexico, since the creation of the System of Social Protection in Health (SPSS) in the year 2003 - whose operational component is the Seguro Popular (SP) - have invested large amounts of resources to increase and improve access to health services. However, in spite of efforts to increase access, the use of services has not been what is desired, in part, due to the negative perceptions of users on quality of services in the public sector, giving rise to people with public coverage of health services seek and make use of services with private providers.

Results of the ENSANUT 2012 show that between 28.4% and 36.6% of the participants in public insurance (including SP) made use of outpatient services from the private sector [2]. A bad perception of quality in terms of care can deter patients to use the available public services, covering then their needs with private suppliers. If users do not trust the public health system to guarantee a minimum quality level of services, services will still be underutilized. On the other hand, bad quality also rises psychological barriers to use the system. Hence, it is important that health service suppliers have as main objective to offer quality services.

The patient's satisfaction is another concept closely linked to health services quality [3] and [4]. There is evidence of greater tendency to pay more for health services in quality institutions, which are focused to meet customer's needs. Health suppliers are improving quality levels in services to meet patients' needs [5]. Satisfaction surveys have been used to deal with access and development problems; also, surveys are key to help decision makers to identify focus groups, clear objectives, define development measures and develop performance information of health services [6]. Other studies suggest that patient's satisfaction is a prevalent concern that entwines with strategic decisions within health services [7]. Therefore, patient's satisfaction must be essential both for evaluating quality and design, and managing the supply of health services [8].

In this study two major objectives are raised: (1) to develop a machine learning model to estimate a satisfaction index of health service users covered by the SPSS, based on the perception of satisfaction with the services they receive in units providing services of the three levels of care; (2) to identify factors that are associated with an increased level of satisfaction of the users.

The work is integrated in the following manner: the first section consists in the approach that will be used for built the semi-supervised learning model using Principal Components and Logistic Model to measuring the perception of health services quality. In the second section, the main results are described; finally, conclusions and discussion are presented.

2   Materials and Methods

2.1   Health Quality

Historically, stablishing rules in terms of quality has been delegated to health professionals. However, recent studies have highlighted the importance of the patient's perspective, the value of perceptions and knowledge that patients can provide to improve health services is recognized. Some authors suggest that "increasingly, the patient's perspective is the one that is seen as a significant quality indicator of health services, and could, in fact, represent the most important point of view" [9].

On its behalf, [10] argue that it is not important if the patient is well or not, what is really important is how the patient feels about the attention he received, despite the perception of health professionals that may be different, and surely it will because they are those who possess knowledge to be able to value quality of the given services. Although, on the other hand, patients may value the development of health professionals, as consumers of this service.

For the latter, it is important to establish criteria in which quality of health services must be assessed. These criteria can be technical or functional [11] and be related to the care process [12] suggests that health service quality may be defined regarding other care technical aspects, interpersonal relation between doctor and patient, and services given during care process. In developing countries such as Mexico, the use of any of these criterion to assess service quality represent additional challenges given the lack of timely and quality information, as well as the diversity of external factors that should be analyzed. As a result, instead of limiting concepts and quality service measures to the theoretical structure and the measures suggested in literature, in this study based on a comprehensive survey, proposes a way of measuring the satisfaction of users of health services covered by the SP and identify which are the factors that most influence satisfaction or perceived quality.

Seven dimensions were used to propose an index of satisfaction and twelve factors that represent the elements that determine the satisfaction perceived by users of health services. These eleven care dimensions were shaped with the general satisfaction index as a target variable. The service quality has been distinguished in which if there are certain things in common, satisfaction is generally seen as a broader concept, while the assessment of service quality focuses on service dimensions'. Given the important bond between service quality and user's satisfaction, this study shapes patient's satisfaction according to the perception of service quality within the units providing health services to the population covered by SPSS in Mexico.

Satisfaction Scale. Previous studies have shown that the expression of overall satisfaction of the patient can be obtained from the recommendation on the visited establishment, but it is also important to consider the direct expression of rating and confidence given to the service [13]. For the satisfaction scale, three questions about the service received on the day of the interview were considered, which on a scale of 1 to 10 established the score of attention, treatment and, in general, the cause of the service; a question about satisfaction with the supply of medicines in the past few months and, finally, a question on general satisfaction at that clinic (which does not refer to a day or specific period of time). The satisfaction scale synthesizes information obtained from the survey about several satisfaction dimensions.

Factors Related to Satisfaction. The quality of health care is based on a set of factors that are not easily measured. The assessment of satisfaction of users of health services is based on a multidimensional approach, which includes various aspects such as information delivery, accessibility, bureaucracy, humanization, attention to psychosocial problems, human talent, financial and physical resources, policies and programs, technology, medical and administrative processes, performance, and service efficiency and interaction with the provision of health services [11], [12] and [14].

2.2 Sources of Information

The study was based on the survey carried out in 2014 to perform the "Study of users' satisfaction of SPSS 2014 (ESATSP 2014)", whose objective is to generate evidence on the satisfaction of health services of population affiliated to SPSS. The ESATSP 2014 survey was conducted based on the collection of primary information from a sample of users of health services covered by the SPSS.

The survey has a probabilistic design and its representative at national level, which included 28,261 interviews with users of first, second and third level of attention throughout the country. The selected sample of health facilities has national and state representativeness, and was stratified in order to generate estimates for the three different care levels. Likewise, the sample was designed to obtain information coming from the total of sanitary jurisdictions dividing the country.

For this study, eleven care aspects were considered, which are decisive elements of the satisfaction of users in terms of the given health service.

- Given disease information.
- Scheduling appointments.
- Waiting time before been attended.
- Consultation duration.
- Information given about the treatment the patient will receive.
- Charges to the patient for receiving attention.
- Perceptions about facilities of the health unit.
- Full supply of prescribed medicines in the same health unit.
- Information given in terms of rights to which SPSS members are entitled.
- Perception about speed of paperwork in the health unit.
- Education level of health service users.

These eleven care aspects were shaped as independent variables and the general satisfaction index as dependent variable in the semi-supervised model developed.

2.3 Semi-supervised Model

A descriptive analysis was performed on the main characteristics of the analyzed sample; also, the corresponding Confidence Intervals are presented with a CI of 95%. In order to generate the satisfaction index, the methodology of Principal Components was used, a multivariate technique that allows to summarize information from a set of correlated variables into a set of smaller size of principal component variables, maintaining the greatest amount of information in the original set. One of the main objectives of estimating Principal Components is reducing the dimension of data. The first component is defined as the lineal combination of original variables that have maximum variance, whose values would be represented by a given vector:

$$Z = \begin{pmatrix} x_{11} & \cdots & x_{1p} \\ \vdots & \ddots & \vdots \\ x_{n1} & \cdots & x_{np} \end{pmatrix} \begin{pmatrix} a_{11} & \cdots & a_{1p} \\ \vdots & \ddots & \vdots \\ a_{n1} & \cdots & a_{np} \end{pmatrix} \qquad (1)$$

Where:

(Z) is the martrix whose columns are values of the (p) components for the (n) individuals and (A) is the matrix of burdens. Therefore, the first component would be result of the following expression:

$$Z_1 = a_1 x_1 + a_2 x_2 + \ldots + a_n x_p \qquad (2)$$

In this case, the set of variables considered as elements that allow to measure quality, would be summarized in a single variable, which would be the built index of quality. The construction of an index by using Principal Components facilitates the analysis in time and between various units of observation, as well as estimating predictive statistical models about the satisfaction of users.

To identify the relationship between the eleven considered factors as determinants of the quality in health care, logistic regression was estimated using the satisfaction index as the dependent variable and the twelve identified factors as independent variables, according to the following approach.

$$E[Y_i = Prob(Y_i = 1)] = M_i \qquad (3)$$

Where:

$$M_i = \frac{1}{1 + e^{-(\alpha + \beta_k X_{ki})}} \qquad (4)$$

is the state probability (i).

$$\frac{M_i}{(1 - M_i)} = \ln\left(e^{\alpha + \beta_k X_{ki}}\right) = \alpha + \beta_k X_{ki} \qquad (5)$$

Are odds ratio and X correspond to the twelve variables identified as determinants of the perception of care quality received by the patient.

The estimates considered the sample design of the survey and in case of dichotomous variables, the coefficient of estimated interest was the Odds Ratio. In addition, it was controlled through confounders (sex, age, socioeconomic status, stratum and schooling) to obtain results that are more robust. Although the survey is representative by level of care, in this study only the results at national level were analyzed. The dependent variable and independent variables were built and codified as dichotomous variables in the following manner (Table 1).

Table 1 Codification of variables for the logistic model

Variable	Value
Satisfaction index	Less than the Median = 0 Greater than or equal to the median = 1
Information of the disease was provided	Yes = 1 No = 0
Waiting time	Less than 30 min = 1 Greater than or equal to 30 min = 0
Scheduling appointments	Scheduled appointment = 1 No scheduled appointment = 0
Duration of the consultation	Greater than or equal to 15 min = 1 Less than 15 min = 0
Information about patient's treatment was provided	Yes = 1 No = 0
Charges were made to the patient at the health unit	Yes = 0 No = 1
Perception of the health unit conditions	Good or very good = 1 Bad, regular or very bad = 0
Full provision of medicines at the same unit	Complete = 1 Partial or not delivered = 0
Information about rights entitled by Seguro Popular affiliates were provided	Yes = 1 No = 0
Perception about speed of paperwork at the health unit	Fast = 1 Regular or slow = 0
Rural or urban stratum	Rural = 1 Urban = 0
Education level of health service users	High school or less = 1 Greater than high school = 0

Source: Own elaboration

3   Results

3.1   Exploratory Data Analysis

The objective of the SPSS is to protect financially the population without social security. The SPSS promotes the inclusion of vulnerable population; under a progressive logic from which people within the four lowest deciles are exempted from the obligation to contribute to their financing. In Table 2 the main characteristics of the surveyed population were presented. The SPSS has more women affiliates than men, 72.2% versus 27.8%. The age distribution of surveyed users was 13.4% of children under 5 years old, 6.0% of children from 5 to 9 years and 10.4% from 10 to 19 years, while the majority group was observed in people of 20 years and older (70.2%). According to the place of residence, the highlighted population is the one of 20 years and more, with an average of 72.6% in rural areas and 69.1% in urban areas.

Table 2 Percentage distribution (CI 95%), 2014

Description	National	
	Proportion	CI (95%)
Sex (N)	28,291	
Man	27.77%	[26.91%, 28.65%]
Woman	72.23%	[71.35%, 73.09%]
Age groups (N)	28,232	
Less than 5 years old	13.36%	[12.49%, 14.28%]
5 to 9 years	5.98%	[5.50%, 6.51%]
10 to 19 years	10.42%	[9.84%, 11.02%]
20 years old or more	70.24%	[68.83%, 71.61%]
Schooling (N)	24,481	
None	12.85%	[11.73%, 14.05%]
Preschool or Kindergarten	2.05%	[1.79%, 2.35%]
Elementary School	41.62%	[40.51%, 42.74%]
Middle School	28.97%	[27.90%, 30.06%]
High School or equivalent	10.52%	[9.85%, 11.24%]
Normal School	0.12%	[0.00%, 0.18%]
Technical or commercial Degree	1.44%	[1.15%, 1.80%]
Professional or Higher education	2.39%	[2.08%, 2.75%]
Master's Degree or PhD	0.04%	[0.02%, 0.09%]
Did you work last week or did any activity from which you get paid? (N)	22,239	
Yes	29.05%	[27.91% ,30.20%]
No	70.94%	[69.78%, 72.07%]
NS	0.01%	[0.00%, 0.04%]
Do you understand any indigenous language? (N)	14,416	
Yes	14.41%	[12.56%, 16.49%]
No	85.29%	[83.23%, 87.13%]
NS/NR	0.30%	[0.17%, 0.52%]
Income quintile (N)	28,267	
First	57.16%	[55.15%, 59.14%]
Second	22.94%	[21.91% ,24.01%]
Third	17.21%	[16.09%, 18.39%]
Fourth	2.26%	[1.98%, 2.57%]
Fifth	0.43%	[0.32%, 0.58%]

Source: Own elaboration based on the ESATSP 2014 survey.

Not surprisingly, the low levels of schooling identified, compared to those observed at national level in other surveys, given that the SPSS is focused toward the most deprived population. At national level, there is a significant difference in the proportion of the heads of households who reported having knowledge and understand some indigenous language, only 14.4% have knowledge of some indigenous language. At national level, a little more than 80% of users participating in the study concentrates on the first two income deciles.

3.2   Satisfaction Index

In order to have an overall measure of satisfaction, an index was estimated by using the Principal Components analysis from eight identified variables earlier in the Materials and Methods section. The criteria to be followed in order to select the number of main components was that the sum of the cumulative variance was greater than 80%, which is why the first three components that build up 85.6% of the variance were selected, the components were standardized. The higher the satisfaction the greater is the value that of the estimated index. The national average index was 0.756. Figure 1 shows the Kernel Density from the global satisfaction index, it is observed that the distribution is biased to the left focusing on positive values which means that a greater proportion of users have a positive perception of the quality of the service received.

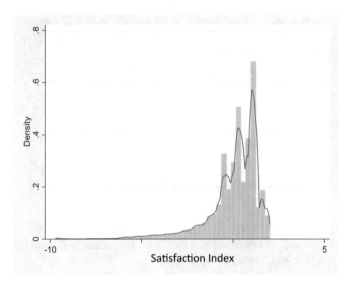

Fig. 1 Kernel density national satisfaction index. Source: Own elaboration based on the ESATSP 2014 survey.

Table 3 summarizes average values of the satisfaction index per place of residence and other characteristics of the surveyed population. At national level, there is better satisfaction among women (0.792) than among men (0.661). The analysis per age groups shows that the greater the age, the greater the satisfaction. On the other hand, if people have study high school or more, the level of satisfaction is lower.

The estimated value of the satisfaction index suggests that the satisfaction levels of indigenous population are lower than the rest of the population: the estimated average index was −0.756. The satisfaction level according to the socioeconomic level is

Table 3. Average value (CI 95%) of the satisfaction index

Description	National	
	Average	IC (95%)
(N)	27,435	
Average value	0.0756	[0.0127, 0.1385]
Sex		
Man	0.0661	[−0.0055, 0.1377]
Woman	0.0792	[0.0118, 0.1466]
Age groups		
Less than 5 years old	−0.0426	[−0.1438, 0.0587]
5 to 9 years	−0.0186	[−0.1218, 0.0845]
10 to 19 years	0.0738	[−0.0211, 0.1688]
20 years old or more	0.1044	[0.0379, 0.1709]
Schooling		
Middle school or less	0.1026	[0.0353, 0.1699]
High school or more	−0.0012	[−0.0727, 0.0703]
Do you understand any indigenous language?		
Yes	−0.0756	[−0.2106, 0.0595]
No	0.0949	[0.0305, 0.1594]
Income Quintile		
First	0.0604	[−0.0236, 0.1443]
Second	0.0662	[−0.0050, 0.1375]
Third	0.0998	[0.0303, 0.1693]
Fourth	0.4037	[0.2832, 0.5242]
Fifth	−0.1221	[−0.6645, 0.4202]

Source: own elaboration based on the ESATSP 2014 survey.

consistent with what is expected, the higher socioeconomic level has the lower level of satisfaction. However, we must read this data carefully, because 80% of health services users are in the first and second income decile.

3.3 Factors Associated with the Satisfaction Level of Population Affiliated to the SPSS

A logistic regression was estimated to identify the relationship between a set of variables considered as determinants of the satisfaction level of the population and as independent variables to semi-supervised model. At national level, the model was statistically significant with an aggregate value of F = 59.36 (p < 0.05). The factors that were identified, as determinants in satisfaction of users of health services were statistically significant at a confidence level of 95% (see Table 4).

The factor that most affects satisfaction is the perception about the infrastructure providing the service, those who reported a good perception on the infrastructure have a level of satisfaction 2.13 (OR = 2.13, 95% CI: 1.91–2.36) times greater than those who have a bad impression. Those who were not charged for the service received have 1.98 (OR = 1.98, 95% CI: 1.43–2.76) times more satisfaction than those who had to pay for the received service (either for the consultation, medication, hospitalization, and/or studies).

Table 4. Logistic regression, national results

Satisfaction (Target) (Y = 1 | Y = 0)	Odds Ratio	Linearized Std. Err.	t	P > |t|	[95% CI]	
1. Info. about the disease	1.4153	0.1271	3.8700	0.0000	1.1866	1.6880
2. Scheduling appointment	1.1753	0.0710	2.6700	0.0080	1.0439	1.3232
3. Waiting time before being attended <=30 min	1.6177	0.0920	8.4600	0.0000	1.4469	1.8086
4. Adequate consultation time >=15 min	1.1704	0.0709	2.6000	0.0100	1.0393	1.3182
5. Information about the treatment	1.6881	0.2418	3.6600	0.0000	1.2745	2.2358
6. Services were not charged	1.9844	0.3349	4.0600	0.0000	1.4251	2.7633
7. Good perception about the infrastructure	2.1258	0.1148	13.9600	0.0000	1.9121	2.3635
8. Full medicine supply	1.8119	0.1399	7.7000	0.0000	1.5572	2.1083
9. Information about rights	1.2880	0.0783	4.1700	0.0000	1.1432	1.4511
10. Adequate time (fast) to perform procedures/paperwork	1.7841	0.1005	10.2800	0.0000	1.5974	1.9925
11. Level of education of middle school or less	1.2531	0.0824	3.4300	0.0010	1.1014	1.4257
12. Urban place of residence	1.1786	0.1179	1.6400	0.1010	0.9686	1.4341
Number of obs = 18236 Population size = 53777847 F (12,1186) = 51.60 Prob > F = 0.0000						

Source: own elaboration based on the ESATSP 2014 survey.

Complying with the regulations according to the magnitude of the OR, full medicine supply is associated with a satisfaction level of 1.81 (OR = 1.81, CI 95%: 1.56–2.11) times greater compared to a partial or null supply. Those who believe that paperwork in the clinic are fast have a satisfaction 1.78 (OR = 1.78, 95% CI: 1.60–1.99) times higher compared to those who consider the time it takes to perform the procedures in the clinic are regular or slow. Table 3 shows that if the user of health services receives information about his treatment, illness and his rights as a member of the SPSS the level of satisfaction is 1.69 (OR = 1.69, CI95%: 1.27–2.24), 1.42 (OR = 1.42; CI95%: 1.19–1.69) and 1.29 (OR = 1.29; CI95%: 1.14–1.45) times higher, respectively, compared to those who did not receive information. Those who report that their consultation lasts 15 min or more have a satisfaction level 1.17 (OR = 1.17; CI96%: 1.04–1.32) times, compared to those who received a consultation with a duration of less than 15 min. On the other hand, those who wait 30 min or less at the health unit to be attended increases 1.62 (OR = 1.62; CI95%: 1.45–1.81) times the satisfaction level versus those who wait more than 30 min.

The aspects that have a lower level of influence on the satisfaction level are: those who have high school or less, their level of satisfaction is 1.25 (OR = 1.25; CI95%: 1.10–1.43) times higher than those patients who have high school or higher level of education. The appointment scheduling is the aspect that least influences the satisfaction of health service users, if they have an appointment schedule they have 1.18 (OR = 1.18; CI95%: 1.04–1.32) times higher levels of satisfaction.

Performance of the Model. The semi-supervised model proposed as an assembly between the analysis of main components and the logistic regression, allows replicating the satisfaction index through the set of independent variables selected. Figure 2 shows the estimated ROC curve with the results obtained; reflecting an acceptable adjustment of 70% in terms of the confusion matrix and this shows an acceptable performance for the model.

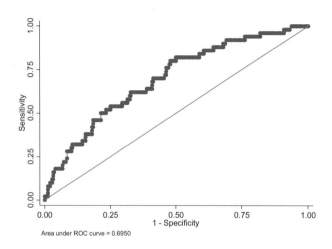

Area under ROC curve = 0.6950

Fig. 2 ROC curve. Source: Own elaboration based on the ESATSP 2014 survey.

4   Conclusions

The last great Health System Reform in Mexico culminated with the creation of the System of Social Protection in Health (SSPS, for its acronym in Spanish) in 2004, as part of this reform public policy actions have been implemented, aimed to increase coverage of population without social security, in addition to increasing the range of interventions financed by the SPSS. Currently, around 50% of the Mexican population is covered by the SPSS and has a set of financed and defined interventions [15]. The natural evolution process of the Health System leads us to a greater emphasis on service quality as counterpart of increasing coverage. Therefore, it is of utmost importance to generate evidence of public policy for the decision-making process, the proposed satisfaction indicator may be an indicator for monitoring the improvement of health care quality of population covered by the SPSS. Additionally, key factors associated with satisfaction of users were identified and that allow defining strategies and lines of action targeted to improve satisfaction levels.

The analysis allows to identify the most relevant aspects in order to explain the satisfaction level of users of health services covered by the SPSS. The perception reported on the facilities conditions was the variable with the greatest effect on the satisfaction level, this result draws attention as a higher relation of satisfaction with aspects directly related to the care process would be expected. Followed by the charge of received services and full medicine supply. The perception of the facilities conditions is a first point of reference for users of health services, with which generate expectations about services to be provided, this first impression could be decisive as it predisposes the user from the beginning of the care process.

The application of a semi-supervised support model was shown, combining the analysis of main components and logistic regression, in the Health Sector. The application of analytical intelligence models in the decision making of different sectors show the great utility of this discipline and the great potential of application to many other areas.

One advantage of estimating a model of this nature lies in a reduction of important costs in the collection of information, since it would be possible to estimate the satisfaction of the user of health services with the collection of a small set of variables. In the same way, the methodology is replicable for other health institutions in Mexico.

References

1. WHO (World Health Organization): Key Components of a Well-Functioning Health System. Ginebra (2010). http://www.who.int/healthsystems/publications/hss_key/en/index.html
2. Gutierrez, J.P., Rivera-Dommarco, J., Shamah-Levy, T., Villalpando-Hernandez, S., Franco, A., Cuevas-Nasu, L., Romero-Martinez, M., Hernandez-Avila, M.: Encuesta Nacional de Salud y Nutrición 2012. Resultados nacionales. 2a. ed. Cuernavaca. México, Instituto Nacional de Salud Pública (MX) (2013)
3. Taylor, S.A., Cronin Jr., J.J.: Modeling patient satisfaction and service quality. J. Health Care Mark. **14**(1), 34–44 (1994)
4. McAlexander, J.H., Kaldenberg, D.O., Koenig, H.F.: Service quality measurement. J. Health Care Mark. **14**(3), 34–40 (1994)
5. Boscarino, J.A.: The public's perception of quality hospitals II: implications for patient surveys. Hosp. Health Serv. Adm. **37**(1), 13–35 (1992)
6. Langseth, P., Langan, P., Talierco, R.: Service delivery survey (SDS): a management tool. The Economic Development Institute of the World Bank, Washington, United States (1995)
7. Gilbert, F.W., Lumpkin, J.R., Dant, R.P.: Adaptation and customer expectation of health care options. J. Health Care Mark. **12**(3), 46–55 (1992)
8. Donabedian, A.: Quality assessment and assurance: unity of purpose, diversity of means. Inquiry **25**, 173–192 (1988)
9. O'Connor, S.J., Shewchuk, R.M., Carney, L.W.: The great gap. J. Health Care Mark. **14**(2), 32–39 (1994)
10. Petersen, M.B.H.: Measuring patient satisfaction: collecting useful data. J. Nurs. Qual. Assur. **2**(3), 25–35 (1988)
11. Babakus, E., Mangold, W.G.: Adapting the SERVQUAL scale to hospital services: an empirical investigation. Health Serv. Res. **26**(6), 767–786 (1992)
12. Weitzman, B.C. In: Kovner, A.R. (ed.) Health care delivery in the United States, 5th edn. Berlin, Springer (2015)
13. Serrano-del-Rosal, R., Loriente-Arín, N.: La anatomía de la satisfacción del paciente. Salud Pública México **50**(2):162–172 (2008)

14. Zeithaml, B.C. In: Kovner, A.R. (ed.) Health care delivery in the United States, 5th edn. Berlin, Springer (2015)
15. Sistema de Protección Social en Salud. Informe de Resultados 2014. SSA. México (2014). http://www.seguro-popular.gob.mx/images/Contenidos/informes/Informe%20de% 20Resultados%202014.pdf

Author Index

© Springer Nature Switzerland AG 2019
P. Vasant et al. (Eds.): ICO 2018, AISC 866, pp. 573–575, 2019.
https://doi.org/10.1007/978-3-030-00979-3

Printed in the United States
By Bookmasters